Molecular Mechanisms of Bioactive Nutrients Promoting Human Health

Molecular Mechanisms of Bioactive Nutrients Promoting Human Health

Guest Editor

Baojun Xu

MDPI

Basel • Beijing • Wuhan • Barcelona • Belgrade • Novi Sad • Cluj • Manchester

Guest Editor
Baojun Xu
Department of Life Sciences
Beijing Normal-Hong Kong
Baptist University
Zhuhai
China

Editorial Office
MDPI AG
Grosspeteranlage 5
4052 Basel, Switzerland

This is a reprint of the Special Issue, published open access by the journal *International Journal of Molecular Sciences* (ISSN 1422-0067), freely accessible at: https://www.mdpi.com/journal/ijms/special_issues/0Y7XV51945.

For citation purposes, cite each article independently as indicated on the article page online and as indicated below:

Lastname, A.A.; Lastname, B.B. Article Title. *Journal Name* **Year**, *Volume Number*, Page Range.

ISBN 978-3-7258-5447-9 (Hbk)
ISBN 978-3-7258-5448-6 (PDF)
https://doi.org/10.3390/books978-3-7258-5448-6

Contents

About the Editor

Baojun Xu

Baojun Xu is a Chair Professor at Beijing Normal-Hong Kong Baptist University (BNBU, a full English teaching university in China), Fellow of the Royal Society of Chemistry, Zhuhai Scholar Distinguished Professor, Department Head of the Department of Life Sciences, Program Director of the Food Science and Technology Program, and author of over 420 peer-reviewed papers. Dr. Xu received Ph.D. in Food Science from Chungnam National University, South Korea. He conducted postdoctoral research work in North Dakota State University (NDSU), Purdue University, and Gerald P. Murphy Cancer Foundation in the USA from 2005 to 2009. He conducted short-term visiting research at NDSU in 2012 and the University of Georgia in 2014, which was followed by visiting research during his sabbatical leave (7 months) at Pennsylvania State University in the USA in 2016. Dr. Xu is serving as the Associate Editor-in-Chief of *Food Science and Human Wellness*, Associate Editor of *Food Research International*, and Associate Editor of *Food Frontiers*, and is an Editorial Board Member of around 10 international journals. He received the inaugural President's Award for Outstanding Research of UIC in 2016 and the President's Award for Outstanding Service of UIC in 2020. Dr. Xu has been listed in the world's top 2% of scientists by Stanford University for the past five consecutive years and was listed as the best scientist in the world in the field of Biology and Biochemistry at Research.com in 2023 and 2024. Prof. Xu was named a Highly Ranked Scholar (top 0.05%) by ScholarGPS^{TM}, ranking at #8 in Food Science and Technology in the world, and ranking at #14 Agricultural and Natural Sciences.

International Journal of *Molecular Sciences*

MDPI

Editorial

Special Issue: Molecular Mechanisms of Bioactive Nutrients Promoting Human Health

Baojun Xu

Food Science and Technology Program, Department of Life Sciences, Beijing Normal-Hong Kong Baptist University, Zhuhai 519087, China; baojunxu@uic.edu.cn

Communicable and non-communicable diseases are major contributors to the global burden of disease. Although their etiologies differ, these diseases share molecular mechanisms, including cellular oxidative stress damage, inflammation, and dysregulated immune signaling. In this context, a growing body of evidence suggests that bioactive nutrients of dietary and microbial origin—such as vitamins, phytochemicals, fatty acids, prebiotics, and probiotics—can mitigate the impact of infectious illnesses and lower the risk of non-communicable diseases by modulating oxidative and inflammatory pathways, shaping the gut microbiota, influencing metabolic and endocrine signaling, and activating immune responses to support human health [1,2]. However, the underlying molecular mechanisms by which these bioactive compounds regulate cellular signaling and adaptively reshape tissue homeostasis in humans remain to be elucidated.

This Special Issue of the *International Journal of Molecular Sciences*, titled "Molecular Mechanisms of Bioactive Nutrients Promoting Human Health," includes eleven contributions spanning in vitro systems, animal models, human studies, and review papers. They offer mechanistic insights into how bioactive nutrients promote human health and propose promising strategies to prevent or treat diseases. For instance, Cocksedge et al. reviewed recent preclinical and clinical findings on superoxide dismutase (SOD)-rich *Tetraselmis chuii*, indicating that it may act as an indirect antioxidant, modulate inflammatory pathways, and prevent immune and mitochondrial dysfunction by activating NRF2 and SIRT1 [Contribution 1]. Yan et al. collected evidence on the dietary flavonoids vitexin and isovitexin, illustrating their cardiovascular, glycemic, anti-obesity, anticancer, antioxidant, anti-inflammatory, and neuroprotective properties. They also emphasize the urgent need to define molecular targets and signaling networks, and to establish clinically effective and physiologically relevant doses [Contribution 2]. Han et al. used the 3T3-L1 adipocytes model to investigate the anti-obesity effect of *Lactobacillus brevis*–fermented γ-aminobutyric acid, showing that it suppresses adipogenesis and lipogenesis, induces lipolysis via fatty acid oxidation, and enhances energy expenditure via UCP1-mediated browning, all of which support its potential as a novel anti-obesity functional food [Contribution 3].

Colon cancer (CRC) is the second leading cause of cancer-related deaths worldwide, driving interest in bioactive nutrient-based interventions with therapeutic potential and fewer side effects. In a comprehensive review, Bentharavithana et al. summarized the anti-cancer effects, particularly the anti-colon cancer properties, of edible medicinal mushrooms. The compounds derived from these medicinal mushrooms, including terpenoids, phenols, polysaccharides, ascorbic acid, glycosides, and organic acids, can inhibit proliferation, induce apoptosis, and reduce inflammation, thereby contributing to the prevention and treatment of colon cancer. The authors emphasize the need for rigorous human clinical trials to establish efficacy and safety before clinical adoption [Contribution 4]. Moreover, studies using an azoxymethane (AOM)-induced CRC rat model showed that low-molar-mass oat β-glucan supplementation improved colonic redox balance, reduced inflammation, and decreased lipid peroxidation. Wilczak et al. further demonstrated that this β-glucan facilitates colonic metabolism remodeling during early carcinogenesis, as evidenced by alterations in the amino acid, purine, biotin, and folate pathways [Contribution 5]. Similarly, Yang et al. reported that chicoric acid alleviated dextran sulfate sodium (DSS)-induced colitis in mice by reducing pro-inflammatory cytokines (TNF-α and IL-6), modifying the gut microbiota (by decreasing *Bacteroidetes* and increasing *Lachnospiraceae*), and improving metabolic homeostasis (by restoring thiamine and lithocholic acid) [Contribution 6]. In an in vivo study using two murine colorectal cancer models (the orthotopic MC-38 cecum injection model and the inflammation-driven AOM/DSS model), probiotic efficacy was found to be model-dependent. *Lactobacillus/Bifidobacterium* mix (CI) reduced tumor burden in the orthotopic MC-38 model, while *Bifidobacterium* alone (CII) suppressed inflammation and tumors in the AOM/DSS model. Collectively, these findings support context-specific, microbiota- and diet-based strategies for CRC prevention and treatment and encourage the development of tailored probiotic strategies [Contribution 7].

Neagu et al. reviewed the interplay between diet, dietary patterns, behaviors, chemical xenobiotic exposures, and breast cancer, highlighting evidence that a healthy diet pattern with appropriate nutritional behaviors activates anti-tumor pathways and suppresses tumor progression, while the accumulation of foodborne contaminants can synergistically promote tumorigenesis [Contribution 8]. Jędrzejewski et al. investigated a combination strategy by pairing *Coriolus versicolor* (CV) with the inhibitor of the phosphatidylinositol 3-kinase (PI3K) signaling pathway, LY294002. In MCF-7, HeLa, and A549 cell models, co-treatment decreased viability and colony formation, induced G0/G1 arrest and apoptosis, and inhibited migration/invasion. While this additive cytotoxic approach shows promise, the authors stress the need to evaluate pharmacokinetics, bioavailability, safety, and the nature of interaction in animal studies [Contribution 9].

In addition to the research on bioactive nutrients in non-communicable diseases, this Special Issue also features two timely studies on antiviral immunity. One study assessed whether an enzymatically liberated salmon oil can increase immune recovery after acute SARS-CoV-2 infection by regulating cytokine, chemokine, and interferon-related gene expression; the other study combined bioinformatics analysis with in vitro testing to evaluate the anti-influenza A effect of *Tagetes erecta* Linn. (TE) extract. In a randomized pilot study of adults with mild to moderate COVID-19 (n = 11), Currie et al. found that participants receiving best supportive care and 4 g/day of full-spectrum, enzymatically liberated salmon oil for 28 days experienced reduced inflammation, improved interferon response, increased lung barrier function, and stronger immune memory [Contribution 10]. However, larger and well-controlled clinical trials are required to determine whether these findings translate into meaningful clinical outcomes. In addition, Kim et al. integrated bioinformatics, molecular docking, antiviral assays, and plaque reduction tests, identifying lutein as a key active component in the TE extract and IL-6 as a central hub target associated with influenza. In conclusion, these contributions demonstrate how nutrition-derived interventions impact immune pathways to enhance human health, though more rigorous clinical validation of efficacy, safety, and dosing is required [Contribution 11].

This Special Issue of the *International Journal of Molecular Sciences* integrates the latest in vitro, animal, and human studies across non-communicable diseases and antiviral immunity, revealing the mechanisms through which bioactive nutrients promote human health. The research contributions to this Special Issue demonstrate the growing potential of dietary interventions in disease prevention and treatment. Although further steps are still necessary for clinical translation, such as optimizing dose and delivery and conducting comprehensive safety assessments, ongoing advances in this rapidly evolving field are promising.

Funding: This research received no external funding.

Institutional Review Board Statement: Not applicable.

Informed Consent Statement: Not applicable.

Data Availability Statement: Not applicable.

Conflicts of Interest: The author declares no conflicts of interest.

List of Contributions:

1. Cocksedge, S.P.; Mantecón, L.; Castaño, E.; Infante, C.; Bailey, S.J. The potential of superoxide dismutase-rich *Tetraselmis chuii* as a promoter of cellular health. *Int. J. Mol. Sci.* **2025**, *26*, 1693.
2. Yan, W.; Cheng, J.; Xu, B. Dietary flavonoids vitexin and isovitexin: New insights into their functional roles in human health and disease prevention. *Int. J. Mol. Sci.* **2025**, *26*, 6997.
3. Han, H.; Song, G.; Kim, J.; Jin, H.; Lee, B.Y. Anti-obesity effects of LB-GABA. *Int. J. Mol. Sci.* **2025**, *26*, 3554.
4. Bentharavithana, J.; Islam, T.; Xu, B. Medicinal mushrooms in colon cancer therapy: Mechanisms of action of bioactive compounds and therapeutic potential. *Int. J. Mol. Sci.* **2025**, *26*, 5304.
5. Wilczak, J.; Prostek, A.; Dziendzikowska, K.; Gajewska, M.; Kopiasz, Ł.; Harasym, J.; Oczkowski, M.; Gromadzka-Ostrowska, J. Oat beta-glucan as a metabolic regulator in early stage of colorectal cancer- A model study on azoxymethane-treated rats. *Int. J. Mol. Sci.* **2024**, *25*, 4635.
6. Yang, J.; Lin, J.; Gu, T.; Sun, Q.; Xu, W.; Peng, Y. Chicoric acid effectively mitigated dextran sulfate sodium (DSS)-induced colitis in BALB/c mice by modulating the gut microbiota and fecal metabolites. *Int. J. Mol. Sci.* **2024**, *25*, 841.
7. Niechcial, A.; Schwarzfischer, M.; Wawrzyniak, P.; Determann, M.; Pöhlmann, D.; Wawrzyniak, M.; Gueguen, E.; Walker, M.R.; Morsy, Y.; Atrott, K.; et al. Probiotic administration modulates gut microbiota and suppresses tumor growth in murine models of colorectal cancer. *Int. J. Mol. Sci.* **2025**, *26*, 4404.
8. Neagu, A.N.; Josan, C.L.; Jayaweera, T.M.; Weraduwage, K.; Nuru, N.; Darie, C.C. Double-edged sword effect of diet and nutrition on carcinogenic molecular pathways in breast cancer. *Int. J. Mol. Sci.* **2024**, *25*, 11078.
9. Jędrzejewski, T.; Sobocińska, J.; Maciejewski, B.; Slovakova, M.; Wrotek, S. Enhanced anti-cancer potential: Investigating the combined effects with *Coriolus versicolor* extract and phosphatidylinositol 3-kinase inhibitor (LY294002) in vitro. *Int. J. Mol. Sci.* **2025**, *26*, 1556.
10. Currie, C.; Myklebust, T.Å.; Bjerknes, C.; Framroze, B. Assessing the potential of an enzymatically liberated salmon oil to support immune health recovery from acute SARS-CoV-2 infection via change in the expression of cytokine, chemokine and interferon-related genes. *Int. J. Mol. Sci.* **2024**, *25*, 6917.
11. Kim, M.; Nowakowska, A.; Kim, J.; Kim, Y.B. Anti-influenza A potential of *Tagetes erecta* Linn. extract based on bioinformatics analysis and in vitro assays. *Int. J. Mol. Sci.* **2024**, *25*, 7065.

References

1. Biswas, S.; Biswas, S.; PadmaPriya, G.; Kaur, J.; Pramanik, A.; Islam, F.; Tariq, F.; Imran, A.; Shah, M.A. Role of nutraceuticals in viral infections as immunomodulators: A comprehensive review. *eFood* **2024**, *5*, e70000. [CrossRef]
2. González-Sarrías, A.; Larrosa, M.; García-Conesa, M.T.; Tomás-Barberán, F.A.; Espín, J.C. Nutraceuticals for older people: Facts, fictions and gaps in knowledge. *Maturitas* **2013**, *75*, 313–334. [CrossRef] [PubMed]

International Journal of
Molecular Sciences

MDPI

Article

Probiotic Administration Modulates Gut Microbiota and Suppresses Tumor Growth in Murine Models of Colorectal Cancer

Anna Niechcial [1], Marlene Schwarzfischer [1], Paulina Wawrzyniak [1], Madita Determann [1], Doris Pöhlmann [1], Marcin Wawrzyniak [1], Emilie Gueguen [1], Maria R. Walker [1], Yasser Morsy [1], Kirstin Atrott [1], Marijn Wilmink [1], Luise Linzmeier [1], Marianne R. Spalinger [1], Sophie Holowacz [2], Anne Leblanc [2] and Michael Scharl [1,*]

[1] Department of Gastroenterology and Hepatology, University Hospital Zurich, University of Zurich, 8091 Zurich, Switzerland; anna.niechcialtsianas@usz.ch (A.N.); marlene.schwarzfischer@usz.ch (M.S.); paulina.wawrzyniak@usz.ch (P.W.); maditaelena.determann@usz.ch (M.D.); doris.poehlmann@usz.ch (D.P.); marcin.wawrzyniak@usz.ch (M.W.); emilie.gueguen@usz.ch (E.G.); maria.walker@usz.ch (M.R.W.); yasser.morsy@usz.ch (Y.M.); kirstin.atrott@usz.ch (K.A.); marijn.wilmink@usz.ch (M.W.); luise.linzmeier@usz.ch (L.L.); marianne.spalinger@usz.ch (M.R.S.)

[2] PiLeJe Laboratoire, 49270 Paris, France; s.holowacz@pileje.com (S.H.); a.leblanc@pileje.com (A.L.)

* Correspondence: michael.scharl@usz.ch

Abstract: Colorectal cancer (CRC) is a leading cause of cancer-related mortality worldwide with limited treatment options for advanced disease stages. Growing evidence implicates the gut microbiota in CRC pathogenesis, prompting interest in probiotics as a potential therapeutic strategy. In this study, we evaluated the effects of two probiotic compositions, CI (a mix of *lactobacilli* and *bifidobacteria*) and CII (*bifidobacteria alone*), in two murine CRC models: the orthotopic MC-38 cecum injection model and the inflammation-driven azoxymethane/dextran sodium sulfate (AOM/DSS) model. CI showed significant anti-tumor effects in the orthotopic model, reducing tumor weight and volume, which was, however, not associated with robust immune activation, suggesting microbiota-driven mechanisms. In contrast, CII was more effective in the AOM/DSS model, reducing colonic inflammation and completely preventing tumor development. Our study demonstrates that probiotics might have great therapeutic potential via modulation of the gut microbiota, and they can exert anti-tumor effects in murine models of CRC with distinct compositions showing differential efficacy depending on the model. CI stabilized the gut microbiome and inhibited pro-tumorigenic taxa in the MC-38 cecum injection model, while CII exhibited anti-inflammatory properties in the AOM/DSS model, highlighting the potential of probiotics as context-specific interventions for CRC. These findings contribute to the growing body of evidence supporting microbiota-targeted strategies in oncology and their relevance for therapeutic applications.

Keywords: colorectal cancer; probiotics; *Lactobacillus*; *Bifidobacterium*; anti-tumor activity

1. Introduction

Colorectal cancer (CRC) is the third most commonly diagnosed cancer worldwide with more than 1.9 million new cases annually [1]. It is also the second leading cause of cancer-related death, accounting for over 900,000 deaths per year [2]. Despite recent advancements in diagnostic and therapeutic strategies, the prognosis for advanced-stage CRC remains poor, underscoring the need for novel preventive and therapeutic approaches [3]. The etiology of CRC is multifactorial, involving genetic, environmental, and lifestyle factors with growing evidence implicating the gut microbiota as a key contributor to its pathogenesis [4]. Emerging evidence suggests that the gut microbiota plays a pivotal role in CRC pathogenesis, influencing carcinogenesis through mechanisms such as inflammation, immune modulation, and the production of carcinogenic metabolites. Consequently, modulating the gut microbiota has been proposed as a potential strategy for CRC prevention and treatment [5–7].

Probiotics, defined as live microorganisms that confer health benefits to the host when administered in adequate amounts [8], have gained considerable attention for their potential to modulate the gut microbiota and exert protective effects against various diseases, including CRC [9]. Mechanistically, probiotics may exert anticancer effects by restoring microbial balance, promoting the mucosal barrier, modulating inflammation, and producing metabolites such as short-chain fatty acids that possess anti-inflammatory and antiproliferative properties [10–12]. Despite these promising mechanisms, the precise role of probiotics in CRC and their impact on tumor development remain poorly understood and are the subject of ongoing research [13].

Animal models, particularly murine models, have proven invaluable in elucidating the complex interactions between probiotics, the gut microbiota, and colorectal carcinogenesis. These models provide a controlled environment to investigate the molecular and cellular mechanisms underlying probiotic effects, offering valuable insights into their potential translational applications in humans [14,15]. In this study, we aimed to evaluate the effects of probiotic administration on cancer development in two well-established murine models of CRC, namely the orthotopic cecum injection model and the azoxymethane/dextran sodium sulfate (AOM/DSS) inflammation-mediated model that mirrors CRC development observed in colitis-associated cancer. By examining tumor incidence and progression, as well as the intestinal microbiome, we sought to shed light on the therapeutic potential of probiotics as a preventive strategy against CRC and contribute to the growing body of evidence supporting microbiota-based interventions in oncology.

2. Results

2.1. Administration of Probiotics Affects the Gut Microbiome of Healthy Mice

In an initial step, the impact of the probiotic compositions (mixes) on the microbiota of healthy mice was investigated. For this purpose, composition I (CI) and composition II (CII) were administered in short-term experiments (two cycles of daily gavages for 3 days within 2 weeks) and long-term experiments (four cycles of daily gavages for 3 days within 14 weeks). In both experiments, control groups received saline gavages only (Figure 1A). The alpha diversity Shannon metric score showed no significant differences either at the beginning or the end of the experiment, neither between the treatment groups nor in relation to the duration of administration (Figure 1B). Weighted UniFrac beta diversity revealed close clustering among all groups at the beginning of the experiment. However, the saline-treated control animals started to diverge over time, while the CI and CII-treated animals remained more closely clustered (Figure 1C, Table S1). At the phylum level, minor changes were observed in CI-treated mice, including an increase in the abundance of *Firmicutes* and *Desulfobacteriota* following long-term administration. Moreover, in CII-treated mice, an effect of the microbial dominance shifted from *Firmicutes* toward *Bacteriota*, which was more pronounced after long-term exposure (Figure 1D). Finally, assessing the abundance of the administered probiotics (Figure 1E) revealed that *Lactobacillus* had already colonized the intestine of mice at the beginning of the experiment with its abundance increasing over time in all groups. Whilst *Bifidobacterium* was completely absent at the beginning of the administration experiments, it evidently increased only following short-term administration of CI and CII. This may be attributed to the 4-week period without treatment after the last gavage in the long-term administration experiment, suggesting that the successful maintenance of intestinal probiotic bacteria requires frequent and continued administration.

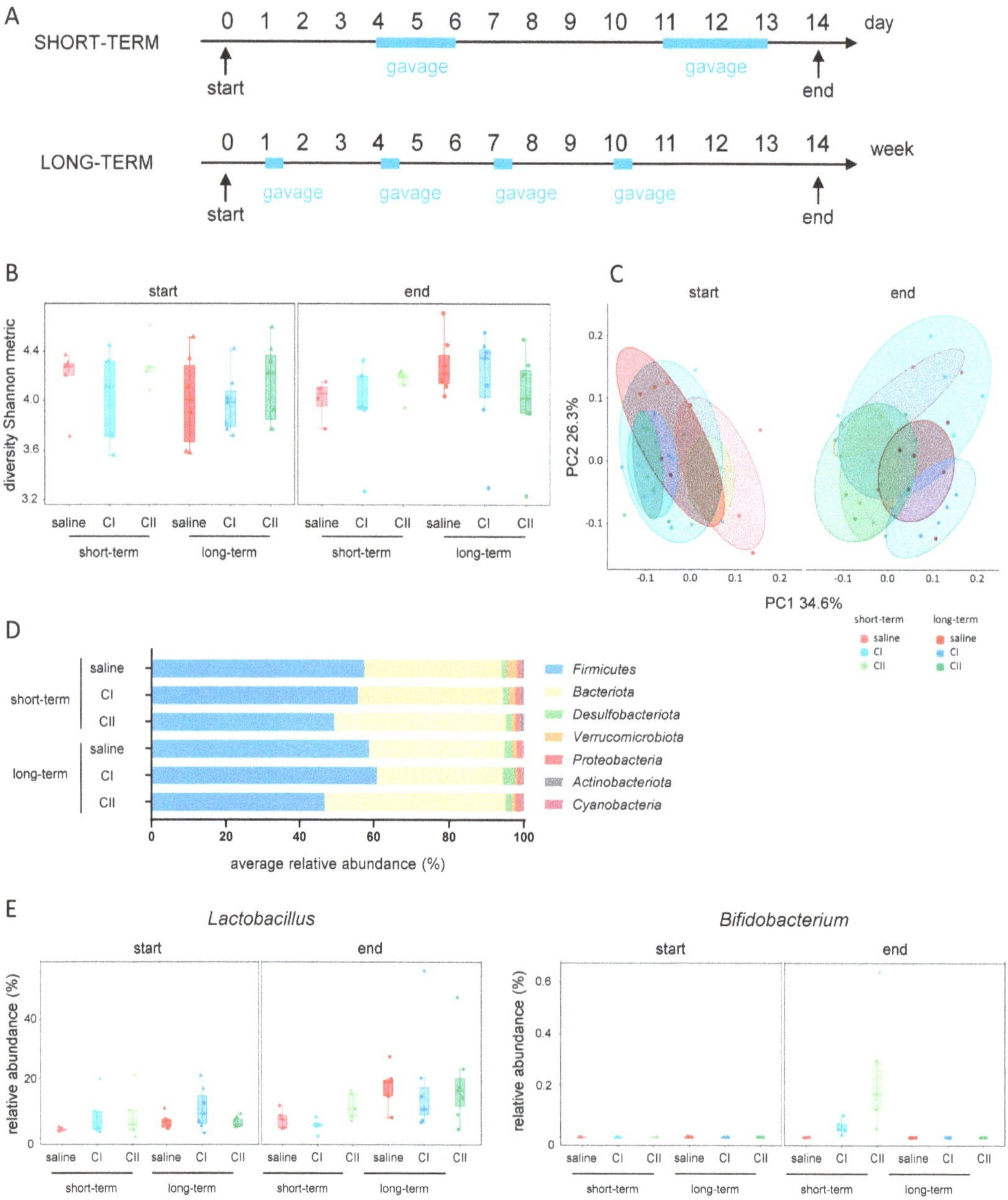

Figure 1. Effects of probiotic administration on intestinal microbiota in healthy mice. (**A**) Experimental design: mice received composition I (CI), composition II (CII), or saline for short-term (2 cycles over 2 weeks) or long-term (4 cycles over 14 weeks), once daily by oral gavage; (**B**) alpha diversity (Shannon metric index); (**C**) beta diversity (weighted UniFrac); (**D**) relative abundance of phyla at the end of experiment; (**E**) probiotic abundance: *Lactobacillus* and *Bifidobacterium* genus.

2.2. Composition I Has Greater Anti-Tumor Potential than Composition II in the Orthotopic CRC Model

Given the observed impact on the intestinal microbiome composition after the administration of our two probiotics compositions, we evaluated their effect on CRC development. For this purpose, the orthotopic MC-38 CRC cecum injection model was used. In this model, MC-38 colon adenocarcinoma cells are injected into the cecum wall of C57BL/6 mice. CI and CII were administered in two cycles of daily gavages—first from day 4 to 6 and secondly from day 11 to 13 (Supplementary Figure S1A). Control animals received gavages of saline only. In two initial independent experiments, both CI and CII impacted tumor development, as evidenced by decreases in tumor weight and volume. However, CI seemed to induce a more pronounced effect (Supplementary Figure S1B,C). Therefore, we further investigated the anti-tumor potential of CI in the cecum injection model in three independent follow-up experiments, using the same administration scheme and experimental design (Figure 2A). Again, a clear reduction in tumor weight and volume (Figure 2B) was observed, which became statistically significant when the data from all three consecutive experiments were pooled (Figure 2C).

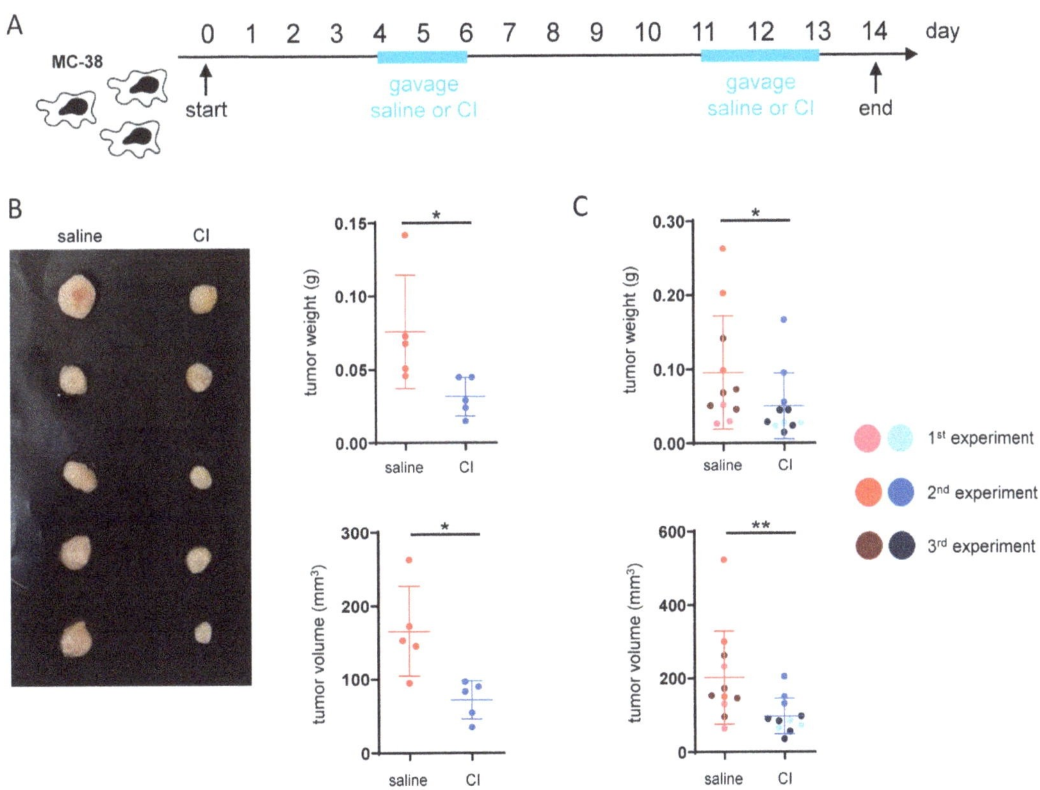

Figure 2. Composition I demonstrates higher anti-tumor potential than composition II in an orthotopic CRC model. (**A**) Experimental design: CRC was induced by implanting the MC-38 cells into the cecum wall. Composition I (CI) was administered in two cycles (days 4–6 and 11–13) once daily by oral gavage, while control mice received saline; (**B**) representative picture and raw tumor data from a single experiment; (**C**) tumor weight and volume—pooled data from three independent experiments; * $p < 0.05$, ** $p < 0.01$.

2.3. The Anti-Tumor Activity of CI Does Not Appear to Be Immune-Mediated

Given that CI demonstrated a clear anti-tumor potential, we aimed to assess whether the observed effect was immune-mediated. To this end, the proportions and activation status of intra-tumoral immune cells was analyzed by flow cytometry (gating strategies are shown in the Supplementary Figure S2). With regard to the overall frequency of B-cells, T-cells and NK cells, only a minor increase in $CD8^+$ cells and $NK1.1^+$ cells in the tumor tissue was visible in the CI-treated group (Figures 3A and S3A). Among the $CD8^+$ cytotoxic T-cells, there was a slight trend toward an increase in perforin-positive (Prf^+) cells. However, no differences were observed in granzyme B (GrB^+) levels (Figure 3B) or T-cell exhaustion, as indicated by PD-1 staining (Figure 3C). In pro-inflammatory $CD4^+$ T-cells, a slight tendency toward an increase in TNF-α but not IFN-γ (Figure 3D) levels was observed. In $CD8^+$ T-cells, no difference in TNF-α or IFN-γ levels was observed upon CI administration. In fact, nearly all $CD8^+$ T-cells were positive for both cytokines regardless of CI administration (Supplementary Figure S3B). No differences in myeloid cell infiltration into the tumor tissue were observed, as the frequency of neutrophils, macrophages, monocytes and dendritic cells did not differ between the treatment groups (Supplementary Figure S3C). However, in CI-treated animals, there was an increase in CD80-expressing macrophages and a significant induction of PDL-1 production (Figure 3E). In monocytes, increased surface expression levels of CD80 and MHCII were observed in CI-treated animals, while PDL-1 expression was slightly reduced (Figure 3F). To further explore potential mechanisms, untargeted metabolomics analysis of the serum was conducted alongside flow cytometry analysis of the immune cells. The relatively close clustering of both treatment groups in the PCA plot suggests that the overall differences in their serum metabolomic profiles were minimal (Supplementary Figure S4, Table S2). The butyrate level was not altered or correlated with tumor load (Supplementary Figure S4). Collectively, these findings suggest that the anti-tumor activity of CI is not primarily driven by immune-mediated or metabolomic alterations.

2.4. Administration of Composition I Modulates the Gut Microbiome to Promote a Healthy Microbial Environment in the MC-38 Tumor Injection Model

To study the impact of CI on the intestinal microbiome in the context of CRC, 16S sequencing of fecal samples (sampled at the start and at the end of each experiment) from control and CI-treated mice was performed, combining data from the three cecal injection experiments with CI. At the start of the experiment, the CI group exhibited slightly lower Shannon diversity indices. However, by the end, alpha diversity increased in CI-treated mice, while it decreased in the saline group (Figure 4A). In terms of beta diversity, there was a substantial overlap between the saline and CI groups at the start of the experiment. However, by the end, the saline group became more distinct, indicating that probiotic administration altered the microbial composition over time (Figure 4B, Table S1). At the phylum level, *Firmicutes* were less abundant in the CI-treated group, while *Bacteroidota* increased, indicating a shift in microbiome composition (Figure 4C). When examining the abundance of the administered probiotics, *Lactobacillales* abundance increased over time in both groups, whereas *Bifidobacteriales* appeared only in the CI-treated animals (Figure 4D). In CI-treated mice, *Bifidobacterium* was clearly present at the genus level, along with the stabilization of *Muribaculum* and the *NK4A214 group* (Figure 4E). The abundance of these genera, which are highly associated with a healthy gut microbiome, notably decreased in the saline-treated group during CRC progression. Furthermore, CI administration inhibited the increase in *Lachnoclostridium*, was associated with pro-tumorigenic processes in CRC, and prevented the rise of pathogenic bacteria such as *Escherichia-Shigella* and *Enterococcus* (Figure 4E).

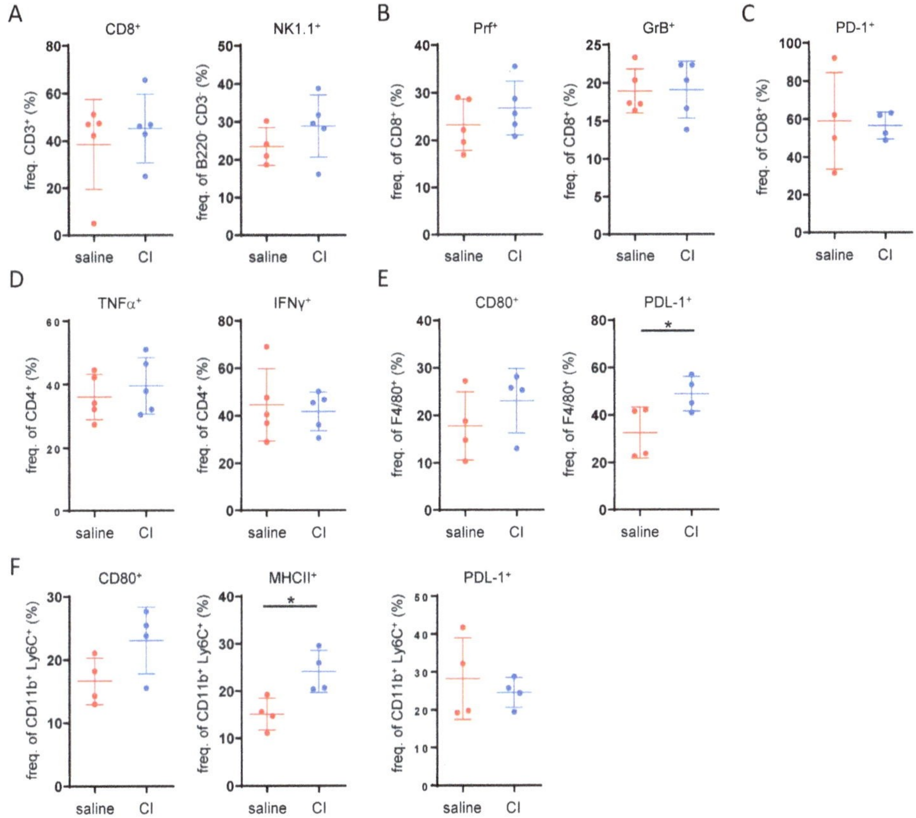

Figure 3. The anti-tumor activity of Composition I is not primarily driven by a clear immune-mediated mechanism. Flow cytometry data showing the frequency on intra-tumoral immune cells: (**A**) $CD8^+$ cytotoxic T-cells, and $NK1.1^+$ NK-cells, (**B**) perforin and granzyme B producing $CD8^+$ T-cells, (**C**) PD-1 expressing $CD8^+$ T-cells, (**D**) TNF-α and IFN-γ producing $CD4^+$ T-cells, (**E**) CD80 and PDL-1 expressing F4/80 macrophages and (**F**) CD80, MHCII and PDL-1 expression in $CD11b^+Ly6C^+$ monocytes. * $p < 0.05$.

2.5. Composition II Exhibits Greater Anti-Tumor Potential in the AOM/DSS Inflammation-Mediated CRC Model

To validate the anti-tumor potential of the probiotics, the therapeutic effects of both compounds were tested in the azoxymethane/dextran sodium sulfate (AOM/DSS) inflammation-dependent model of colitis-associated cancer. In this model, tumor development is induced by administering the chemical carcinogen AOM in combination with DSS, which triggers chronic intestinal inflammation. The experiment involved four cycles of DSS administration in drinking water, each lasting 7 days, which was followed by a 14-day recovery period with regular drinking water (no DSS). After the final DSS cycle, the animals underwent a 4-week recovery period. In each cycle, AOM was administered via intraperitoneal injection on days 1 and 9 (Figure 5A). At the end of each DSS cycle, mice received daily gavage of saline, CI or CII for 3 consecutive days. Throughout the experiment, mice in each group showed weight loss in response to DSS and AOM treatments but quickly regained weight during phases with regular drinking water (Figure 5B). To evaluate the impact of probiotics on intestinal inflammation, colonoscopy was performed at the end of the experiment. The murine endoscopic index of colitis severity (MEICS) score indicated that CI had a limited effect, while CII reduced colonic inflammation (Figure 5C). While both compositions had only little effect on colon length (which typically shortens upon the presence of intestinal inflammation), mice treated with CII did not develop tumors in the colon (Figure 5D).

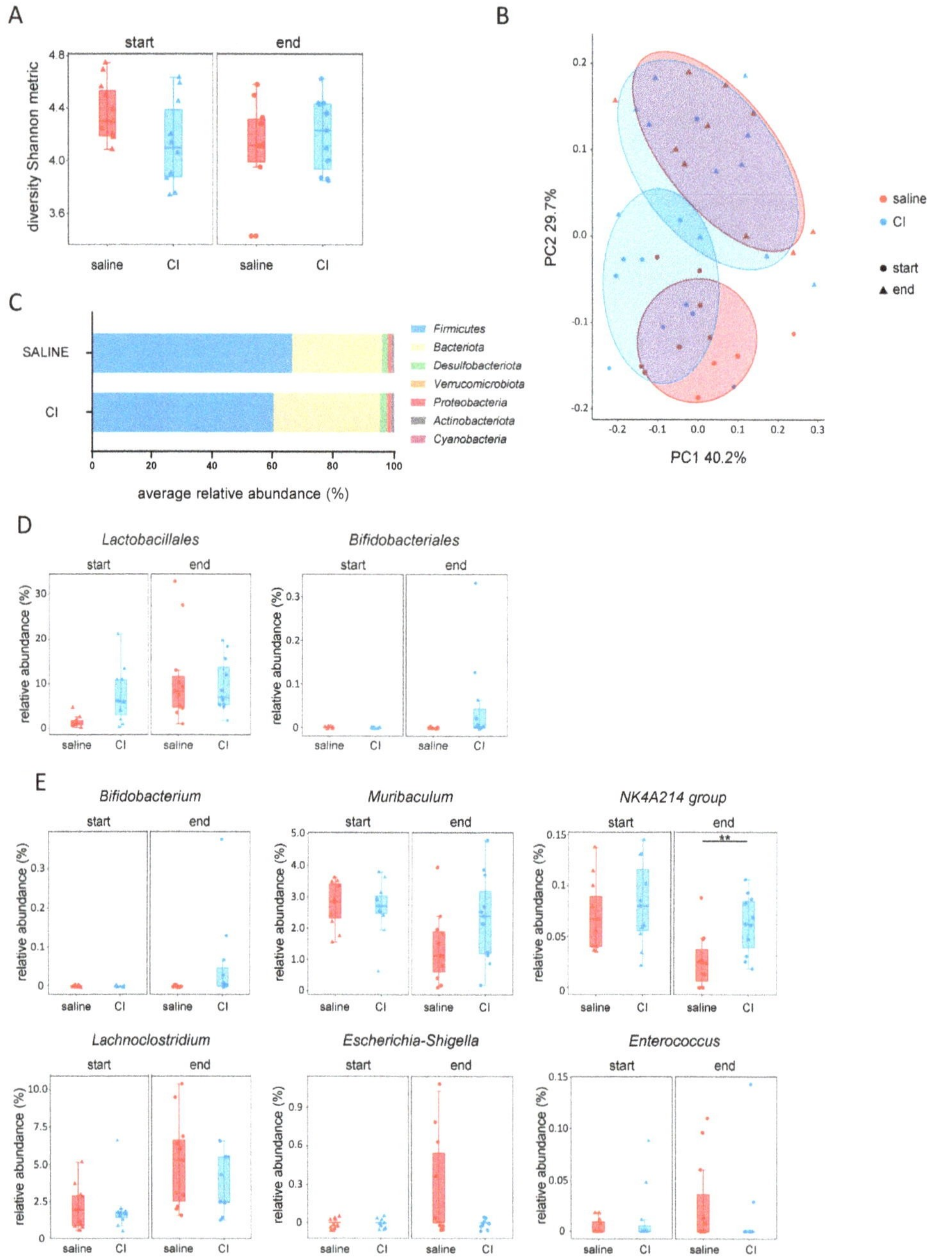

Figure 4. Composition I administration shapes the gut microbiome to foster a healthy microbial environment in the MC-38 tumor model. (**A**) Alpha diversity (Shannon metric index); (**B**) beta diversity (weighted UniFrac); (**C**) relative abundance of phyla at the end of experiment; (**D**) probiotic abundance: *Lactobacillus* and *Bifidobacterium*; (**E**) genus-level abundance of *Bifidobacterium*, *Muribaculum*, *NK4A214 group*, *Lachnoclostridium*, *Escherichia-Shigella*, and *Enterococcus*.

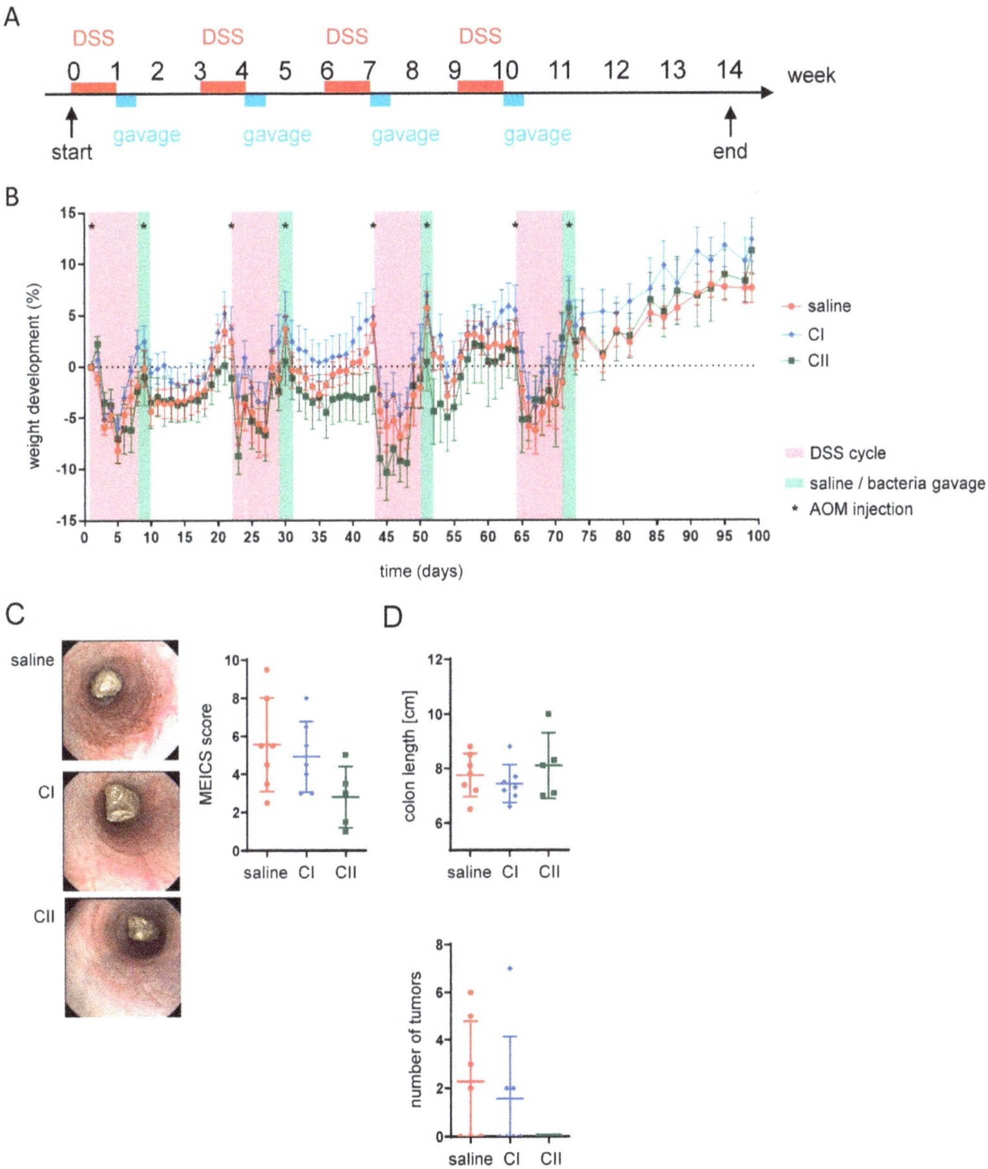

Figure 5. Composition II demonstrates higher anti-tumor potential in the AOM/DSS inflammation-mediated CRC model. (**A**) Experimental design: colon tumors were induced by co-administration of DSS in drinking water (four 1-week cycles followed by 2 weeks of only water recovery period) and AOM intraperitoneal injections on days 1 and 9 of each cycle; mice received daily gavages of saline, composition I (CI), or composition II (CII) for 3 days at the end of each DSS cycle; (**B**) weight development of mice throughout the experiment; (**C**) representative pictures from mouse colonoscopy performed at the end of the experiment and the respective murine endoscopic index of colitis severity (MEICS) score; (**D**) colon length and number of tumors within the colon.

When examining the fecal microbiome composition after probiotics treatment by 16s rRNA sequencing, no significant differences were observed in the abundance of *Lactobacil-*

lus. In contrast, *Bifidobacterium* increased with the administration of both compositions (Figure 6).

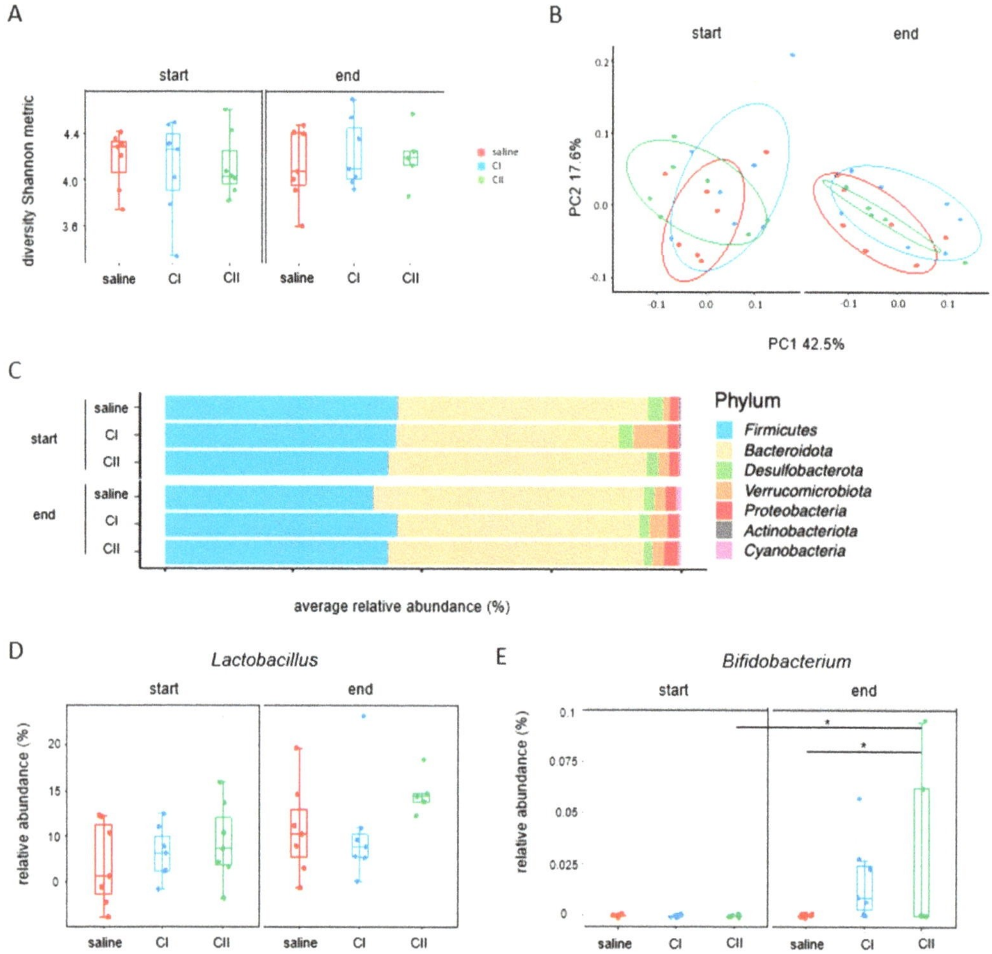

Figure 6. Impact of composition I and composition II on the fecal microbiome in the AOM/DSS inflammation-mediated CRC model. (**A**) Alpha diversity (Shannon metric index); (**B**) beta diversity (weighted UniFrac); (**C**) relative abundance of phyla at the start and at the end of the experiment; (**D**,**E**) relative abundance of probiotics: *Lactobacillus* and *Bifidobacterium*. * $p < 0.05$.

3. Discussion

Our study highlights the potential of probiotics in the prevention of CRC through the modulation of the gut microbiota while also revealing key differences in their effectiveness depending on microbial composition and disease model. The probiotic composition I (CI, *lactobacilli* and *bifidobacteria* mix) and II (CII, *bifidobacteria*) demonstrated distinct effects in altering the microbial composition and influencing tumor progression in the two well-established murine CRC models that were used here. The alterations observed in the gut microbiota following probiotic administration indicate a potential to promote a potentially healthier microbial environment.

The administration of both CI and CII resulted in distinct microbial shifts. CI demonstrated the ability to increase the abundance of *Bifidobacterium* and other health-associated genera such as *Muribaculum* and *Lachnospiraceae NK4A214 group* while reducing pathogenic taxa including *Escherichia-Shigella* and *Enterococcus*. Together with the tumor-reducing effects that we observed upon the administration of CI, this aligns with the hypothesis that a balanced microbiota might mitigate CRC-associated dysbiosis [16,17]. Furthermore, the enhanced abundance of beneficial taxa in CI-treated mice, combined with the inhibition of pro-tumorigenic genera such as *Lachnoclostridium*, further supports the role of probiotics in reshaping the gut microbiome to counteract carcinogenesis [16]. These findings are particularly relevant given the pressing need for novel therapeutic intervention strategies and the growing interest in microbiota-based interventions for CRC prevention [18].

Notably, CI treatment was associated with an increased abundance of *Lachnospiraceae NK4A214*, which is a known butyrate-producing group. While our untargeted metabolomics did not show significant differences in systemic metabolite profiles, we acknowledge that targeted short-chain fatty acid (SCFA) quantification, particularly of butyrate in fecal or cecal samples, could provide more specific insight. This remains an important avenue for future investigation, as local SCFA production may contribute to the tumor-suppressive effects observed with CI.

In addition, our data demonstrate a clear distinction in the anti-tumor potential of the two compositions depending on the CRC model employed. In the orthotopic tumor injection model, CI exhibited significant anti-tumor activity, which was evidenced by reduced tumor weight and volume. In contrast, CII demonstrated superior efficacy in the AOM/DSS inflammation-mediated CRC model, effectively reducing colonic inflammation and preventing tumor development. The differential efficacy of CI and CII across CRC models highlights the importance of tailoring probiotic interventions to specific disease contexts. The ability of CI to promote a healthy microbial environment and reducing pathogenic bacteria growth is particularly relevant in contexts where microbiota dysbiosis is associated with tumor progression. Conversely, the anti-inflammatory properties of CII render it a promising candidate for addressing inflammation-driven CRC. Although we did not perform in-depth immune or metabolite profiling in the CII group, the rapid clinical improvement suggests early anti-inflammatory effects. Whether these effects are mediated by microbial changes or through host responses to probiotic-derived components remains to be clarified. We recognize that the lack of mechanistic data for CII is a limitation and have now outlined this as a priority for future research. Overall, these findings emphasize the importance of further investigation into the mechanistic underpinnings of probiotic action, particularly the role of microbial metabolites and their interaction with host pathways.

Our results suggest that the observed anti-tumor effects of CI in the MC-38 injection model may be mediated through microbiota-driven pathways rather than direct immune or metabolomic modulation. Despite slight trends in the induction of cytotoxic $CD8^+$ T-cells and macrophage activation, no significant changes in cytokine production, cytotoxicity or immune activation were observed. Furthermore, serum metabolomics analysis revealed no significant differences between the CI-treated and control groups with close clustering of the two groups in PCA plots, indicating subtle or minimal differences in the metabolomic profiles.

While our study provides evidence of the anti-tumor potential of probiotics, there are also some limitations. One notable limitation was the relatively small group size in the orthotopic MC-38 cecum injection model which was accounted to logistical challenges related to surgical cell implantation procedures, such as the availability of trained personnel and the need to ensure consistent, precise operations, which restricted the number of animals that could be treated simultaneously. While the results observed in this model are promising and consistent across experiments, future studies should aim to include larger cohorts to strengthen statistical power and enhance the robustness of the findings. Secondly, the reduced tumor size observed in treated animals, while indicative of therapeutic efficacy, resulted in less material available for subsequent analyses, which limited the extent of downstream investigations, such as immune cell characterization and more comprehensive molecular profiling. Additionally, variability in immune responses between experiments was observed, highlighting the complexity of probiotic–host interactions and the need for more in-depth investigations into their molecular mechanisms. The long-term effects of probiotics on CRC progression and recurrence remain to be explored. Lastly, translating these findings to human CRC requires clinical validation, as murine models may not fully recapitulate the complexities of human disease.

Importantly, the anti-tumor effects of both compositions were observed alongside distinct microbial patterns, supporting the hypothesis that probiotics may function through microbiota-driven mechanisms, which are shaped by both composition and disease context. However, our microbiota analyses were limited to initial and final time points. While this allowed us to assess global shifts, it likely missed dynamic or transient microbial changes during disease progression, particularly in the long-term AOM/DSS model. We now acknowledge this as a limitation and emphasize the need for intermediate time-point sampling in future longitudinal studies to better capture microbiota–host interactions over time.

Despite the limitations mentioned, this study provides several novel contributions. First, to our knowledge, this is among the first direct comparisons of two compositionally distinct probiotic blends across two CRC models, revealing model-specific efficacy. Second, we challenge the common assumption that probiotics exert anti-tumor effects primarily through immune activation. In the case of CI, our data suggest a microbiota-centric mode of action, which may operate independently of systemic immune responses. Finally, the differential response to CI and CII across models underscores the need for more personalized and context-aware strategies in microbiota-based therapy.

Future research should investigate the combinatorial effects of probiotics with other therapeutic modalities, such as chemotherapy and immunotherapy, since the integration of probiotics into existing treatment regimens may enhance therapeutic efficacy while minimizing side effects.

In conclusion, our study reinforces the therapeutic potential of probiotics in CRC prevention and treatment, highlighting the importance of model-specific strategies. It contributes to the growing evidence supporting microbiota-targeted interventions in oncology, encourages further research into the mechanistic pathways involved, including microbial metabolites and host interactions, and lays the groundwork for integrating probiotics into personalized medicine frameworks. More personalized approaches to probiotic therapy, guided by individual microbiome profiles, hold promise for optimizing outcomes in the prevention and treatment of CRC in the future.

4. Materials and Methods

4.1. Animal Experiments

All animal experiments were performed according to the Swiss Animal Welfare Legislation and approved by the local veterinary office, the Veterinary Office of the Canton Zurich (Licenses ZH153-2020 and ZH154-2020). Mice: In all studies, 12-week-old C57BL/6J female wild-type (WT) mice purchased from Janvier Labs, Le Genest-Saint-Isle, France, were used for the experiment. Upon arrival at our facility, the mice were given 2 weeks of acclimatization time. The mice were kept in a specific pathogen-free (SPF) facility with chow and water ad libitum. Study design: The primary outcome of the study was either the tumor volume measured in mm^3 or number of colon tumors (AOM/DSS model). Randomization: Prior to the experiment, the mice were randomized using a random number generator. Blinding: Treatment administration was conducted blindly by using allocation groups to ensure that all animals in the experiment were handled, monitored, and treated the same way. Sample collection at the end of the experiment and sample analysis were conducted in a blinded manner. Statistics: Analysis was performed using the GraphPad Prism 10 software. Sample size calculations: The tumor cells were injected, and the bacteria were administered via oral gavage for each mouse individually; thus, each mouse was considered an experimental unit. Student's *t*-test was performed when comparing only two conditions, while a one-way analysis of variance (ANOVA) test was chosen when comparing more than 3 conditions. Data are presented as means ± standard deviation. *p*-values lower than 0.05 were considered statistically significant.

4.2. Orthotopic Cecum Injection Model

MC-38 cells (isolated and provided from Prof. Dr. Lubor Borsig, University Zürich, Zürich, Switzerland): 300,000 cells in high-glucose DMEM (Gibco, Waltham, MA, USA) medium mixed 1:1 with Matrigel (Corning, Corning, NY, USA) were injected into the cecum wall during a surgical procedure under isoflurane anesthesia. Prior to the surgery, mice received a subcutaneous injection of the painkiller Finadyne (MSD Animal Health, Rahway, NJ, USA). After the surgery, an analgesia mix consisting of tramadol (Tramal, Grünenthal, Aachen, Germany) and paracetamol (Dafalgan, UPSA, Rueil Malmaison, France) was administered in the drinking water for 2–3 days. Mice were sacrificed 14 days post tumor cells injection.

4.3. AOM/DSS Model

Four cycles of dextran sodium sulfate (DSS, Fisher Scientific/MP Biomedicals, Irvine, CA, USA) were applied; in each, 1–1.5% DSS was administered in the drinking water for 7 days, which was followed by 14 days of recovery with drinking water only. Azoxymethane (AOM, Sigma-Aldrich/Merck, St. Louis, MO, USA) was injected intraperitoneally (1 mg/kg in 100 µL PBS) on the 1st and 9th day of each DSS cycle. After the final DSS cycle, mice received normal drinking water for 4 weeks before sacrifice.

4.4. Probiotics

Composition I (*Bifidobacterium lactis PI61, Bifidobacterium bifidum PI62, Bifidobacterium lactis PI63, Bifidobacterium breve PI64, Lactobacillus acidophilus PI1, Lactobacillus rhamnosus PI28, Lactobacillus gasseri PI17*, and *Lactobacillus acidophilus PI4* mix, abbreviated as CI) and composition II (*Bifidobacterium lactis PI61, Bifidobacterium bifidum PI62, Bifidobacterium lactis PI63*, and *Bifidobacterium breve PI64 mix*, abbreviated as CII) were provided by PiLeJe (Paris, France) and administered by oral gavage (10×10^8 CFU per dose). For every administration, probiotics were freshly prepared by reconstituting frozen lyophilizates in saline. In the short-term administration experiment and in cecum injection experiments, bacteria were applied once daily in two periods: days 4–6 and days 11–13. In the long-term administration experiment and in the AOM/DSS experiment, bacteria were applied once daily for the first 3 days of weeks 1, 4, 7 and 10 of the experiment. In the control groups, mice received gavages of saline instead of bacteria.

4.5. Clinical Readouts

In every cecum injection experiment, the obtained tumors were weighted and measured. The tumor volume was calculated according to the following formula: $V = \frac{4}{3} \times 3.14 \times \frac{a}{2} \times \frac{b}{2} \times \frac{c}{2}$, where "V" is the tumor volume, and "a", "b", "c" are the tumor dimensions. In the AOM/DSS experiment colon length was measured, number of tumors within the colon was counted and colonoscopy was performed using a mouse endoscope (Karl Storz, Tuttlingen, Germany) and evaluated using the murine endoscopic index of colitis severity (MEICS score) as described by Becker et al. [19], which consists of scoring for the thickening and granularity of the mucosal wall as well as alterations in vascularity, stool consistency and the presence of fibrin.

4.6. Tumor Processing for Flow Cytometry

Upon removal, tumors were placed in PBS and kept on ice prior to further processing. Once all the tumors were collected, each tumor was cut into pieces using small size surgical scissors. Tumors were then enzymatically digested using the Tumor Dissociation Kit Mouse (Miltenyi Biotec, Bergisch Gladbach, Germany) according to the manufacturer's instructions.

4.7. Flow Cytometry

Single-cell suspensions obtained from each tumor were divided into two different sets for flow cytometry staining. One set was re-stimulated using a cocktail for immune cell activation and subsequently stained for cytokine's evaluation, while the other set was stained immediately after tissue processing. Re-stimulation was performed using 50 ng/mL PMA (Sigma Aldrich, St. Louis, MO, USA), 1 µg/mL ionomycin (Sigma Aldrich) and 1 µg/mL brefeldin A (Sigma Aldrich) in RPMI 1640 medium (Thermo Fisher Scientific, Waltham, MA, USA). Cells were re-stimulated in 12-well cell culture plates (TPP) for 3 h in 5% CO_2. First, viability staining was performed using Zombie NIR (BioLegend) for 30 min at room temperature in the dark. Cells where then washed with PBS and stained with TruStain FcS™ PLUS (BioLegend) for 5 min at 4 °C in the dark. Further, cells were washed with PBS and stained with anti-CD4-BUV563 (BD Biosciences, Franklin Lakes, NJ, USA), anti-NK1.1-BUV395 (BD Biosciences, Franklin Lakes, NJ, USA), anti-CD3-BV785 (BioLegend, San Diego, CA, USA), anti-CD45-PB (BioLegend, San Diego, CA, USA), anti-CD45R(B220)-PE-Cy5 (Thermo Fisher Scientific, Waltham, MA, USA), and anti-CD8-PE-CF594 (BD Biosciences, Franklin Lakes, NJ, USA) antibodies for 15 min at 4 °C in the dark. After washing with MACS buffer (autoMACS™ Running Buffer, Miltenyi Biotec, Bergisch Gladbach, Germany), cells were permeabilized for 20 min at 4 °C in the dark using the BD Cytofix/Cytoperm (BD Biosciences, Franklin Lakes, NJ, USA) kit and stained intracellularly using anti-CD4-BUV563 (BD Biosciences, Franklin Lakes, NJ, USA), anti-CD3-BV785 (BioLegend), anti-TNFα-BV650 (BioLegend), anti-IL-17-BV510 (BioLegend), anti-Granzyme-B-PerCP-Cy5.5 (BioLegend), anti-Perforin-FITC (e-Bioscience), anti-IFN-γ-PE-Cy7 (BioLegend), and anti-CD8A-PE-CF594 (BD Biosciences, Franklin Lakes, NJ, USA) antibodies for 30 min at 4 °C in the dark. Finally, cells were washed with BD Perm Wash (BD Biosciences) and kept at 4 °C in 0.1% paraformaldehyde (PFA, Sigma-Aldrich, St. Louis, MO, USA) until acquisition. For the second set of cells without the re-stimulation, cells were processed in the same way; however, the surface staining consisted of the following antibodies: anti-CD80-BUV737 (BD Biosciences, Franklin Lakes, NJ, USA), anti-Ly6G-BUV563 (BD Biosciences, Franklin Lakes, NJ, USA), anti-CD4-BUV496 (BD Biosciences, Franklin Lakes, NJ, USA), anti-CD3-BV785 (BioLegend, San Diego, CA, USA), anti-CD45-BV650 (BioLegend, San Diego, CA, USA), anti-CD11b-BV605 (BioLegend, San Diego, CA, USA), anti-Ly6C-BV570 (BioLegend, San Diego, CA, USA), anti-CD274(PD-L1)-BV480 (BD Biosciences, Franklin Lakes, NJ, USA), anti-F4/80-FITC (BioLegend, San Diego, CA, USA), anti-CD11c-PE-Cy7 (BioLegend, San Diego, CA, USA), anti-CD45R(B220) (Thermo Fisher Scientific, Waltham, MA, USA), anti-CD8A-PE-CD594 (BD Biosciences, Franklin Lakes, NJ, USA), anti-I-A/I-E(MHCII)-AF700 (BioLegend, San Diego, CA, USA) and anti-CD279(PD-1)-APC (BioLegend, San Diego, CA, USA), and anti-CD274(PD-L1)-BV480 (BD Biosciences, Franklin Lakes, NJ, USA). Cells were acquired using the FACSymphony (BD

Biosciences, Franklin Lakes, NJ, USA) and the analysis was performed using the FlowJo software (Version 10, FlowJo LLC/BD, Ashland, OR, USA).

4.8. Microbiota 16S Sequencing

Fecal samples were collected at the beginning and at the end of each experiment and subsequently stored at −80 °C prior to analysis. DNA extraction and sequencing of the V4 region of 16S ribosomal DNA was performed by Microsynth AG (Balgach, Switzerland). The QIIME2 pipeline was utilized for analyzing 16S rRNA [20]. After assessing data quality, denoising with DADA2 was conducted to merge the paired reads, producing amplicon sequence variants (ASVs) [21]. The alpha rarefaction module was employed to ensure sufficient depth for capturing most features. We then calculated both alpha and beta diversity using the core-metrics-phylogenetic module. For the alpha analysis, we utilized the faith phylogenetic matrix to compute richness and incorporate phylogenetic relationships. For beta diversity, we employed the weighted UniFrac distance matrix to quantify the dissimilarity between communities. Principal coordinate analysis (PCoA) was conducted to enhance the visualization of beta diversity. Taxonomy was assigned to ASVs using the q2-feature-classifier classify-sklearn naïve Bayes taxonomy classifier against the pre-trained Naïve Bayes silva-132-99-nb-classifier, which was trained on Silva (release 132) full-length sequences [22]. The Taxa barplot module was used to visualize the different taxonomy compositions.

4.9. Serum Metabolomics

On the last day of the experiment, blood was collected into BD Microtainer tubes with a serum separator and centrifuged at the maximum speed for 5 min at room temperature. The obtained serum was transferred into a 0.5 mL Eppendorf tube and immediately snap frozen in liquid nitrogen, and all samples were stored at −80 °C prior to analysis. Metabolomics analysis was carried out with a high-resolution mass spectrometer (Agilent QTOF 6550, Agilent, Santa Clara, CA, USA), and analysis was performed by MetaboAnalystR package version 4.2 for the statistical analysis [23]. The normalized intensities by the median for each sample were mean-centered and divided by the standard deviation of each variable for all annotated features. A feteroscedastic (two-tail, unequal variance) *t*-test was used to determine significant differences between the groups. In addition, *p*-values were adjusted according to Benjamini–Hochberg, and q-values were adjusted according to Storey and Tibshirani. Principle component analysis (PCA) was used to visualize sample variance. Heatmap visually presented hierarchical clustering using the Euclidean method for distance calculation and Ward's linkage for clustering.

4.10. Statistical Analysis

Analysis was performed using the GraphPad Prism 10 software. Student's *t*-test was performed when comparing only two conditions, while an ANOVA test with Tukey's HSD was chosen when comparing more than 3 conditions. Data are presented as means ± standard deviation. *p*-values lower than 0.05 were considered statistically significant.

Supplementary Materials: The following supporting information can be downloaded at: https://www.mdpi.com/article/10.3390/ijms26094404/s1.

Author Contributions: Conceptualization: M.S. (Michael Scharl), S.H. and A.L.; research overview: M.S. (Michael Scharl) and A.N.; data acquisition: A.N. and M.S. (Marlene Schwarzfischer), P.W., M.D., D.P., M.W. (Marcin Wawrzyniak), E.G., M.R.W., K.A., M.W. (Marijn Wilmink), L.L. and M.R.S.; formal analysis: A.N., P.W., M.W. (Marcin Wawrzyniak) and Y.M.; data interpretation: A.N.; writing—original draft preparation: A.N.; writing—review and editing: M.S. (Michael Scharl). All authors have read and agreed to the published version of the manuscript.

Funding: The study was supported by a research grant from PiLeJe SA, France.

Institutional Review Board Statement: Not applicable.

Informed Consent Statement: Not applicable.

Data Availability Statement: The original contributions presented in this study are included in the article. Further inquiries can be directed to the corresponding author.

Acknowledgments: Flow cytometry data were recorded using the instruments of the Flow Cytometry Facility at the University of Zurich in Zurich, Switzerland.

Conflicts of Interest: M.S. (Michael Scharl) has shares and is co-founder of Recolony AG, Zurich, CH and has shares in PharmaBiome AG, Zurich, CH. M.S. (Michael Scharl) served as Advisor for Abbvie, Gilead, Fresenius, Topadur, Takeda, Roche and Celltrion. M.S. (Michael Scharl) received speaker's honoraria from Janssen, Falk Pharma, Vifor Pharma, Pileje, Phytolis and Bromatech. M.S. (Michael Scharl) received research grants from Abbvie, Takeda, Gilead, Gnubiotics, Roche, Axalbion, Pharmabiome, Topadur, Basilea, MBiomics, Storm Therapeutics, LimmatTech, Zealand Pharma, NodThera, Calypso Biotech, Pileje, Herbodee, Vifor, Menarini. Anne Leblanc and Sophie Holowacz are employees of PiLeJe. The remaining authors have no conflicts of interest to declare.

References

1. Morgan, E.; Arnold, M.; Gini, A.; Lorenzoni, V.; Cabasag, C.J.; Laversanne, M.; Vignat, J.; Ferlay, J.; Murphy, N.; Bray, F. Global burden of colorectal cancer in 2020 and 2040: Incidence and mortality estimates from GLOBOCAN. *Gut* **2023**, *72*, 338–344. [CrossRef]
2. World Health Organization (WHO). Cancer: Fact Sheets. Available online: https://www.who.int/news-room/fact-sheets/detail/cancer (accessed on 5 January 2025).
3. Zajkowska, M.; Mroczko, B. A Novel Approach to Staging and Detection of Colorectal Cancer in Early Stages. *J. Clin. Med.* **2023**, *12*, 3530. [CrossRef] [PubMed]
4. Marmol, I.; Sanchez-de-Diego, C.; Pradilla Dieste, A.; Cerrada, E.; Rodriguez Yoldi, M.J. Colorectal Carcinoma: A General Overview and Future Perspectives in Colorectal Cancer. *Int. J. Mol. Sci.* **2017**, *18*, 197. [CrossRef]
5. Dougherty, M.W.; Jobin, C. Intestinal bacteria and colorectal cancer: Etiology and treatment. *Gut Microbes* **2023**, *15*, 2185028. [CrossRef] [PubMed]
6. Wong, C.C.; Yu, J. Gut microbiota in colorectal cancer development and therapy. *Nat. Rev. Clin. Oncol.* **2023**, *20*, 429–452. [CrossRef] [PubMed]
7. Rebersek, M. Gut microbiome and its role in colorectal cancer. *BMC Cancer* **2021**, *21*, 1325. [CrossRef]
8. Hill, C.; Guarner, F.; Reid, G.; Gibson, G.R.; Merenstein, D.J.; Pot, B.; Morelli, L.; Canani, R.B.; Flint, H.J.; Salminen, S.; et al. Expert consensus document. The International Scientific Association for Probiotics and Prebiotics consensus statement on the scope and appropriate use of the term probiotic. *Nat. Rev. Gastroenterol. Hepatol.* **2014**, *11*, 506–514. [CrossRef]
9. Chong, E.S. A potential role of probiotics in colorectal cancer prevention: Review of possible mechanisms of action. *World J. Microbiol. Biotechnol.* **2014**, *30*, 351–374. [CrossRef]
10. Latif, A.; Shehzad, A.; Niazi, S.; Zahid, A.; Ashraf, W.; Iqbal, M.W.; Rehman, A.; Riaz, T.; Aadil, R.M.; Khan, I.M.; et al. Probiotics: Mechanism of action, health benefits and their application in food industries. *Front. Microbiol.* **2023**, *14*, 1216674. [CrossRef]
11. Markowiak-Kopec, P.; Slizewska, K. The Effect of Probiotics on the Production of Short-Chain Fatty Acids by Human Intestinal Microbiome. *Nutrients* **2020**, *12*, 1107. [CrossRef]
12. Torres-Maravilla, E.; Boucard, A.S.; Mohseni, A.H.; Taghinezhad, S.S.; Cortes-Perez, N.G.; Bermudez-Humaran, L.G. Role of Gut Microbiota and Probiotics in Colorectal Cancer: Onset and Progression. *Microorganisms* **2021**, *9*, 1021. [CrossRef] [PubMed]
13. Khaleel, S.M.; Shanshal, S.A.; Khalaf, M.M. The Role of Probiotics in Colorectal Cancer: A Review. *J. Gastrointest. Cancer* **2023**, *54*, 1202–1211. [CrossRef] [PubMed]
14. Neto, I.; Rocha, J.; Gaspar, M.M.; Reis, C.P. Experimental Murine Models for Colorectal Cancer Research. *Cancers* **2023**, *15*, 2570. [CrossRef]
15. Li, C.; Lau, H.C.; Zhang, X.; Yu, J. Mouse Models for Application in Colorectal Cancer: Understanding the Pathogenesis and Relevance to the Human Condition. *Biomedicines* **2022**, *10*, 1710. [CrossRef]
16. Li, Q.; Liu, D.; Liang, M.; Zhu, Y.; Yousaf, M.; Wu, Y. Mechanism of probiotics in the intervention of colorectal cancer: A review. *World J. Microbiol. Biotechnol.* **2024**, *40*, 306. [CrossRef] [PubMed]
17. Mahooti, M.; Abdolalipour, E.; Sanami, S.; Zare, D. Inflammatory Modulation Effects of Probiotics: A Safe and Promising Modulator for Cancer Prevention. *Curr. Microbiol.* **2024**, *81*, 372. [CrossRef]

18. Hu, Y.; Zhou, P.; Deng, K.; Zhou, Y.; Hu, K. Targeting the gut microbiota: A new strategy for colorectal cancer treatment. *J. Transl. Med.* **2024**, *22*, 915. [CrossRef]
19. Becker, C.; Fantini, M.C.; Wirtz, S.; Nikolaev, A.; Kiesslich, R.; Lehr, H.A.; Galle, P.R.; Neurath, M.F. In vivo imaging of colitis and colon cancer development in mice using high resolution chromoendoscopy. *Gut* **2005**, *54*, 950–954. [CrossRef] [PubMed]
20. Bolyen, E.; Rideout, J.R.; Dillon, M.R.; Bokulich, N.A.; Abnet, C.C.; Al-Ghalith, G.A.; Alexander, H.; Alm, E.J.; Arumugam, M.; Asnicar, F.; et al. Reproducible, interactive, scalable and extensible microbiome data science using QIIME 2. *Nat. Biotechnol.* **2019**, *37*, 852–857. [CrossRef]
21. Callahan, B.J.; McMurdie, P.J.; Rosen, M.J.; Han, A.W.; Johnson, A.J.; Holmes, S.P. DADA2: High-resolution sample inference from Illumina amplicon data. *Nat. Methods* **2016**, *13*, 581–583. [CrossRef]
22. Bokulich, N.A.; Kaehler, B.D.; Rideout, J.R.; Dillon, M.; Bolyen, E.; Knight, R.; Huttley, G.A.; Gregory Caporaso, J. Optimizing taxonomic classification of marker-gene amplicon sequences with QIIME 2's q2-feature-classifier plugin. *Microbiome* **2018**, *6*, 90. [CrossRef] [PubMed]
23. Pang, Z.; Chong, J.; Li, S.; Xia, J. MetaboAnalystR 3.0: Toward an Optimized Workflow for Global Metabolomics. *Metabolites* **2020**, *10*, 186. [CrossRef] [PubMed]

International Journal of
Molecular Sciences

MDPI

Article

Anti-Obesity Effects of LB-GABA

Hyein Han, Gunju Song, Jongwon Kim, Heegu Jin and Boo-Yong Lee *

Department of Food Science and Biotechnology, College of Life Science, CHA University, Seongnam 13488, Republic of Korea; hyeinoo@naver.com (H.H.); juhun022188@naver.com (G.S.); nanananakim@naver.com (J.K.); heegu94@hanmail.net (H.J.)
* Correspondence: bylee@cha.ac.kr

Abstract: Obesity is characterized by an excessive imbalance in energy metabolism and is associated with metabolic syndrome. Mammals have two types of adipose tissue: white adipose tissue (WAT) and brown adipose tissue (BAT). These are key factors in regulating the energy balance. Strategies aimed at reducing obesity should encompass not only the prevention of lipid accumulation but also the stimulation of browning in both WAT and BAT, with the aim of enhancing energy expenditure. In this study, the mechanism by which Lactobacillus brevis-fermented gamma-aminobutyric acid (LB-GABA) prevents obesity was investigated, as well as whether it induces lipolysis and browning in WAT using 3T3-L1 adipocytes. The expression of proteins involved in signaling pathways regulating lipid accumulation and degradation, as well as browning, was measured using Western blotting analysis. We demonstrated that LB-GABA significantly inhibited lipid accumulation by suppressing adipogenesis and lipogenesis. In addition, the microscopic analysis of WAT demonstrated that LB-GABA reduced the adipocyte size and the number of lipid droplets. Moreover, Western blot analysis revealed that GABA increased lipolysis and activated the protein kinase A (PKA) signaling pathway, which promotes uncoupling protein 1 (UCP1)-mediated WAT browning. In conclusion, these results suggest that LB-GABA activates energy expenditure through lipid metabolism regulation and exerts anti-obesity effects.

Keywords: obesity; lipid metabolism; energy expenditure; 3T3-L1 cells; LB-GABA

1. Introduction

Obesity is defined as an imbalance between dietary energy intake and energy expenditure and is also associated with metabolic syndrome [1]. Obesity is a major contributor to the development of metabolic syndrome, a condition that predisposes individuals to several health complications, such as type 2 diabetes, cardiovascular disease, and nonalcoholic fatty liver disease [2]. The prevalence of obesity has increased very rapidly over the last 30 years, posing a serious threat to human health and socio-economic development [3,4]. In recent years, the epidemiology of obesity, predisposing factors, and prevention mechanisms have been the focus of an increasing number of studies [5].

White adipose tissue (WAT) is an important regulator of lipid metabolism and the energy balance [6]. Two types of adipose tissue, WAT and brown adipose tissue (BAT), have been identified as the key regulators of the energy balance in mammals [7]. Obesity is caused by excessive WAT accumulation, where excess energy is stored in the form of triglycerides (TGs) [8]. BAT, composed of adipocytes containing numerous small multifunctional fat droplets, exhibits high levels of mitochondrial biosynthesis and dissipates energy as heat because it expresses uncoupling protein 1 (UCP1) on its inner mitochondrial membrane [9]. Studies have shown that cold exposure, adrenaline stimulation, and other conditions can trigger browning, which results in a shift from WAT to brown-like (beige or light) fat cells [10].

The 3T3-L1 cell line is one of the most well-established models for the study of the differentiation of preadipocytes to mature adipocytes (adipogenesis). The 3T3-L1 cell line was utilized in the study to examine the processes of browning and obesity-related properties. This particular cell line is a well-established preadipocyte cell line developed in mouse embryos. These 3T3-L1 cells manifest an adipocyte-like phenotype when cultivated under the appropriate conditions [11].

Adipocyte differentiation and subsequent lipid accumulation in mature adipocytes occur through a well-organized process of hyperplasia and hypertrophy, respectively [12]. During preadipocyte differentiation, the expression of peroxisome proliferator-activated receptor gamma (PPARγ) and fatty acid binding protein 4 (FABP4) and the elevated expression of lipogenic factors, such as lysophosphatidic acid acyltransferase theta (LPAATθ), Lipin 1, and fatty acid synthase (FAS), lead to increased lipid accumulation and the development of obesity [13]. Lipolysis is defined as the catabolic mechanism that facilitates the breakdown of TGs stored in WAT, resulting in the release of free fatty acids (FFAs) and glycerol [14]. TG hydrolysis is the result of protein phosphorylation of protein kinase A (PKA), which in turn regulates several key lipolytic proteins. In adipocytes, these lipolytic proteins control TG lipase activity [15]. These are transported across the membrane by carnitine palmitoyl transferase 1 (CPT1) and undergo mitochondrial β-oxidation [16,17]. The major lipases involved in the lipolytic process are adipocyte triglyceride lipase (ATGL), diglyceride to monoglyceride lipase, and hormone-sensitive lipase (HSL) [18,19]. Glycerol and FFAs released from WAT are transported through the blood and penetrate into other tissues, where they distribute lipids and regulate the energy balance [18]. In order to address the issue of obesity, there exist methodologies that not only prevent fat accumulation but also promote energy expenditure by activating the browning of BAT and WAT [19]. The expression of UCP1 uncouples electron transfer in the mitochondria to generate energy as heat [20]. WAT and BAT are readily interconvertible [21]. Similarly, the browning of WAT is known to require the expression of the thermogenic protein UCP1 [22], as these phenotypic changes activate thermogenesis, promoting the browning of WAT, which may be a therapeutic approach to treat obesity [23].

LB-GABA consists of about 20% gamma-aminobutyric acid (GABA) (*w/v*) and about 80% modified food starch [24]. GABA is a non-proteinogenic amino acid that is widely found in microorganisms, animals, and plants and is naturally produced in the human organism [25–27]. GABA is produced from glutamate within the central nervous system and functions as a significant inhibitory neurotransmitter in the human brain [28,29]. GABA is transported and used by the transporter [30,31]. The levels of GABA, which have been shown to be significantly reduced in cases of insomnia, anxiety, depression, panic disorder, and the aging process, have been identified as a chronic disease of modern people. Consequently, the supplementation of GABA has emerged as a promising therapeutic approach aimed at mitigating symptoms such as excitement, irritation, and stress while also counteracting the physical and psychological effects of aging [32–35]. The known effects of GABA are as follows: the body's energy balance is regulated by the inhibition of the activation of these neurons through the release of the neurotransmitter GABA, and GABA has been reported to have a variety of physiological effects, including anti-inflammatory and anti-diabetic [36–40], in addition to its role as an endogenous inhibitory neurotransmitter in the mammalian central nervous system [35]. Due to these effects, GABA is used as an active ingredient in pharmaceuticals and food products, and its demand is increasing. However, the supply of foods containing GABA has proven inadequate in meeting this demand [41]. Consequently, GABA is predominantly produced and utilized through fermentation processes involving yeast, fungi, and bacteria [42–45]. The LB-GABA employed in this study was produced using Lactobacillus brevis fermentation, a method which we hypothesize can counterbalance the limited GABA content in food.

However, in 3T3-L1 adipocytes, it is not known whether LB-GABA acts by decreasing the expression of adipogenic proteins, such as PPARγ and FABP4; lipogenic proteins, such as LPAAT θ and Lipin 1; and FA synthesis proteins such as FAS, thereby inhibiting lipogenesis. Its ability to inhibit obesity by increasing the expression of pPKA and its downstream lipolytic proteins, increasing the expression of UCP1 proteins, and increasing the browning of white fat has not been reported. The hypothesis under investigation is that given the proven effects of GABA contained in LB-GABA on oxidative stress and associated inflammation, it would be expected to have potential benefits in terms of efficacy against obesity. In addition, the anti-obesity effect of GABA has already been revealed in previous studies [34], and it is expected that the same effect can be expected from LB-GABA, a fermented GABA. Therefore, this study focused on the effects of LB-GABA on the metabolism of 3T3-L1 cells and investigated the overall anti-obesity signaling of LB-GABA in 3T3-L1 cells.

2. Results

2.1. Effects of LB-GABA on the Viability of 3T3-L1 Adipocytes

To determine the cytotoxic effect of LB-GABA on 3T3-L1 adipocytes and to determine the appropriate concentration to use in our experiments, we measured the effect of LB-GABA on lipid metabolism using the MTT (3-(4,5-dimethylthiazol-2yl)-2,5-diphenyltetrazolium bromide) assay. As shown in Figure 1 and Table S1, treatment with 200 μg/mL LB-GABA resulted in a significant difference, suggesting that it was cytotoxic; therefore, we used low and high concentrations of 25 and 100 μg/mL in the 3T3-L1 cell medium.

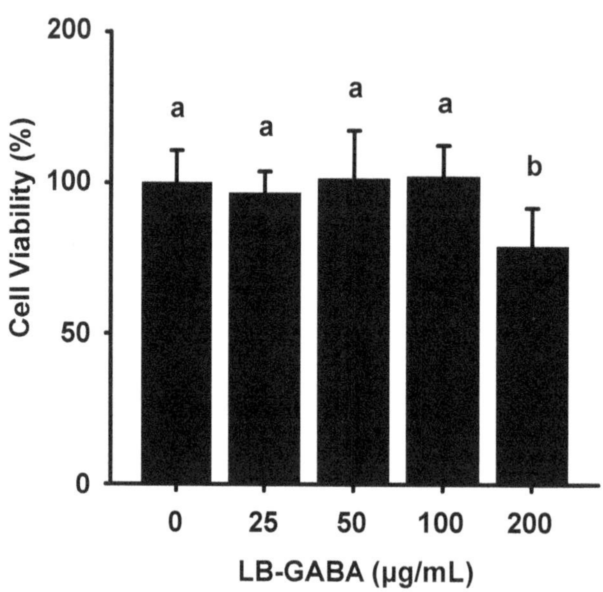

Figure 1. LB-GABA effects on 3T3-L1 cell cytotoxicity. The viability of 3T3-L1 preadipocytes treated with LB-GABA for 24 h was assessed using an MTT assay. Data are expressed as the mean ± SEM (n = 8). Values indicated by different letters are significantly different; $p < 0.05$ (a > b).

2.2. LB-GABA Inhibits Lipid Accumulation

To determine the effect of LB-GABA on lipid accumulation in adipocytes, we induced differentiation with IBMX (3-Isobutyl-1-methylxanthine), dexamethasone, and insulin for 8 days and treated them with two concentrations (25 and 100 μg/mL) of LB-GABA and then used oil red O staining to determine the level of lipid accumulation. As shown in Figure 2A and Table S2, differentiation induced intracellular lipid accumulation, but LB-GABA inhibited lipid accumulation in a dose-dependent manner. Microscopic examination showed that the MDI-treated group had larger lipid droplets, but the LB-GABA-treated group had smaller lipid droplets and fewer number of droplets (Figure 2B,C). Consequently, the findings of this study demonstrate that LB-GABA exerts a regulatory function on adipose tissue accumulation by reducing the magnitude and quantity of adipose tissue droplets in differentiated 3T3-L1 cells.

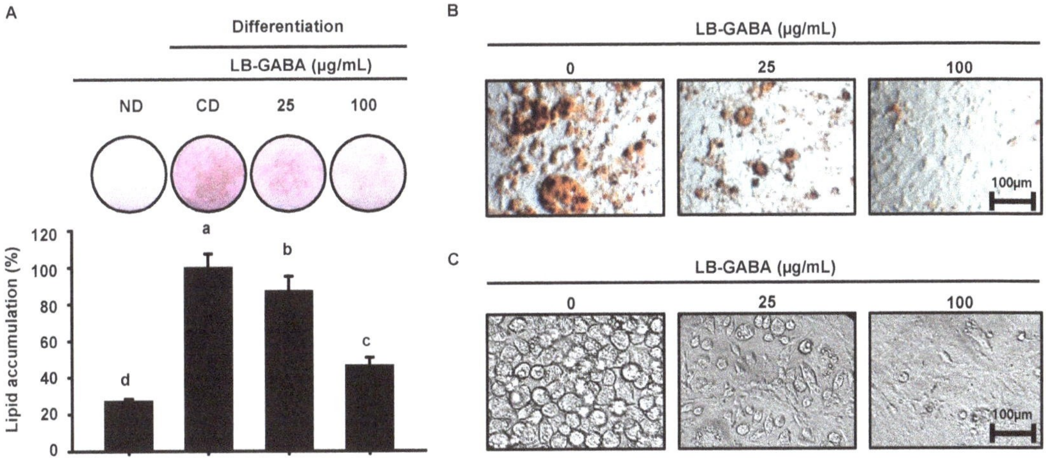

Figure 2. LB-GABA inhibits lipid accumulation in 3T3-L1 adipocytes. (**A**) The effect of LB-GABA on lipid accumulation, as determined by oil red O staining. A differentiation-inducing cocktail, with or without LB-GABA, was added to 3T3-L1 adipocytes. (**B**,**C**) Microscopic images of fully differentiated 3T3-L1 cells were obtained at day 8 (D8). The scale bar represents 100 μm. Data are expressed as the mean ± SEM (n = 12). Values indicated by different letters are significantly different; $p < 0.05$ (a > b > c > d).

In order to investigate the mechanism for the lipid accumulation inhibitory effect of LB-GABA, the expression of key transcription factors and biomarkers of adipocyte differentiation was measured by means of Western blot analysis. As shown in Figure 3A, Tables S3–S5 and S9–S11, the expression of key adipogenesis proteins (PPARγ and FABP4) was increased in the differentiation inducer group, but LB-GABA decreased the expression of these proteins. And as shown in Figure 3B,C, lipogenesis proteins (LPAATθ and Lipin1) and fatty acid synthesis (FAS) proteins were upregulated in the differentiation group, but LB-GABA downregulated their expression. These results suggest that LB-GABA inhibits adipogenesis and lipogenesis in 3T3-L1 cells.

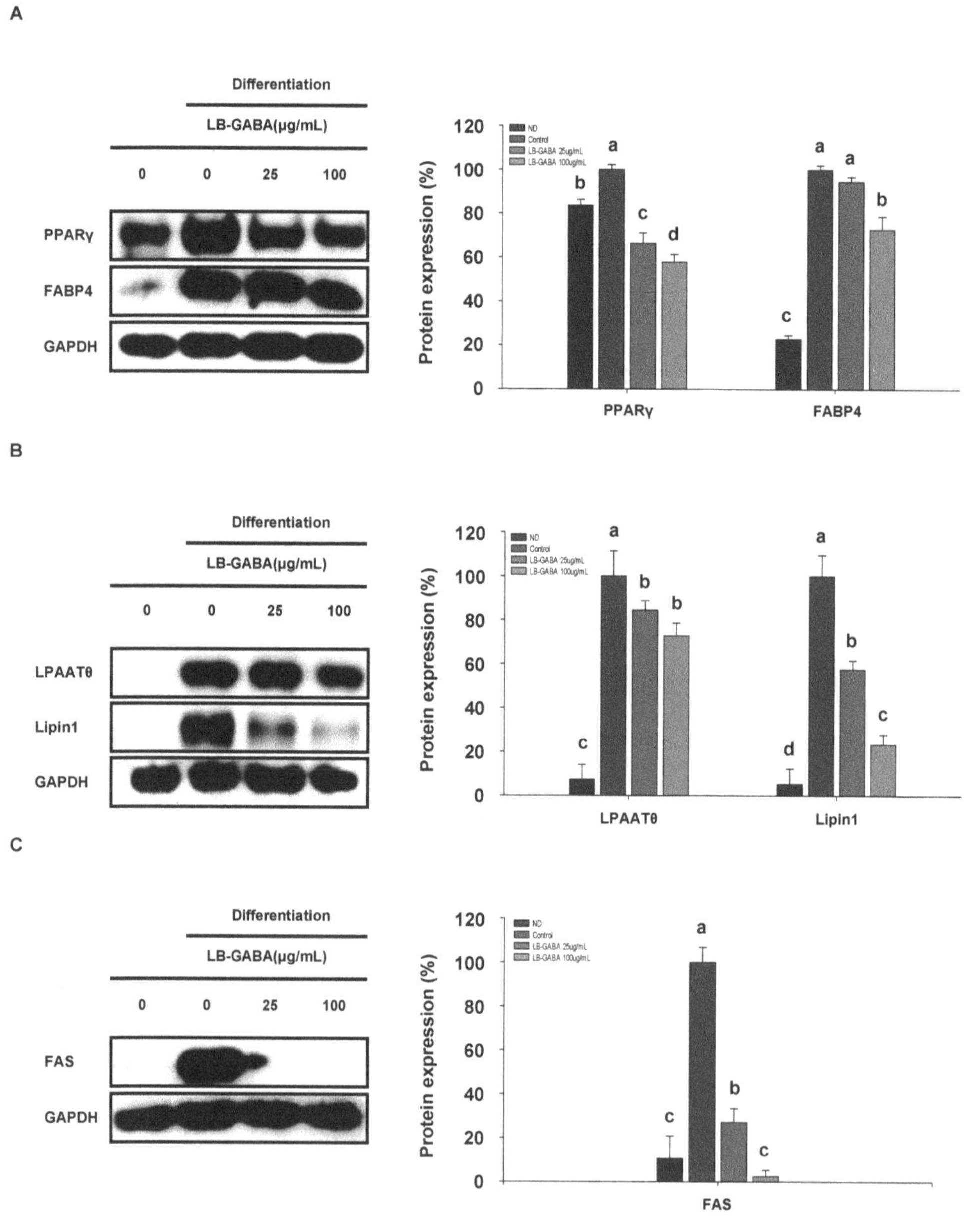

Figure 3. LB-GABA reduces adipogenic and lipogenic proteins in 3T3-L1 cells. Cells were incubated with various concentrations of LB-GABA for 8 days. The expression of (**A**) adipogenic proteins (PPARγ and FABP4), (**B**) lipogenic proteins (LPAAT θ and Lipin 1), and (**C**) proteins involved in fatty acid synthesis (FAS) was determined via Western blot. ND, undifferentiated; control, differentiated. Data are expressed as the mean ± SEM. Values indicated by different letters are significantly different; $p < 0.05$ (a > b > c > d).

2.3. LB-GABA Increases Lipolysis in 3T3-L1 Adipocytes

The process of lipolysis is characterized by the sequential activity of lipolytic enzymes (ATGL and pHSL), which are induced by PKA activation [35]. Therefore, the present study employed Western blot analysis to determine whether LB-GABA increased lipolysis through the upregulation of the expression of lipolytic enzymes. As shown in Figure 4, Tables S6 and S12, treatment with LB-GABA resulted in a dose-dependent increase in the level of phosphorylation of PKA and the expression levels of its sub-proteins (ATGL and pHSL) in 3T3-L1 cells. This result suggests that LB-GABA increases lipolysis in 3T3-L1 cells.

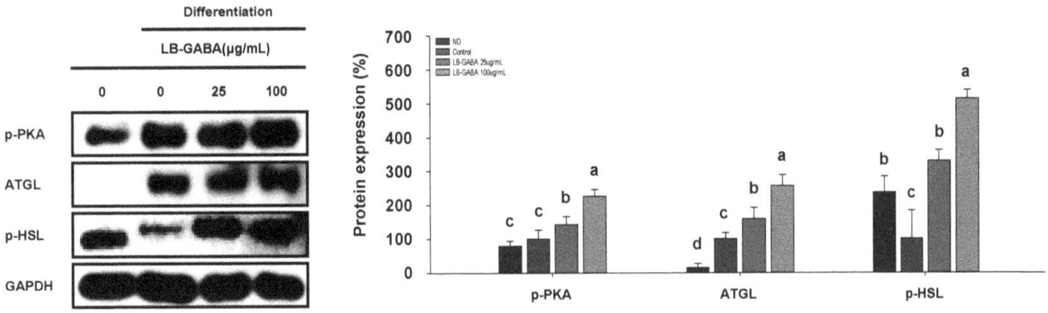

Figure 4. LB-GABA increases lipolysis in 3T3-L1 adipocytes. Adipocytes were cultivated in the presence of a series of concentrations of LB-GABA. The expression of lipolysis proteins (pPKA, ATGL, and pHSL) was determined using Western blot. ND, undifferentiated; control, differentiated. Data are expressed as the mean ± SEM. Values indicated by different letters are significantly different; $p < 0.05$ (a > b > c > d).

2.4. LB-GABA Activates Energy Metabolism by Promoting the Browning of 3T3-L1 Adipocytes

White adipocytes can be browned to resemble brown adipocytes [36]. The presence of UCP1 expression is a defining feature of brown adipocytes, which are known to consume energy through the process of inefficient oxidation of FFAs within the mitochondria. Western blot analysis was therefore performed to measure the expression of the brown adipocyte-specific protein UCP1. As shown in Figure 5, Tables S7 and S13, the administration of LB-GABA resulted in a dose-dependent augmentation in the expression of thermogenic proteins. The present research demonstrates that LB-GABA promotes browning by increasing the expression of UCP1 in mitochondria.

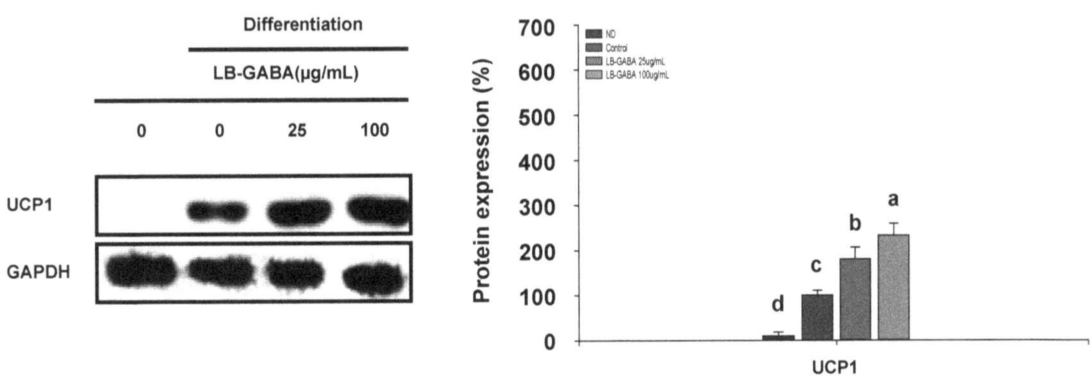

Figure 5. LB-GABA increases browning in 3T3-L1 adipocytes. Adipocytes were cultivated in a series of concentrations of LB-GABA. The expression of a browning protein (UCP1) was determined via Western blot. ND, undifferentiated; control, differentiated. Data are expressed as the mean ± SEM. Values indicated by different letters are significantly different; $p < 0.05$ (a > b > c > d).

2.5. Investigation of LB-GABA's Effect on Reducing Fatty Acid Oxidation in 3T3-L1 Adipocytes

The activities of CPT1 and UCP1 in the mitochondria are linked [37]. The pharmacological ligand of peroxisome proliferator-activated receptor alpha (PPARα) has been shown to activate the expression of genes involved in the processes of fatty acid and glucose oxidation, including carnitine palmitoyltransferase-1A (CPT-1A). At the initial stage of mitochondrial oxidation of long-chain fatty acids, CPT1 functions as a catalyst, facilitating the transfer of long-chain fatty acids from acyl-CoA to carnitine across the mitochondrial membrane [38]. Therefore, we performed a Western blot analysis to determine whether LB-GABA increased the expression of proteins involved in fatty acid oxidation. As shown in Figure 6, Tables S8 and S14, the LB-GABA treatment increased the expression of PPARα and CPT1, which are pivotal in the process of fatty acid oxidation. These findings indicate that LB-GABA enhances the oxidation of fatty acids in 3T3-L1 cells.

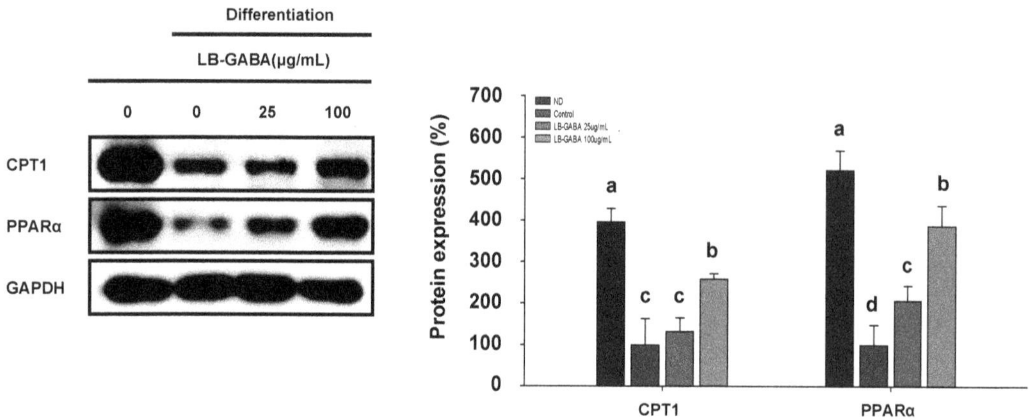

Figure 6. LB-GABA increases fatty acid oxidation in 3T3-L1 adipocytes. Adipocytes were cultured with a series of concentrations of LB-GABA. The expression of fatty acid oxidation proteins (CPT1 and PPARα) was determined via Western blot. ND, undifferentiated; control, differentiated. Data are expressed as the mean ± SEM. Values indicated by different letters are significantly different; $p < 0.05$ (a > b > c > d).

3. Discussion

Obesity and its associated diseases, including diabetes and hypertension, have emerged as significant global health concerns. In addition, oxidative stress has been demonstrated to induce inflammatory responses, which can result in extensive damage to bodily functions and, in severe cases, even death [39]. Diagnosing and timely managing metabolic syndrome, which is not a disease itself but a risk factor which significantly increases the risk of developing chronic illnesses, can lead to substantial savings—both in terms of time/life quality and financial resources—at the individual and healthcare system levels. Consequently, this study was conducted to ascertain a sustainable method of preventing obesity, a condition that has been demonstrated to diminish the quality of life and reduce life expectancy. The anti-obesity effect of LB-GABA was confirmed in differentiated 3T3-L1 adipocytes.

In this study, we demonstrated that LB-GABA reduces adipogenesis by decreasing the expression of adipogenic proteins, including PPARγ and FABP4, in 3T3-L1 adipocytes. These adipogenic factors play important roles after differentiation in white adipocytes. We also showed that LB-GABA can reduce adipogenesis by decreasing the expression of adipogenic factors, an enzyme that is stimulated by TG biosynthesis, and by decreasing PPARγ expression by decreasing LPAATθ expression.

In adipocytes, lipolysis is activated by p-PKA and carried out by enzymes including HSL [40]. In particular, previous studies have shown that the phosphorylation of HSL is induced following PKA activation, releasing FFAs that can be oxidized in the mitochondria [41,42]. The process of lipolysis is defined as the hydrolysis of TGs, leading to the consequent release of fatty acids and glycerol from adipocytes. TG hydrolysis requires the action of lipases such as ATGL and phosphorylated HSL, resulting in the release of free fatty acids. HSL in particular is the key enzyme responsible for the hydrolysis of triglycerides stored in adipose tissue [43]. PKA activation in adipocytes was confirmed, as would be expected to increase lipolysis and mitochondrial activity. Furthermore, it would be expected to increase the expression of ATGL and p-HSL, leading to further hydrolysis to glycerol and free fatty acids. Consistent with this, in this study, LB-GABA treatment effectively increased lipolytic biomarkers, including ATGL and p-HSL, and its downstream factors by increasing the expression of phosphorylated PKA. In summary, LB-GABA not only reduces adipogenesis and lipogenesis but also upregulates lipolytic pathways in 3T3-L1 adipocytes, thereby reducing lipid accumulation.

There has been a significant increase in interest in inducing a switch in differentiation from white adipocytes to BAT-like adipocytes or increasing energy expenditure in BATs. In addition, a current research trend is the use of natural dietary compounds to prevent and treat obesity through lipid metabolism, which is involved in fat browning and energy expenditure [46–50]. Increasing energy expenditure has the potential to prevent metabolic syndrome from obesity, as obesity occurs when energy intake consistently exceeds energy expenditure. A mechanism called 'browning' is critically associated with the induction of UCP1 expression in adipocytes [51,52]. Several small molecules, such as berberine and curcumin, appear to induce browning by activating thermogenic transcription factors or modulating key signaling pathways in adipocytes [53,54]. Therefore, a promising therapeutic strategy to increase energy expenditure is the use of substances to stimulate the induction of beige adipocytes or activate brown adipocytes. The process of thermogenesis in BAT-like adipocytes is initiated by the activation of UCP1 [51]. Therefore, this study focused on the differentiation of adipocytes and the inhibition of lipogenesis, as well as the promotion of lipolysis of stored fat through energy consumption and energy metamorphosis. In this study, we found that LB-GABA increased the expression of UCP1, which is an important regulator of this process, thereby inducing the browning of adipocytes and promoting energy metabolism. Furthermore, the results obtained suggest that lipolysis releases FFAs through β-oxidation via PPARa and CPT1, indicating that the β-oxidation of FFAs is also important in brown adipocytes when UCP1 is expressed [52].

In summary, the study utilized 3T3-L1 adipocytes, a model that has been extensively validated for its application in the research domain. The 3T3-L1 model is particularly well suited for investigating the process of adipogenesis, which refers to the transformation of preadipocytes into fully mature adipocytes. This transformation occurs during the developmental stage of mouse embryos. The findings from this study have the potential to inform and guide future experimental research, including the design of animal and clinical trials. Consequently, this cell experimentation confirmed the concentration of LB-GABA administration and its inhibitory effect on obesity. Subsequent investigations will address the anti-obesity impact of LB-GABA in animals. The findings of this study demonstrate that PKA activation by LB-GABA has the potential to enhance metabolic health by inhibiting lipid accumulation and increasing energy expenditure. The positive effects on metabolic health are presumably attributable to lower adiposity, smaller adipocytes, and increased energy expenditure. Consequently, LB-GABA has the potential to serve as a therapeutic agent in the treatment of obesity, with the aim of enhancing metabolic health.

4. Materials and Methods

4.1. Materials

Dulbecco's modified Eagle's medium (DMEM), bovine calf serum (BS), fetal bovine serum (FBS), penicillin streptomycin (P/S), phosphate-buffered saline (PBS), trypsin-EDTA (T/E), and insulin were purchased from Gibco (Gaithersburg, MD, USA). IBMX, dexamethasone, isopropanol, oil red O, phosphatase inhibitors, and the protease inhibitor cocktail were purchased from Sigma-Aldrich (St. Louis, MO, USA). Primary antibodies specific for PPARγ (sc7273, 55 kDa), FABP4 (sc-30088, 15 kDa), LPAATθ (sc-68372, 42 kDa), FAS (sc-20140, 270 kDa), PPARα (sc-9000, 55 kDa), p-PKA (sc-136460, 53 kDa), and glyceraldehyde 3-phosphate dehydrogenase (GAPDH, sc365062, 37 kDa) were purchased from Santa Cruz Biotechnology (Dallas, TX, USA). Antibodies specific to ATGL (cs-#1238, 54 kDa) and pHSL (cs-#4139, 81 kDa) were purchased from Cell Signaling Technology (Danvers, MA, USA), and antibodies specific to Lipin 1 (ab70138, 110 kDa), CPT1 (ab128568, 37 kDa), and UCP1 (ab23841, 33 kDa) were purchased from Abcam (Cambridge, UK).

4.2. Preparation and Analysis of LB-GABA

LB-GABA production was performed using Lactobacillus brevis fermentation from monosodium L-glutamate. Lactobacillus brevis PF-1, originating from Kimchi, a Republic of Korean fermented vegetable, was cultured in a broth containing monosodium l-glutamate at 27–30 °C for 96–120 h. Culture broth was filter-sterilized, concentrated, and mixed with modified starch. Afterward, it underwent another round of filter sterilization before being subjected to the spray-drying process. LB-GABA comprises approximately 20% GABA (*w*/*w*) and 80% modified food starch [46]. The LB-GABA product contained 323.44 ± 29.48 µg of GABA per milligram (*w*/*w*) in modified food starch (AMOREPACIFIC Co., Seoul, Republic of Korea).

The main peak was detected (Figure 7). The elution time of GABA was 23.95 min (Figure 7B), which was the same as the elution time of synthetic GABA (Figure 7C). Therefore, the LB-GABA sample was confirmed as GABA when compared against the standard synthetic sample [47].

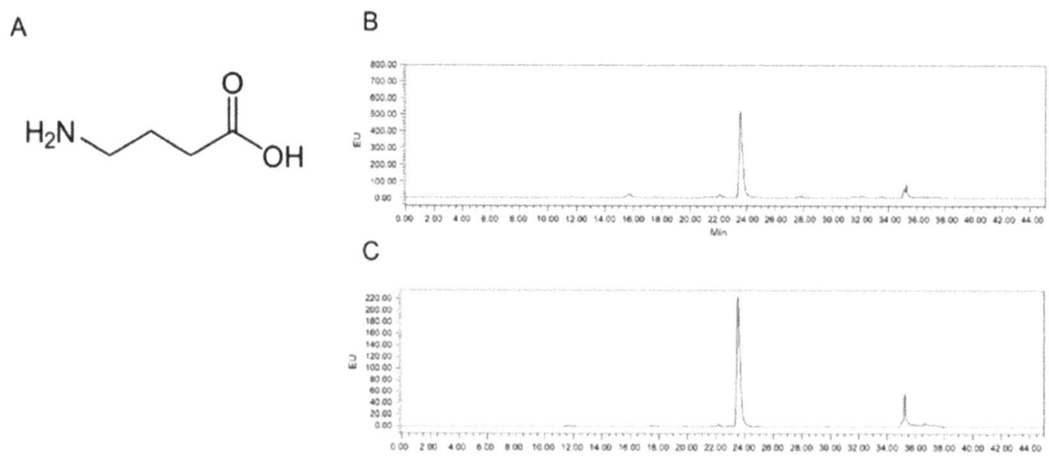

Figure 7. (**A**) Chemical structure and (**B**) high-performance liquid chromatography (HPLC) chromatograms of Lactobacillus-fermented γ-aminobutyric acid (LB-GABA) and the (**C**) GABA standard.

4.3. Cell Culture

The 3T3-L1 preadipocytes were purchased from the American Type Culture Collection (Manassas, VA, USA) and maintained in the growth medium (DMEM containing 10% BS, 1% P/S, and 3.7 g/L $NaHCO_3$) at 37 °C in a 5% CO_2 incubator. The 3T3-L1 cells were cultured while replacing the growth medium every 2 days until they became 100% confluent. Two days after they became confluent (D0), the growth medium was replaced with the differentiation medium (DMEM containing 10% FBS, 1% P/S, and 3.7 g/L $NaHCO_3$ containing 0.5 mM IBMX, 1 µM dexamethasone, and 4 µg/mL insulin [MDI]) for 2 days. On day 2 (D2), the differentiation medium was removed and replaced with the maintenance medium (DMEM containing 10% FBS, 1% P/S, and 3.7 g/L $NaHCO_3$ containing 4 µg/mL insulin) every 2 days up to day 8 (D8).

4.4. Cell Viability

The 3T3-L1 preadipocytes were seeded (5×10^3 cells/well) in 96-well plates and incubated overnight in the growth medium. The cells were then treated with LB-GABA (0, 25, 50, 100, or 200 µM) and incubated for a further 24 h. After this, 20 µL of the MTT solution was added to each well, and the cells were incubated for 3 h; then, the MTT-containing medium was removed, and 100 µL of DMSO was added to elute the formazan crystals. The absorbance of each well was then measured at 570 nm (BioTek, Winooski, VT, USA) [48].

4.5. Oil Red O Staining

Fully differentiated 3T3-L1 adipocytes were fixed in 10% formaldehyde for 1 h at room temperature and then washed twice with 60% isopropanol. The fixed cells were stained with the oil red O solution (ratio of 6:4 with distilled water) for 20 min at room temperature and then washed with distilled water. After drying, the stained cells were imaged; then, the dye was eluted using 100% isopropanol, and its absorbance was measured at 490 nm.

4.6. Western Blotting

The 3T3-L1 adipocytes were washed twice with PBS and then lased in lysis buffer (iNtRON Biotechnology, Seoul, Republic of Korea) containing phosphatase and protease inhibitors. The lysate protein concentrations were determined using a protein assay reagent (Bio-Rad, Hercules, CA, USA). Equal amounts of protein (20 µg) were diluted in 5× sample buffer (50 mM Tris pH 6.8, 2% sodium dodecyl sulfate (SDS), 10% glycerol, 5% β-mercaptoethanol, and 0.1% bromophenol blue) and heated for 5 min at 90 °C. After cooling, the proteins were separated using 8–12% SDS–polyacrylamide gel electrophoresis and transferred to polyvinylidene fluoride membranes. Then, the membranes were blocked with 5% skim milk for 1 h at room temperature. After blocking, the membranes were washed with Tris-buffered saline containing Tween 20 (TBST) and incubated with primary antibodies (1:1000) overnight at 4 °C, followed by incubation with secondary antibodies conjugated with horseradish peroxidase (1:5000) (Santa Cruz Biotechnology) in TBST containing 5% skim milk for 2 h at room temperature [46,47]. Specific protein bands were detected using enhanced chemiluminescence and then imaged using an Amersham Imager 680 (GE Healthcare Life Sciences, Chicago, IL, USA). The localization of proteins was determined using the protein markers of Tri-Glycine 4–20%.

4.7. Statistical Analysis

Data are expressed as the mean ± standard deviation (SD). One-way ANOVA with Duncan's test (SPSS, Chicago, IL, USA) was used to analyze differences among multiple groups. Statistically significant differences were accepted when $p < 0.05$.

All data are expressed as the mean ± standard deviation (SD) and are the results of experiments performed at least three times. Prior to conducting subsequent analyses, normality tests were performed. Differences between groups were analyzed using one-way analysis of variance (ANOVA) with Duncan's test (SPSS, version 20, Chicago, IL, USA) to identify significant differences.

5. Conclusions

The results of this study showed that LB-GABA inhibits adipogenesis and lipogenesis in 3T3-L1, induces lipolysis through FA oxidation, and promotes the browning of adipocytes, confirming the biomarkers of overall obesity through the activation of energy metabolism. In conclusion, LB-GABA exhibits anti-obesity effects within adipocytes. But the long-term use of substances that inhibit adipogenesis may result in a condition called lipodystrophy, which is of genetic or autoimmune origin. It is imperative to be cognizant of the potential issues that may emerge. Further research is required to ascertain the precise anti-obesity effect of LB-GABA in animal models. The present study thus sought to confirm the overall anti-obesity effect of LB-GABA, and it is planned that animal experiments using the established concentration will be conducted in the future. It is hypothesized that this will facilitate the development of new anti-obesity health functional foods.

Supplementary Materials: The following supporting information can be downloaded at: https://www.mdpi.com/article/10.3390/ijms26083554/s1.

Author Contributions: Conceptualization, H.H. and B.-Y.L.; methodology, H.H.; software, H.H.; validation, H.H. and G.S.; formal analysis, H.H.; investigation, H.H.; resources, B.-Y.L.; data curation, H.H., G.S. and J.K.; writing—original draft preparation, H.H.; writing—review and editing, H.H. and H.J.; visualization, H.H.; supervision, B.-Y.L.; project administration, B.-Y.L.; funding acquisition, B.-Y.L. All authors have read and agreed to the published version of the manuscript.

Funding: This study received no funding support.

Institutional Review Board Statement: Not applicable.

Informed Consent Statement: Not applicable.

Data Availability Statement: All data generated or analyzed during this study are included in this published article.

Conflicts of Interest: The authors declare no conflicts of interest.

Abbreviations

WAT: white adipose tissue; BAT: brown adipose tissue; LB-GABA: Lactobacillus brevis-fermented gamma-aminobutyric acid; GABA: gamma-aminobutyric acid; PKA: protein kinase A; UCP1: uncoupling protein 1; TG: triglycerides; PPARγ: peroxisome proliferator-activated receptor gamma; FABP4: fatty acid binding protein 4; LPAATθ: lysophosphatidic acid acyltransferase theta; FAS: fatty acid synthase; DGAT1: diacylglycerol acyltransferase 1; FFAs: free fatty acids; CPT1: carnitine palmitoyl transferase 1; ATGL: adipocyte triglyceride lipase; HSL: hormone-sensitive lipase; MGL: monoglyceride lipase; LB-GABA: Lactobacillus-fermented gamma-aminobutyric acid; Lipin1: Phosphatidate phosphatase LPIN1; FA: fatty acid; MTT: 3-(4,5-dimethylthiazol-2yl)-2,5-diphenyltetrazolium bromide; IBMX: 3-Isobutyl-1-methylxanthine; MDI: 0.5 mM 3-isobutyl-1-methylxanthine, 1 μM dexamethasone, and 4 μg/mL insulin; D8: day 8; ND: non-differentiation; PPARα: peroxisome proliferator-activated receptor alpha; CPT-1A: carnitine palmitoyltransferase-1A; DMEM: Dulbecco's modified Eagle's medium; BS: bovine calf serum; FBS: fetal bovine serum; P/S: penicillin streptomycin; GAPDH: glyceraldehyde 3-phosphate dehydrogenase; HPLC: high-performance liquid chromatography; DMSO: dimethyl sulfoxide; SDS: sodium dodecyl sulfate; TBST: Tris-buffered saline containing Tween 20; SD: standard deviation; ANOVA: analysis of variance.

References

1. Dietrich, M.O.; Horvath, T.L. Limitations in anti-obesity drug development: The critical role of hunger-promoting neurons. *Nat. Rev. Drug Discov.* **2012**, *11*, 675–691. [CrossRef] [PubMed]
2. Yang, M.; Liu, S.; Zhang, C. The Related Metabolic Diseases and Treatments of Obesity. *Healthcare* **2022**, *10*, 1616. [CrossRef] [PubMed]
3. Hu, C.; Jia, W. Diabetes in China: Epidemiology and Genetic Risk Factors and Their Clinical Utility in Personalized Medication. *Diabetes* **2018**, *67*, 3–11. [CrossRef] [PubMed]
4. Zhou, M.; Wang, H.; Zeng, X.; Yin, P.; Zhu, J.; Chen, W.; Li, X.; Wang, L.; Wang, L.; Liu, Y.; et al. Mortality, morbidity, and risk factors in China and its provinces, 1990–2017: A systematic analysis for the Global Burden of Disease Study 2017. *Lancet* **2019**, *394*, 1145–1158. [CrossRef]
5. Letukienė, A.; Hendrixson, V.; Ginevičienė, V. Current knowledge and scientific trends in myokines and exercise research in the context of obesity. *Front. Med.* **2024**, *11*, 1421962. [CrossRef]
6. Choe, S.S.; Huh, J.Y.; Hwang, I.J.; Kim, J.I.; Kim, J.B. Adipose Tissue Remodeling: Its Role in Energy Metabolism and Metabolic Disorders. *Front. Endocrinol.* **2016**, *7*, 30. [CrossRef]
7. Cannon, B.; Nedergaard, J. Brown Adipose Tissue: Function and Physiological Significance. *Physiol. Rev.* **2004**, *84*, 277–359. [CrossRef]
8. Sørensen, T.I.A. Obesity Defined as Excess Storage of Inert Triglycerides—Do We Need a Paradigm Shift? *Obes. Facts* **2011**, *4*, 91–94. [CrossRef]
9. Nedergaard, J.; Golozoubova, V.; Matthias, A.; Asadi, A.; Jacobsson, A.; Cannon, B. UCP1: The only protein able to mediate adaptive non-shivering thermogenesis and metabolic ine§ciency. *Biochim. Biophys. Acta (BBA)—Bioenerg.* **2001**, *1504*, 82–106. [CrossRef]
10. Wu, J.; Cohen, P.; Spiegelman, B.M. Adaptive thermogenesis in adipocytes: Is beige the new brown? *Genes Dev.* **2013**, *27*, 234–250. [CrossRef]
11. Green, H.; Meuth, M. An established pre-adipose cell line and its differentiation in culture. *Cell* **1974**, *3*, 127–133. [CrossRef] [PubMed]
12. Zubiría, M.; Alzamendi, A.; Moreno, G.; Portales, A.; Castrogiovanni, D.; Spinedi, E.; Giovambattista, A. Relationship between the Balance of Hypertrophic/Hyperplastic Adipose Tissue Expansion and the Metabolic Profile in a High Glucocorticoids Model. *Nutrients* **2016**, *8*, 410. [CrossRef] [PubMed]
13. Zhou, Y.-T.; Grayburn, P.; Karim, A.; Shimabukuro, M.; Higa, M.; Baetens, D.; Orci, L.; Unger, R.H. Lipotoxic heart disease in obese rats: Implications for human obesity. *Proc. Natl. Acad. Sci. USA* **2000**, *97*, 1784–1789. [CrossRef] [PubMed]
14. Grabner, G.F.; Xie, H.; Schweiger, M.; Zechner, R. Lipolysis: Cellular mechanisms for lipid mobilization from fat stores. *Nat. Metab.* **2021**, *3*, 1445–1465. [CrossRef]
15. Cerk, I.K.; Wechselberger, L.; Oberer, M. Adipose Triglyceride Lipase Regulation: An Overview. *CPPS* **2017**, *19*, 221–233. [CrossRef]
16. Henique, C.; Mansouri, A.; Fumey, G.; Lenoir, V.; Girard, J.; Bouillaud, F.; Prip-Buus, C.; Cohen, I. Increased Mitochondrial Fatty Acid Oxidation Is Sufficient to Protect Skeletal Muscle Cells from Palmitate-induced Apoptosis. *J. Biol. Chem.* **2010**, *285*, 36818–36827. [CrossRef]
17. Tardelli, M. Monoacylglycerol lipase reprograms lipid precursors signaling in liver disease. *WJG* **2020**, *26*, 3577–3585. [CrossRef]
18. Luo, L.; Liu, M. Adipose tissue in control of metabolism. *J. Endocrinol.* **2016**, *231*, R77–R99. [CrossRef]
19. Schirinzi, V.; Poli, C.; Berteotti, C.; Leone, A. Browning of Adipocytes: A Potential Therapeutic Approach to Obesity. *Nutrients* **2023**, *15*, 2229. [CrossRef]
20. Kuryłowicz, A.; Puzianowska-Kuźnicka, M. Induction of Adipose Tissue Browning as a Strategy to Combat Obesity. *Int. J. Mol. Sci.* **2020**, *21*, 6241. [CrossRef]
21. Timmons, J.A.; Wennmalm, K.; Larsson, O.; Walden, T.B.; Lassmann, T.; Petrovic, N.; Hamilton, D.L.; Gimeno, R.E.; Wahlestedt, C.; Baar, K.; et al. Myogenic gene expression signature establishes that brown and white adipocytes originate from distinct cell lineages. *Proc. Natl. Acad. Sci. USA* **2007**, *104*, 4401–4406. [CrossRef] [PubMed]
22. Machado, S.A.; Pasquarelli-do-Nascimento, G.; Da Silva, D.S.; Farias, G.R.; De Oliveira Santos, I.; Baptista, L.B.; Magalhães, K.G. Browning of the white adipose tissue regulation: New insights into nutritional and metabolic relevance in health and diseases. *Nutr. Metab.* **2022**, *19*, 61. [CrossRef] [PubMed]
23. Kim, S.H.; Plutzky, J. Brown Fat and Browning for the Treatment of Obesity and Related Metabolic Disorders. *Diabetes Metab. J.* **2016**, *40*, 12. [CrossRef] [PubMed]
24. Roberts, E.; Frankel, S. γ-Aminobutyric Acid in Brain: Its Formation from Glutamic Acid. *J. Biol. Chem.* **1950**, *187*, 55–63. [CrossRef]
25. Luo, H.; Liu, Z.; Xie, F.; Bilal, M.; Liu, L.; Yang, R.; Wang, Z. Microbial production of gamma-aminobutyric acid: Applications, state-of-the-art achievements, and future perspectives. *Crit. Rev. Biotechnol.* **2021**, *41*, 491–512. [CrossRef]

26. Puts, N.A.J.; Edden, R.A.E. In vivo magnetic resonance spectroscopy of GABA: A methodological review. *Prog. Nucl. Magn. Reson. Spectrosc.* **2012**, *60*, 29–41. [CrossRef]
27. Braga, J.D.; Thongngam, M.; Kumrungsee, T. Gamma-aminobutyric acid as a potential postbiotic mediator in the gut–brain axis. *npj Sci. Food* **2024**, *8*, 16. [CrossRef]
28. Park, C.; Chen, X.; Tian, C.-L.; Park, G.N.; Chenouard, N.; Lee, H.; Yeo, X.Y.; Jung, S.; Bi, G.; Tsien, R.W.; et al. Inhibitory synaptic vesicles have unique dynamics and exocytosis properties. *bioRix* **2020**. [CrossRef]
29. Park, C.; Chen, X.; Tian, C.-L.; Park, G.N.; Chenouard, N.; Lee, H.; Yeo, X.Y.; Jung, S.; Tsien, R.W.; Bi, G.-Q.; et al. Unique dynamics and exocytosis properties of GABAergic synaptic vesicles revealed by three-dimensional single vesicle tracking. *Proc. Natl. Acad. Sci. USA* **2021**, *118*, e2022133118. [CrossRef]
30. Yoto, A.; Murao, S.; Motoki, M.; Yokoyama, Y.; Horie, N.; Takeshima, K.; Masuda, K.; Kim, M.; Yokogoshi, H. Oral intake of γ-aminobutyric acid affects mood and activities of central nervous system during stressed condition induced by mental tasks. *Amino Acids* **2012**, *43*, 1331–1337. [CrossRef]
31. Simmonite, M.; Carp, J.; Foerster, B.R.; Ossher, L.; Petrou, M.; Weissman, D.H.; Polk, T.A. Age-Related Declines in Occipital GABA are Associated with Reduced Fluid Processing Ability. *Acad. Radiol.* **2019**, *26*, 1053–1061. [CrossRef] [PubMed]
32. Tu, J.; Jin, Y.; Zhuo, J.; Cao, X.; Liu, G.; Du, H.; Liu, L.; Wang, J.; Xiao, H. Exogenous GABA improves the antioxidant and anti-aging ability of silkworm (*Bombyx mori*). *Food Chem.* **2022**, *383*, 132400. [CrossRef] [PubMed]
33. Zuppichini, M.D.; Hamlin, A.M.; Zhou, Q.; Kim, E.; Rajagopal, S.; Beltz, A.M.; Polk, T.A. GABA levels decline with age: A longitudinal study. *Imaging Neurosci.* **2024**, *2*, 1–15. [CrossRef]
34. Jin, H.; Han, H.; Song, G.; Oh, H.-J.; Lee, B.-Y. Anti-Obesity Effects of GABA in C57BL/6J Mice with High-Fat Diet-Induced Obesity and 3T3-L1 Adipocytes. *Int. J. Mol. Sci.* **2024**, *25*, 995. [CrossRef]
35. Zhang, X.; Zhang, C.C.; Yang, H.; Soni, K.G.; Wang, S.P.; Mitchell, G.A.; Wu, J.W. An Epistatic Interaction between Pnpla2 and Lipe Reveals New Pathways of Adipose Tissue Lipolysis. *Cells* **2019**, *8*, 395. [CrossRef]
36. Lee, Y.-H.; Mottillo, E.P.; Granneman, J.G. Adipose tissue plasticity from WAT to BAT and in between. *Biochim. Biophys. Acta (BBA)—Mol. Basis Dis.* **2014**, *1842*, 358–369. [CrossRef]
37. Calderon-Dominguez, M.; Sebastián, D.; Fucho, R.; Weber, M.; Mir, J.F.; García-Casarrubios, E.; Obregón, M.J.; Zorzano, A.; Valverde, Á.M.; Serra, D.; et al. Carnitine Palmitoyltransferase 1 Increases Lipolysis, UCP1 Protein Expression and Mitochondrial Activity in Brown Adipocytes. *PLoS ONE* **2016**, *11*, e0159399. [CrossRef]
38. Madsen, M.S.; Siersbæk, R.; Boergesen, M.; Nielsen, R.; Mandrup, S. Peroxisome Proliferator-Activated Receptor γ and C/EBPα Synergistically Activate Key Metabolic Adipocyte Genes by Assisted Loading. *Mol. Cell. Biol.* **2014**, *34*, 939–954. [CrossRef]
39. Richter, K.; Haslbeck, M.; Buchner, J. The Heat Shock Response: Life on the Verge of Death. *Mol. Cell* **2010**, *40*, 253–266. [CrossRef]
40. Jin, H.; Oh, H.-J.; Kim, J.; Lee, K.-P.; Han, X.; Lee, O.-H.; Lee, B.-Y. Effects of Ecklonia stolonifera extract on the obesity and skeletal muscle regeneration in high-fat diet-fed mice. *J. Funct. Foods* **2021**, *82*, 104511. [CrossRef]
41. Calderon-Dominguez, M. Fatty acid metabolism and the basis of brown adipose tissue function. *Adipocyte* **2016**, *5*, 98–118. [CrossRef] [PubMed]
42. Seo, Y.-J.; Jin, H.; Lee, K.; Song, J.-H.; Chei, S.; Oh, H.-J.; Oh, J.-H.; Lee, B.-Y. Cardamonin suppresses lipogenesis by activating protein kinase A-mediated browning of 3T3-L1 cells. *Phytomedicine* **2019**, *65*, 153064. [CrossRef] [PubMed]
43. Langin, D. Adipose tissue lipolysis as a metabolic pathway to define pharmacological strategies against obesity and the metabolic syndrome. *Pharmacol. Res.* **2006**, *53*, 482–491. [CrossRef] [PubMed]
44. Löffler, M.C.; Betz, M.J.; Blondin, D.P.; Augustin, R.; Sharma, A.K.; Tseng, Y.-H.; Scheele, C.; Zimdahl, H.; Mark, M.; Hennige, A.M.; et al. Challenges in tackling energy expenditure as obesity therapy: From preclinical models to clinical application. *Mol. Metab.* **2021**, *51*, 101237. [CrossRef]
45. Song, N.-J.; Chang, S.-H.; Li, D.Y.; Villanueva, C.J.; Park, K.W. Induction of thermogenic adipocytes: Molecular targets and thermogenic small molecules. *Exp. Mol. Med.* **2017**, *49*, e353. [CrossRef]
46. Kim, H.; Kim, H.; Suh, H.J.; Choi, H.-S. *Lactobacillus brevis*-Fermented Gamma-Aminobutyric Acid Ameliorates Depression-and Anxiety-Like Behaviors by Activating the Brain-Derived Neurotrophic Factor-Tropomyosin Receptor Kinase B Signaling Pathway in BALB/C Mice. *J. Agric. Food Chem.* **2024**, *72*, 2977–2988. [CrossRef]
47. Jeong, A.-H.; Hwang, J.; Jo, K.; Kim, S.; Ahn, Y.; Suh, H.J.; Choi, H.-S. Fermented Gamma Aminobutyric Acid Improves Sleep Behaviors in Fruit Flies and Rodent Models. *Int. J. Mol. Sci.* **2021**, *22*, 3537. [CrossRef]
48. Khan, I.; Preeti, K.; Kumar, R.; Khatri, D.K.; Singh, S.B. Activation of SIRT1 by silibinin improved mitochondrial health and alleviated the oxidative damage in experimental diabetic neuropathy and high glucose-mediated neurotoxicity. *Arch. Physiol. Biochem.* **2024**, *130*, 420–436. [CrossRef]
49. Khan, I.; Kaur, S.; Rishi, A.K.; Boire, B.; Aare, M.; Singh, M. Cannabidiol and Beta-Caryophyllene Combination Attenuates Diabetic Neuropathy by Inhibiting NLRP3 Inflammasome/NFκB through the AMPK/sirT3/Nrf2 Axis. *Biomedicines* **2024**, *12*, 1442. [CrossRef]

50. Khan, I.; Preeti, K.; Kumar, R.; Kumar Khatri, D.; Bala Singh, S. Piceatannol promotes neuroprotection by inducing mitophagy and mitobiogenesis in the experimental diabetic peripheral neuropathy and hyperglycemia-induced neurotoxicity. *Int. Immunopharmacol.* **2023**, *116*, 109793. [CrossRef]
51. Fedorenko, A.; Lishko, P.V.; Kirichok, Y. Mechanism of Fatty-Acid-Dependent UCP1 Uncoupling in Brown Fat Mitochondria. *Cell* **2012**, *151*, 400–413. [CrossRef] [PubMed]
52. Kalinovich, A.V.; De Jong, J.M.A.; Cannon, B.; Nedergaard, J. UCP1 in adipose tissues: Two steps to full browning. *Biochimie* **2017**, *134*, 127–137. [CrossRef] [PubMed]
53. Zhang, Z.; Zhang, H.; Li, B.; Meng, X.; Wang, J.; Zhang, Y.; Yao, S.; Ma, Q.; Jin, L.; Yang, J.; et al. Berberine activates thermogenesis in white and brown adipose tissue. *Nat. Commun.* **2014**, *5*, 5493. [CrossRef] [PubMed]
54. Lone, J.; Choi, J.H.; Kim, S.W.; Yun, J.W. Curcumin induces brown fat-like phenotype in 3T3-L1 and primary white adipocytes. *J. Nutr. Biochem.* **2016**, *27*, 193–202. [CrossRef]

International Journal of
Molecular Sciences

MDPI

Article

Enhanced Anti-Cancer Potential: Investigating the Combined Effects with *Coriolus versicolor* Extract and Phosphatidylinositol 3-Kinase Inhibitor (LY294002) In Vitro

Tomasz Jędrzejewski [1,*], Justyna Sobocińska [1], Bartosz Maciejewski [1], Marcela Slovakova [2] and Sylwia Wrotek [1]

[1] Department of Immunology, Faculty of Biological and Veterinary Sciences, Nicolaus Copernicus University, Lwowska 1 Str., 87-100 Toruń, Poland; j.sobocinska@umk.pl (J.S.); bartosz.maciejewski@umk.pl (B.M.); wrotek@umk.pl (S.W.)

[2] Department of Biological and Biochemical Sciences, Faculty of Chemical Technology, University of Pardubice, Studentska 573 Str., 532 10 Pardubice, Czech Republic; marcela.slovakova@upce.cz

* Correspondence: tomaszj@umk.pl

Abstract: *Coriolus versicolor* (CV), known in traditional Chinese medicine for over 2000 years, is currently used in China and Japan to reduce chemotherapy or radiotherapy side effects in cancer patients. Despite extensive research, its effects still need improvement. This study aimed to determine if combining CV extract with LY294002, an inhibitor of the phosphatidylinositol-3-kinase (PI3K) signalling pathway, enhances cancer cell treatment, potentially leading to a novel therapeutic approach. Three human cancer cell lines (MCF-7, HeLa, and A549) were treated with CV extract alone or combined with LY294002. Cell viability was assessed using MTT assays. Then, HeLa and MCF-7 cells most sensitive to the co-treatment were used to evaluate colony formation, apoptosis, cell cycle, cell migration and invasion, and phospho-PI3K expression. The results demonstrated that LY294002 enhanced the CV extract's anti-tumour effects by reducing cell viability and colony formation. The combined treatment with CV extract and LY294002 more effectively induced G0/G1 cell cycle arrest, promoted apoptosis, reduced cell invasion and migration, and inhibited phospho-PI3K expression compared to each agent alone. This study highlights the potent cytotoxic enhancement between CV extract and LY294002 on cancer cells, primarily by inhibiting phospho-PI3K expression. These findings suggest promising avenues for developing novel combination therapies targeting cancer.

Keywords: *Coriolus versicolor*; phosphatidylinositol 3-kinase; LY294002; cancer

1. Introduction

Coriolus versicolor (L.) Quél. (1886) (CV), also known as *Trametes versicolor* (L.) Lloyd (1920) mushroom, is one of the most investigated species, and has been used in traditional Chinese herbal medicine for over 2000 years. Ancient Chinese CV formulation is believed to promote health, strength, and longevity. Within traditional Chinese medicine, CV mushroom is believed to aid in detoxification, bolstering strength, optimizing liver and spleen functionality, and augmenting the immune system's response, particularly when it's dried, ground, and brewed into tea [1].

Nowadays, it is well-established that CV mushroom possesses various beneficial properties, including anti-oxidant, hypoglycaemic, anti-inflammatory, anti-viral, immunostimulating, and liver protection properties. These indicate its potential application value in treating liver disease, diabetes, arteriosclerosis, Alzheimer's disease, cardiovascular and cerebrovascular diseases, osteoarthritis, inflammatory bowel disease, and many others [2–5]. However, the most widely studied properties of the compounds derived from CV mushroom are their anti-cancer activity. Numerous in vitro and in vivo studies have shown that CV compounds induce anti-tumoricidal effects against multiple cancer cell lines, inhibit tumour growth and metastasis in animal models, and display specific anti-angiogenic properties, which have been widely discussed in several review papers [3,5–7]. Clinical trials have reported both the indirect anti-cancer properties of CV through immunostimulating mechanisms and its direct anti-cancer activity, leading to the adoption of CV compounds as an adjunct therapy for cancer treatment in Japan and China. The bioactive compounds derived from CV, recommended for patients undergoing or post-radiation and chemotherapy, enhance their chances of survival, mitigate the immunosuppressive impacts of standard treatments, and alleviate symptoms associated with cancer therapy, such as fatigue, decreased appetite, vomiting, and pain [8–12]. Although CV extract is helpful in cancer treatment, it is not potent enough to be used as monotherapy. Hence, it is necessary to identify a substance that can enhance its effectiveness.

The present study investigated the effect of the combined treatment of cancer cells with CV extract and LY294002, a chemical inhibitor of the phosphatidylinositol-3-kinase (PI3K) signalling pathway. It is well-established that the PI3K pathway is frequently over-activated in human cancers, playing a significant role in carcinogenesis, tumour cell proliferation, invasion, and metastasis [13]. Over recent decades, research has concentrated on crafting PI3K inhibitors, aiming at individual or multiple proteins, progressing from preclinical tool compounds to specific medications for cancer patients [14]. Among them, LY294002 is a morpholine-containing compound which causes the induction of apoptosis in tumour cells, inhibits the invasiveness of cancer cells, and has anti-angiogenic properties [15,16]. The treatment with a combination of LY294002 and other drugs, such as tamoxifen [17], talazoparib [18], sorafenib [19], or rapamycin [20], is still under investigation, since this combination could both overcome the toxicity associated with LY294002 and may sensitise cancer cells to these drugs.

This study aimed to determine if combining CV extract with LY294002 can enhance cancer cell treatment, potentially leading to a novel therapeutic approach. For the experiments, we selected cell lines representing the most prevalent cancers worldwide in 2022, based on the latest World Health Organization (WHO) data. A549 cells were chosen as a model for lung cancer, the most commonly diagnosed cancer, while MCF-7 cells represent female breast cancer, which ranks second in global incidence [21]. Moreover, we included HeLa cells, one of the most extensively studied and well-documented cancer cell lines, facilitating comparative analysis and reproducibility across different studies.

2. Results

2.1. *LY294002 and CV Extract Have Additive Cytotoxic Effects Against Cancer Cells*

The effect of LY294002 and CV extract on cancer cells was assessed through MTT and colony-forming assays. As shown in Figure 1, the viability of HeLa (Figure 1A–C), MCF-7 (Figure 1D–F), and A549 (Figure 1G–I) cells was decreased following the treatment with the CV extract in a dose- and time-dependent manner. The co-stimulation of cells with CV extract and LY294002 significantly reduced cell survival compared to the CV extract-treated cells. It was mainly observed after 72 h of stimulation when this additive effect was noticed for all tested doses of the CV extract and all used cell lines, except for a dose of 50 μg/mL. Since the CV extract at a 100 μg/mL concentration was the highest non-toxic dose according to the International Organization for Standardization (ISO) 10993-5 norm [22], it was used for further experiments. Moreover, LY294002 at a concentration of 10 μM was selected for the cell co-treatment, since this dose did not reduce cell viability below 70% (Figure S1 in the Supplementary Materials).

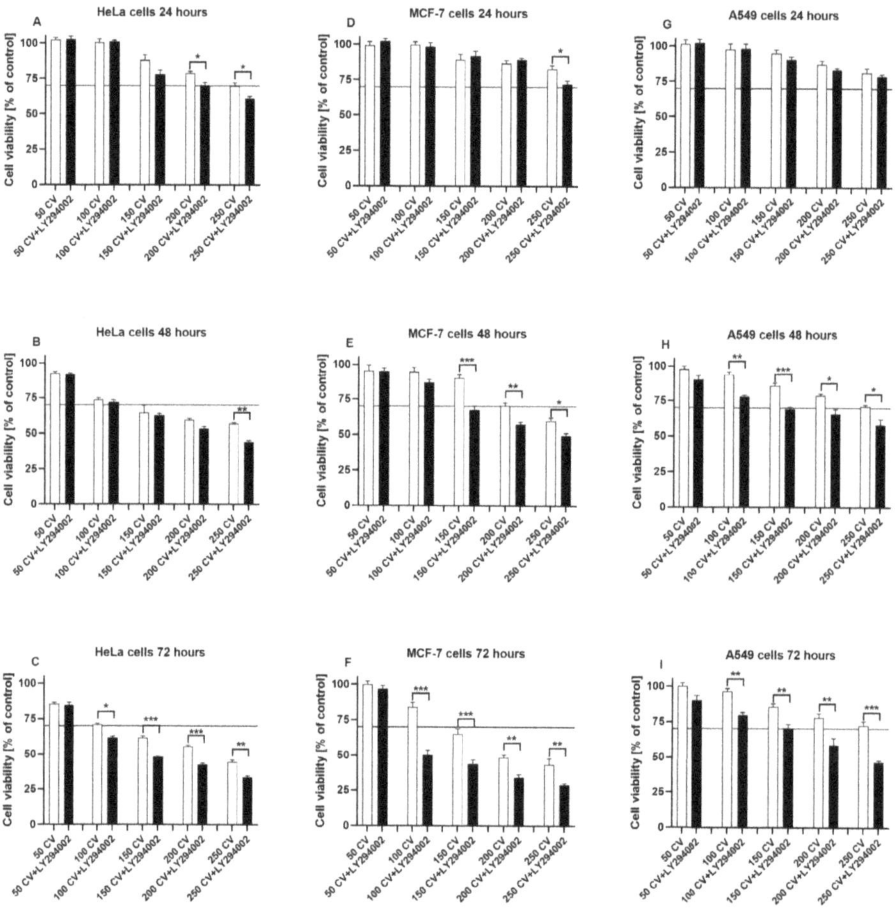

Figure 1. The viability of HeLa (**A–C**), MCF-7 (**D–F**), and A549 (**G–I**) cells stimulated with CV extract (50–250 μg/mL) or co-treated with the CV extract and LY294002 (10 μM) for 24, 48, and 72 h. The results are expressed as means ± S.E.M. of untreated cells from three independent experiments. Asterisks indicate significant differences between CV extract-treated and co-treated cells (* $p < 0.05$; ** $p < 0.01$; *** $p < 0.001$). Horizontal lines mark the 70% viability threshold for cytotoxicity.

The half inhibitory concentrations (IC_{50} values) calculated based on the results from the MTT assay provided more detailed information about the differences in the cytotoxicity of the tested compounds (Table 1). These results show that the co-treatment of HeLa and MCF-7 cells with the CV extract and LY294002 (at a non-toxic dose) more effectively inhibited cancer cell survival than stimulation of cells only with the extract. This effect was observed for HeLa cells after 48 and 72 h of stimulation ($p < 0.01$ and $p < 0.001$, respectively) and for MCF-7 cell treatment after 72 h ($p < 0.001$). In contrast, the IC_{50} values calculated for A549 cells revealed that stimulation of these cells with CV extract in the presence of LY294002 did not significantly improve the cytotoxic effect compared with the cells treated only with CV extract. Therefore, the rest of the experiments were only conducted using HeLa and MCF-7 cells.

Table 1. The time-dependent inhibitory concentration of the CV extract and LY29402 (10 μM) against different cell lines in the MTT assay.

Time	Sample	IC_{50}		
		HeLa	**MCF-7**	**A549**
24 h	CV	>250	>250	>250
	CV + LY294002	>250	>250	>250
48 h	CV	>250	>250	>250
	CV + LY294002	215.5 ± 7.6 **	239.4 ± 9.9	>250
72 h	CV	225.2 ± 11.2	208.6 ± 7.8	>250
	CV + LY294002	153.0 ± 4.5 ***	133.4 ± 3.9 ***	244.7 ± 6.7

Data are presented as mean ± S.E.M. IC_{50}: half-maximal inhibitory concentration (μg/mL). Asterisks show significant differences between the co-stimulated cells and those treated only with CV extract (*** $p < 0.001$; ** $p < 0.01$).

The colony-forming assay was performed to evaluate the survival of cancer cells, measuring their capacity to form colonies from single cells. As shown in Figure 2, the stimulation of both HeLa and MCF-7 cells with the CV extract alone or LY294002 alone significantly inhibited colony formation compared to control cells. However, combining the CV extract and LY294002 provoked the lowest percentage of colony formation noticed for HeLa cells and MCF-7 cells.

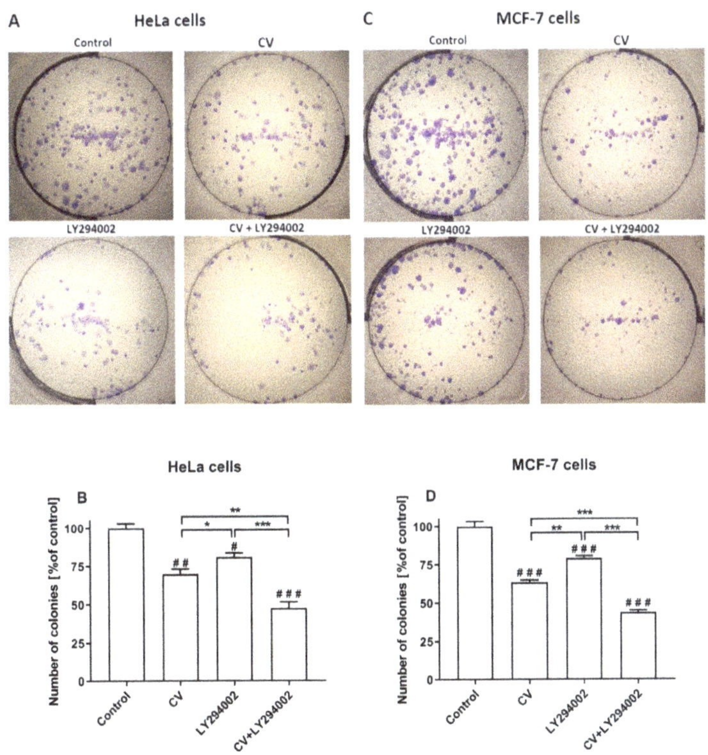

Figure 2. Colony numbers of HeLa (**A**,**B**) and MCF-7 (**C**,**D**) cells stimulated with CV extract (100 μg/mL), LY294002 (10 μM), or their combination. Colony formation was quantified as a percentage of untreated cells and presented as the means ± S.E.M. from three independent experiments. Asterisks show differences between the variants of cell treatment as indicated (* $p < 0.05$; ** $p < 0.01$; *** $p < 0.001$). Hashes show differences between control and treated cells (# $p < 0.05$; ## $p < 0.01$; ### $p < 0.001$).

2.2. Co-Treatment of Cancer Cells with LY294002 and CV Extract Increases Cell Cycle Arrest at the G0/G1 Phase

Cell cycle analysis was performed to further clarify the mechanism of additive cytotoxic properties of CV extract and LY294002 inhibitor against cancer cells. The results showed that, in comparison with control cells, the single treatments led to a significant increase in the percentage of both MCF-7 and HeLa cells in the G0/G1 phase both after stimulation with CV extract ($p < 0.01$ and $p < 0.05$, respectively) as well as LY294002 ($p < 0.001$ and $p < 0.05$, respectively). However, the use of the combination of these agents significantly enhanced cell cycle arrest at the G0/G1 phase in comparison with either CV extract alone ($p < 0.001$ for MCF-7 cells and $p < 0.01$ for HeLa cells, respectively) or LY294002 alone ($p < 0.05$ for MCF-7 cells and $p < 0.01$ for HeLa cells, respectively). At the same time, the stimulation of both cell lines with CV extract alone or LY294002 alone resulted in a significant decrease in the S cell population phase compared to control cells ($p < 0.05$ and $p < 0.001$, respectively). However, this effect was the most noticeable in the co-treated cells. Finally, the decreased percentage of MCF-7 cells in the G2 phase was observed after stimulation with CV extract alone ($p < 0.05$) or a combination of CV and LY294002 ($p < 0.05$). In contrast, the G2 HeLa cell population was increased after CV extract treatment ($p < 0.001$) and decreased after cell co-stimulation (Figure 3).

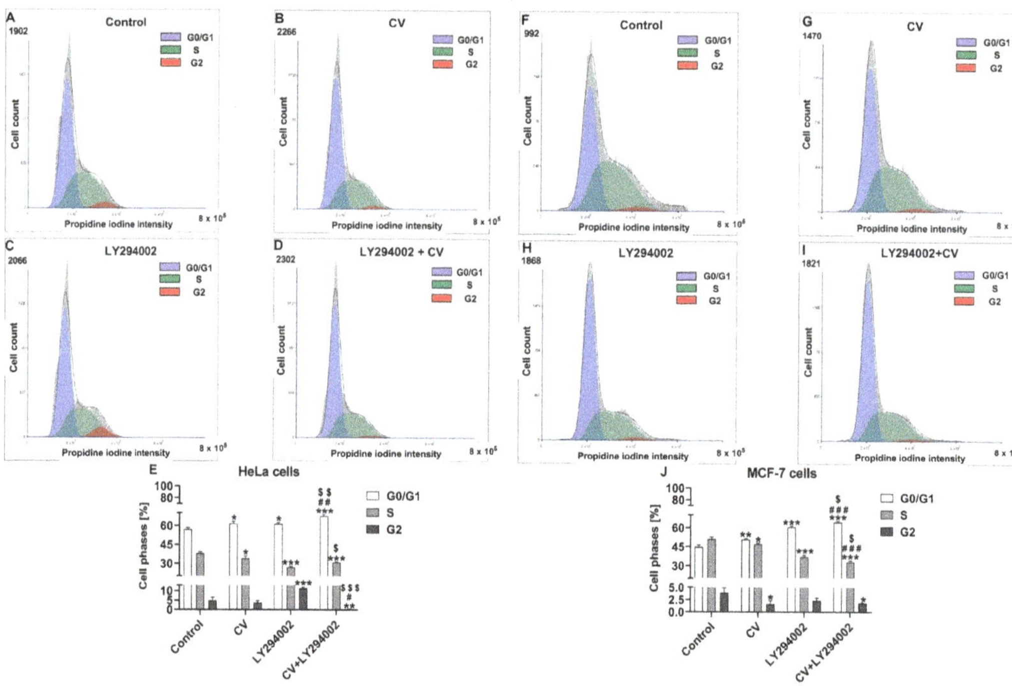

Figure 3. Cell cycle distribution of HeLa cells (**A–E**) and MCF-7 cells (**F–J**) treated with the CV extract (100 µg/mL), LY294002 (100 µM), or their combination compared with untreated cells (control). Representative histograms in (**A–D**) and (**F–I**) show the percentage of cells in G0/G1, S, and G2 phases. (**E**,**J**) present the cell cycle distribution as means ± S.E.M. from three independent experiments. Asterisks indicate differences between control and treated cells (* $p < 0.05$; ** $p < 0.01$; *** $p < 0.001$). Hashes show differences between CV extract and co-treated cells (# $p < 0.05$; ## $p < 0.01$; ### $p < 0.001$). $ marks present differences between LY294002 and co-treated cells ($ $p < 0.05$; $$ $p < 0.01$; $$$ $p < 0.001$).

2.3. Co-Treatment of Cancer Cells with LY294002 and CV Extract Increases Cell Apoptosis

The apoptosis analysis was performed using the cell death detection ELISAPLUS assay, which allows for quantifying histone-complexed DNA fragments out of the cytoplasm of the apoptotic cells. This analysis elucidated the underlying mechanism of the additive inhibition of the growth of HeLa and MCF-7 cancer cells co-stimulated with CV extract and LY294002. The results showed that the treatments of both cell lines only with the CV extract significantly increased the level of apoptosis, which was observed after 48 and 72 h (Figure 4). A similar effect was noticed for the cells stimulated with LY294002 alone, except for the HeLa cells treated with this inhibitor for 48 h. Moreover, the increase in nucleosome release in LY294002-treated cells was lower than the values calculated for the cells stimulated with the CV extract only. However, combining the CV extract and LY294002 induced the highest level of apoptosis, which was observed in HeLa cells after 72 h and MCF-7 cells after 48 h.

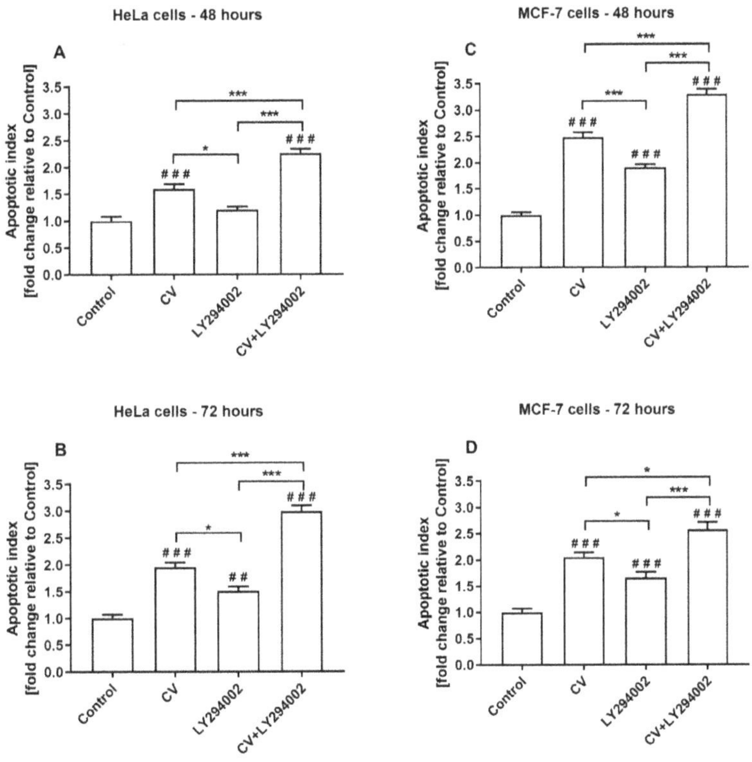

Figure 4. The induction of apoptosis in HeLa (**A**,**B**) and MCF-7 (**C**,**D**) treated with the CV extract (100 μg/mL), LY294002 (10 μM), or co-stimulated with both agents for 48 and 72 h. The level of the apoptotic index was expressed as a fold change relative to untreated cells (control; served as 1) and is presented as the means ± S.E.M. Asterisks indicate differences between the variants of cell stimulation as indicated (* $p < 0.05$; *** $p < 0.001$). Hashes present differences between control cells and the cells treated with the agents (## $p < 0.01$; ### $p < 0.001$).

2.4. LY294002 Increases the Anti-Migratory Activity of the CV Extract

To investigate the anti-migratory activity of the tested agents, scratch assays were performed to estimate the cell migration levels. As shown in Figure 5, although the use of CV extract alone or LY294002 alone decreased cell migration, the combined cell treatment with these two agents most effectively inhibited the migration of HeLa cells and MCF-7 cells. Moreover, both HeLa and MCF-7 cells stimulated only with LY294002 demonstrated higher levels of scratch closure than the CV extract-treated cells.

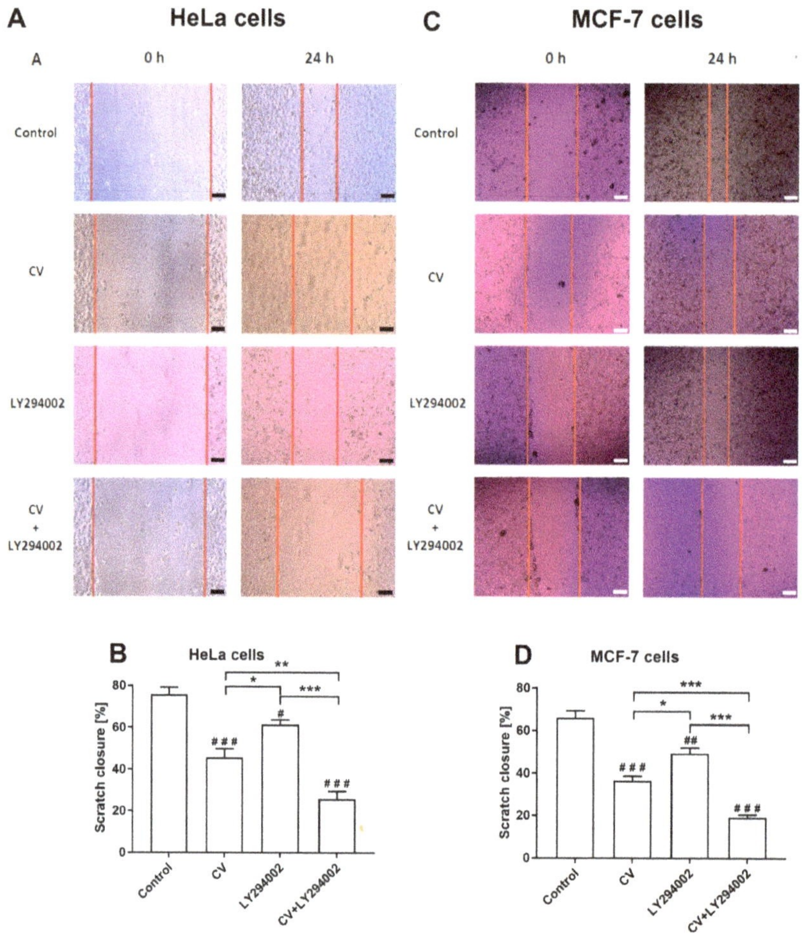

Figure 5. Anti-migratory effect of CV extract (100 μg/mL), LY294002 (10 μM), or their combination on HeLa (**A**,**B**) and MCF-7 (**C**,**D**) cells compared to control cells. (**A**,**C**) show the representative cell images at 0 h and after 24 h, respectively. Black and white lines in the lower corners represent a scale of 200 μm. (**B**,**D**) present the quantitative scratch closure (%) from 0 h to 24 h. Data are presented as mean ± S.E.M. from three independent experiments. Asterisks indicate significant differences between treatments (* $p < 0.05$; ** $p < 0.01$; *** $p < 0.001$). Hashes show differences between control and treated cells (# $p < 0.05$; ## $p < 0.01$; ### $p < 0.001$).

2.5. LY294002 Decreases the Invasive Ability of the Cancer Cells Stimulated with the CV Extract

The results of the transwell tumour cell invasion assays showed that the CV extract at a concentration of 100 μg/mL significantly decreased the invasive ability of only MCF-7 cells ($p < 0.05$) compared to control cells, without an effect on HeLa cells. In contrast, the number of HeLa and MCF-7 cells invaded after the stimulation only with LY294002 was lower than control cells ($p < 0.01$). However, the combined treatment with the CV extract and LY294002 most significantly reduced the invasive ability of HeLa cells and MCF-7 cells ($p < 0.001$) (Figure 6).

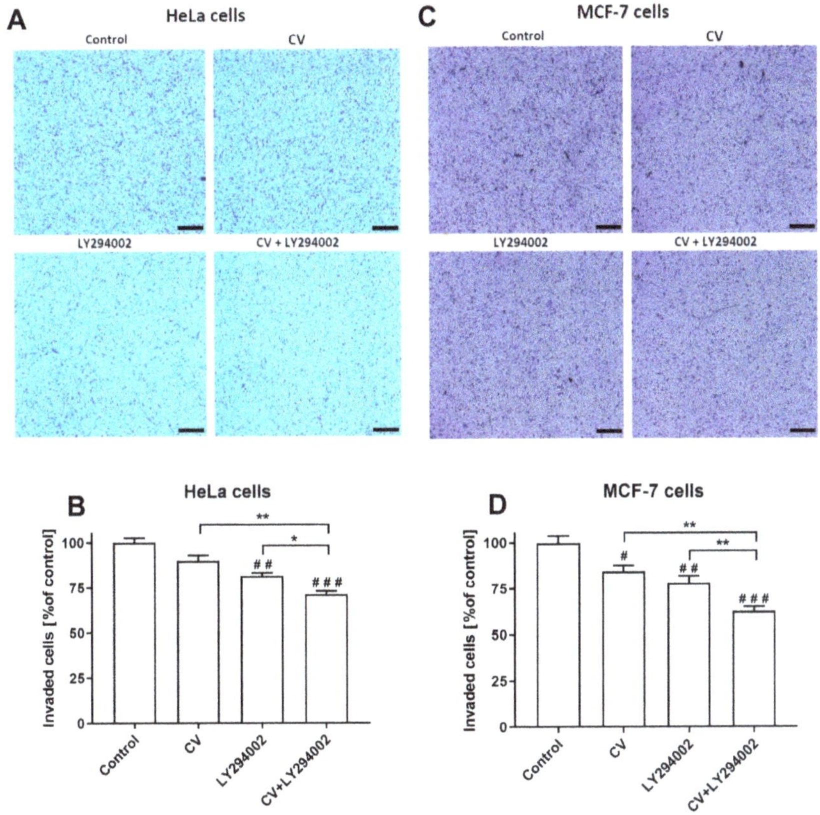

Figure 6. Invasive ability of HeLa cells (**A**,**B**) and MCF-7 cells (**C**,**D**) stimulated with CV extract (100 µg/mL), LY294002 (10 µM), or their combination compared to control cells. (**A**,**C**) show representative images of invaded cells, while (**B**,**D**) display quantitative evaluations (% of control). Black lines in the lower corners represent a scale of 200 µm. Data are presented as mean ± S.E.M. from three independent experiments. Asterisks indicate significant differences between treatments (* $p < 0.05$; ** $p < 0.01$). Hashes show differences between control and treated cells (# $p < 0.05$; ## $p < 0.01$; ### $p < 0.001$).

2.6. Co-Treatment of Cancer Cells with LY294002 and CV Extract Potentiated Inhibition of Phospho-PI3K Expression

The Western blot analysis was applied to check whether the bilateral effect of cancer cell co-stimulation with the CV extract and LY294002 results from changes in the phospho-PI3K (p-PI3K) expression. The results showed that the CV extract decreased the expression of p-PI3K in both cancer cell lines compared to control cells ($p < 0.01$ for HeLa cells and $p < 0.05$ for MCF-7 cells). However, this effect was weaker than in the cells stimulated only with LY294002 ($p < 0.05$ for both cancer cell lines). The lowest level of p-PI3K expression was observed in the cancer cells co-treated with both compounds for 24 h (Figure 7).

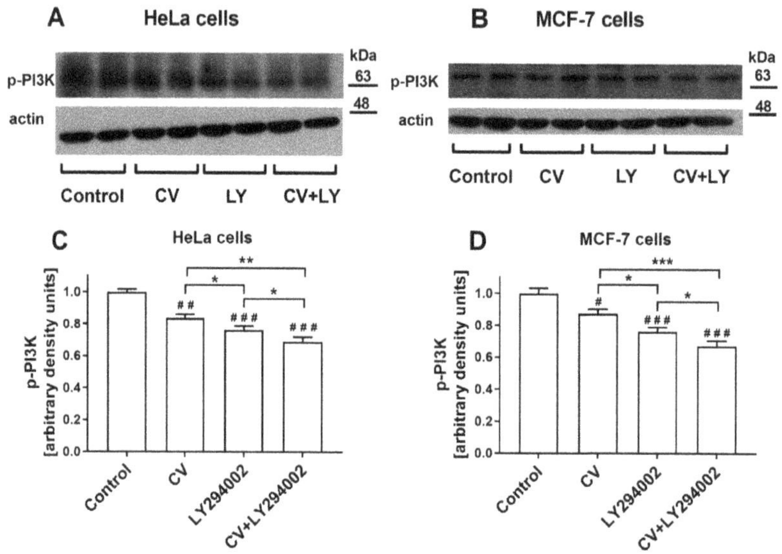

Figure 7. Representative Western blot images of p-PI3K and actin expression in untreated (control) HeLa (**A**) and MCF-7 cells (**C**) or cells stimulated with CV extract (100 μg/mL), LY294002 (10 μM), or both for 24 h. Densitometric bar graphs show p-PI3K levels in cell lysates (**B**,**D**). Data are presented as mean ± S.E.M. Asterisks indicate significant differences between treatments (* $p < 0.05$; ** $p < 0.01$; *** $p < 0.001$). Hashes show differences between control and treated cells (# $p < 0.05$; ## $p < 0.01$; ### $p < 0.001$).

3. Discussion

Out of over 270 species of mushrooms known for their immunomodulatory properties, 50 are identified as non-toxic and have undergone testing in animal models. However, human cancer studies have investigated only six of these species. Notably, the *Coriolus versicolor* mushroom stands out as the sole species studied in phase I, II, and III clinical trials involving patients with stomach, colorectal, oesophageal, and breast cancers [23]. The promising application of CV mushroom in cancer therapy stems from its capacity to discriminate between cancerous and normal cells, induce different modes of cancer cell death, such as apoptosis, necrosis, or necroptosis, and enhance the host immune response to cancer, which has been widely discussed in several review papers [3,5–7]. Moreover, numerous in vitro and in vivo studies show that active compounds derived from CV can induce anti-tumour effects when used as a monotherapy, and the cancer cell sensitization to various chemotherapeutic agents by CV has also been widely studied.

Despite promising results, there is still a need to refine therapy using CV by introducing an additional component because its anti-cancer effectiveness was observed only at high doses, which is manifested by high IC_{50} values observed for certain cancer cell lines, such as melanoma, breast cancer, leukaemia, stomach cancer, and colorectal cancer [5,24–27]. On the other hand, the use of CV in combination with chemotherapy has disadvantages related to side effects induced by chemotherapeutics that, unlike CV, do not differentiate cancerous cells and normal cells. Therefore, solutions capable of potentiating the anti-tumour value of CV mushroom in combined treatment in oncology patients, are still being searched for.

This study aimed to explore the therapeutic efficacy of co-treating cancer cells with the PI3K inhibitor (LY294002) and CV extract. The PI3K signalling pathway encompasses a range of biological processes pertinent to malignant cancers, such as cell proliferation, differentiation, invasion, and metastasis. The aberrant activation of this pathway also holds significant sway over the emergence of multidrug resistance in various types of neoplasms, representing a primary hurdle in chemotherapy efficacy [28]. Although several PI3K signalling inhibitors have been influential in the inhibition of tumour progressions during clinical trials and approved by the Food and Drug Administration in the USA, most of them have demonstrated only modest clinical efficiency as monotherapies because of drawbacks in their pharmacokinetics and tolerability [29,30]. We have chosen to explore the impact of a PI3K inhibitor on CV-induced effects for two main reasons. Our previous findings suggested that CV extract mediates immunostimulatory properties related to cytokine production in macrophages through the PI3K signalling pathway [31]. Secondly, there is a lack of research examining the potential impact of CV extract on cancer cells via this pathway.

Previous results from MTT assays showed that CV extract decreased cancer cell viability, confirming the results widely discussed in many papers [3,5,7,32]. In the present paper, we have also demonstrated that this extract reduced cancer cell colony formation, indicating that CV extract may suppress the regrowth and recurrence of tumours after treatment [33]. Significantly, the use of LY294002 inhibitor at a non-cytotoxic dose improved the anti-tumour properties of the CV extract compared to the anti-cancer activity of the CV extract alone, which was manifested by the decreased cell survival rate (IC_{50} values) and reduced ability of single cancer cells to survive and reproduce to form colonies.

Additionally, we investigated cell cycle distributions to further explain the underlying mechanism of the additive toxic effect resulting from treating cancer cells with the CV extract and LY294002 combination. The results showed that, compared to control cells, the treatments with CV extract alone or LY294002 alone led to a significant increase in the percentage of cancer cells in the G0/G1 while causing a significant decrease in the S cell population phase. These discoveries align with prior research, which has shown that CV extract disrupts the progression of the cancer cell cycle and induces cell cycle arrest at the G0 [34], G0/G1 [35], G1/S [36], and G2/M [37] phases. Likewise, LY294002 is regarded as a promising therapeutic agent due to its ability to inhibit the growth of different cancer cell types, primarily by inducing cell cycle arrest at both the G0/G1 [38] and G1/M [17] phases. However, our results demonstrated that combined treatment of cancer cells with CV extract and LY294002 notably intensified cell cycle arrest at the G0/G1 phase in cancer cells compared to using each compound alone.

The G0/G1-phase arrest of cell cycle progression is widely recognised as a crucial juncture where cells can either engage in repair mechanisms or proceed along the apoptotic pathway. Apoptosis, a significant form of cell death triggered by cytotoxic effects, is thus facilitated by this arrest [39]. Numerous experiments also show that CV extract [35,40–42] and LY294002 [17,43] can induce the apoptosis of cancer cells. Based on these findings and the results from the cell cycle distributions, we decided to check how combined treatment with CV extract and LY294002 affects the apoptotic death of cancer cells. Our results showed that co-treatment with the CV extract and LY294002 more effectively induced the apoptosis of cancer cells than the stimulation with each of these agents alone. This effect can be partially associated with the more effective arrest of the cancer cell cycle at the G0/G1 phase observed during the combined treatment of cancer cells [17].

Cancer cells' migration and invasion capabilities facilitate their movement and infiltration into lymphatic and blood vessels, enabling the spread of neoplastic cells and the development of metastatic growth in distant organs. Therefore, controlling cancer cell migration and invasion becomes a pivotal focus of cancer treatment [44]. The aggressive nature of cancer cells is governed by an intricate network of signalling pathways that oversee crucial functions, including growth, survival, migration, and invasion. The PI3K signalling pathway has been causally linked to these four responses. Moreover, overexpression of PI3K in a wide range of human tumours is detected and often linked with poor prognosis [13]. Accumulating evidence suggests that a PI3K inhibitor named LY294002 can significantly attenuate the invasion and migration of leukaemia cells [45], prostate cancer cells [46], oral squamous carcinoma cells [47], osteosarcoma cells [48], and pancreatic cancer cells [49]. Here, we illustrated that LY294002 also exhibits these properties when tested against breast and cervical cancer cells. There are also some experimental data showing that CV extract has anti-migratory and anti-invasive potentials against numerous tumour cells, including triple-negative breast cancer [42], oestrogen receptor (ER)-positive breast cancer [26], colon cancer [25], pancreatic and gastric cancer [50], and melanoma [51]. Our results demonstrated that CV extract also inhibits the invasiveness and migration of cervical cancer cells. The combined treatment with the CV extract and LY294002 significantly reduced the invasive and migratory activities of ER-positive breast and cervical cancer cells more considerably than the cell stimulation with each compound alone. Additional investigations are imperative to clearly elucidate the mechanism behind this enhanced interaction. However, we suppose that it can result from the effect of both compounds on the simultaneous inhibition of the PI3K pathway. Our assumption is confirmed by the results from the Western blot analysis, which showed that co-treatment of cells with the CV extract and LY294002 more significantly inhibited the protein expression of p-PI3K in the cancer cells than LY294002 alone. We also observed that CV extract decreased p-PI3K expression in cancer cells for the first time compared to control cells.

Additional studies are required to assess the effects of combined stimulation with CV extract and LY294002 on normal cells, providing a deeper understanding of their impact on healthy tissues. Numerous studies, including ours, have shown that CV extract not only lacks toxic effects on normal cells but that it has also been found to enhance the viability and proliferation of these cells. This mitogenic effect has been particularly observed in various immune cell types, including B and T lymphocytes, splenocytes, monocytes, macrophages, and dendritic cells [3,5]. Moreover, CV extract can increase the resistance of normal cells to chemotherapy-induced cytotoxicity, mainly by enhancing antioxidant activity and immune response. Studies investigating the co-treatment of CV extract and chemotherapeutics have revealed several beneficial effects, including protective effects on bone marrow cells and hepatocytes, as well as the enhancement of immune cell functions. These findings suggest that CV extract may help mitigate the side effects of chemotherapy while supporting the body's natural defence mechanisms [5,10]. In contrast, treatment of normal cells with LY294002 can exhibit cytotoxic effects, particularly at high concentrations. This is primarily due to the inhibition of the PI3K pathway, which plays a crucial role in the survival, proliferation, and metabolic regulation of various normal cells [52]. Therefore, it is crucial to use relatively low doses of LY294002 to minimise toxicity in healthy cells while simultaneously enhancing the sensitivity of cancer cells to the cytotoxic effects of natural compounds, like CV extract.

4. Materials and Methods

4.1. Cell Culture

The MCF-7 breast cancer cells and lung cancer cell line A549 were procured from the European Collection of Authenticated Cell Cultures (ECACC; Salisbury, UK), while cervical cancer cells (HeLa) were sourced from the American Type Culture Collection (ATCC; Manassas, VA, USA). Each cell line was cultured in DMEM supplemented with 10% foetal bovine serum (FBS) at 37 °C in an atmosphere with 5% CO_2. The reagents used for cell cultures were purchased from Merck KGaA (Darmstadt, Germany).

4.2. Preparation of CV Extract and LY294002 Solution

The CV extract was provided by the MycoMedica Company (Police nad Metují, Czech Republic). Previously, we demonstrated that CV extract contains polysaccharide peptide (PSP) and polysaccharide krestin (PSK), which are considered the most potent compounds for anti-tumour and immunomodulatory effects. Additionally, we have identified various low-molecular-weight compounds in the extract, including vitamins (D3, K, retinyl acetate), monosaccharides (arabinose, glucuronic acid), amino acids (histidine), fatty acids (palmitic acid, oleic acid, linolenic acid), and phenolic compounds (p-hydroxy benzoic acid) [53].

The culture medium was utilised to dissolve the CV extract, forming a 4 mg/mL stock solution, which was received by continuous agitation for 48 h at room temperature. The soluble supernatant, which held 1 mg of polysaccharide peptides, underwent sterilization via a 0.22 μm filter. Subsequently, it was diluted with culture medium to the specified concentrations as indicated.

The LY294002, an inhibitor of PI3K, was purchased from Cell Signalling Technology (Leiden, The Netherlands) and reconstituted in dimethyl sulfoxide (DMSO; Merck KGaA, Darmstadt, Germany). Before cell stimulation, the stock solution of the inhibitor was diluted to the desired concentration using a culture medium. Throughout the cell treatment, the ultimate concentration of DMSO remained below 0.1%.

4.3. Cell Viability

The cancer cells were seeded in a density of 3×10^3 cells/well and pre-incubated overnight. Then, the cells were treated with the CV extract (50, 100, 150, 200, and 250 μg/mL) alone or co-stimulated with the CV extract and LY294002 (10 μM) for 24, 48, and 72 h. For the cell co-treatment, a 10 μM concentration of inhibitor was selected as it maintained cell viability above 70% (Supplemental Figure S1) and, therefore, it cannot be considered a cytotoxic agent according to the ISO 10993-5 norm [22]. After treatment, the MTT solution (0.5 mg/mL; Merck KGaA, Darmstadt, Germany) was added for 3 h, followed by dissolving formazan crystals with DMSO and measuring the optical density at 570 nm using a microplate spectrophotometer (Synergy HT; BioTek Instruments, Winooski, VT, USA). The results were presented as a percentage of the control cells. Using data from the MTT assays, the IC_{50} values were determined using GraphPad Prism 7.0 (GraphPad Software Inc., San Diego, CA, USA). For subsequent experiments, HeLa and MCF-7 cells were used, since their viability was decreased during co-stimulation with CV extract and LY294002 compared with the treatment with the extract alone. Furthermore, a 100 μg/mL CV extract concentration was selected, as it maintained viability above 70%.

4.4. Colony Formation Assay

The cells were seeded in 6-well plates (3×10^5 cells/well) overnight, followed by the stimulation with the CV extract (100 μg/mL), LY294002 (10 μM), or their combination for 48 h. After trypsinization, the live HeLa and MCF-7 cells were seeded in 6-well plates (in triplicates) at a density of 2×10^2 or 1×10^3 cells/well. They were cultivated in 2 mL media for 10 or 14 days, respectively. Every three days, the media were exchanged with fresh culture medium. After the designated culture period, the cells were washed with PBS and fixed with 100% *v*/*v* methanol for 20 min at room temperature. Next, staining was carried out using a 0.5% *v*/*v* crystal violet solution (prepared in 25% *v*/*v* methanol) for 25 min; the colonies were then washed with water and we air-dried the plates before performing the colony counting. Colonies containing more than 50 cells were counted in triplicate.

4.5. Apoptosis Assay

The cell death detection ELISAPLUS assay (Roche Diagnostics, Mannheim, Germany) was used to quantify histone-complexed DNA fragments (mono- and oligonucleosomes) out of the apoptotic cells' cytoplasm after the induction of apoptosis. HeLa and MCF-7 (1×10^4 cells/well) were pre-incubated for 24 h, followed by treatment with the CV extract (100 μg/mL), LY294002 (10 μM), or co-stimulated with both agents for 48 and 72 h. The subsequent procedures of the sandwich enzyme-linked immunoassay followed the instructions provided by the manufacturer. The colour development, indicative of the number of nucleosomes captured in the antibody sandwich, was assessed at 405 nm (with a reference wavelength of 490 nm) using a Synergy HT Microplate Reader (BioTek Instruments, Winooski, VT, USA). The results were expressed as a fold change relative to untreated cells, normalised to 1.

4.6. Cell Cycle Analysis

The cell cycle was assessed following the manufacturer's guidelines using the Cell-CycleFlowEx Kit (EXBIO Praha, a.s., Vestec, Czech Republic), which facilitates the quantification of DNA content through propidium iodide staining, followed by analysis using flow cytometry. The cells were seeded at a density of 3×10^5 cells in a 25 cm^2 tissue culture flask (in triplicates), cultured overnight, followed by starving with serum-free medium for 6 h. Then, the cells were treated with the CV extract (100 μg/mL), LY294002 (10 μM), or their combination for 48 h. For flow cytometry analysis, the cells were stained for 30 min with propidium iodide, and RNA was digested using RNAse A (EXBIO Praha, a.s., Vestec, Czech Republic). Finally, the cell cycle distributions were analysed using a BriCyte E6 flow cytometer (Mindray, Shenzhen, China) and FCS Express 7 Image Cytometry Software (DeNovo Software, Pasadena, CA, USA).

4.7. Scratch Assay

The cells (2×10^5/well) were seeded in 12-well plates and pre-incubated in DMEM with 10% FBS until they reached 100% confluency. The cell monolayer was scratched (in triplicate) using a 10 μL pipette tip, and cells were treated with CV extract, LY294002, or both agents in DMEM with 1% FBS for 24 h. Cell migration into the scratched area was monitored using a Leica DMi1 inverted microscope with a digital camera (Leica Microsystems, Wetzlar, Germany), both at the initial moment and 24 h later. Migration scratch closure was measured with ImageJ software v1.54d (National Institutes of Health, Bethesda, MD, USA) and expressed as a percentage of the initial distance.

4.8. Invasion Assay

Cell invasion was assessed using the CHEMICON Cell Invasion Assay Kit (Merck KGaA, Darmstadt, Germany). Cells (3×10^5/well) were seeded in 6-well plates overnight (in triplicates) and treated with CV extract, LY294002, or both agents for 48 h. Then, 5×10^4 live cells were suspended in 300 μL serum-free DMEM and added to the upper chambers, while 500 μL DMEM with 20% FBS was added to the lower chambers. After 24 h, non-invasive cells were removed using a cotton swab, and invasive cells were stained with a staining solution provided by the manufacturer and imaged (40× magnification). The level of cell invasion was also quantitated by dissolving stained cells in 10% acetic acid (200 μL/well) in addition to taking colorimetric absorbance readings at 560 nm (Synergy HT Microplate Reader, BioTek Instruments, Winooski, VT, USA). The invasion levels were expressed as a percentage of untreated cells.

4.9. Western Blot Analysis

Western blotting was used to determine p-PI3K protein levels of in HeLa and MCF-7 cells after 24 h treatment with CV extract, LY294002, or both agents. After stimulation, the cells were washed with ice-cold PBS and lysed in RIPA buffer supplemented with a protease and phosphatase inhibitor cocktail (all reagents sourced from Merck KGaA, Darmstadt, Germany). The protein concentration within the lysates was assessed using a Pierce™ BCA Protein Assay Kit (Thermo Fisher Scientific, Waltham, MA, USA). The lysates were mixed with sample buffer and subjected to electrophoresis. Then, proteins were transferred to nitrocellulose membranes, and incubated with rabbit anti-phospho-PI3K p85 (cat. no. 4292; Lot: 1; Cell Signalling Technology, Leiden, The Netherlands) and mouse anti-actin (cat no. CP01; Lot: D00164515; Merck KGaA, Darmstadt, Germany) antibodies. Subsequently, the membranes were exposed to either goat anti-rabbit (cat. no. 612657; Lot: 0296526 MP Biomedicals; Santa Ana, CA, USA) or goat anti-mouse (cat no. 115-035-003; Lot no.: 152341; Jackson ImmunoResearch, Cambridge, United Kingdom) antibodies conjugated with horseradish peroxidase. Protein bands were visualised with SuperSignalWest Pico substrate (Thermo Fisher Scientific, Waltham, MA, USA) and analysed densitometrically using ImageJ software v1.54d.

4.10. Statistical Analyses

Statistical analyses were performed using GraphPad Prism 7.0 software (GraphPad Software Inc., San Diego, CA, USA). Data are presented as mean $\pm$ standard error of the mean (S.E.M.) and were assessed using one-way ANOVA followed by Tukey's multiple comparisons test, with statistical significance set at $p < 0.05$.

5. Conclusions

In conclusion, while compounds derived from CV extract are utilised in complementary cancer treatment, there remains a quest for novel therapies to enhance its efficacy. Our study illustrates that combining CV extract with a chemical inhibitor of the PI3K signalling pathway induces an additive cytotoxic effect on cancer cells. This suggests a promising avenue for developing more effective anti-cancer therapies incorporating CV extract within conventional treatment protocols in China, Japan, and in Western countries. However, this study has certain limitations, including the need to compare the effects of CV extract and LY294002 not only on cancer cells but also on healthy cells to assess the potential cytotoxicity of this kind of cell treatment. Additionally, further research is required to examine the interactions between the two compounds and determine their nature, such as their synergistic or additive effects, using isobolographic analysis. Finally, future studies should incorporate animal models to evaluate the pharmacokinetics, bioavailability, and systemic effects of the CV extract and LY294002 combination.

Supplementary Materials: The following supporting information can be downloaded at: https://www.mdpi.com/article/10.3390/ijms26041556/s1.

Author Contributions: Conceptualization, T.J.; methodology, T.J.; validation, T.J., J.S., and B.M.; formal analysis, T.J., J.S., B.M., M.S., and S.W.; investigation, T.J. and J.S.; resources, M.S. and S.W.; Writing—original draft preparation, T.J.; Writing—review and editing, T.J., M.S., and S.W.; Visualization, T.J. and B.M.; Supervision, T.J.; Project administration, T.J.; Funding acquisition, T.J. All authors have read and agreed to the published version of the manuscript.

Funding: This work in the field of a post-doctoral research internship of Tomasz Jędrzejewski at the University of Pardubice was financed under the "Excellence Initiative—Research University" program (Mobility for employees, 9th edition) by Nicolaus Copernicus University.

Institutional Review Board Statement: Not applicable.

Informed Consent Statement: Not applicable.

Data Availability Statement: The original contributions presented in this study are included in the article/Supplementary Material. Further inquiries can be directed to the corresponding author.

Conflicts of Interest: The authors declare no conflicts of interest.

Abbreviations

The following abbreviations are used in this manuscript:

CV	Coriolus versicolor
LY294002	phosphatidylinositol-3-kinase inhibitor
PI3K	phosphatidylinositol-3-kinase
p-PI3K	phospho- phosphatidylinositol-3-kinase

References

1. Jing, Y.; Zhang, S.; Li, M.; Ma, Y.; Zheng, Y.; Zhang, D.; Wu, L. Research Progress on the Extraction, Structure, and Bioactivities of Polysaccharides from *Coriolus versicolor*. *Foods* **2022**, *11*, 2126. [CrossRef]
2. Chen, Y.; Yao, F.; Ming, K.; Wang, D.; Hu, Y.; Liu, J. Polysaccharides from Traditional Chinese Medicines: Extraction, Purification, Modification, and Biological Activity. *Molecules* **2016**, *21*, 1705. [CrossRef] [PubMed]
3. Habtemariam, S. *Trametes versicolor* (Synn. *Coriolus versicolor*) Polysaccharides in Cancer Therapy: Targets and Efficacy. *Biomedicines* **2020**, *8*, 135. [CrossRef] [PubMed]
4. Bains, A.; Chawla, P.; Kaur, S.; Najda, A.; Fogarasi, M.; Fogarasi, S. Bioactives from Mushroom: Health Attributes and Food Industry Applications. *Materials* **2021**, *14*, 7640. [CrossRef] [PubMed]
5. Jędrzejewski, T.; Pawlikowska, M.; Sobocińska, J.; Wrotek, S. COVID-19 and Cancer Diseases-The Potential of *Coriolus versicolor* Mushroom to Combat Global Health Challenges. *Int. J. Mol. Sci.* **2023**, *24*, 4864. [CrossRef]
6. Saleh, M.H.; Rashedi, I.; Keating, A. Immunomodulatory Properties of *Coriolus versicolor*: The Role of Polysaccharopeptide. *Front. Immunol.* **2017**, *8*, 1087. [CrossRef]
7. He, Z.; Lin, J.; He, Y.; Liu, S. Polysaccharide-Peptide from *Trametes versicolor*: The Potential Medicine for Colorectal Cancer Treatment. *Biomedicines* **2022**, *10*, 2841. [CrossRef]
8. Maehara, Y.; Tsujitani, S.; Saeki, H.; Oki, E.; Yoshinaga, K.; Emi, Y.; Morita, M.; Kohnoe, S.; Kakeji, Y.; Yano, T.; et al. Biological mechanism and clinical effect of protein-bound polysaccharide K (KRESTIN(®)): Review of development and future perspectives. *Surg. Today* **2012**, *42*, 8–28. [CrossRef]
9. Torkelson, C.J.; Sweet, E.; Martzen, M.R.; Sasagawa, M.; Wenner, C.A.; Gay, J.; Putiri, A.; Standish, L.J. Phase 1 Clinical Trial of *Trametes versicolor* in Women with Breast Cancer. *ISRN Oncol.* **2012**, *2012*, 251632. [CrossRef]
10. Chang, Y.; Zhang, M.; Jiang, Y.; Liu, Y.; Luo, H.; Hao, C.; Zeng, P.; Zhang, L. Preclinical and clinical studies of Coriolus versicolor polysaccharopeptide as an immunotherapeutic in China. *Discov. Med.* **2017**, *23*, 207–219.
11. Pilkington, K.; Wieland, L.S.; Teng, L.; Jin, X.Y.; Storey, D.; Liu, J.P. *Coriolus* (*Trametes*) *versicolor* mushroom to reduce adverse effects from chemotherapy or radiotherapy in people with colorectal cancer. *Cochrane Database Syst. Rev.* **2022**, *11*, CD012053. [CrossRef] [PubMed]
12. Sivanesan, I.; Muthu, M.; Gopal, J.; Oh, J.W. Mushroom Polysaccharide-Assisted Anticarcinogenic Mycotherapy: Reviewing Its Clinical Trials. *Molecules* **2022**, *27*, 4090. [CrossRef] [PubMed]
13. Rascio, F.; Spadaccino, F.; Rocchetti, M.T.; Castellano, G.; Stallone, G.; Netti, G.S.; Ranieri, E. The Pathogenic Role of PI3K/AKT Pathway in Cancer Onset and Drug Resistance: An Updated Review. *Cancers* **2021**, *13*, 3949. [CrossRef] [PubMed]
14. Castel, P.; Toska, E.; Engelman, J.A.; Scaltriti, M. The present and future of PI3K inhibitors for cancer therapy. *Nat. Cancer* **2021**, *2*, 587–597. [CrossRef]
15. Wang, Y.; Kuramitsu, Y.; Baron, B.; Kitagawa, T.; Tokuda, K.; Akada, J.; Maehara, S.I.; Maehara, Y.; Nakamura, K. PI3K inhibitor LY294002, as opposed to wortmannin, enhances AKT phosphorylation in gemcitabine-resistant pancreatic cancer cells. *Int. J. Oncol.* **2017**, *50*, 606–612. [CrossRef]
16. Meng, D.; He, W.; Zhang, Y.; Liang, Z.; Zheng, J.; Zhang, X.; Zheng, X.; Zhan, P.; Chen, H.; Li, W.; et al. Development of PI3K inhibitors: Advances in clinical trials and new strategies (Review). *Pharmacol. Res.* **2021**, *173*, 105900. [CrossRef]
17. Abdallah, M.E.; El-Readi, M.Z.; Althubiti, M.A.; Almaimani, R.A.; Ismail, A.M.; Idris, S.; Refaat, B.; Almalki, W.H.; Babakr, A.T.; Mukhtar, M.H.; et al. Tamoxifen and the PI3K Inhibitor: LY294002 Synergistically Induce Apoptosis and Cell Cycle Arrest in Breast Cancer MCF-7 Cells. *Molecules* **2020**, *25*, 3355. [CrossRef]

18. Guney Eskiler, G.; Ozturk, M. Therapeutic potential of the PI3K inhibitor LY294002 and PARP inhibitor Talazoparib combination in BRCA-deficient triple negative breast cancer cells. *Cell. Signal.* **2022**, *91*, 110229. [CrossRef]
19. Huang, A.; Zeng, P.; Li, Y.; Lu, W.; Lai, Y. LY294002 Is a Promising Inhibitor to Overcome Sorafenib Resistance in FLT3-ITD Mutant AML Cells by Interfering with PI3K/Akt Signaling Pathway. *Front. Oncol.* **2021**, *11*, 782065. [CrossRef]
20. Badinloo, M.; Esmaeili Mahani, S. Combined Application of Phosphoinositide 3-Kinase and Mammalian Target of Rapamycin Inhibitors Suppresses Cell Growth and Promotes Apoptosis in Human Lung Cancer Cell Lines. *Int. J. Cancer. Manag.* **2016**, *9*, e3433. [CrossRef]
21. Bray, F.; Laversanne, M.; Sung, H.; Ferlay, J.; Siegel, R.L.; Soerjomataram, I.; Jemal, A. Global cancer statistics 2022: GLOBOCAN estimates of incidence and mortality worldwide for 36 cancers in 185 countries. *CA Cancer J. Clin.* **2024**, *74*, 229–263. [CrossRef] [PubMed]
22. *ISO 10993-5:2021*; Biological Evaluation of Medical Devices. Part 5: Tests for in Vitro Cytotoxicity. International Organization for Standardization: Geneva, Switzerland, 2021.
23. Standish, L.J.; Wenner, C.A.; Sweet, E.S.; Bridge, C.; Nelson, A.; Martzen, M.; Novack, J.; Torkelson, C. Trametes versicolor mushroom immune therapy in breast cancer. *J. Soc. Integr. Oncol.* **2008**, *6*, 122–128. [PubMed]
24. Roca-Lema, D.; Martinez-Iglesias, O.; Fernández de Ana Portela, C.; Rodríguez-Blanco, A.; Valladares-Ayerbes, M.; Díaz-Díaz, A.; Casas-Pais, A.; Prego, C.; Figueroa, A. In Vitro Anti-proliferative and Anti-invasive Effect of Polysaccharide-rich Extracts from Trametes Versicolor and Grifola Frondosa in Colon Cancer Cells. *Int. J. Med. Sci.* **2019**, *16*, 231–240. [CrossRef] [PubMed]
25. Jędrzejewski, T.; Sobocińska, J.; Pawlikowska, M.; Dzialuk, A.; Wrotek, S. Extract from the *Coriolus versicolor* Fungus as an Anti-Inflammatory Agent with Cytotoxic Properties against Endothelial Cells and Breast Cancer Cells. *Int. J. Mol. Sci.* **2020**, *21*, 9063. [CrossRef]
26. Winder, M.; Bulska-Będkowska, W.; Chudek, J. The use of *Hericium erinaceus* and *Trametes versicolor* extracts in supportive treatment in oncology. *Acta Pharm.* **2021**, *71*, 1–16. [CrossRef]
27. Lowenthal, R.; Taylor, M.; Gidden, J.A.; Heflin, B.; Lay, J.O., Jr.; Avaritt, N.; Tackett, A.J.; Urbaniak, A. The mycelium of the *Trametes versicolor* synn. *Coriolus versicolor* (Turkey tail mushroom) exhibit anti-melanoma activity in vitro. *Biomed. Pharmacother.* **2023**, *161*, 114424. [CrossRef]
28. Liu, R.; Chen, Y.; Liu, G.; Li, C.; Song, Y.; Cao, Z.; Li, W.; Hu, J.; Lu, C.; Liu, Y. PI3K/AKT pathway as a key link modulates the multidrug resistance of cancers. *Cell Death Dis.* **2020**, *11*, 797. [CrossRef]
29. Curigliano, G.; Shah, R.R. Safety and Tolerability of Phosphatidylinositol-3-Kinase (PI3K) Inhibitors in Oncology. *Drug. Saf.* **2019**, *42*, 247–262. [CrossRef]
30. Coleman, N.; Moyers, J.T.; Harbery, A.; Vivanco, I.; Yap, T.A. Clinical Development of AKT Inhibitors and Associated Predictive Biomarkers to Guide Patient Treatment in Cancer Medicine. *Pharmgenomics. Pers. Med.* **2021**, *14*, 1517–1535. [CrossRef]
31. Jędrzejewski, T.; Sobocińska, J.; Pawlikowska, M.; Dzialuk, A.; Wrotek, S. Dual Effect of the Extract from the Fungus *Coriolus versicolor* on Lipopolysaccharide-Induced Cytokine Production in RAW 264.7 Macrophages Depending on the Lipopolysaccharide Concentration. *J. Inflamm. Res.* **2022**, *15*, 3599–3611. [CrossRef]
32. Blagodatski, A.; Yatsunskaya, M.; Mikhailova, V.; Tiasto, V.; Kagansky, A.; Katanaev, V.L. Medicinal mushrooms as an attractive new source of natural compounds for future cancer therapy. *Oncotarget* **2018**, *9*, 29259–29274. [CrossRef] [PubMed]
33. Gomes, N.P.; Frederick, B.; Jacobsen, J.R.; Chapnick, D.; Su, T.T. A High Throughput Screen with a Clonogenic Endpoint to Identify Radiation Modulators of Cancer. *Radiat. Res.* **2023**, *199*, 132–147. [CrossRef] [PubMed]
34. Ricciardi, M.R.; Licchetta, R.; Mirabilii, S.; Scarpari, M.; Parroni, A.; Fabbri, A.A.; Cescutti, P.; Reverberi, M.; Fanelli, C.; Tafuri, A. Preclinical Antileukemia Activity of Tramesan: A Newly Identified Bioactive Fungal Metabolite. *Oxid. Med. Cell. Longev.* **2017**, *2017*, 5061639. [CrossRef] [PubMed]
35. Jiménez-Medina, E.; Berruguilla, E.; Romero, I.; Algarra, I.; Collado, A.; Garrido, F.; Garcia-Lora, A. The immunomodulator PSK induces in vitro cytotoxic activity in tumour cell lines via arrest of cell cycle and induction of apoptosis. *BMC Cancer* **2008**, *8*, 78. [CrossRef]
36. Wan, J.M.; Sit, W.H.; Yang, X.; Jiang, P.; Wong, L.L. Polysaccharopeptides Derived from *Coriolus Versicolor* Potentiate the S-Phase Specific Cytotoxicity of Camptothecin (CPT) on Human Leukemia HL-60 Cells. *Chin. Med.* **2010**, *5*, 16. [CrossRef]
37. Hsieh, T.C.; Wu, P.; Park, S.; Wu, J.M. Induction of cell cycle changes and modulation of apoptogenic/anti-apoptotic and extracellular signaling regulatory protein expression by water extracts of I'm-Yunity (PSP). *BMC Complement. Altern. Med.* **2006**, *6*, 30. [CrossRef]
38. Lu, X.X.; Cao, L.Y.; Chen, X.; Xiao, J.; Zou, Y.; Chen, Q. PTEN Inhibits Cell Proliferation, Promotes Cell Apoptosis, and Induces Cell Cycle Arrest via Downregulating the PI3K/AKT/hTERT Pathway in Lung Adenocarcinoma A549 Cells. *Biomed. Res. Int.* **2016**, *2016*, 2476842. [CrossRef]
39. Pucci, B.; Kasten, M.; Giordano, A. Cell cycle and apoptosis. *Neoplasia* **2000**, *2*, 291–299. [CrossRef]
40. Lau, C.B.; Ho, C.Y.; Kim, C.F.; Leung, K.N.; Fung, K.P.; Tse, T.F.; Chan, H.H.; Chow, M.S. Cytotoxic activities of *Coriolus versicolor* (Yunzhi) extract on human leukemia and lymphoma cells by induction of apoptosis. *Life Sci.* **2004**, *75*, 797–808. [CrossRef]

41. Hirahara, N.; Fujioka, M.; Edamatsu, T.; Fujieda, A.; Sekine, F.; Wada, T.; Tanaka, T. Protein-bound polysaccharide-K (PSK) induces apoptosis and inhibits proliferation of promyelomonocytic leukemia HL-60 cells. *Anticancer. Res.* **2011**, *31*, 2733–2738.
42. Luo, K.W.; Yue, G.G.; Ko, C.H.; Lee, J.K.; Gao, S.; Li, L.F.; Li, G.; Fung, K.P.; Leung, P.C.; Lau, C.B. In vivo and in vitro anti-tumor and anti-metastasis effects of *Coriolus versicolor* aqueous extract on mouse mammary 4T1 carcinoma. *Phytomedicine* **2014**, *21*, 1078–1087. [CrossRef] [PubMed]
43. Jiang, H.; Fan, D.; Zhou, G.; Li, X.; Deng, H. Phosphatidylinositol 3-kinase inhibitor(LY294002) induces apoptosis of human nasopharyngeal carcinoma in vitro and in vivo. *J. Exp. Clin. Cancer Res.* **2010**, *29*, 34. [CrossRef] [PubMed]
44. Novikov, N.M.; Zolotaryova, S.Y.; Gautreau, A.M.; Denisov, E.V. Mutational drivers of cancer cell migration and invasion. *Br. J. Cancer* **2021**, *124*, 102–114. [CrossRef] [PubMed]
45. Liu, P.; Xu, B.; Li, J.; Lu, H. LY294002 inhibits leukemia cell invasion and migration through early growth response gene 1 induction independent of phosphatidylinositol 3-kinase-Akt pathway. *Biochem. Biophys. Res. Commun.* **2008**, *377*, 187–190. [CrossRef]
46. Shukla, S.; Maclennan, G.T.; Hartman, D.J.; Fu, P.; Resnick, M.I.; Gupta, S. Activation of PI3K-Akt signaling pathway promotes prostate cancer cell invasion. *Int. J. Cancer* **2007**, *121*, 1424–1432. [CrossRef]
47. Duarte, A.; Silveira, G.G.; Soave, D.F.; Costa, J.P.O.; Silva, A.R. The Role of the LY294002—A Non-Selective Inhibitor of Phosphatidylinositol 3-Kinase (PI3K) Pathway- in Cell Survival and Proliferation in Cell Line SCC-25. *Asian. Pac. J. Cancer. Prev.* **2019**, *20*, 3377–3383. [CrossRef]
48. Zhou, Y.; Zhu, L.B.; Peng, A.F.; Wang, T.F.; Long, X.H.; Gao, S.; Zhou, R.P.; Liu, Z.L. LY294002 inhibits the malignant phenotype of osteosarcoma cells by modulating the phosphatidylinositol 3-kinase/Akt/fatty acid synthase signaling pathway in vitro. *Mol. Med. Rep.* **2015**, *11*, 1352–1357. [CrossRef]
49. Teranishi, F.; Takahashi, N.; Gao, N.; Akamo, Y.; Takeyama, H.; Manabe, T.; Okamoto, T. Phosphoinositide 3-kinase inhibitor (wortmannin) inhibits pancreatic cancer cell motility and migration induced by hyaluronan in vitro and peritoneal metastasis in vivo. *Cancer Sci.* **2009**, *100*, 770–777. [CrossRef]
50. Zhang, H.; Morisaki, T.; Matsunaga, H.; Sato, N.; Uchiyama, A.; Hashizume, K.; Nagumo, F.; Tadano, J.; Katano, M. Protein-bound polysaccharide PSK inhibits tumor invasiveness by down-regulation of TGF-beta1 and MMPs. *Clin. Exp. Metastasis* **2000**, *18*, 343–352. [CrossRef]
51. Matsunaga, K.; Ohhara, M.; Oguchi, Y.; Iijima, H.; Kobayashi, H. Antimetastatic effect of PSK, a protein-bound polysaccharide, against the B16-BL6 mouse melanoma. *Invasion Metastasis* **1996**, *16*, 27–38.
52. Esposito, A.; Viale, G.; Curigliano, G. Safety, Tolerability, and Management of Toxic Effects of Phosphatidylinositol 3-Kinase Inhibitor Treatment in Patients with Cancer: A Review. *JAMA Oncol.* **2019**, *5*, 1347–1354. [CrossRef] [PubMed]
53. Jędrzejewski, T.; Sobocińska, J.; Maciejewski, B.; Spisz, P.; Walczak-Skierska, J.; Pomastowski, P.; Wrotek, S. In vitro treatment of triple-negative breast cancer cells with an extract from the *Coriolus versicolor* mushroom changes macrophage properties related to tumorigenesis. *Immunol. Res.* **2025**, *73*, 14. [CrossRef] [PubMed]

International Journal of
Molecular Sciences

MDPI

Article

Anti-Influenza A Potential of *Tagetes erecta* Linn. Extract Based on Bioinformatics Analysis and In Vitro Assays

Minjee Kim, Aleksandra Nowakowska, Jaebum Kim and Young Bong Kim *

Department of Biomedical Science and Engineering, Konkuk University, Seoul 05029, Republic of Korea; mj0411@konkuk.ac.kr (M.K.); szalonystudent@gmail.com (A.N.); jbkim@konkuk.ac.kr (J.K.)
* Correspondence: kimera@konkuk.ac.kr; Tel.: +82-2-450-4208

Abstract: *Tagetes erecta* Linn. (TE) is traditionally used to treat cardiovascular, renal, and gastrointestinal diseases. In this study, we investigated the active compounds and targets of TE extract that may exert antiviral effects against influenza A. Active compounds and targets of TE extract were identified using the Traditional Chinese Medicine Systems Pharmacology database (TCSMP). The influenza A-related gene set was screened using GeneCards and the Kyoto Encyclopedia of Genes and Genomes (KEGG). A protein–protein interaction (PPI) network was built to establish the hub targets. Pathway and target studies were conducted using Gene Expression Omnibus (GEO). The interactions between active compounds and potential targets were assessed by molecular docking. An in vitro study was performed using antiviral and plaque reduction assays. From the compound and target search, we identified 6 active compounds and 95 potential targets. We retrieved 887 influenza-associated target genes and determined 14 intersecting core targets between TE and influenza. After constructing a compound–target network, we discovered lutein and beta-carotene to be the key compounds. Next, PPI network analysis identified the top three hub genes associated with influenza (IL-6, HIF1A, and IL-1β). Similarly, GEO analysis revealed IL-6, TGFB1, and CXCL8 to be the top three target genes. In our docking study, we identified that lutein and IL-6 had the strongest bindings. Our in vitro experimental results revealed that the TE extract exhibited therapeutic rather than prophylactic effects on influenza disease. We identified lutein as a main active compound in TE extract, and IL-6 as an important target associated with influenza, by using data mining and bioinformatics. Our in vitro findings indicated that TE extract exerted protective properties against the influenza A virus. We speculated that lutein, as a key active component in TE extract, is largely responsible for its antiviral effects. Therefore, we suggest TE extract as an alternative in the treatment of influenza.

Keywords: network pharmacology; molecular docking; influenza; antiviral treatment; *Tagetes erecta*; bioinformatics

1. Introduction

Influenza is a respiratory disease that occurs due to continuous antigenic drift and sporadic shift [1,2]. The influenza virus has a negative-sense RNA genome without a proofreading mechanism, leading to constant mutations over the years. It is reported that vaccines are the first line of defense against influenza; therefore, sequence analyses are commonly used to develop effective global vaccines [1]. Therapeutic interventions are also important, particularly when vaccines are unavailable or ineffective. However, to date, only one class of drugs has been approved for antiviral treatment. These are neuraminidase (NA) inhibitors (NAIs), virus-specific drugs that bind to the active site of the virus. However, a major drawback of NAIs is that they must be administered within 48 h of onset, a challenging requirement in many countries. Moreover, the NAI-resistant strains lower the efficacy of available antiviral agents, necessitating the development of novel influenza drugs [3].

Natural products are commonly used in medicine because they contain bioactive compounds that control cellular targets in many diseases [4]. Their main benefits include few adverse effects, affordability, and accessibility. Currently, high-throughput screening is the primary technique used to evaluate the pharmacological effects of natural products with potential to be used as herbal medicines [5]. However, with the rapidly increasing need for new drugs, the traditional drug-discovery strategy of "one drug–one target–one disease" has become inefficient and time-consuming [6]. Instead, the concept of "network pharmacology" has emerged, the idea of developing multi-target drugs for complex diseases by predicting genes controlled by bioactive compounds [7].

Tagetes erecta Linn. (TE; family Asteraceae) exerts several biological benefits, including antioxidant, antiproliferative, and antidiabetic effects [8]. In South America, TE flower extracts were commonly used to treat cardiovascular, renal, and gastrointestinal diseases [9]. Research in rodents suggests that TE flower extracts reduce inflammation and improve gastric diseases [10]. TE extracts contain high levels of lutein that are used as nutritional supplements to prevent aging or disease-related loss of visual acuity [11].

In this study, we aimed to determine the action of TE extract against influenza infection, in addition to identifying the major associated compounds, targets, and associated pathways. First, we constructed a compound–target network to identify active compounds, their targets, and pathways associated with the influenza virus. Second, we performed molecular docking analysis to identify the potential compound and target. Finally, we evaluated in vitro cytotoxicity and antiviral effects of the TE extract.

2. Results

2.1. In Silico Findings

2.1.1. Active Compounds and Target Prediction

We selected six active compounds from TCMSP based on the criteria of OB $\geq$ 20 and DL $\geq$ 0.1, following the database suggestions (Table 1). The plant extract composition is given in Supplementary Table S1. The targets regulated by these compounds in the TE extract were queried against the TCMSP and NCBI databases to obtain 95 targets.

Table 1. Active compounds in the *Tagetes erecta* (TE) extract.

Compound	Classification	MW	OB	DL
Alpha-carotene	Carotenoid	536.96	34.51	0.58
Beta-carotene	Carotenoid	536.96	37.18	0.58
Beta-sitosterol	Phytosterol	414.79	36.91	0.75
Campesterol	Phytosterol	400.76	37.58	0.71
Lutein	Carotenoid	568.96	22.59	0.55
Stigmasterol	Phytosterol	412.77	43.83	0.76

MW: molecular weight; OB: oral bioavailability; DL: drug-likeness.

2.1.2. Influenza Target Prediction

We obtained 160 and 720 targets with the search terms "influenza" and "Homo sapiens" from KEGG and GeneCards, respectively. Removing duplicates yielded 14 intersecting target genes associated with both the TE extract and influenza. These 14 targets were further analyzed using a compound–disease target network and four compounds (lutein, beta-carotene, beta-sitosterol, and alpha-carotene). The results indicated that lutein had the highest degree of connection, followed by beta-carotene (Figure 1).

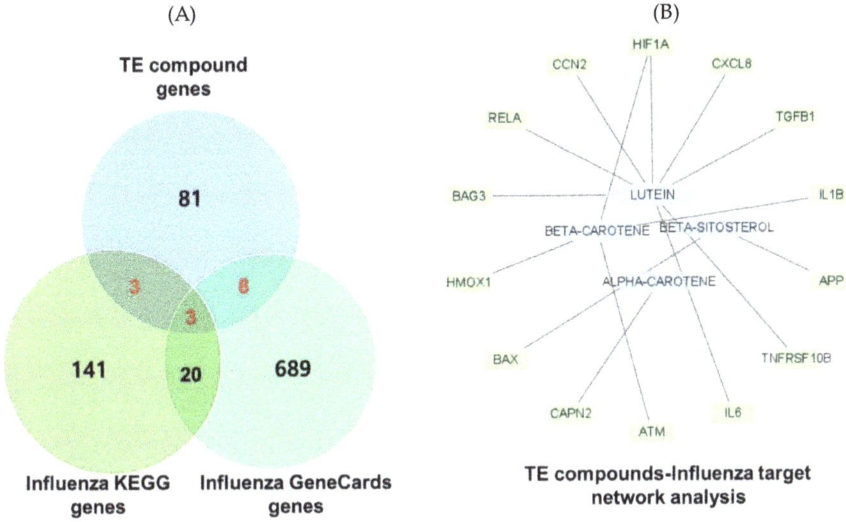

Figure 1. Target prediction. (**A**) Intersecting active targets. (**B**) Compound–disease target network analysis. Fourteen TE–influenza common targets were retrieved and constructed as a compound-disease target network.

2.1.3. Construction of the Protein–Protein Interaction (PPI) Network of Potential Targets

We analyzed 14 intersection targets of TE–active compounds associated with influenza using the Search Tool for the Retrieval of Interacting Genes/Proteins (STRING) database (https://string-db.org/ accessed on 2 February 2024) and imported them into Cytoscape to construct a compound–target network (Figure 2).

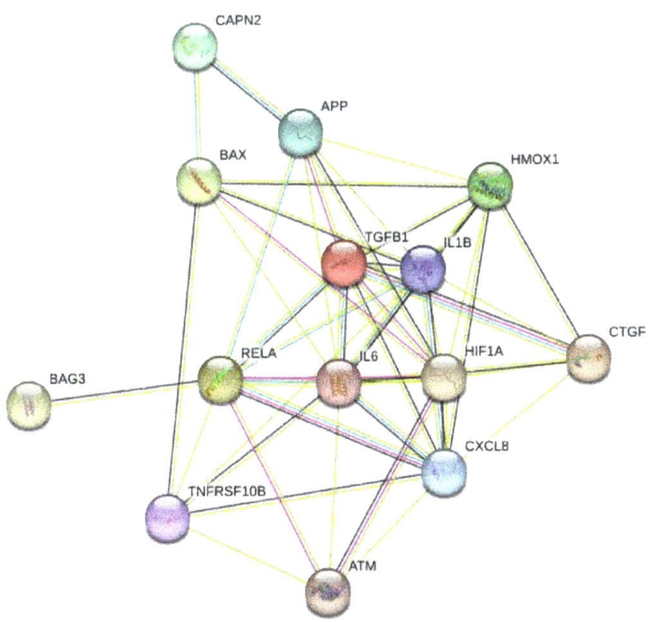

Figure 2. Target protein–protein interaction (PPI) network. Fourteen hub targets of TE and influenza were used as inputs for the STRING database. The network contained 14 nodes and 49 edges.

The network contained 14 nodes linked to 49 edges (see Table 2 for targets). The PPI network analysis revealed that IL-6 exhibited the highest degree of connectivity.

Table 2. Intersection target prediction.

Target	Description	Degree
IL-6	Interleukin-6	11
HIF1A	Hypoxia-inducible factor 1 subunit alpha	10
IL-1β	Interleukin-1 beta	10
CXCL8	Interleukin-8	9
RELA	RELA proto-oncogene, NF-κB subunit	9
HMOX1	Heme oxygenase 1	8
TGFB1	Transforming growth factor beta 1	8
APP	Amyloid beta precursor protein	7
BAX	Apoptosis regulator BAX	6
CTGF	Cellular communication network factor 2	6
TNFRSF10B	Tumor necrosis factor receptor superfamily member 10B	6
ATM	ATM serine/threonine kinase	5
CAPN2	Calpain 2	2
BAG3	BAG cochaperone 3	1

2.1.4. GEO Analysis

We obtained 20,396 H1N1 influenza-related disease genes from the GSE48466 dataset, then characterized the expression data of seven overlapping targets from GEO case samples. Volcano and box plots of the gene expression data are given in Supplementary Figures S1 and S2. A gene-difference heatmap (Figure 3) revealed that the average expression levels of H1N1 patients compared to healthy controls were ranked as IL6 > TGFB1 > CXCL8 > RELA > TNFRSF10B > BAX > ATM. Details of the expression levels are given in Supplementary Table S2.

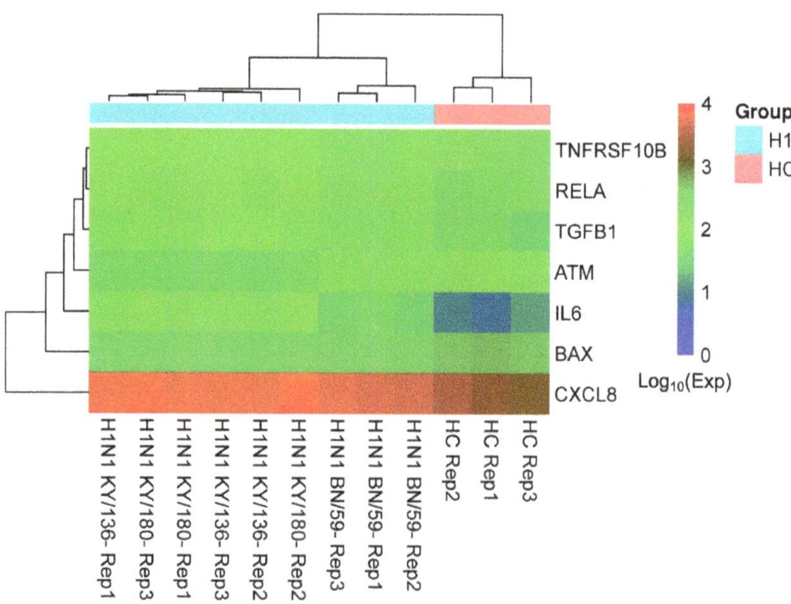

Figure 3. Gene difference heatmap of seven target genes. The Log_{10} expression levels of seven target genes (rows in the heatmap) are shown for nine H1N1 samples (columns with the prefix "H1N1" in the heatmap) and three healthy controls (columns with the prefix "HC" in the heatmap).

2.1.5. KEGG Pathway and GO Analysis

The KEGG analysis revealed 81 enriched signaling pathways, and the top 10 pathway enrichment rankings were plotted in bubble diagrams (Figure 4). Influenza was the disease with the third highest correlation after diabetic complications and non-alcoholic fatty liver disease.

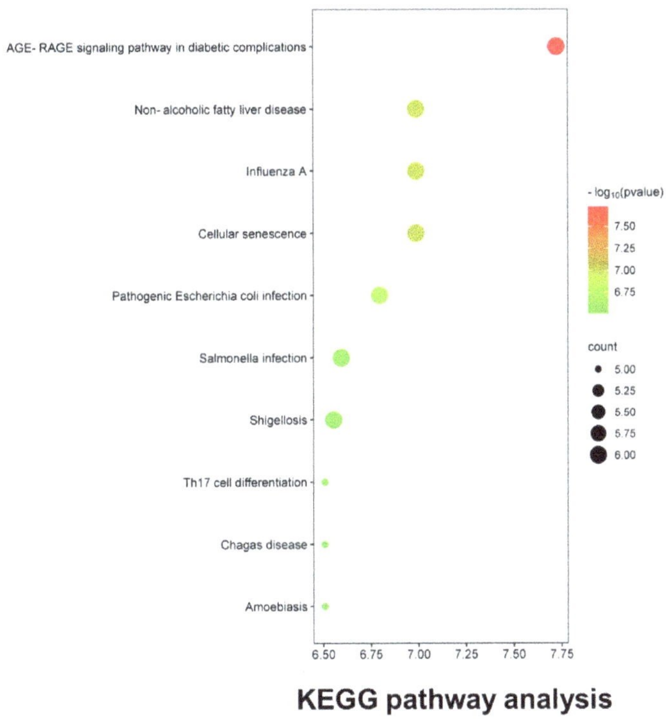

Figure 4. KEGG pathway enrichment analysis. TE and influenza−associated pathways and disease were screened and plotted in a bubble diagram according to the enriched gene counts.

For GO biological process (BP) analysis, we used seven targets to draw a chord diagram containing four biological regulatory processes (Figure 5). Six targets were enriched in a GO chord. IL6, TGFB1, CXCL8, and RELA were associated with inflammatory response (GO:0006954), while IL6, TGFB1, and BAX were involved in cellular response to the virus (GO:0098586).

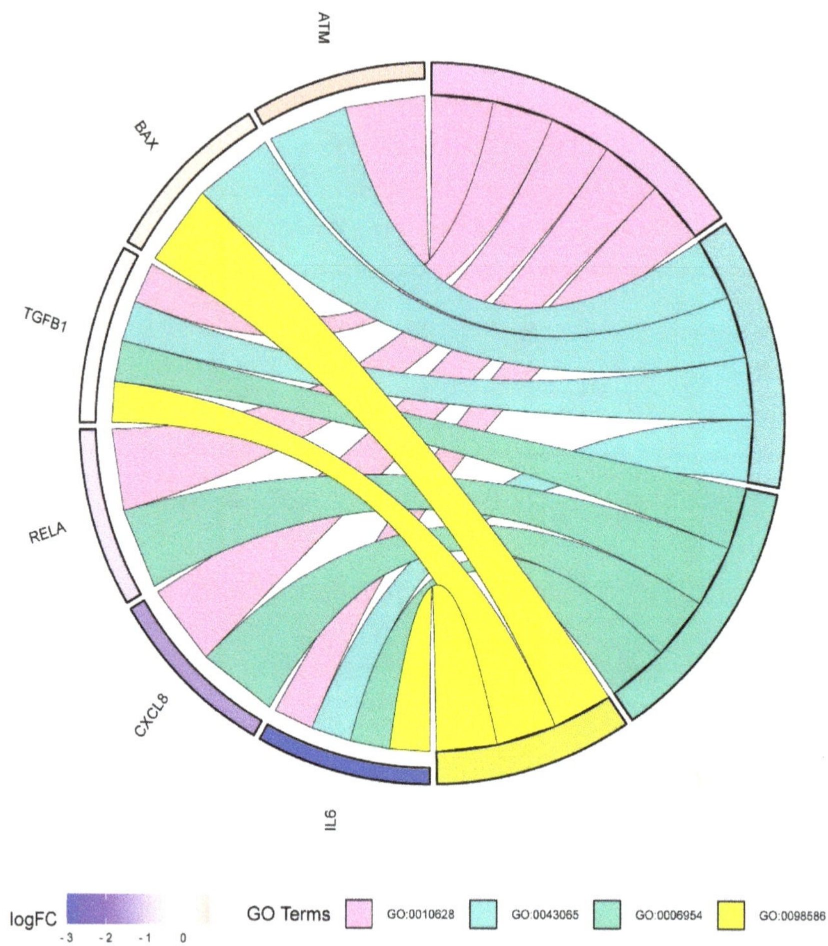

Figure 5. Gene ontology (GO) biological process (BP) analysis. The BP of the final screen targets was analyzed based on logFC, and six targets were enriched in a GO chord.

2.1.6. Molecular Docking of Active Compounds with Hub Target Genes

We conducted molecular docking to evaluate the potential target bindings with lutein and beta-carotene in the TE extract (Table 3). Three main targets retrieved from GEO (IL-6, TGFB1, and CXCL8) were docked to measure the binding affinity with the two compounds. IL-6 exhibited the strongest binding affinity of the three targets with lutein and beta-carotene. Between the two compounds, lutein bound more strongly to IL-6, whereas beta-carotene bound more strongly to TFGB1 and CXCL8. We then investigated the 2D binding of lutein and IL-6 in Biovia Discovery Studio (Figure 6). The key bonds were alkyl bonds with LYS66, PHE74, CYS73, ALA180, and PHE78 as well as pi-alkyl bonds with PHE74 and ARG179.

Table 3. Binding affinities of the active compounds of the TE extract with influenza-associated genes.

Compound	Binding Affinity (kcal/mol)		
	IL-6	TGFB1	CXCL8
Lutein	−8.6	−7.3	−7.3
Beta-carotene	−8.2	−7.6	−7.6

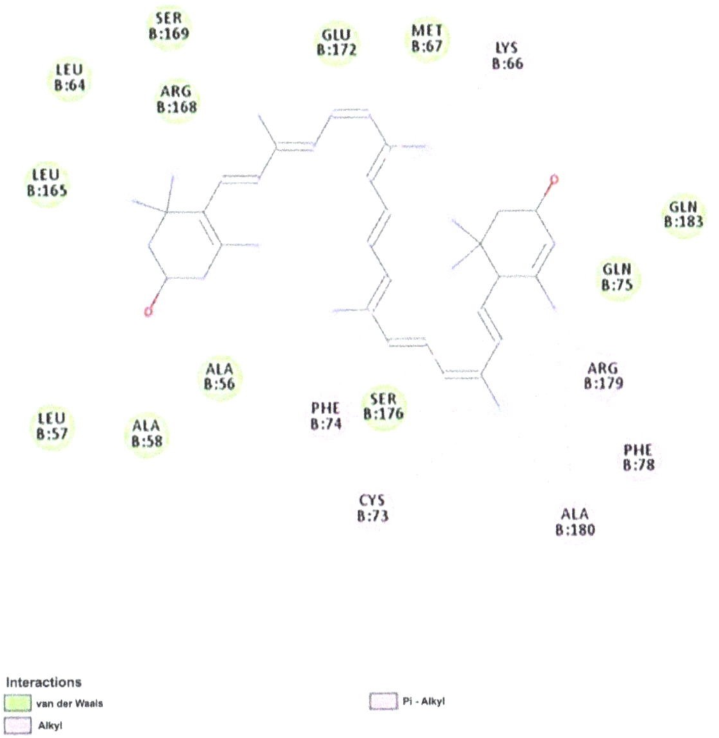

Figure 6. The 2D molecular docking of lutein and IL-6. The ligand–target interaction was visualized by using Discovery Studio Visualizer. The key bonds were alkyl bonds with LYS66, PHE74, CYS73, ALA180, and PHE78 as well as pi-alkyl bonds with PHE74 and ARG179.

2.2. *In Vitro Findings*

2.2.1. Cytotoxicity and Antiviral Activity of TE Extract

A cytotoxicity test was conducted to determine the dose and toxicity at the cell level (Figure S3). The cell viability levels with TE extract at the lowest (500 μg/mL) to highest concentration (2000 μg/mL) were between 60–95%. The mean cytotoxicity concentration of the 50% (CC50) value was higher than 2000 μg/mL (Table 4). The IC50 value was defined based on a post-treatment assay. The IC50 value was higher than 70 μg/mL and the SI value of >30. Four separate experiments were performed to determine whether the TE extract exhibits anti-influenza activity at different stages of the viral life cycle (Figure 7). Our in vitro results revealed that the TE extract had effects in the attachment inhibition and co-treatment assays (Figure 7B,C). In both cases, the extract at doses of 1000 and 500 μL/mL significantly improved cell viability from untreated conditions. However, antiviral activity was absent in the pre-treatment assay (Figure 7A), indicating that TE extract is unlikely to elicit protective effects against influenza infection. Instead, the extract exhibited antiviral properties later in infection. Post-treatment yielded the best results among the four assays, with cell viability increasing to 31 μg/mL. Interestingly, incubation of the extract with infected cells inhibited viral activity and increased cell growth, leading to cell viability of >100% from the control levels.

Table 4. CC_{50}, IC_{50}, and SI values of the TE extract.

CC_{50} (μg/mL)	IC_{50} (μg/mL)	SI
>2000	76 ± 18	>30

CC_{50}: mean cytotoxicity concentration of 50% compared with the control. IC_{50}: mean half-maximal inhibitory concentration determined with WST method in the post-treatment assay. SI: selectivity index (CC_{50}/IC_{50}).

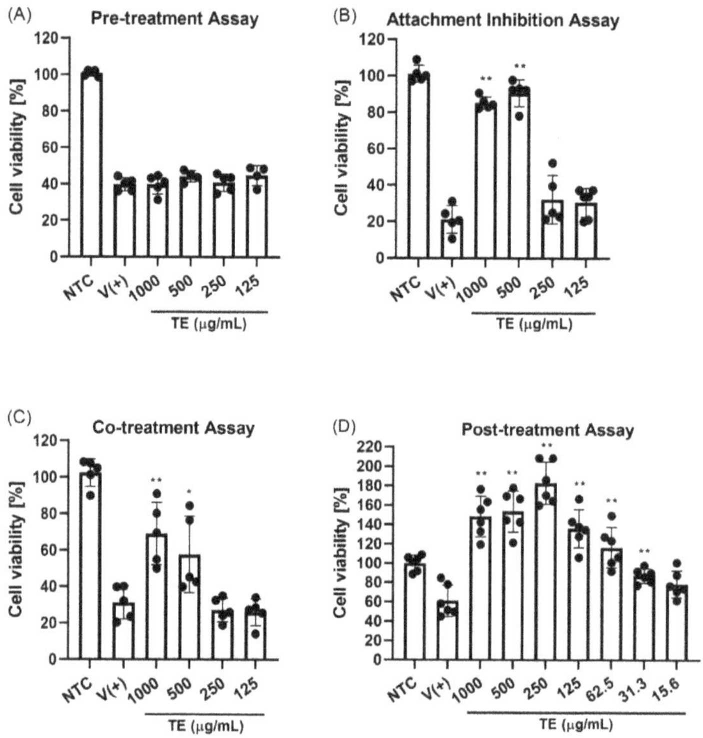

Figure 7. Antiviral properties of TE extract. (**A**) Pre-treatment assay: MDCK cells were treated with different concentrations of TE extract for 1 h at 37 °C, infected with influenza virus (100 50% tissue culture infectious dose [TCID50]/well), and incubated with viral inoculum for 1 h at 37 °C. (**B**) Influenza H1N1 virus attachment inhibition: MDCK cells were infected with influenza virus (100 TCID50/well) and co-treated with different concentrations of TE extract for 1 h at 4 °C. (**C**) Co-treatment assay: MDCK cells were co-incubated with influenza virus (100 TCID50/well) and different concentrations of TE extract for 1 h at 37 °C. (**D**) Post-treatment: MDCK cells were infected with influenza virus (100 TCID50/well) and incubated for 1 h at 37 °C. Virus was then removed, and cells were treated with different concentrations of TE extract. NTC: non-infected and non-treated control group; V (+): infected and non-treated group; ** $p < 0.01$, and * $p < 0.05$.

2.2.2. Time-of-Addition Assay

To further investigate the therapeutic effect of the TE extract in a post-treatment test and to note the stage of viral life cycle inhibition, a time-of-additional assay was conducted. The results showed a decreasing trend with delayed treatment initiation

(Figure 8), suggesting that the best antiviral effects are assured with treatment administered shortly after (2 h) infection.

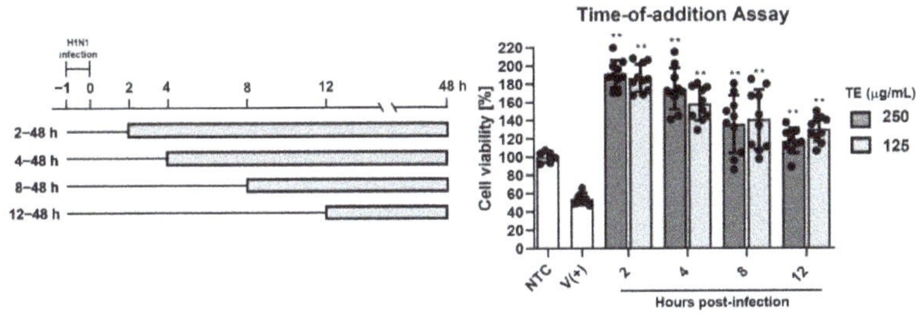

Figure 8. Time-of-addition assay: MDCK cells were infected with influenza virus (100 TCID50/well) and incubated for 1 h at 37 °C (−1 to 0 h). Virus was removed, and cells were treated with 250 or 125 μg/mL of TE extract starting 2, 4, 8, or 12 h after infection. NTC: non-infected and non-treated control group; V (+): infected and non-treated group; ** $p < 0.01$ (compared with virus-only control).

2.2.3. Plaque Reduction Assay

The antiviral activity of the TE extract was confirmed using a plaque reduction assay (Figure 9), comparing the number of plaques in treated infected (50 PFU/well) cells to the number in untreated infected (50 PFU/well) cells. Compared with the infection-only control group, the plaque count decreased to approximately 60% post-treatment with 1000 μL/mL of the TE extract and stayed at approximately 70% for other groups.

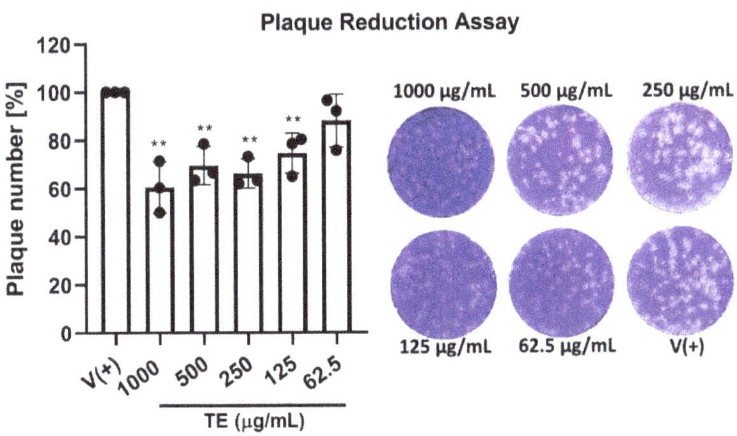

Figure 9. Plaque reduction assay. MDCK cells were infected with influenza H1N1 virus (50 plaque-forming units [PFUs]/well) and incubated for 1 h at 37 °C. Next, cells were treated with different concentrations of TE extract and incubated in the same conditions for 1 h. Compound mixtures were then removed, and 3 mL of overlay medium was added to each well. After 2 d, cells were stained with a crystal violet solution and plaques were calculated. Plaque reduction was determined via comparison with untreated infected wells ($n = 3$). ** $p < 0.01$ (compared with virus-only control).

3. Discussion

Influenza is a serious threat to public health [12]. Due to antigenic shift and the complications of influenza infection, influenza vaccines are formulated annually to match the circulating strains [13]. Antiviral agents that target influenza have been developed, but the use of these agents have many limitations [2].

Although TE was traditionally used to treat gastrointestinal diseases, and is considered to have medicinal properties such as anthelmintic, diuretic, and sedative effects, there are no studies on the effectiveness of TE against influenza virus [10]. In this study, we aimed to find active compounds and targets of TE associated with influenza for the first time by using data mining, bioinformatics, and systems biology. We identified two core active compounds—lutein and beta-carotene—via compound–disease target network analysis. For the target search, we applied a PPI network to identify the top three hub genes associated with influenza (IL-6, HIF1A, and IL-1β). From our PPI analysis, we identified HIF1A as one of the key targets associated with influenza. HIF1A is a key mediator in inflammation and is involved in the transcriptional regulation of cytokines, such as IL-6 and TNF-α [14]. Recent research reported that influenza infection induced the nuclear translocation of HIF1A, which may promote the production of proinflammatory cytokines [15]. Influenza infection also induces IL-1β, resulting in lung inflammation [16]. It was reported that significantly higher levels of IL-1β and IL-6 were detected in influenza patients with more severe conditions [17]. Similarly, GEO analysis revealed IL-6, TGFB1, and CXCL8 to be the top three target genes. TGF-beta is reported to control the immune responses—both inflammatory and regulatory [18]. Additionally, influenza infection stimulated CXCL8 secretion by primary human alveolar epithelial cells [19]. In influenza, disease severity is strongly associated with high levels of circulating IL-6-induced proteins, and monocyte chemotactic protein [20,21]. In our study, we identified IL-6 as a hub target in both PPI and GEO analysis. IL-6 levels were reported to be significantly increased in the sera of patients with uncomplicated influenza [20]. Moreover, elevated levels of the pro-inflammatory cytokine IL-6 are correlated with a high number of hospital admissions [22]. These findings correspond to our study result on potential targets associated with influenza infection. Herein, we conducted molecular docking to assess target binding of the two main compounds in TE extract. IL-6 exhibited the strongest binding affinity among the three selected targets with lutein and beta-carotene. We found that lutein binds more strongly than beta-carotene to IL-6. We speculated that lutein, as a key component in TE extract, is largely responsible for its antiviral effects by possibly reducing IL-6 level.

Virus invasion causes oxidative stress, resulting in inflammation and an exaggerated immune response, also known as cytokine storm [23]. Antioxidant therapy in combination with antiviral drugs has been suggested to reduce the lethal effects of influenza infection [24]. Our two main compounds, lutein and beta-carotene, belong to a carotenoid family that comprises natural lipid-soluble antioxidants. We speculate that the antioxidant potential of lutein and beta-carotene may include a reduction in inflammatory mediators and the prevention of viral mutations [25]. Antioxidants represent a potential therapeutic option to prevent influenza infection [26]. Therefore, the development of novel antioxidant compounds for influenza virus research and clinical use is needed to reinforce their potential.

Finally, we validated the anti-influenza effects of TE extract in MDCK cells and assessed its effects on the viral cycle. The post-treatment assay yielded the most potent effects and was effective up to a concentration of 31 μg/mL; this value was very low compared to the concentrations obtained in the other three assays. The time-of-addition assay revealed that the treatment applied shortly after infection (2 h) was most effective. Moreover, infected cell growth increased under incubation with TE extract, an improvement that may be attributable to the antioxidant activity of the natural compounds we identified.

With the continued advancement of big data, it is possible to screen and develop drugs from natural products using computer science [27]. Bioinformatics-driven drug screenings are cost- and time-effective research tools in developing useful pharmacological agents [28]. The primary focus of this study was to retrieve a list of active compounds of TE and target proteins of influenza infection through extensive data mining. The novelty of this study is that it is the first study of the effect of TE extract on influenza infection, and it was mainly conducted with computational work. The results of this study differ from the previously known effects of TE extract. However, further studies involving oxidative stress with influenza and its underlying mechanism will be required to provide a therapeutic option for the prevention and the control of influenza virus infection.

4. Materials and Methods

4.1. In Silico Analyses

4.1.1. Identification of Potential Active Compounds in TE Extract

Known TE compounds were searched in Dr. Duke's Phytochemical and Ethnobotanical database (version 1.10.07; https://phytochem.nal.usda.gov/phytochem accessed on 2 February 2024) and the Traditional Chinese Medicine Systems Pharmacology (TCMSP) database (version 2.3; http://tcmspw.com/tcmsp.php accessed on 2 February 2024). Based on the oral bioavailability (OB) and drug-likeness (DL) results from TCMSP, active compounds were then identified with an in silico absorption, distribution, metabolism, and excretion screening model.

4.1.2. Potential Targets of TE Extract and Influenza Virus

Regulatory target genes associated with the identified TE compounds were investigated using TCMSP and DrugBank (version 5.1.10; https://www.drugbank.ca accessed on 2 February 2024). Influenza virus-associated target genes (Homo sapiens) were obtained from Kyoto Encyclopedia of Genes and Genomes (KEGG) and GeneCards (version 5.14.0; https://www.genecards.org/ accessed on 2 February 2024).

4.1.3. Network Construction and Pathway Analysis

Compounds and intersecting core target networks were constructed using Cytoscape (version 3.9.0), and then analyzed to determine hub compounds and targets. Gene Ontology (GO) and pathway analyses of predicted targets were performed with the Database for Annotation, Visualization, and Integrated Discovery (DAVID, https://david.ncifcrf.gov accessed on 4 February 2024) and Kyoto Encyclopedia of Genes and Genomes (KEGG) [29], respectively.

4.1.4. Gene Difference and Gene Enrichment Analyses

The Gene Expression Omnibus (GEO, https://www.ncbi.nlm.nih.gov/geo/ accessed on 6 February 2024) from NCBI was used to further screen influenza targets from clinical data. For preprocessing and normalization, we used the GEO2R analysis tool provided by NCBI GEO. Influenza clinical case samples were retrieved from the GSE48466 dataset, containing twelve clinical samples (nine from patients infected with human influenza H1N1 and three from healthy controls). An analysis of gene differences across GSE clinical samples allowed for the construction of an intuitive heatmap that further improved target screening.

4.1.5. Molecular Docking Analysis

The major compounds in the TE extract were docked into three main targets retrieved from the network analysis to understand target specificity. The three-dimensional (3D) structures of the active compounds (lutein and beta-carotene) were downloaded from the PubChem database (https://pubchem.ncbi.nlm.nih.gov/ accessed on 24 February 2024) and converted to PDB files using Biovia Discovery Studio (version 20.1.0.19295). The 3D structures of the main targets were retrieved from the Research Collaboratory for Structural Bioinformatics (RCSB) Protein Data Bank (for the targets, IL-6 [PDB ID: 1P9M], transforming growth factor beta 1 [TGFB1; PDB ID: 5VQP], and IL-8 [CXCL8, PDB ID: 3IL8]). All the targets were converted to PDB files using Biovia Discovery Studio (ver. 4.5). Ligand-target dockings were conducted with AutoDock Vina (ver. 1.1.2) using PyRx (ver. 0.9.6) based on a scoring system. The CASTp (Computer Atlas of Surface Topology of proteins) server (ver. 3.0) was used to locate the active sites of the following targets: IL-6, TGFB1, and IL-6. Two-dimensional ligand–target interaction was performed using Biovia Discovery Studio.

4.2. In Vitro Experiment

4.2.1. Plant Material and Extraction

TE flowers were grown at Woori Bio Smart farm (Gyeongsan, Republic of Korea). The plant materials (5 g) were ground and extracted using 100% ethanol (100 mL, reflux extraction for 3 h). Ethanol was used as it is safe for infused edibles and provides consistent results [30]. The extracts were filtered and evaporated at a low temperature under reduced pressure. After the extraction, the sample (450 mg) was provided by Woori Bio for the in vitro experiment. This research on plants complies with relevant institutional, national, and international guidelines and legislation. The extracted material has been deposited at Konkuk University (Seoul, Republic of Korea) and Woori Bio Smart farm.

4.2.2. Cells and Viral Infection

Madin–Darby canine kidney (MDCK) cells were obtained from the American Type Culture Collection (ATCC, Manassas, VA, USA) and cultured in Eagle's minimum essential medium (EMEM; ATCC) containing 10% fetal bovine serum (Gibco, Carlsbad, CA, USA) and 1% penicillin/streptomycin at 37 °C and 5% CO_2. Human influenza type A/California/07/09 (H1N1) was provided by the Centers for Disease Control and Prevention (Korea). Infected MDCK cells were maintained in virus growth medium (EMEM) supplemented with 0.3% bovine serum albumin, 1% penicillin/streptomycin, and 0.5 µg/mL trypsin-TPCK (Sigma, Saint Louis, MO, USA). The virus titer was defined as the 50% tissue culture infectious dose (TCID50), according to the Reed–Muench endpoint method. All experiments involving virus-related work were performed in the Biosafety level 2 (BL2) facility using a biosafety cabinet.

4.2.3. Cytotoxicity and Antiviral Assay

To determine the cytotoxicity of the TE extract, cell viability was determined by the water-soluble tetrazolium salt (WST) method using an EZ-Cytox kit (Daeil Lab Service, Seoul, South Korea) according to the manufacturer's instructions. Briefly, MDCK cells were seeded in a 96-well plate at a density of 1×10^4 cells/mL. After 24 h, cells were treated with a serial dilution of extract and incubated at 37 °C for 2 days. After this time the EZ-Cytox solution was added to each well and incubated in the dark for 2 h, followed by spectrophotometric measurements of optical density (OD) at 540 nm. The CC50 value of extract concentration inducing 50% cell death was calculated via regression analysis.

The antiviral activity of the TE extract against H1N1 infection was determined using the following four methods: pretreatment, attachment inhibition, co-treatment, and post-treatment assays. For all assays, MDCK cells were seeded in a 96-well plate and incubated for 24 h. In the pre-treatment assay, cells were first subjected to different concentrations of TE extract and incubated for 1 h at 37 °C. Next, samples were removed and infected with the virus (100 TCID50/well) for 1 h. In the attachment inhibition assay, the virus (100 TCID50/well) was pre-absorbed into cells before they were treated with different concentrations of TE extract for 1 h at 4 °C. The co-treatment assay involved incubating the cells with the virus (100 TCID50/well) and extract for 1 h at 37 °C. Lastly, for the post-treatment assay, cells were pre-incubated with the virus (100 TCID50/well) for 1 h at 37 °C, and then incubated for 48 h with various concentrations of the TE extract prepared in virus growth media. After 48 h, cell viability was measured using the WST method with the EZ-Cytox kit.

4.2.4. Time-of-Addition Assay

To define the timeline of viral replication in which TE extract shows anti-influenza properties, a time-of-addition assay was performed. MDCK cells were seeded in a 96-well plate and incubated for 24 h. Next, cells were pre-incubated with the virus (100 TCID50/well) for 1 h at 37 °C. Later, the viral inoculum was removed, and cells were treated with 250 or 125 µg/mL of TE extract at different times post-infection. The experiment included treatment for 2–48 h, 4–48 h, 8–48 h, and 12–48 h post-infection. At the end of the experiment, cell viability was measured using the WST method with the EZ-Cytox kit.

4.2.5. Plaque Reduction Assay

MDCK cells were seeded in a six-well plate and cultured for 24 h. Cells were infected with the virus at a dose of 50 plaque-forming units (PFUs) per well, and then incubated at 37 °C and 5% CO_2 for 1 h. After removing the inoculum, the cells were treated for 1 h with different concentrations of TE extract. The mixture was then removed, and 3 mL of overlay medium containing 0.6% low-melting agar was added. The overlaid plates were incubated for 48 h at 37 °C and 5% CO_2. Finally, the overlay medium was removed and the cells were subjected to crystal violet staining for plaque counting.

4.3. Statistical Analysis

The results are presented as the average of at least three replications. Differences between groups were analyzed using one-way analysis of variance and Dunnett's test. Significance was set at $p < 0.05$. Data processing and statistical analyses were performed in GraphPad Prism 8.0.2 (GraphPad Software Inc., San Diego, CA, USA).

5. Conclusions

In conclusion, we identified lutein as a key active component in TE extract, largely responsible for its antiviral effects. Through an in silico study, we identified IL-6 as a hub target associated with influenza infection. Our research suggests that TE flower extract could be important as an alternative supplementation or treatment option for influenza in the future.

Supplementary Materials: The following supporting information can be downloaded at: https://www.mdpi.com/article/10.3390/ijms25137065/s1.

Author Contributions: M.K.: analysis and interpretation of in silico data, drafting the article; A.N.: acquisition of in vitro data; J.K.: acquisition of in silico data; Y.B.K.: conception and design of the study. All authors have read and agreed to the published version of the manuscript.

Funding: This research was supported by grants from the Ministry of Food and Drug Safety (22183MFDS443) and the Korean Health Technology R&D Project (Nos. HV23C1857/RS-2023-KH143167 and HV23C0074/RS-2023-KH138023).

Institutional Review Board Statement: Not applicable.

Informed Consent Statement: Not applicable.

Data Availability Statement: Requests for data sharing should be addressed to the corresponding author.

Conflicts of Interest: The authors declare no conflict of interest.

References

1. Webster, R.G.; Govorkova, E.A. Continuing challenges in influenza. *Ann. N. Y. Acad. Sci.* **2014**, *1323*, 115–139. [CrossRef] [PubMed]
2. Krammer, F.; Smith, G.J.; Fouchier, R.A.; Peiris, M.; Kedzierska, K.; Doherty, P.C.; Palese, P.; Shaw, M.L.; Treanor, J.; Webster, R.G.; et al. Influenza. *Nat. Rev. Dis. Primers* **2018**, *4*, 3. [CrossRef] [PubMed]
3. Gubareva, L.V. Molecular mechanisms of influenza virus resistance to neuraminidase inhibitors. *Virus Res.* **2004**, *103*, 199–203. [CrossRef]
4. Atanasov, A.G.; Zotchev, S.B.; Dirsch, V.M.; Supuran, C.T. Natural products in drug discovery: Advances and opportunities. *Nat. Rev. Drug Discov.* **2021**, *20*, 200–216. [CrossRef] [PubMed]
5. Tan, N.; Gwee, K.A.; Tack, J.; Zhang, M.; Li, Y.; Chen, M.; Xiao, Y. Herbal medicine in the treatment of functional gastrointestinal disorders: A systematic review with meta-analysis. *J. Gastroenterol. Hepatol.* **2020**, *35*, 544–556. [CrossRef] [PubMed]
6. Casas, A.I.; Hassan, A.A.; Larsen, S.J.; Gomez-Rangel, V.; Elbatreek, M.; Kleikers, P.W.; Guney, E.; Egea, J.; López, M.G.; Baumbach, J.; et al. From single drug targets to synergistic network pharmacology in ischemic stroke. *Proc. Natl. Acad. Sci. USA* **2019**, *116*, 7129–7136. [CrossRef]
7. Hopkins, A.L. Network pharmacology: The next paradigm in drug discovery. *Nat. Chem. Biol.* **2008**, *4*, 682–690. [CrossRef]

8. Burlec, A.F.; Pecio, Ł.; Kozachok, S.; Mircea, C.; Corciovă, A.; Vereştiuc, L.; Cioancă, O.; Oleszek, W.; Hăncianu, M. Phytochemical Profile, Antioxidant Activity, and Cytotoxicity Assessment of *Tagetes erecta* L. Flowers. *Molecules* **2021**, *26*, 1201. [CrossRef] [PubMed]
9. Zanovello, M.; Mariano, L.N.; Cechinel-Zanchett, C.C.; Boeing, T.; Tazinaffo, G.C.; da Silva, L.M.; Silva, D.B.; Junior, A.G.; de Souza, P. *Tagetes erecta* L. flowers, a medicinal plant traditionally used to promote diuresis, induced diuretic and natriuretic effects in normotensive and hypertensive rats. *J. Ethnopharmacol.* **2021**, *279*, 114393. [CrossRef]
10. Meurer, M.; de Oliveira, B.M.; Cury, B.J.; Jerônimo, D.T.; Venzon, L.; França, T.C.; Mariott, M.; Silva-Nunes, R.; Santos, A.C.; Roman-Junior, W.A.; et al. Extract of *Tagetes erecta* L., a medicinal plant rich in lutein, promotes gastric healing and reduces ulcer recurrence in rodents. *J. Ethnopharmacol.* **2022**, *293*, 115258. [CrossRef]
11. Bernstein, P.S.; Li, B.; Vachali, P.P.; Gorusupudi, A.; Shyam, R.; Henriksen, B.S.; Nolan, J.M. Lutein, zeaxanthin, and meso-zeaxanthin: The basic and clinical science underlying carotenoid-based nutritional interventions against ocular disease. *Prog. Retin. Eye Res.* **2016**, *50*, 34–66. [CrossRef] [PubMed]
12. Kim, M.; Choi, H.; Kim, S.; Kang, L.W.; Kim, Y.B. Elucidating the Effects of Curcumin against Influenza Using In Silico and In Vitro Approaches. *Pharmaceuticals* **2021**, *14*, 880. [CrossRef] [PubMed]
13. Javanian, M.; Barary, M.; Ghebrehewet, S.; Koppolu, V.; Vasigala, V.; Ebrahimpour, S. A brief review of influenza virus infection. *J. Med. Virol.* **2021**, *93*, 4638–4646. [CrossRef] [PubMed]
14. Chen, S.; Okahara, F.; Osaki NShimotoyodome, A. Increased GIP signaling induces adipose inflammation via a HIF-1α-dependent pathway and impairs insulin sensitivity in mice. *Am. J. Physiol. Endocrinol. Metab.* **2015**, *308*, 23. [CrossRef] [PubMed]
15. Guo, X.; Zhu, Z.; Zhang, W.; Meng, X.; Zhu, Y.; Han, P.; Zhou, X.; Hu, Y.; Wang, R. Nuclear translocation of HIF-1α induced by influenza A (H1N1) infection is critical to the production of proinflammatory cytokines. *Emerg. Microbes Infect.* **2017**, *6*, 21. [CrossRef] [PubMed]
16. Kim, K.S.; Jung, H.; Shin, I.K.; Choi, B.R.; Kim, D.H. Induction of interleukin-1 beta (IL-1β) is a critical component of lung inflammation during influenza A (H1N1) virus infection. *J. Med. Virol.* **2015**, *87*, 1104–1112. [CrossRef] [PubMed]
17. Chiaretti, A.; Pulitanò, S.; Barone, G.; Ferrara, P.; Romano, V.; Capozzi, D.; Riccardi, R. IL-1 beta and IL-6 upregulation in children with H1N1 influenza virus infection. *Mediat. Inflamm.* **2013**, *2013*, 495848. [CrossRef] [PubMed]
18. Denney, L.; Branchett, W.; Gregory, L.G.; Oliver, R.A.; Lloyd, C.M. Epithelial-derived TGF-β1 acts as a pro-viral factor in the lung during influenza A infection. *Mucosal Immunol.* **2018**, *11*, 523–535. [CrossRef] [PubMed]
19. Ito, Y.; Correll, K.; Zemans, R.L.; Leslie, C.C.; Murphy, R.C.; Mason, R.J. Influenza induces IL-8 and GM-CSF secretion by human alveolar epithelial cells through HGF/c-Met and TGF-α/EGFR signalinag. *Am. J. Physiol. Lung Cell. Mol. Physiol.* **2015**, *308*, 10. [CrossRef]
20. Davey, R.T., Jr.; Lynfield, R.; Dwyer, D.E.; Losso, M.H.; Cozzi-Lepri, A.; Wentworth, D.; Lane, H.C.; Dewar, R.; Rupert, A.; Metcalf, J.A.; et al. The association between serum biomarkers and disease outcome in influenza A(H1N1)pdm09 virus infection: Results of two international observational cohort studies. *PLoS ONE* **2013**, *8*, e57121. [CrossRef]
21. Betakova, T.; Kostrabova, A.; Lachova, V.; Turianova, L. Cytokines Induced During Influenza Virus Infection. *Curr. Pharm. Des.* **2017**, *23*, 2616–2622. [CrossRef] [PubMed]
22. Lee, N.; Wong, C.K.; Chan, P.K.; Chan, M.C.; Wong, R.Y.; Lun, S.W.; Ngai, K.L.; Lui, G.C.; Wong, B.C.; Lee, S.K.; et al. Cytokine response patterns in severe pandemic 2009 H1N1 and seasonal influenza among hospitalized adults. *PLoS ONE* **2011**, *6*, e26050. [CrossRef] [PubMed]
23. Liu, M.; Chen, F.; Liu, T.; Liu, S.; Yang, J. The role of oxidative stress in influenza virus infection. *Microbes Infect.* **2017**, *19*, 580–586. [CrossRef] [PubMed]
24. Uchide, N.; Toyoda, H. Antioxidant therapy as a potential approach to severe influenza-associated complications. *Molecules* **2011**, *16*, 2032–2052. [CrossRef]
25. Johra, F.T.; Bepari, A.K.; Bristy, A.T.; Reza, H.M. A Mechanistic Review of β-Carotene, Lutein, and Zeaxanthin in Eye Health and Disease. *Antioxidants* **2020**, *9*, 1046. [CrossRef] [PubMed]
26. Kang, H.; Hu, H.; Park, S.K. Serum antioxidant status and mortality from influenza and pneumonia in US adults. *Public Health Nutr.* **2022**, *25*, 3466–3475. [CrossRef] [PubMed]
27. Silva, D.R.; Sardi, J.D.C.O.; Freires, I.A.; Silva AC, B.; Rosalen, P.L. In silico approaches for screening molecular targets in Candida albicans: A proteomic insight into drug discovery and development. *Eur. J. Pharmacol.* **2019**, *842*, 64–69. [CrossRef] [PubMed]
28. Gill, S.K.; Christopher, A.F.; Gupta, V.; Bansal, P. Emerging role of bioinformatics tools and software in evolution of clinical research. *Perspect. Clin. Res.* **2016**, *7*, 115–122. [PubMed]
29. Kanehisa, M.; Furumichi, M.; Sato, Y.; Kawashima, M.; Ishiguro-Watanabe, M. KEGG for taxonomy-based analysis of pathways and genomes. *Nucleic Acids Res.* **2023**, *51*, D587–D592. [CrossRef]
30. Hikmawanti, N.P.; Fatmawati, S.; Asri, A.W. The Effect of Ethanol Concentrations as The Extraction Solvent on Antioxidant Activity of Katuk (*Sauropus androgynus* (L.) Merr.) Leaves Extracts. *IOP Sci.* **2020**, *755*, 012060. [CrossRef]

International Journal of
Molecular Sciences

MDPI

Article

Assessing the Potential of an Enzymatically Liberated Salmon Oil to Support Immune Health Recovery from Acute SARS-CoV-2 Infection via Change in the Expression of Cytokine, Chemokine and Interferon-Related Genes

Crawford Currie [1,*], Tor Åge Myklebust [2,3], Christian Bjerknes [1] and Bomi Framroze [1]

1 Hofseth BioCare, Keiser Wilhelms Gate 24, 6003 Alesund, Norway; chbj@hofsethbiocare.no (C.B.); bf@hofsethbiocare.no (B.F.)
2 Department of Research and Innovation, More og Romsdal Hospital Trust, 6026 Ålesund, Norway; tor.age.myklebust@helse-mr.no
3 Department of Registration, Cancer Registry of Norway, 0379 Oslo, Norway
* Correspondence: cc@hofsethbiocare.no

Abstract: Cytokines, chemokines, and interferons are released in response to viral infection with the ultimate aim of viral clearance. However, in SARS-CoV-2 infection, there is an imbalanced immune response, with raised cytokine levels but only a limited interferon response with inefficient viral clearance. Furthermore, the inflammatory response can be exaggerated, which risks both acute and chronic sequelae. Several observational studies have suggested a reduced risk of progression to severe COVID-19 in subjects with a higher omega-3 index. However, randomized studies of omega-3 supplementation have failed to replicate this benefit. Omega-3 fats provide important anti-inflammatory effects; however, fatty fish contains many other fatty acids that provide health benefits distinct from omega-3. Therefore, the immune health benefit of whole salmon oil (SO) was assessed in adults with mild to moderate COVID-19. Eleven subjects were randomized to best supportive care (BSC) with or without a full spectrum, enzymatically liberated SO, dosed at 4 g daily, for twenty-eight days. Nasal swabs were taken to measure the change in gene expression of markers of immune response and showed that the SO provided both broad inflammation-resolving effects and improved interferon response. The results also suggest improved lung barrier function and enhanced immune memory, although the clinical relevance needs to be assessed in longer-duration studies. In conclusion, the salmon oil was well tolerated and provided broad inflammation-resolving effects, indicating a potential to enhance immune health.

Keywords: COVID-19; long COVID; gene expression; immune health; cytokine; chemokine; interferon; lipid mediators

1. Introduction

The successful development of pharmaceutical interventions for SARS-CoV-2 infection has dramatically reduced the associated morbidity and mortality of COVID-19. Nevertheless, COVID-19 continues to impose significant health challenges, not least from the post-acute sequelae of COVID-19 (PASC), commonly referred to as long COVID.

Marked elevations in the pro-inflammatory cytokines IL-6 and TNFα, predominantly secreted by lung macrophages, appear central to both the effects of acute infection and long COVID [1,2]. Serum CXCL-10, which activates macrophages via the CXCR3 receptor, is also typically elevated and further increases the inflammatory response. Higher CXCL-10 levels are also associated with an increased risk of long COVID [3,4].

Although the lungs are the first organ to be affected by SARS-CoV-2 infection, the virus can directly infect many organs, including the gut, kidney, brain, myocardium, and vasculature. This is likely a significant driver of the common long-term sequelae of COVID-19, such as shortness of breath, fatigue, brain fog, and sleep disturbances [5,6].

SARS-CoV-2′s ability to evade recognition by the innate immune system results in lower levels of interferon-I and -III in the lungs and peripheral blood compared to respiratory viral infections [7]. The innate immune system uses pattern recognition receptors (PRRs) to detect the presence of viruses and mount an immune response to clear the virus (8). This detection system includes toll-like receptors (TLRs) and ultimately results in the release of inflammatory cytokines and chemokines, type I and III interferons, and the activation of immune effector cells, especially natural killer (NK) cells [8–10]. In contrast, COVID-19 infection results in an imbalanced response with a significant cytokine response but a muted interferon response and hence reduced cellular antiviral activity. Compounding this further is an increase of TGFβ seen with SARS-CoV-2 infection, which reduces NK cell anti-viral activity and viral clearance [11].

However, whilst viral persistence is observed in up to 50% of long COVID sufferers, in others, the virus is undetectable but the inflammatory response, characterized by IL-6 and TNFα, persists [12]. Immune dysregulation with elevated inflammatory markers therefore appears central to the pathology and symptoms of long COVID [13,14].

The management of long COVID remains a significant challenge, with estimates of at least 65 million individuals worldwide affected by this condition [13]. Furthermore, whilst the absolute risk of long COVID is lower in those suffering mild infections, this cohort represents the largest single group of COVID-19 infections and, therefore, a significant contributor to long COVID cases. Indeed, COVID-19 results in a prolonged inflammatory signature even after apparent recovery from acute infection.

Observational data, including from the United Kingdom Biobank, has suggested that subjects with a higher omega-3 index (O3i) have a lower risk of hospitalization and severe COVID-19 [15]. The regular consumption of fresh fish provides anti-inflammatory and antioxidant benefits from the fatty fraction of the fish. The metabolism of omega-3 fatty acids results in the production of specialist pro-resolving mediators (SPMs) of inflammation [16]. However, other fractions within fish oil appear to provide anti-inflammatory benefits distinct from omega-3. These effects include the regulation of nuclear factor-kappa beta (NF-kB) activation and the moderation of inflammatory responses, including the reduced expression of the pro-inflammatory cytokines IL-6 and IL-8 [17–19]. A non-fatty fraction of microcolin lipopeptides, derived from fish-eating blue-green algae, has also been described as moderating type II inflammation by modulating eosinophil effector function [20]. These wide-ranging inflammation-resolving effects of marine fatty acids might, therefore, explain how subjects with higher O3i levels were able to fare better when infected by COVID-19.

OmeGo is enzymatically liberated from Norwegian Atlantic salmon and has the same lipid profile as contained in whole salmon. Our hypothesis was that OmeGo might, therefore, aid recovery from acute COVID-19 infection and thereby reduce the risk of progression to severe disease. To this end, our study recruited patients with milder forms of COVID-19 and subjects were randomized to receive standard-of-care treatment with or without OmeGo. This study assessed both clinical and biomarker outcomes. The biomarker work, assessing the immune response to COVID-19 infection and recovery along with biomarkers associated with long COVID, is the focus of this paper.

2. Results

This study ran from 10 August 2021 to 19 May 2022. The gene expression analysis set was comprised of eleven subjects, six in the OmeGo group and five in the BSC group. Four of the six subjects in the OmeGo group were female and two were male, with an average age of 33.8 years, and in the BSC group, three were female and two were male, with an average age of 29.4 years (Table 1). As noted, nasal swabs were taken on days zero, fourteen, and twenty-eight.

Table 1. Age distribution and ethnicity of the study subjects. Southeast Asian denotes persons having origin from any of the following countries: Thailand, Singapore, Indonesia, and Taiwan.

Age Range (Years)	Number of Subjects	Subjects on OmeGo and BSC	Subjects on BSC Only
20–29	6	2	4
30–39	3	3	0
40–49	0	0	0
50–59	2	1	1
Mean age (years)		33.8	29.4
Ethnicity			
Caucasian		6	3
Southeast Asian		0	2

The two groups were balanced in terms of sex, age, and BMI, whereas there was an imbalance in terms of ethnicity. Mean blood leukocyte count, CRP, and glucose were all normal at the baseline and on day twenty-eight, as were blood electrolytes, creatinine, and liver enzymes. There were no statistically significant differences between treatment arms on any clinical endpoints. No serious adverse events were reported during this study. Overall, two subjects (one in the BSC group and one in the OmeGo group) reported three adverse events that were judged not to be related to the study treatment.

2.1. Change in Cytokine Gene Expression Levels

The gene expression analysis focused on the cytokines closely associated with acute SARS-CoV-2 infection and recovery and those associated with long COVID. The gene expression of the core pro-inflammatory cytokines in COVID-19, IL-6, and TNFα, were significantly lower with OmeGo than in the BSC group ($p < 0.01$). The expression of IL-6 halved from the baseline with OmeGo, whereas it declined by around 25% in the BSC group, resulting in an inter-group difference of three fold on day 28 (Figure 1). TNFα expression was reduced by around 20% with OmeGo, whereas it increased by almost 25% in the BSC group, resulting in a 3.4-fold difference by the end of this study (Figure 1). Elevations of IL-1β are also noted in both acute and long COVID, as well as IL-8, IL-13, GM-CSF (granulocyte–monocyte stimulatory factor), and TGFβ. In our study, IL-1β expression was just over 20% higher in the OmeGo subjects at the baseline compared to the BSC group, and whilst there was a 1.6-fold decline in IL-1β with OmeGo (and no change in the BSC arm), there was ultimately no difference between the two arms at the end of this study (Figure 1).

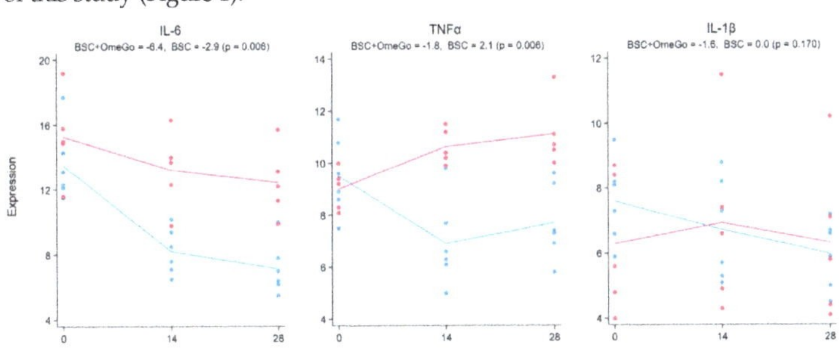

Figure 1. Fold change in the gene expression of cytokines IL-6, TNFα, and IL-1β from the baseline to day twenty-eight. The numeric change from day zero to day twenty-eight is given for both groups at the top of the chart along with the *p*-value to assess for significance of the difference between the two groups. The significance level was set at 5% ($p \leq 0.05$). The blue lines represent the BSC and OmeGo groups and the red lines the BSC-only group. The dots represent the gene expression level for the individual subjects. The lines represent a line of best fit and do not go through the mean scores.

There was also no difference seen in IL-8, with expression levels unchanged over the duration of the trial in either arm. In contrast, IL-13 expression was reduced by over 40% with OmeGo, whilst it increased by 25% in the BSC group, resulting in a 4-fold difference on day 28 ($p < 0.01$). The expression of IL-4 remained unchanged with BSC and showed a limited 0.9-fold increase with OmeGo, which was significant ($p < 0.01$). GM-CSF also declined more in the OmeGo arm; however, the 2.4-fold difference was not significant, and a similar outturn was seen with TGFβ, with a 1.6-fold difference in favor of OmeGo, which again was non-significant ($p = 0.07$ for both) (Figure 2).

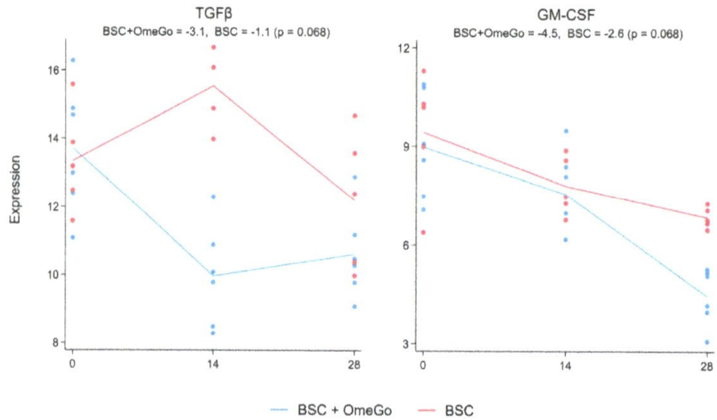

Figure 2. Fold change in the gene expression of TGFβ and GM-CSF from the baseline to day twenty-eight. The numeric change is given for both groups at the top of the chart along with the *p*-value to assess for significance of the difference between the two groups. The significance level was set at 5% ($p \leq 0.05$). The blue lines represent the BSC and OmeGo groups and the red lines the BSC-only group. The dots represent the gene expression level for the individual subjects. The lines represent a line of best fit and do not go through the mean scores.

2.2. *Change in Chemokine Gene Expression Levels*

The chemokine analysis showed a more than halving of CXCL-10 expression from the baseline to the end of this study in the OmeGo arm, with a small decline in the control arm, which resulted in an almost four-fold significant difference in expression between the two arms ($p < 0.01$). The associated chemokines, CXCL-9 and CXCL-11, showed more limited changes (Figure 3). CXCL-9 trended up in the OmeGo arm and down in the control arm, resulting in a 0.6-fold difference ($p = 0.01$), whereas the reduction in CXCL-11 with an almost 2-fold expression difference compared to the control missed significance ($p = 0.1$). CCL-19 showed a significant 2-fold greater reduction in expression with OmeGo, and levels in the control group were basically unchanged ($p < 0.01$). CCL-22 and CXCL-17, which are often raised alongside CCL-19 in COVID-19, showed little difference between the two arms (Figure 4). Changes in the expression of CCL-3 and CCL-5, chemokines associated with acute viral infection, were non-significant and showed little difference between the two arms. Finally, CXCL-13 and CCL-21, chemokines associated with immune memory, were significantly increased with OmeGo by 2.4 and 3.6 folds, respectively (both $p < 0.01$) (Figure 5).

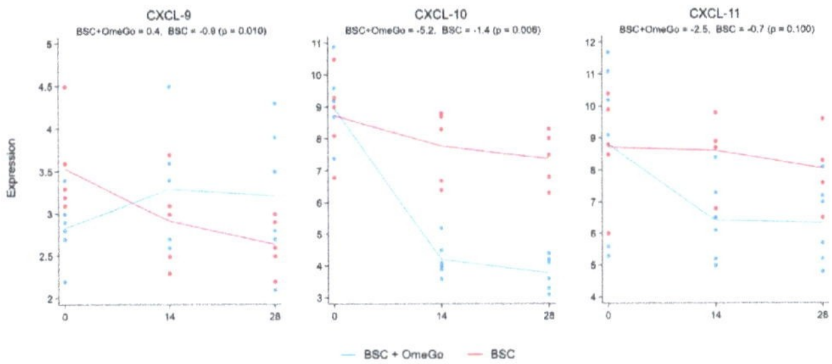

Figure 3. Fold change in the gene expression of chemokines CXCL-9, CXCL-10, and CXCL-11 from the baseline to day twenty-eight. The numeric change is given for both groups at the top of the chart along with the *p*-value to assess for significance of the difference between the two groups. The significance level was set at 5% ($p \leq 0.05$). The blue lines represent the BSC and OmeGo groups and the red lines the BSC-only group. The dots represent the gene expression level for the individual subjects. The lines represent a line of best fit and do not go through the mean scores.

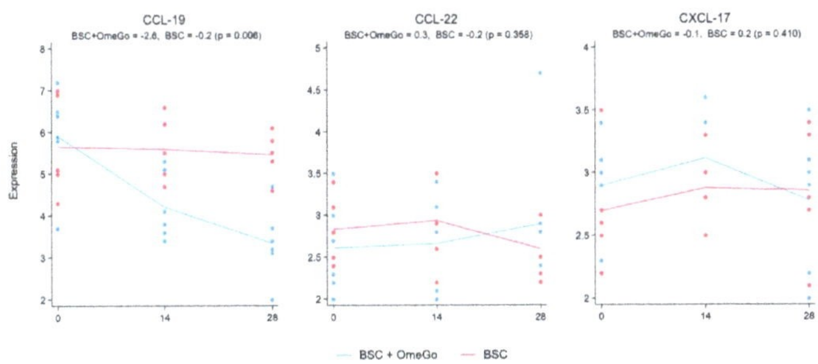

Figure 4. Fold change in the gene expression of chemokines CCL-19, CCL-22, and CXCL-17 from the baseline to day twenty-eight. The numeric change is given for both groups at the top of the chart along with the *p*-value to assess for significance of the difference between the two groups. The significance level was set at 5% ($p \leq 0.05$). The blue lines represent the BSC and OmeGo groups and the red lines the BSC-only group. The dots represent the gene expression level for the individual subjects. The lines represent a line of best fit and do not go through the mean scores.

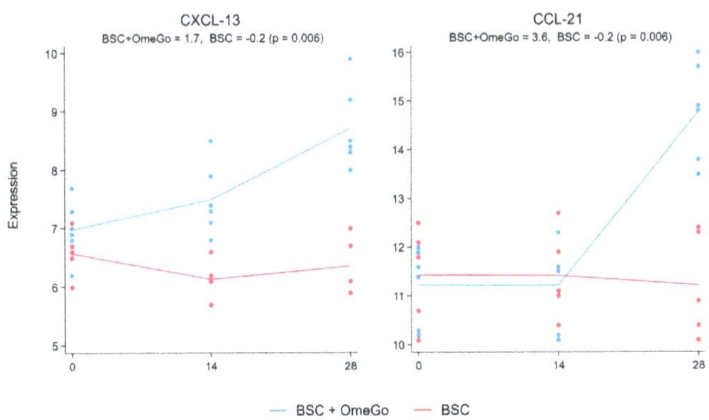

Figure 5. Fold change in the gene expression of chemokines CXCL-13 and CCL-21 from the baseline to day twenty-eight. The numeric change is given for both groups at the top of the chart along with the *p*-value to assess for significance of the difference between the two groups. The significance level was set at 5% ($p \leq 0.05$). The blue lines represent the BSC and OmeGo groups and the red lines the BSC-only group. The dots represent the gene expression level for the individual subjects. The lines represent a line of best fit and do not go through the mean scores.

2.3. Change in Interferon and Interferon-Related Factors Gene Expression Levels

In contrast to pro-inflammatory cytokine and chemokine elevations observed in COVID-19, interferon and interferon-related factors are typically under-expressed compared to responses typically seen with other viral infections. OmeGo resulted in a more than doubling of the expression of IFN1β compared to little change in the control arm, with a 6.4-fold difference between the two arms on day 28 ($p < 0.01$). Interestingly, the difference was even more marked on day 14, at which point the expression in the control arm had declined by 1.6 fold, whereas it had increased in the OmeGo arm by almost 12 fold. A very similar pattern was also seen with the expression of IFNγ, with the largest difference between the arms on day 14 and a six-fold difference declining to an almost three-fold significant difference ($p < 0.05$) (Figure 6). The interferon-stimulated gene, IFIT1, showed a sustained increase throughout the study period in the OmeGo group, with an almost 4-fold difference on day 28 versus BSC ($p < 0.01$), whereas the 1.6-fold difference in interferon regulatory factor 7 (IRF7) on day 28 did not reach significance ($p = 0.07$) (Figure 7).

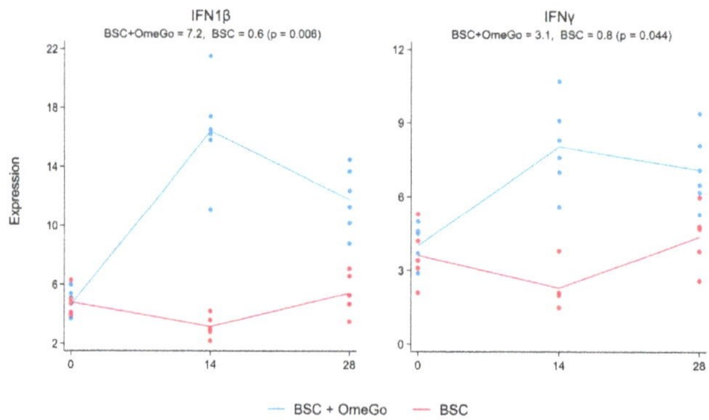

Figure 6. Fold change in the gene expression of interferons IFN1β and IFNγ from the baseline to day twenty-eight. The numeric change is given for both groups at the top of the chart along with the p-value to assess for significance of the difference between the two groups. The significance level was set at 5% ($p \leq 0.05$). The blue lines represent the BSC and OmeGo groups and the red lines the BSC-only group. The dots represent the gene expression level for the individual subjects. The lines represent a line of best fit and do not go through the mean scores.

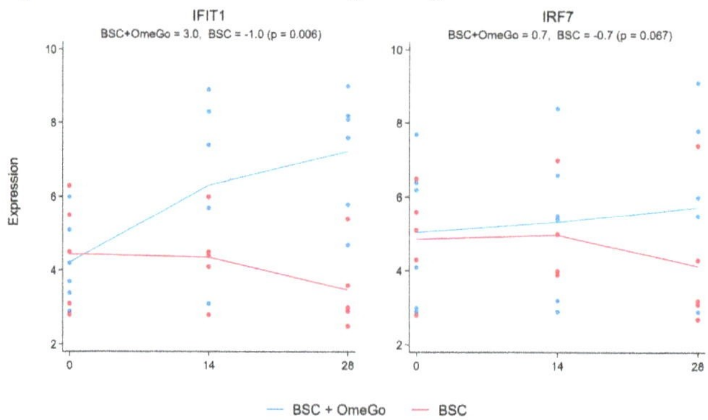

Figure 7. Fold change in the gene expression of the interferon-related factors, IFIT1 and IRF7, from the baseline to day twenty-eight. The numeric change is given for both groups at the top of the chart along with the p-value to assess for significance of the difference between the two groups. The significance level was set at 5% ($p \leq 0.05$). The blue lines represent the BSC and OmeGo groups and the red lines the BSC-only group. The dots represent the gene expression level for the individual subjects. The lines represent a line of best fit and do not go through the mean scores.

3. Discussion

This twenty-eight-day study analyzed changes in gene expression in the nasal mucosa of subjects with milder forms of COVID-19 to assess the impact of OmeGo, an enzymatically liberated salmon oil, on the immune response and recovery from SARS-CoV-2 infection. The subjects were all ambulatory with mild symptoms and, therefore, the response to SARS-CoV-2 infection in the nasal epithelium was proposed as a more sensitive means to assess immune response compared to serum markers. Whilst this is a relatively short-duration study, the twenty-eight-day timeframe does cover the critical time points of the period of peak viral load, the risk period for hospitalization, and the average time to viral clearance with SARS-CoV-2 infection [21–23]. Hence, this enabled the assessment of both the initial immune response and early post-infection recovery along with insights into the potential to moderate the risk of long COVID, a syndrome that has been associated with a prolonged inflammatory profile post-SARS-CoV-2 infection [24].

A hallmark of early COVID-19 is an increase in pro-inflammatory cytokines, especially IL-6 and TNFα, as well as IL-1 β, IL-8, GM-CSF, and the type 2 inflammatory cytokine IL-13 [1,25,26]. Whilst such a response is important initially to help control viral replication and spread, excessive or prolonged levels of inflammation can result in damage to the body and ill health.

By day fourteen, subjects in the OmeGo arm showed marked reductions in the expression of IL-6 and TNFα, as well as IL-13, compared to BSC, and this difference persisted to day twenty-eight, and these were all statistically significant. However, the 2.4-fold lower expression of GM-CSF seen with OmeGo did not reach significance, which was perhaps a result of the small number of subjects in our study. IL-1β and IL-8 showed little change in either group. Whilst IL-4 showed a 75% significant relative increase compared to the control, the absolute 0.9-fold difference in expression would question any benefit in terms of reducing the risk of long COVID. Indeed, a minimum fold change of 1.5 is frequently taken as physiologically relevant [27,28].

Chemokines are also central to the early immune response to infection, helping to recruit immune cells to infected areas. Stimulatory loops between cytokines and chemokines help ramp up inflammation and viral clearance. However, as per cytokines, a balanced response with progressive resolution is required for optimal recovery from the infection.

There was an almost 50% greater decrease in the gene expression of the chemokine CXCL-10, an important driver of inflammation in COVID-19, in the OmeGo-treated group [29]. The chemokines, CXCL-9, -10, and -11, are closely related and act through a common receptor, CXCR3, expressed on a number of immune cells including monocytes, T-cells, and NK cells. CXCL-9 and -11 have also been noted to be raised with SARS-CoV-2 infection [30]. OmeGo appeared to have a more limited impact on both of these, and whilst the 1.3-fold increase of CXCL-9 compared to BSC was statistically significant, the physiological impact would likely be limited. CCL-19, CCL-22, and CXCL-17 are also frequently raised in acute COVID-19 cases and are involved in T-cell, dendritic cell, and macrophage chemotaxis. CCL-19 showed a significant 2-fold greater reduction in expression with OmeGo with levels basically unchanged in the control group, and minimal changes were seen with CCL-22 and CXCL-17 in either group of subjects. Limited changes were seen in the expression of CCL-3 and CCL-5, chemokines that are often raised in viral infections and help recruit both NK cells and macrophages. Finally, the significant increases in CXCL-13 and CCL-21 are noteworthy. These play a part in antigen-specific B-cell maturation, T-cell activation, and immune self-tolerance. Whether this suggests a potential for a more sustained immunity against SARS-CoV-2 is unclear as we did not assess other markers, such as the expression of CD27 and CD21 on B-cells and plasma cells, to provide greater weight to this potential effect [31,32]. Furthermore, excessive amounts of CXCL-13 have been associated with worse outcomes in COVID-19 [33]. As always, a balance of activity is required with the immune system to support ongoing health.

Taken together, these cytokine and chemokine changes could suggest that OmeGo helped restore a healthier level of inflammation compared to the control. IL-6 and TNFα are central to the inflammatory response in COVID-19, and IL-13 has also been linked to disease severity, driving increased levels of eosinophils and macrophages in the lungs and stimulating smooth muscle hypertrophy and fibrosis [1,25–27]. CXCL-10 is associated with IL-6 secretion with raised levels seen in COVID-19 patients with pulmonary immune cell infiltration and cytokine storm [29]. The 2.4-fold reduction in GM-CSF is consistent with a decreased inflammatory profile, with GM-CSF being involved in the differentiation of alveolar macrophages and changes in lung barrier function [26]. As such, this reduction might suggest a faster resolution of lung inflammation; however, the change versus BSC did not reach statistical significance. Finally, whilst the three-fold decline from the baseline in TGFβ with OmeGo is consistent with a potentially improved immune profile, the 1.6-fold greater decrease in TGFβ versus BSC missed statistical significance.

In contrast to the inflammatory cytokine response to acute SARS-CoV-2 infection, limited interferon production can blunt the overall immune response with an increased risk of severe disease. This effect is linked to the production of inhibitory proteins encoded by SARS-CoV-2 [6,34,35]. Subjects in the OmeGo arm showed a significant, greater than six-fold increase in interferon-1 (IFNβ1). Type 1 interferon response is typically viewed as helping drive the most robust anti-viral immune response and helping to improve barrier function in the lungs and gut, along with type III interferons [9,36–38].

Treatment with OmeGo also resulted in a near doubling expression of IFNγ (type II interferon) and a significant difference compared to the limited change in the BSC arm. Whilst IFNγ is not typically involved in the natural course of SARS-CoV-2 infection, it does appear to augment the immune response against SARS-CoV-2 infection in the lungs. A study of IFNγ in almost 2000 outpatients presenting with COVID-19 within seven days of infection showed a halving of subsequent hospitalization compared to those receiving a placebo despite high vaccination rates [39]. Further, a study in hospitalized patients with moderate, new SARS-CoV-2 infection showed a faster recovery in those receiving IFNγ; however, a small case series in critically ill patients did not show a benefit for IFNγ [40,41]. In total, these studies suggest that the greatest impact of IFNγ on COVID-19 is in the earlier stages of infection. Consistent with this, subjects who suffered from respiratory infections prior to contracting SARS-CoV-2 seem to have a more robust immune response (including an IFNγ response), resulting in milder COVID-19 [42].

The interferon response in this study for both IFN1β and IFNγ peaked on day 14 and progressively declined from there onwards. This might suggest that OmeGo helped support a better interferon response; however, we did not assess the impact on viral clearance or longer-term outcomes, such as long COVID. We did, however, see consistent effects in the interferon-related immune system. The action of interferon includes the induction of interferon-stimulated genes (ISGs), which target different parts of the viral replication cycle [43]. Interferon regulatory factors (IRFs) also induce the expression of ISGs in response to TLR (toll-like receptors) and RLR (RIG-1-like receptors) activation. One important group of ISGs is the IFIT proteins (interferon-induced protein with tetratricopeptide repeats), which target viral protein production [44]. IFIT1 is a central part of this defense mechanism, and a muted IFIT1 response is seen in SARS-CoV-2 infection, which is probably a result of direct viral mechanisms and a limited interferon response. Mirroring the increase seen with IFN-1β and -γ, OmeGo showed a significant two-fold increase in IFIT1 expression. Whilst this could suggest a potential for more efficient clearing of the viral infection, further work is needed to explore this potential benefit.

The increase in the gene expression of IRF7 (interferon-regulatory factor-7) was less marked at 1.6 fold compared to BSC, and this change missed significance. Interferon regulatory factors (IRFs) help sustain the production of type I and III interferons and help reduce their expression after viral clearance, and IRF7 is considered a key IRF [45–47]. The limited change in IRF7 could reflect the relatively mild cases of COVID-19 in this study, which would not require prolonged IFN production. Consistent with this, the expression of both IFNβ and IFNγ declined by 50% from day 14 to day 28, which suggests a resolving infection.

Overall, these changes suggest that the subjects in the OmeGo arm had a better immune health profile by day 28. The interferon response appears improved and is combined with a progressive reduction in the expression of pro-inflammatory cytokines and chemokines. Even in mild cases of COVID-19, the immune inflammatory gene signature can remain exaggerated for many months after the resolution of symptoms [48]. Monocyte-derived macrophages continue to express a number of inflammatory genes, in particular, lipid mediators of immune function, which is, in essence, an upregulation of eicosanoid (prostaglandins, leukotrienes, and thromboxane) production with a downregulation of lipid mediators involved in inflammation resolution [48–50].

Lipid mediators are an important part of the immune response. Eicosanoids, metabolites of 5-lipoxygenase, have numerous actions, including white cell chemotaxis, as well as increasing vascular permeability to enable immune cells to reach infected areas. They also help sustain the immune response. However disordered lipid metabolism with increased eicosanoid production and reduced SPM expression can result in fibrosis with organ damage, including airway remodeling and reduced lung function. Raised IL-6 and TNFα levels appear central to PASC/long COVID and complications, such as the increased risk of cardiovascular events, persist long after the acute infection, as does the neuroinflammation and associated cognitive deficits ("brain fog") [51,52]. IL-1, IL-13, CXCL-10, and TGFβ are also raised in long COVID, especially in those with respiratory complications and suppressed levels of IL-4 and IL-10 [1,2,53].

OmeGo not only contains the omega-3 polyunsaturated fatty acids eicosapentaenoic acid (EPA), docosapentaenoic acid (DHA), and docosahexaenoic acid (DPA) but also omega-5, omega-6, omega-7, omega-9, and omega-11. The health benefits of omega-3 fatty acids derive from their metabolism into three different classes of SPMs, namely, protectins, resolvins, and maresins, which support immune and overall health [54]. For instance, specialized pro-resolving mediators (SPMs) reduce the production of inflammatory cytokines, including IL-6, modify macrophage function to a more protective role, and balance neutrophil and T-cell activity with other SPMs, appearing to suppress viral replication [55–57].

These combined effects could enhance recovery from viral infections, consistent with the observation of a higher omega-3 index appearing protective against severe COVID-19 [17]. However, studies of omega-3 supplements have delivered somewhat equivocal results. A large Norwegian study of cod liver oil (containing EPA, DHA, and vitamin D) showed no benefit in preventing COVID-19 or reducing its severity and, overall, other clinical trials have not shown a conclusive benefit for omega-3 oils containing EPA and DHA in COVID-19 [58,59].

The importance of DPA in immune health has emerged more recently than that of EPA and DHA, and DPA is now recognized as the source of a distinct family of SPMs. These SPMs have a range of actions, including the inhibition of viral replication and reduced platelet aggregation [57,60–62]. Recent publications have also shown the health benefits of non-omega-3 fatty acids contained in fish oil. For example, omega-9 (oleic acid) and pentadecanoic acid provide non-overlapping anti-inflammatory effects with cardiometabolic health benefits [63,64]. For oleic acid, at least part of this effect appears to be via a reduction in the inflammatory action of advanced glycation end products (AGEs) by raising levels of soluble receptors for advanced glycation end products (sRAGEs), a decoy receptor that has no transcriptional activity [65]. This reduces the AGEs available to bind to membrane RAGE (mRAGE) to trigger cellular inflammation and oxidative stress. Lower levels of sRAGE occur with aging and in chronic diseases, such as obesity, and have been correlated to COVID-19 severity and also implicated in long COVID [66].

The array of polyunsaturated and monounsaturated fats contained in OmeGo could, therefore, provide broader inflammation-resolving and immune health benefits than an intervention focused on two omega-3 fatty acids, namely, EPA and DHA, and more closely mimic the health benefits of regularly eating whole fish. Beyond the anti-inflammatory effects of oleic acid and pentadecanoic acid noted above, OmeGo also contains a lipopeptide fraction composed of microcolins. These originate from blue-green algae, a natural part of the fish's diet, and have been shown to moderate type 2 eosinophilic inflammation, including a reduction in eosinophilic activation and migration and associated reductions in IL-13 levels [20,67,68].

We did not measure plasma cytokine levels and note that changes in mRNA expression and protein levels can show a variable correlation due to various biological factors, such as post-translational modifications. The overall correlation between mRNA and protein expression is generally positive but moderate, as observed in similar studies of human monocytes [69]. Correlations likely depend on the gene categories in question, as the transcription and translation of various genes will be variably modified prior to the final protein synthesis. Quantification methods employed also have some inherent variability, adding some noise to the correlation. The strength of the correlations also appears context dependent. For instance, the differential expression of mRNA versus the control shows better correlations with their protein products compared to non-differentially expressed ones. This would inherently make sense that a differential mRNA expression should have a functional impact, such as a change in protein expression. If not, it would question the purpose of the differential expression. This was demonstrated in an ovarian cancer xenograft model, where differentially expressed mRNAs showed significant protein correlation, increasing confidence in using mRNA data for biological discovery in specific contexts [70]. The generally positive strength of the correlations was the basis for investigating differential mRNA expression in this research.

Our study has some weaknesses. We only followed the subjects up for twenty-eight days, and we have no data on whether subjects in either group subsequently suffered from long COVID. Nevertheless, this time period does include important time points relevant to acute infection and recovery. It should also be noted that the biomarker study was a sub-study of only eleven subjects with an average age of 32 years and only two subjects in their sixth decade. The effectiveness of the immune system against infectious diseases, including COVID-19, declines with age. A nutritional supplement to better support immune health in the elderly would be an important finding; however, our study does not provide any insights into elderly subjects. We did not measure the change in viral load, which would have further supported the immune health benefits indicated by the changes in gene expression seen in this study. Delayed viral clearance has been correlated to higher levels of TNFα and IL-6 and lower levels of neutralizing antibodies with a greater risk of severe disease and long COVID [51,52,71]. Whilst we assessed the change in TNFα and IL-6 expression, we did not assess blood antibody levels, which would have strengthened the data into a potential for OmeGo to support efficient viral clearance. Finally, we did not measure the change in the omega-3 index, which could have provided further insights into the relative drivers of inflammation resolution with OmeGo. Nevertheless, the biomarker changes seen in the OmeGo group are consistent with previously published research, both with OmeGo and on the modulation of immune signaling by lipid mediators. Further clinical trial work is needed to delineate OmeGo's potential role in supporting immune health, such as the prevention or treatment of long COVID, a condition of immune dysregulation and inflammation. Pollution-related cough, resulting from underlying particulate matter inflammation, is another area of unmet need with a significant impact on quality of life, including sleep quality. It also has a broad inflammatory signature with similarities to that seen in COVID-19, albeit at a significantly lower level. A low-cost intervention derived from fresh fish would be an appealing approach for this common symptom, and clinical trials should be considered in this setting.

4. Materials and Methods

4.1. Study Design and Study Subjects

This randomized, open-label, parallel study was conducted in five sites in three countries: Hungary, Serbia, and Brazil. It was designed by the sponsor, Hofseth BioCare, with input from Eurofins (Luxembourg City, Luxembourg), the Clinical Research Organization that managed the trial sites. Institutional review boards approved this study at each site, and it was run in accordance with the principles of the Declaration of Helsinki, International Conference on Harmonization (ICH) guidelines for current Good Clinical Practice (GCP), and applicable regulatory requirements. All subjects freely provided their written, informed consent.

Subjects were eligible if they were aged between 18 and 75 years with confirmed mild to moderate COVID-19 infection. Patients with a known fish allergy or hypersensitivity were excluded. Enrolled patients were randomized on day 0 to either OmeGo plus the best standard of care (BSC) or BSC alone. Patients randomized to the OmeGo arm were instructed to take two capsules (equivalent to 2 g of salmon oil) twice daily. They were also requested to save all unused and open packages to determine treatment compliance. A standard assessment, including physical, laboratory, and radiographic investigations, was undertaken to assess the patient's clinical condition at the baseline, and the same assessments were undertaken periodically through the twenty-eight days of this study. On days zero, fourteen, and twenty-eight, nasal swabs were taken for gene expression analysis (Figure 8). The selected genes were those involved in the immune response to COVID-19 infection, recovery from COVID-19, and markers associated with long COVID.

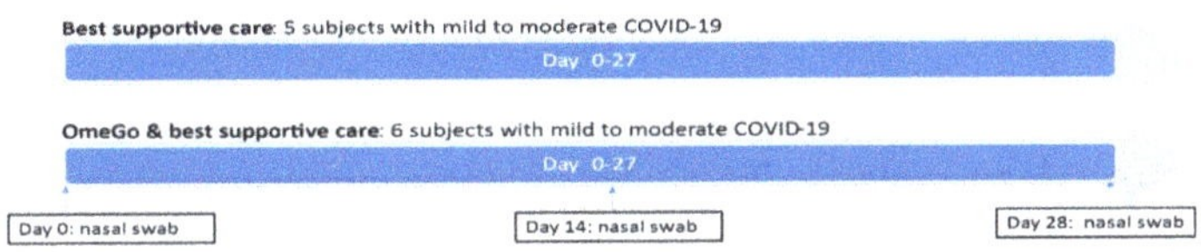

Figure 8. Trial design. Subjects with mild to moderate COVID-19 infection were randomized to best supportive care (BSC) with or without OmeGo for 28 days. Voluntary nasal swabs to analyze changes in the gene expression of markers of immune response and recovery were taken on days zero, fourteen, and twenty-eight.

4.2. Gene Expression Analysis

Following the collection of the nasal swab samples, each swab was vortexed while still in the tube with the transport medium for fifteen seconds. Using sterile forceps, the swab was removed from the transport tube and squeezed by pressing it against the side of the tube to wring out any remaining fluid and to ensure that most of the cells were in the liquid medium.

The RNeasy UCP Micro Kit (Qiagen, Venlo, Netherlands) and the standard protocol were used to purify mRNA from the nasal epithelial cells. The fluid collected from each of the 3 × 11 patient swabs was pelleted by centrifugation for five minutes at 1000 rpm in a centrifuge tube. All the supernatant was carefully removed by aspiration, making sure that all the cell media had been removed thoroughly. The cells were disrupted by adding a 350 µL buffer RULT, taking care to loosen the cell pellet from the tube, vortexed to mix thoroughly, and the mixture was homogenized by passing the lysate five times through a 20-gauge needle fitted to an RNase-free syringe. Next, 350 µL of 70% ethanol was added to the lysate and mixed again by pipetting (some precipitate was visible but was not a problem for the assays).

The sample was transferred, including any precipitate that may have formed, to an RNeasy UCP MinElute spin column placed in a 2 mL collection tube, centrifuged for fifteen seconds at 10,000 rpm, and washed as per the instructions. Next, 10 µL DNase I stock solution was added to 70 µL buffered RDD, mixed gently, and 80 µL of the DNase I mix was added directly to the spin column membrane and vortexed for 15 min. Then, 350 L buffer RUWT was added to the spin column and centrifuged for 15 s at 10,000 rpm to wash it, and the flow through was discarded. The spin column was placed in a new 2 mL collection tube, 500 µL of buffer RUPE was added and centrifuged for fifteen seconds at 10,000 rpm, and the flow through was discarded.

The spin column was washed again with 500 µL of 80% ethanol, placed in a new 2 mL collection tube, and centrifuged for 5 min at full speed with the lid open to make sure all the ethanol was removed. The spin column was placed in a new 1.5 mL collection tube, 14 μL ultra-clean water was added directly to the center of the spin column membrane, the lid was closed, and it was centrifuged for 1 min at full speed to elute the RNA. The dead volume of the spin column was 4 μL, which was eluted with 16 μL of ultra-clean water to yield a ~20 μL (4 μg) RNA eluate.

For cDNA preparation for RT-PCR, the RNA samples from each of the 3 × 11 patients were added to 40 μL of buffer GE2 (gDNA elimination buffer) and RNase-free water to make a final volume of 60 μL. This was incubated at 37 °C for five minutes and immediately placed on ice for two minutes, and 62 μL of the BC5 Reverse Transcriptase Mix was added to each 60 μL RNA sample for a final volume of 102 μL. This was incubated at 42 °C for exactly fifteen minutes, and then the reaction was immediately stopped by heating at 95 °C for five minutes (then held on ice until qPCR as required).

The Human Oxidative Stress RT2 Profiler PCR Array (Qiagen, Venlo, Netherlands) was used, as RT-PCR is a highly sensitive and reliable method for gene expression analysis. The assay was used to analyze expression levels of nineteen genes related to the immune response in nasal epithelial cells from SARS-CoV-2-positive subjects treated with and without CARDIO salmon oil capsules for twenty-eight days. The cDNA from above was mixed with the RT2 SYBR Green fluor Mastermix (Qiagen, Venlo, Netherlands) and aliquoted into the wells of the PCR Array. RT-PCR was performed on an iCycler (Bio-Rad Labs, Hercules, CA, USA). Gene expression was compared using Ct values, and the results were calculated using the ΔΔ Ct. method with normalization to the average expression levels of the five common genes (ACTB, B2M, GAPDH, HPRT, and RPL13A).

5. Conclusions

In summary, in subjects with milder forms of SARS-CoV-2 infection, twenty-eight days of treatment with OmeGo, an enzymatically liberated whole salmon oil, resulted in changes in the cytokine, chemokine, and interferon gene expression, which suggests potential for a faster return to immune homeostasis. This is consistent with the published literature on the inflammation-resolving effects of the range of lipid mediators contained in whole fish. Further clinical trials are needed to assess how this translates into outcomes, such as reducing the complications of infection, such as long COVID.

Author Contributions: Conceptualization, C.C. and B.F.; methodology, C.C., B.F., T.Å.M. and C.B.; software, T.Å.M.; formal analysis, T.Å.M.; resources, B.F.; data curation, B.F.; writing—original draft preparation, C.C.; writing—review and editing, T.Å.M., C.B. and B.F.; visualization, T.Å.M. All authors have read and agreed to the published version of the manuscript.

Funding: The research received no external funding or grants. This work was funded in its entirety by Hofseth BioCare ASA.

Institutional Review Board Statement: The study was conducted in accordance with the Declaration of Helsinki and was approved by the Independent Ethics Committees of Markhot Ferenc Oktatókórház és Rendelőintézet, Eger, Hungary (29.06.2001), Albert Schweitzer Kórház-Rendelőintézet, Hatvan, Hungary (13.07.2021), and General Hospital Paracin, Serbia (3 August 2021). Protocol number 20CCHH/OP110921.HBC.

Informed Consent Statement: Informed consent was obtained from all subjects involved in the study.

Data Availability Statement: Data requests can be directed to the authors of this paper.

Conflicts of Interest: C.C., C.B., and B.F. are employees or consultants for Hofseth BioCare ASA, the sponsor of this study. T.Å.M. collaborates with Hofseth BioCare ASA to support statistical planning and analysis and is paid for this on an ad hoc basis.

References

1. Queiroz, M.A.F.; Moura da Neves, P.F.; Lima, S.S.; Lopes, J.D.C.; Torres, M.K.D.S.; Vallinoto, I.M.V.C.; Bichara, C.D.A.; Santos, E.F.D.; de Brito, M.T.F.M.; da Silva, A.L.S.; et al. Cytokine Profiles Associated With Acute COVID-19 and Long COVID-19 Syndrome. *Front. Cell Infect. Microbiol.* **2022**, *12*, 922422. [CrossRef] [PubMed]
2. Scultheis, C.; Wilscher, E.; Paschold, L.; Gottschick, C.; Klee, B.; Henkes, S.S.; Bosurgi, L.; Dutzmann, J.; Sedding, D.; Frese, T.; et al. The IL-1, IL-6 and TNF cytokine triad is associated with post-acute sequelae of COVID-19. *Cell Rep. Med.* **2022**, *6*, 100663. [CrossRef]
3. Gudowska-Sawczuk, M.; Mroczko, B. What is Currently Known about the Role of CXCL10 in SARS-CoV-2 Infection? *Int. J. Mol. Sci.* **2022**, *7*, 3673. [CrossRef] [PubMed]
4. Unterman, A.; Sumida, T.S.; Nouri, N.; Yan, X.; Zhao, A.Y.; Gasque, V.; Schupp, J.C.; Asashima, H.; Liu, Y.; Cosme, C., Jr.; et al. Single cell multi-omics reveals dyssynchrony of the innate and adaptive immune system in progressive COVID-19. *Nat. Commun.* **2022**, *13*, 440. [CrossRef] [PubMed]
5. Liu, J.; Li, Y.; Liu, Q.; Yao, Q.; Wang, X.; Zhang, H.; Chen, R.; Ren, L.; Min, J.; Deng, F.; et al. SARS-CoV-2 cell tropism and multiorgan infection. *Cell Discov.* **2021**, *1*, 17. [CrossRef] [PubMed]
6. Eberhardt, N.; Noval, M.G.; Kaur, R.; Amadori, L.; Gildea, M.; Sajja, S.; Das, D.; Cilhoroz, B.; Stewart, O.J.; Fernandez, D.M.; et al. SARS-CoV-2 infection triggers pro-atherogenic inflammatory responses in human coronary vessels. *Nat. Cardiovasc. Res.* **2023**, *2*, 889–916. [CrossRef] [PubMed]
7. Blanco-Meol, D.; Nilsson-Payant, E.; Liu, W.C.; Uhl, S.; Hoagland, D.; Møller, R.; Jordan, T.X.; Oishi, K.; Panis, M.; Sachs, D.; et al. Imbalanced Host Response to SARS-CoV-2 Drives Development of COVID-19. *Cell* **2020**, *5*, 1036–1045. [CrossRef]
8. Janeway, C.A.; Medzihitov, R. Innate Immune Recognition. *Ann. Rev. Immunol.* **2002**, *20*, 197–216. [CrossRef]
9. Lazaer, H.M.; Schoggins, J.W.; Diamond, M.S. Shared and Distinct Functions of Type I and III Interferons. *Immunity* **2019**, *4*, 907–923. [CrossRef]
10. Vivier, E.; Tomasello, E.; Baratin, M.; Walzer, T.; Ugolini, S. Functions of natural killer cells. *Nat. Immunol.* **2008**, *9*, 503–510. [CrossRef]
11. Witkowski, M.; Tizian, C.; Ferreira-Gomes, M.; Niemeyer, D.; Jones, T.C.; Heinrich, F.; Frischbutter, S.; Angermair, S.; Hohnstein, T.; Mattiola, I.; et al. Untimely TGFβ responses in COVID-19 limit antiviral functions of NK cells. *Nature* **2021**, *600*, 295–300. [CrossRef]
12. Terjerina, F.; Catalan, P.; Rodriguez-Grande, C.; Adan, J.; Rodriguez-Gonzalez, C.; Muñoz, P.; Aldamiz, T.; Diez, C.; Perez, L.; Fanciulli, C.; et al. Post-COVID-19 syndrome. SARS-CoV-2 RNA detection in plasma, stool, and urine in patients with persistent symptoms after COVID-19. *BMC Infect. Dis.* **2022**, *22*, 211.
13. Davis, H.E.; McCorkell, L.; Vogel, J.M.; Topol, E.J. Long COVID: Major findings, mechanisms and recommendations. *Nat. Rev. Microbiol.* **2023**, *21*, 133–146. [CrossRef] [PubMed]
14. Merad, M.; Blish, C.A.; Sallusto, F.; Iwasaki, A. The immunology and immunopathology of COVID-19. *Science* **2022**, *375*, 1122–1127. [CrossRef]
15. Harris, W.S.; Tintle, N.L.; Sathyanarayanan, S.P.; Westra, J. Association between blood N-3 fatty acid levels and the risk of coronavirus disease 2019 in the UK Biobank. *Am. J. Clin. Nutr.* **2023**, *117*, 357–363. [CrossRef] [PubMed]
16. Chiang, N.; Serhan, C.N. Specialized pro-resolving mediator network: An update on production and actions. *Essays Biochem.* **2020**, *3*, 443–462.
17. Chang, C.S.; Sun, H.L.; Lii, C.K.; Chen, H.W.; Chen, P.Y.; Liu, K.L. Gamma-linolenic acid inhibits inflammatory responses by regulating NF-kappaB and AP-1 activation in lipopolysaccharide-induced RAW 264.7 macrophages. *Inflammation* **2010**, *1*, 46–57. [CrossRef] [PubMed]
18. De Souza, C.O.; Valenzuela, C.A.; Baker, E.J.; Miles, E.A.; Rosa Neto, J.C.; Calder, P.C. Palmitoleic acid has stronger anti-inflammatory potential in human endothelial cells compared to oleic and palmitic acids. *Mol. Nutr. Food Res.* **2018**, *20*, e1800322. [CrossRef]
19. Venn-Watson, S.; Lumpkin, R.; Dennis, E.A. Efficacy of dietary odd-chain saturated fatty acid pentadecanoic acid parallels broad associated health benefits in humans: Could it be essential? *Sci. Rep.* **2020**, *10*, 8161. [CrossRef]
20. Framroze, B.; Heggdal, H. An in vitro study to explore the modulation of eosinophil effector function in human allergic peripheral blood eosinophils using enzymatically extracted salmonid oil. *Funct. Foods Health Dis.* **2020**, *10*, 357. [CrossRef]
21. Frediani, J.K.; Parsons, R.; McLendon, K.B.; Westbrook, A.L.; Lam, W.; Martin, G.; Pollock, N.R. The New Normal: Delayed Peak SARS-CoV-2 Viral Loads Relative to Symptom Onset and Implications for COVID-19 Testing Programs. *Clin. Infect Dis.* **2024**, *78*, 301–307. [CrossRef] [PubMed]
22. Faes, C.; Abrams, S.; van Beckhoven, D.; Meyfroidt, G.; Vlieghe, E.; Hens, N.; Belgian Collaborative Group on COVID-19 Hospital Surveillance. Time between Symptom Onset, Hospitalisation and Recovery or Death: Statistical Analysis of Belgian COVID-19 Patients. *Int. J. Environ. Res. Public Health* **2020**, *17*, 7560. [CrossRef] [PubMed]
23. Yang, Y.; Guo, L.; Yuan, J.; Xu, Z.; Gu, Y.; Zhang, J.; Guan, Y.; Liang, J.; Lu, H.; Liu, Y. Viral and antibody dynamics of acute infection with SARS-CoV-2 omicron variant (B.1.1.529): A prospective cohort study from Shezhen, China. *Lancet* **2023**, *4*, e632–e641. [CrossRef] [PubMed]
24. Woodruff, M.C.; Bonham, K.S.; Anam, F.A.; Walker, T.A.; Faliti, C.E.; Ishii, Y.; Kaminski, C.Y.; Ruunstrom, M.C.; Cooper, K.R.; Truong, A.D.; et al. Chronic inflammation, neutrophil activity, and autoreactivity splits long COVID. *Nat. Commun.* **2023**, *14*, 4201. [CrossRef] [PubMed]

25. Donlan, A.N.; Sutherland, T.E.; Marie, C.; Preissner, S.; Bradley, B.T.; Carpenter, R.M.; Sturek, J.M.; Ma, J.Z.; Moreau, G.B.; Donowitz, J.R.; et al. IL-13 is a driver of COVID-19 severity. *JCI Insight* **2021**, *15*, e150107. [CrossRef] [PubMed]
26. Bonaventura, A.; Vecchie, A.; Wang, T.S.; Lee, E.; Cremer, P.C.; Carey, B.; Rajendram, P.; Hudock, K.M.; Korbee, L.; Van Tassell, B.W.; et al. Targeting GM-CSF in COVID-19 Pneumonia: Rationale and Strategies. *Front. Immunol.* **2020**, *11*, 01625. [CrossRef] [PubMed]
27. Kim, D.M.; Kim, Y.; Seo, J.W.; Lee, J.; Park, U.; Ha, N.Y.; Koh, J.; Park, H.; Lee, J.W.; Ro, H.J.; et al. Enhanced eosinophil-mediated inflammation associated with antibody and complement-dependent pneumonic insults in critical COVID-19. *Cell. Rep.* **2021**, *1*, 109798. [CrossRef]
28. Peart, M.J.; Smyth, G.K.; van Laar, R.K.; Bowtell, D.D.; Richon, V.M.; Marks, P.A.; Holloway, A.J.; Johnstone, R.W. Identification and functional significance of genes regulated by structurally different histone deacetylase inhibitors. *Proc. Natl Acad. Sci. USA* **2005**, *102*, 3697–3702. [CrossRef] [PubMed]
29. McCarthy, D.J.; Smyth, G.K. Testing significance relative to a fold-change threshold is a TREAT. *Bioinformatics* **2009**, *6*, 765–771. [CrossRef]
30. Cperchini, F.; Chiovato, L.; Rotondo, M. Interleukin-6, CXCL10 and Infiltrating Macrophages in COVID-19-Related Cytokine Storm: Not One for All But All for One! *Front. Immunol.* **2021**, *12*, 668507. [CrossRef]
31. Noto, A.; Joo, V.; Mancarella, A.; Suffiotti, M.; Pellaton, C.; Fenwick, C.; Perreau, M.; Pantaleo, G. CXCL-12 and CXCL-13 Cytokine Serum Levels Are Associated with the Magnitude and the Quality of SARS-CoV-2 Humoral Responses. *Viruses* **2022**, *14*, 2665. [CrossRef] [PubMed]
32. Kozai, M.; Kubo, Y.; Katakai, T.; Kondo, H.; Kiyonari, H.; Schaeuble, K.; Luther, S.A.; Ishimaru, N.; Ohigashi, I.; Takahama, Y. Essential role of CCL21 in establishment of central self-tolerance in T cells. *J. Exp. Med.* **2017**, *7*, 1925–1935. [CrossRef] [PubMed]
33. Perreau, M.; Suffiotti, M.; Marques-Vidal, P.; Wiedemann, A.; Levy, Y.; Laouénan, C.; Ghosn, J.; Fenwick, C.; Comte, D.; Roger, T.; et al. The cytokines HGF and CXCL13 predict the severity and the mortality in COVID-19 patients. *Nat. Commun.* **2021**, *12*, 4888. [CrossRef] [PubMed]
34. Celik, N.; Celik, O.; Laloglu, E.; Özkaya, A. The CXCL9/10/11-CXCR3 axis as a predictor of COVID-19 progression: A prospective case-control study. *Rev. Soc. Bras. Med. Trop.* **2023**, *56*, e0128–e2023. [CrossRef]
35. Xia, H.; Cao, Z.; Xie, X.; Zhang, X.; Chen, J.Y.C.; Wang, H.; Menachery, V.D.; Rajsbaum, R.; Shi, P.Y. Evasion of Type I Interferon by SARS-CoV-2. *Cell. Rep.* **2020**, *33*, 08234. [CrossRef] [PubMed]
36. Lei, X.; Dong, X.; Ma, R.; Wang, W.; Xiao, X.; Tian, Z.; Wang, C.; Wang, Y.; Li, L.; Ren, L.; et al. Activation and evasion of type I interferon responses by SARS-CoV-2. *Nat. Commun.* **2020**, *1*, 3810. [CrossRef] [PubMed]
37. Le Messurier, K.S.; Häcker, H.; Chi, L.; Tuomanen, E.; Redecke, V. Type I interferon protects against pneumococcal invasive disease by inhibiting bacterial transmigration across the lung. *PLoS Pathog.* **2013**, *9*, e1003727.
38. Odendall, C.; Voak, A.A.; Kagan, J.C. Type III IFNs are commonly induced by bacteria-sensing TLRs and reinforce epithelial barriers during infection. *J. Immunol.* **2017**, *199*, 3270–3279. [CrossRef] [PubMed]
39. Reis, G.; Moreira Silva, E.A.S.; Meideros Silva, D.C.; Thabane, L.; Campos, V.H.; Ferreira, T.S.; Santos, C.V.; Nogueira, A.M.; Almeida, A.P.; Savassi, L.C.; et al. Early treatment with pegylated interferon lambda for COVID-19. *N. Engl. J. Med.* **2023**, *388*, 518–528. [CrossRef]
40. Myasnikov, A.L.; Berns, S.A.; Talyzin, P.A.; Ershov, F.I. Interferon gamma in the treatment of patients with moderate COVID-19. *Vopr. Virusol.* **2021**, *66*, 47–54. [CrossRef]
41. Van Laarhoven, A.; Kurver, L.; Overheul, G.J.; Kooistra, E.J.; Abdo, W.F.; van Crevel, R.; Duivenvoorden, R.; Kox, M.; Oever, J.T.; Schouten, J.; et al. Interferon gamma immunotherapy in five critically ill COVID-19 patients with impaired cellular immunity: A case series. *Med* **2021**, *2*, 1163–1170. [CrossRef] [PubMed]
42. Hilligan, K.L.; Namasivayam, S.; Clancy, C.S.; Baker, P.J.; Old, S.I.; Peluf, V.; Amaral, E.P.; Oland, S.D.; O'Mard, D.; Laux, J.; et al. Bacterial-induced or passively administered interferon gamma conditions the lung for early control of SARS-CoV-2. *Nat. Commun.* **2023**, *14*, 8829. [CrossRef] [PubMed]
43. Schoggins, J.W. ISGs: What do they do? *Ann. Rev. Virol.* **2019**, *1*, 567–584. [CrossRef] [PubMed]
44. Vladimer, G.I.; Gorner, M.W.; Superti-Furga, G. IFITS: Emerging roles as key antiviral proteins. *Front. Immunol.* **2014**, *5*, 94. [CrossRef] [PubMed]
45. Honda, K.; Takaoka, A.; Taniguchi, T. Type I interferon gene induction by the interferon regulatory factor family of transcription factors. *Immunity* **2006**, *25*, 349–360. [CrossRef] [PubMed]
46. Ma, W.; Huang, G.; Wang, Z.; Wang, L.; Gao, Q. IRF7: Role and regulation in immunity and autoimmunity. *Front. Immunol.* **2023**, *14*, 1236923. [CrossRef] [PubMed]
47. Honda, K.; Yanai, H.; Negishi, H.; Asagiri, M.; Sato, M.; Mizutani, T.; Shimada, N.; Ohba, Y.; Takaoka, A.; Yoshida, N.; et al. Irf-7 is the master regulator of type-I interferon-dependent immune responses. *Nature* **2005**, *7034*, 772–777. [CrossRef]
48. Bohnacker, S.; Hartung, F.; Henkel, F.; Quaranta, A.; Kolmert, J.; Priller, A.; Ud-Dean, M.; Giglberger, J.; Kugler, L.M.; Pechtold, L.; et al. Mild COVID-19 imprints a long-term inflammatory eicosanoid-and chemokine memory in monocyte-derived macrophages. *Mucosal Immunol.* **2022**, *15*, 515–524. [CrossRef]
49. Palmas, F.; Clarke, J.; Colas, R.A.; Gomez, E.A.; Keogh, A.; Boylan, M.; McEvoy, N.; McElvaney, O.J.; McElvaney, O.; Alalqam, R.; et al. Dysregulated plasma lipid mediator profiles in critically ill COVID-19 patients. *PLoS ONE* **2021**, *8*, e0256226. [CrossRef]

50. Koenis, D.S.; Beegun, I.; Jouvene, C.C.; Aguirre, G.A.; Souza, P.R.; Gonzalez-Nunez, M.; Ly, L.; Pistorius, K.; Kocher, H.M.; Ricketts, W.; et al. Disrupted Resolution Mechanisms Favor Altered Phagocyte Response in COVID-19. *Circ. Res.* **2021**, *129*, e54–e71. [CrossRef]
51. Xie, Y.; Xu, E.; Bowe, B.; Al-Aly, Z. Long-term cardiovascular outcomes of COVID-19. *Nat. Med.* **2022**, *28*, 583–590. [CrossRef] [PubMed]
52. Monje, M.; Iwasaki, A. The neurobiology of long COVID. *Neuron* **2022**, *110*, 3484–3496. [CrossRef]
53. Lai, Y.J.; Liu, S.H.; Manachevakul, S.; Lee, T.A.; Kuo, C.T.; Bello, D. Biomarkers in long COVID-19: A systematic review. *Front. Med.* **2023**, *10*, 1085988. [CrossRef]
54. Calder, P.C. Marine omega-3 fatty acids and inflammatory processes: Effects, mechanisms and clinical relevance. *Biochem. Biophys. Acta.* **2015**, *1851*, 469–484. [CrossRef]
55. Regidor, P.-A.; Mueller, A.; Sailer, M.; Gonzalez Santos, F.; Rizo, J.M.; Egea, F.M. Chronic inflammation in PCOS: The potential benefits of specialized pro-resolving lipid mediators (SPMs) in the improvement of the resolutive response. *Int. J. Mol. Sci.* **2020**, *22*, 384. [CrossRef]
56. Chiang, N.; Serhan, C.N. Structural elucidation and physiologic functions of specialized pro-resolving mediators and their receptors. *Mol. Aspects Med.* **2017**, *58*, 114–129. [CrossRef]
57. Morita, M.; Kuba, K.; Ichikawa, A.; Nakayama, M.; Katahira, J.; Iwamoto, R.; Watanebe, T.; Sakabe, S.; Daidoji, T.; Nakamura, S.; et al. The lipid mediator protectin D1 inhibits influenza virus replication and improves severe influenza. *Cell* **2013**, *1*, 112–125. [CrossRef] [PubMed]
58. Brunvoll, S.H.; Nygaard, B.; Ellingjord-Dale, M.; Holland, P.; Istre, M.S.; Kalleberg, K.T.; Søraas, C.L.; Holven, K.B.; Ulven, S.M.; Hjartåker, A.; et al. Prevention of COVID-19 and other acute respiratory infections with cod liver oil supplementation, a low dose vitamin D supplement: Quadruple blinded, randomised placebo-controlled trial. *BMJ* **2022**, *378*, e071245. [CrossRef]
59. Velotti, F.; Costantini, L.; Merendino, N. Omega-3 Polyunsaturated Fatty Acids (n-3 PUFAs) for Immunomodulation in COVID-19 Related Acute Respiratory Distress Syndrome (ARDS). *J. Clin. Med.* **2022**, *1*, 304. [CrossRef] [PubMed]
60. Hansen, T.V.; Dalli, J.; Serhan, C.N. The novel lipid mediator PD1n-3 DPA: An overview of the structural elucidation, synthesis, biosynthesis and bioactions. *Prostaglandins* **2017**, *133*, 103–110. [CrossRef]
61. Pistorius, K.; Souza, P.R.; De Matteis, R.; Austin-Williams, S.; Primdahl, K.G.; Vik, A.; Mazzacuva, F.; Colas, R.A.; Marques, R.M.; Hansen, T.V.; et al. PDn-3 DPA Pathway Regulates Human Monocyte Differentiation and Macrophage Function. *Cell Chem. Biol.* **2018**, *6*, 749–760 e9. [CrossRef] [PubMed]
62. Droin, G.; Rioux, V.; Legrand, P. The n-3 docosapentaenoic acid (DPA): A new player in the n-3 long chain polyunsaturated fatty acid family. *Biochemie* **2019**, *159*, 36–48. [CrossRef]
63. Santa-María, C.; López-Enríquez, S.; la Paz, S.M.-D.; Geniz, I.; Reyes-Quiroz, M.E.; Moreno, M.; Palomares, F.; Sobrino, F.; Alba, G. Update on Anti-Inflammatory Molecular Mechanism Induced by Oleic Acid. *Nutrients* **2023**, *1*, 224. [CrossRef]
64. Venn-Watson, S.K.; Butterworth, C.N. Broader and safer clinically-relevant activities of pentadecanoic acid compared to omega-3: Evaluation of emerging essential fatty acid across twelve primary human cell-based disease systems. *PLoS ONE.* **2022**, *5*, e0268778. [CrossRef] [PubMed]
65. Deo, P.; Dhillon, V.S.; Thomas, P.; Fenech, M. Oleic Acid Status Positively Correlates with the Soluble Receptor for Advanced Glycation End-Products (sRAGE) in Healthy Adults Who Are Homozygous for G Allele of RAGE G82S Polymorphism. *Cells* **2023**, *12*, 1662. [CrossRef]
66. Holms, R.D. Long COVID (PASC) Is Maintained by a Self-Sustaining Pro-Inflammatory TLR4/RAGE-Loop of S100A8/A9 >TLR4/RAGE Signalling, Inducing Chronic Expression of IL-1b, IL-6 and TNFa: Anti-Inflammatory Ezrin Peptides as Potential Therapy. *Immuno* **2022**, *2*, 512–533. [CrossRef]
67. Currie, C.; Framroze, B.; Singh, D.; Sharma, D.; Bjerknes, C.; Hermansen, E. Pharmacological evaluation of the effects of enzymatically liberated fish oil on eosinophilic inflammation in animal models. *Biotechnol. Appl. Biochem.* **2023**, *1*, 157–163. [CrossRef]
68. Currie, C.; Framroze, B.; Singh, D.; Lea, S.; Bjerknes, C.; Hermansen, E. Assessing the Anti-Inflammatory Effects of an Orally Dosed Enzymatically Liberated Fish Oil in a House Dust Model of Allergic Asthma. *Biomedicines* **2022**, *10*, 2574. [CrossRef]
69. Guo, Y.; Xiao, P.; Lei, S.; Deng, F.; Xiao, G.G.; Liu, Y.; Chen, X.; Li, L.; Wu, S.; Chen, Y. How is mRNA expression predictive for protein expression? A correlation study on human circulating monocytes. *Acta Biochim. Biophys. Sin.* **2008**, *40*, 426–436. [CrossRef] [PubMed]
70. Koussounadis, A.; Langdon, S.P.; Um, I.H.; Harrison, D.J.; Smith, V.A. Relationship between differentially expressed mRNA and mRNA-protein correlations in a xenograft mouse model system. *Sci. Rep.* **2015**, *5*, 10775. [CrossRef]
71. Zlei, M.; Sidorov, I.A.; Joosten, S.A.; Heemskerk, M.H.M.; Myeni, S.K.; Pothast, C.R.; de Brouwer, C.S.; der Zanden, A.L.B.-V.; van Meijgaarden, K.E.; Morales, S.T.; et al. Immune Determinants of Viral Clearance in Hospitalised COVID-19 Patients: Reduced Circulating Naïve CD4+ Cell Counts Correspond with Delayed Viral Clearance. *Cells* **2022**, *11*, 2743. [CrossRef] [PubMed]

International Journal of *Molecular Sciences*

MDPI

Article

Oat Beta-Glucan as a Metabolic Regulator in Early Stage of Colorectal Cancer—A Model Study on Azoxymethane-Treated Rats

Jacek Wilczak [1,*], Adam Prostek [1], Katarzyna Dziendzikowska [2], Małgorzata Gajewska [1], Łukasz Kopiasz [2], Joanna Harasym [3], Michał Oczkowski [2] and Joanna Gromadzka-Ostrowska [2]

[1] Department of Physiological Sciences, Institute of Veterinary Medicine, Warsaw University of Life Sciences, 02-776 Warsaw, Poland; adam_prostek@sggw.edu.pl (A.P.); malgorzata_gajewska@sggw.edu.pl (M.G.)

[2] Department of Dietetics, Institute of Human Nutrition Sciences, Warsaw University of Life Sciences, 02-776 Warsaw, Poland; katarzyna_dziendzikowska@sggw.edu.pl (K.D.); lukasz_kopiasz@sggw.edu.pl (Ł.K.); michal_oczkowski@sggw.edu.pl (M.O.); joanna_gromadzka-ostrowska@sggw.edu.pl (J.G.-O.)

[3] Department of Biotechnology and Food Analysis, Wroclaw University of Economics and Business, 53-345 Wroclaw, Poland; joanna.harasym@ue.wroc.pl

* Correspondence: jacek_wilczak@sggw.edu.pl

Abstract: Factors that reduce the risk of developing colorectal cancer include biologically active substances. In our previous research, we demonstrated the anti-inflammatory, immunomodulatory, and antioxidant effects of oat beta-glucans in gastrointestinal disease models. The aim of this study was to investigate the effect of an 8-week consumption of a diet supplemented with low-molar-mass oat beta-glucan in two doses on the antioxidant potential, inflammatory parameters, and colonic metabolomic profile in azoxymethane(AOM)-induced early-stage colorectal cancer in the large intestine wall of rats. The results showed a statistically significant effect of AOM leading to the development of neoplastic changes in the colon. Consumption of beta-glucans induced changes in colonic antioxidant potential parameters, including an increase in total antioxidant status, a decrease in the superoxide dismutase (SOD) activity, and a reduction in thiobarbituric acid reactive substance (TBARS) concentration. In addition, beta-glucans decreased the levels of pro-inflammatory interleukins (IL-1α, IL-1β, IL-12) and C-reactive protein (CRP) while increasing the concentration of IL-10. Metabolomic studies confirmed the efficacy of oat beta-glucans in the AOM-induced early-stage colon cancer model by increasing the levels of metabolites involved in metabolic pathways, such as amino acids, purine, biotin, and folate. In conclusion, these results suggest a wide range of mechanisms involved in altering colonic metabolism during the early stage of carcinogenesis and a strong influence of low-molar-mass oat beta-glucan, administered as dietary supplement, in modulating these mechanisms.

Keywords: azoxymethane; colorectal cancer; interleukins; metabolome; oat beta-glucan; redox status

1. Introduction

Colorectal cancer (CRC) ranks as the third most diagnosed type of cancer and the second leading cause of cancer-related deaths worldwide [1]. Primary risk factors for CRC development include intestinal inflammation (typically non-specific), older age, and environmental factors, including alcohol consumption, smoking, obesity, and diet type. Although early stages of CRC often remain asymptomatic, early diagnosis is associated with a good prognosis of approximately 90% of 5-year survival rate [2]. Recent studies suggest that diet (its quality and the content of potentially carcinogenic ingredients) plays a crucial role in the initiation of inflammatory processes that lead to carcinogenesis. Meta-analysis performed by Shivappa and co-workers has shown a direct link between the consumption of certain dietary products and an increased risk of CRC development [3]. On the other hand, various factors, including regular physical activity a diet rich in fruits, vegetables, fibre, folic acid, calcium, dairy products, vitamin D, vitamin B6, magnesium, garlic, and regular fish intake, have been associated with a reduction in CRC rates [4]. Specific dietary components, such as polyphenols, prebiotic fibre fractions, polyunsaturated n-3 fatty acids, known for their antioxidative and anti-inflammatory properties, may exert an inhibitory effect on CRC development [5].

Among the dietary components that show potential direct anti-inflammatory and indirect antioxidative effects are beta-glucans. Beta-glucans are a group of compounds composed of D-glucose molecules linked by beta-glycosidic bonds, forming non-starch polysaccharides that belong to the dietary fibre fraction. They serve as a structural components in the cell walls of fungi, yeasts, and marine algae, as well as the aleurone layer and endosperm of certain cereals. The structure of beta-glucans varies depending on their source of origin, which in turn leads to their different biological functions. Cereal beta-glucans (including those isolated from oats), unlike mushroom, yeast, and seaweed beta-glucans, form long, slightly branched chains consisting of β-D-glucopyranose monomers linked mainly via β-1,4-glycosidic bonds and, to a lesser extent, β-1,3-glycosides. This unique structure renders oat beta-glucans well soluble in water, forming sticky gels. Among the numerous properties of beta-glucans that directly affect gastrointestinal tract functions, their anti-inflammatory and prebiotic properties are of most interest. Our previous research demonstrated the inhibitory effect of oat beta-glucans on LPS-induced inflammatory process in rats' colon [5,6]. In these studies, the effect of high- and low-molar-mass oat beta-glucans was investigated, and our results showed that the low-molar-mass beta-glucan had stronger anti-inflammatory properties [5,6]. Therefore, in the current study, we focused on oat beta-glucan with a low molar mass (5.2×10^4 g/mol) and purity of 99.3%.

To demonstrate the anti-carcinogenic properties of the low-molar-mass oat beta-glucan, we used an experimental model of azoxymethane (AOM)-induced early-stage colorectal cancer in rats. AOM and its final metabolites of 1,2-dimethylhydrazine (DMH) are responsible for the methylation of the DNA nitrogenous bases in various cells, including the epithelial cells of colonic crypts. This process leads to colon cell apoptosis and an increased mutagenicity of colonic epithelial cells [7–10]. In the present study, we determined the levels of selected pro- and anti-inflammatory cytokines (IL-1α, IL-1β, IL-12, IL-10, as well as C reactive protein), markers of the redox status (glutathione reductase, glutathione peroxidase, superoxide dismutase, and total antioxidant status), as well as the metabolomic profile in the large intestine tissue of rats with AOM-induced early stage of colorectal cancer. The rats were fed either a feed supplemented with the low-molar-mass oat beta-glucan (OBG) at the level of 1% or 3% or a non-supplemented feed. This comprehensive approach was adopted to explore the possible mechanisms underlying the protective effect of oat beta-glucans against carcinogenesis in the colon.

2. Results

2.1. Colorectum Histopathological Changes

Histopathological examination of colorectal specimens showed pronounced changes in the epithelium, including the presence of aberrant crypts and hyperplasia (Figure 1B), observed mainly in rats from the CRC group without dietary intervention, which confirms the effectiveness of the research model used. In control animals, regardless of the type of diet consumed, and in most of the rats from groups CRC_BG_1 and CRC_BG_3, the colorectal epithelium showed normal morphology without microscopic signs of abnormalities (Figure 1A). A detailed description of the analysis of histological changes can be found in our previous publication [11].

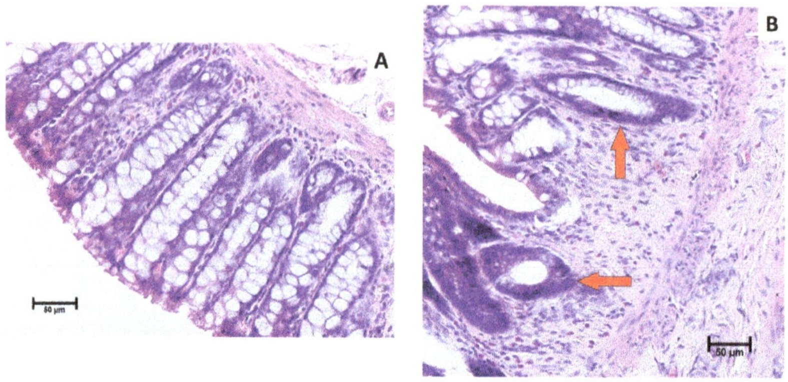

Figure 1. Microscopic images of colon epithelium of experimental rats. (**A**) normal colon epithelium; (**B**) colon epithelium with aberrant crypts and hyperplasia (red arrows).

2.2. The Redox Balance of the Colorectal Tissue

The analyses indicated significant changes in the redox balance in the colorectum of the examined animals. AOM treatment resulted in a statistically significant decrease in TAS level along with a simultaneous increase in SOD activity and TBARS concentration in rats from the CRC group in comparison to animals from the CON group. On the other hand, in the CON_BG_1 and CON_BG_3 groups, the SOD activity and TBARS concentration were at the same levels as those in the corresponding control group, although TAS levels were increased in comparison to those in the control group (CON). In animals with CRC, the 8-week consumption of feed supplemented with 1% or 3% OBG (CRC_BG_1 and CRC_BG_3 groups) significantly altered all parameters when compared with corresponding CON groups. TAS values in CRC_BG_1 and CRC_BG_3 groups reached levels observed in animals from the CON group (Figure 2).

An analysis of all markers related to glutathione metabolism (Table 1) showed that AOM administration caused a decrease in reduced gluthatione (GSH) and an increase in the oxidized gluthatione (GSSG) concentration, which resulted in the lowest GSH/GSSG ratio in the colorectal tissue of rats from the CRC group in comparison to other tested groups. In the colorectal tissue of rats from the CON group, the value of GSH/GSSG ratio was higher (0.13 ± 0.02 vs. 0.68 ± 0.07). Dietary intervention with OBG, regardless of the dose used, increased the GSH/GSSG ratio to the value of approximately 0.31 in the case of animals from the CRC_BG_1 and CRC_BG_3 groups. The rate of conversion of GSSG to GSH is a consequence of the activity of glutathione reductase (GR). Its activity did not change in control animals fed a diet containing OBG, regardless of its amount in the diet. However, the activity of GR was decreased during AOM-induced CRC. Supplementation of diet with OBG significantly increased the GR activity in comparison to the CRC group, but this effect was observed only in the case of 1% OBG dietary dose (group CRC_BG_1). In addition, a significant increase in glutathione peroxidase (GPx) activity was observed in CRC groups. Only in rats from the CRC_BG_1 group was a significant decrease in the activity of GPx was observed; however, this value did not reach the level characteristic for control animals (CON_BG_1 and CON_BG_3 groups).

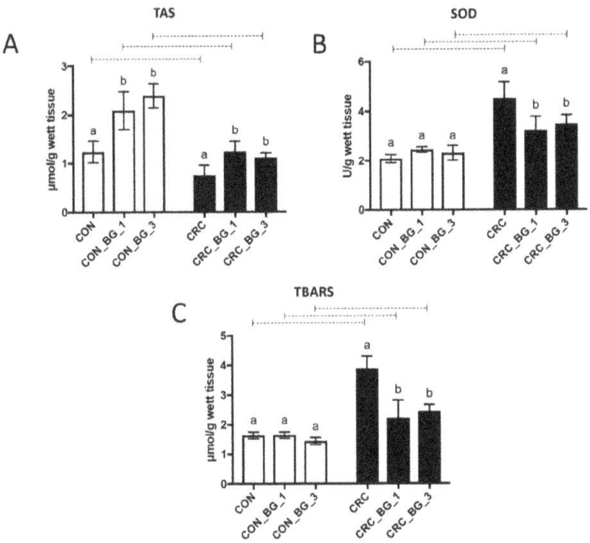

Figure 2. Redox status of the colorectal tissue: total antioxidant status (TAS) (**A**), superoxide dismutase activity (SOD) (**B**), and thiobarbituric acid reactive substances (TBARSs) (**C**) in colorectal tissue of rats from control (CON) and AOM-induced colorectal cancer (CRC) groups fed a diet with 1% or 3% of oat beta-glucan (BG_1 and BG_3, respectively) or control diet. Results are presented as means ± standard deviation. Statistically significant differences between groups and within groups were evaluated using two-way ANOVA with Tukey's post hoc test. Means followed by a common letter are not significantly different at the 5% level of significance ($p < 0.05$) when the comparison is made among dietary subgroups in the control/CRC group. Connecting dotted line is placed above bars that represent statistically different means between the control and CRC groups on the same feed at $p < 0.05$.

Table 1. Concentration of reduced (GSH) and oxidized glutathione (GSSG), GSH/GSSG ratio, activity of glutathione reductase (GR) and glutathione peroxidase (GPx) in colorectal tissue of rats from control (CON) and AOM-induced colorectal cancer (CRC) groups fed a diet with 1% or 3% addition of oat beta-glucan (BG_1 and BG_3, respectively) or control diet.

Parameters	CON Group			CRC Group		
	CON (n = 7)	CON_BG_1 (n = 7)	CON_BG_3 (n = 7)	CRC (n = 8)	CRC_BG_1 (n = 8)	CRC_BG_3 (n = 8)
GSH	269.89 ± 11.2 ^{a,*}	223.72 ± 12.6 ^{b,*}	230.26 ± 18.5 ^{b,*}	109.05 ± 13.3 ^{a,*}	165.82 ± 13.4 ^{b,*}	172.99 ± 17.6 ^{b,*}
GSSG	346.5 ± 25.7 *	357.5 ± 12.1 *	314.3 ± 17.6 *	712.9 ± 72.9 ^{a,*}	473.4 ± 47.0 ^{ab,*}	588.5 ± 73.4 ^{a,*}
GSH/GSSG	0.68 ± 0.07 ^{a,*}	0.56 ± 0.03 ^{b*}	0.67 ± 0.08 ^{a,*}	0.13 ± 0.02 ^{a,*}	0.31 ± 0.05 ^{b,*}	0.31 ± 0.07 ^{b,*}
GR	2.60 ± 0.22 *	2.55 ± 0.07 *	2.521 ± 0.10 *	1.71 ± 0.14 ^{a,*}	2.07 ± 0.17 ^{b,*}	1.78 ± 0.15 ^{ab,*}
GPx	150.85 ± 11.5 *	175.59 ± 6.8 *	170.97 ± 10.9 *	331.22 ± 30.1 ^{a,*}	246.43 ± 12.4 ^{b,*}	296.63 ± 25.2 ^{a,*}

All values were expressed as means ± standard deviation. Statistically significant differences between groups and within groups were evaluated using two-way ANOVA with Tukey's post hoc test; ^{a,b}—different letters denote significant differences among dietary subgroups in the control/CRC group at $p < 0.05$; *—denotes significantly different results between the control and CRC groups on the same feed at $p < 0.05$. GSH and GSSG concentration is expressed in µmol/g of wet tissue; GR and GPx activity is expressed in U/g of wet tissue.

2.3. Immunological Parameters in Colorectal Tissue Samples

The induction of the early stage of CRC in rats by AOM (CRC group) caused an increase in the concentration of pro-inflammatory interleukins and CRP protein, accompanied by a statistically significant decrease in the concentration of the anti-inflammatory cytokine IL-10 (Figure 3). The intake of feed supplemented with OBG had the greatest effect on changing the inflammatory profile of the colorectal tissue, especially in animals from the CRC groups. The concentrations of both pro-inflammatory interleukins IL-1α and IL-1β, in CRC_BG_1 and CRC_BG_3 groups, reached values similar to those in the control group (CON), suggesting that the consumption of feed with OBG lowered the concentration of those interleukins to physiological values (Figure 3A,B). A similar trend was observed in the case of IL-10. The level of anti-inflammatory cytokine IL-10 was increased in the colorectal tissue of CRC rats receiving diet with OBG in comparison to corresponding CRC group (Figure 3E). Consumption of feed supplemented with 1% or 3% of OBG (CRC_BG_1 and CRC_BG_3) caused a significant reduction in the IL-12 and CRP levels; however, the concentration of these proteins did not reach the values observed in control animals (Figure 3C,D). In the case of CRP, the strongest effect was noted when rats were fed a diet supplemented with 3% OBG (Figure 3D).

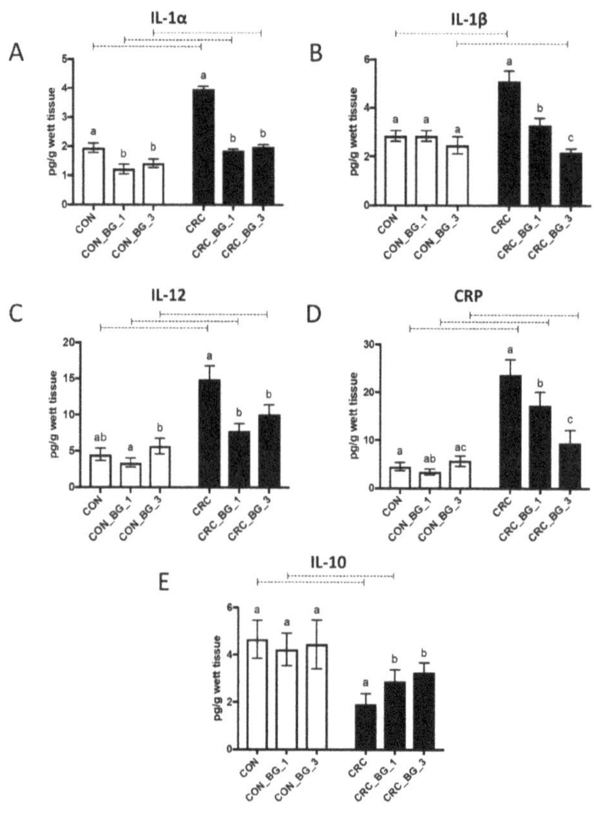

Figure 3. Immunological parameters in colorectal tissue: levels of interleukin 1alpha (IL-1α) (**A**), Interleukin 1beta (IL-1β) (**B**), interleukin 12 (IL-12) (**C**), C-reactive protein (CRP) (**D**), and interleukin 10 (IL-10) (**E**) in colorectal tissue of rats from control (CON) and AOM-induced colorectal cancer (CRC) groups fed a diet with 1% or 3% of oat beta-glucan (BG_1 and BG_3, respectively) or control diet. Results are presented as means ± standard deviation. Statistically significant differences between groups and within groups were evaluated using two-way ANOVA with Tukey's post hoc test. Means followed by a common letter are not significantly different at the 5% level of significance ($p < 0.05$) when the comparison is made among dietary subgroups in the control/CRC group. Connecting dotted line is placed above bars that represent statistically different means between the control and CRC groups on the same feed at $p < 0.05$.

2.4. The Metabolomic Profile of Colorectal Tissue

To determine the changes in metabolomic profiles of CRC rats under the influence of consumption of feed supplemented with two different doses of OBG, metabolomic data from the following group comparisons were analysed: CRC vs. CON, CRC vs. CRC_BG_1 and CRC vs. CRC_BG_3. This approach allowed a precise comparison of the metabolomes of the large intestine under the influence of tested factors, including AOM effects and dietary intervention. In addition, it facilitated the determination of the metabolic pathways connected with the above factors. The final effect of such comparison was the selection of metabolites that generated the most pronounced and statistically significant differences between the experimental groups.

Data were mean-centred using scaling and principal component analysis (PCA) to analyse the metabolomic data. A p value at $\leq$0.05 was considered statistically significant. Metabolites that differed among tested groups were subjected to pathway analysis by MetaboAnalyst 5.0, which combined results from powerful pathway enrichment analysis with the pathways' topology analysis. The identified metabolites were then mapped to the KEGG pathway for biological interpretation to achieve a higher-level understanding of systemic functions.

A PCA analysis (Figure 4) showed that the metabolomic profile in the colorectal tissue samples of rats was statistically significantly affected by both the AOM treatment (CON vs. CRC) and by consumption the feed supplemented with different doses of OBG. The blocked diagrams show the separation of metabolomes of groups CON and CRC (Figure 4). Homogeneity within the control groups (non-supplemented and supplemented with OBG) can be noticed. The consumption of 1% or 3% OBG with the feed did not cause pronounced differences in the metabolomes of control animals.

This difference in metabolomic profiles was also reflected in the number of differentiating metabolites detected in the investigated groups. Within all CON groups, 465 differential metabolites were identified, whereas in three CRC groups, the number of characteristic metabolites amounted to 1435. Therefore, a more detailed analysis was conducted to examine the differences in metabolomes of the following experimental groups: CRC vs. CON, CRC vs. CRC_1BG, CRC vs. CRC_3BG (Supplementary Data, Tables S1–S3). Supplementary Table S1 presents metabolic pathways in which the metabolites differed between the CRC and CON groups. The following metabolic pathways were found: metabolism of butanoate, valine, leucine and isoleucine degradation, metabolism of lysine and tryptophan, fatty acids degradation and metabolism, purine metabolism, sphingolipid metabolism, one carbon pool by folate, and others.

Metabolites that differed significantly between CRC and CRC_BG_1 groups (fed non-supplemented versus supplemented with 1% OBG diets) represented the following metabolic pathways: arachidonic acid, linoleate, vitamin A (retinol), glycerophospholipid, porphyrin metabolism, prostaglandin formation from arachidonate, leukotriene metabolism, and C21-steroid hormone biosynthesis and metabolism (Supplementary Table S2a). When the amount of OBG added to the feed was increased to 3%, the metabolites that differed significantly between CRC and CRC_BG_3 groups represented the following metabolic pathways: aspartate and asparagine metabolism, methionine and cysteine metabolism, mono- and di-unsaturated fatty acid beta-oxidation, leukotriene, linoleate metabolism, omega-3 fatty acids metabolism, prostaglandin formation from arachidonate, C21-steroid hormone biosynthesis and metabolism, androgen and oestrogen biosynthesis and metabolism, porphyrin metabolism, and others (Supplementary Table S3a). Significantly ($p < 0.05$) regulated metabolites detected in the colorectal tissues of rats from analysed experimental groups are presented in Supplementary Tables S1b, S2b, and S3b.

Additional analysis was performed to confirm the presence of chosen specific metabolites, selected from a list of metabolites that differed significantly between the CRC group and other experimental groups included in the metabolomic analysis (CON, CRC_BG_1, CRC_BG_3). After the initial verification of metabolites using the ChemSpider platform, a LC-MS/MS based targeted metabolomic analysis was performed with a thorough analysis of the fragmentation spectra. Only the confirmation of these spectra ensured that the metabolites were correctly selected for further research. Out of the initially selected 12 metabolites, 4 were conclusively confirmed: kynurenine, methionine, citrulline and tryptophan. These compounds showed significantly higher levels in the CRC group in comparison to the control (CON) groups and CRC groups fed the diet supplemented with OBG, regardless of the dose used (CRC_BG_1 and CRC_BG_3) (Figure 5). The consumption of the feed supplemented with OBG led to a reduction in the levels of kynurenine, methionine, citrulline, and tryptophan, bringing them in line with the values observed in the colorectal samples from rats in the control groups (Figure 5). The role of these compounds is described in more detail in the discussion section.

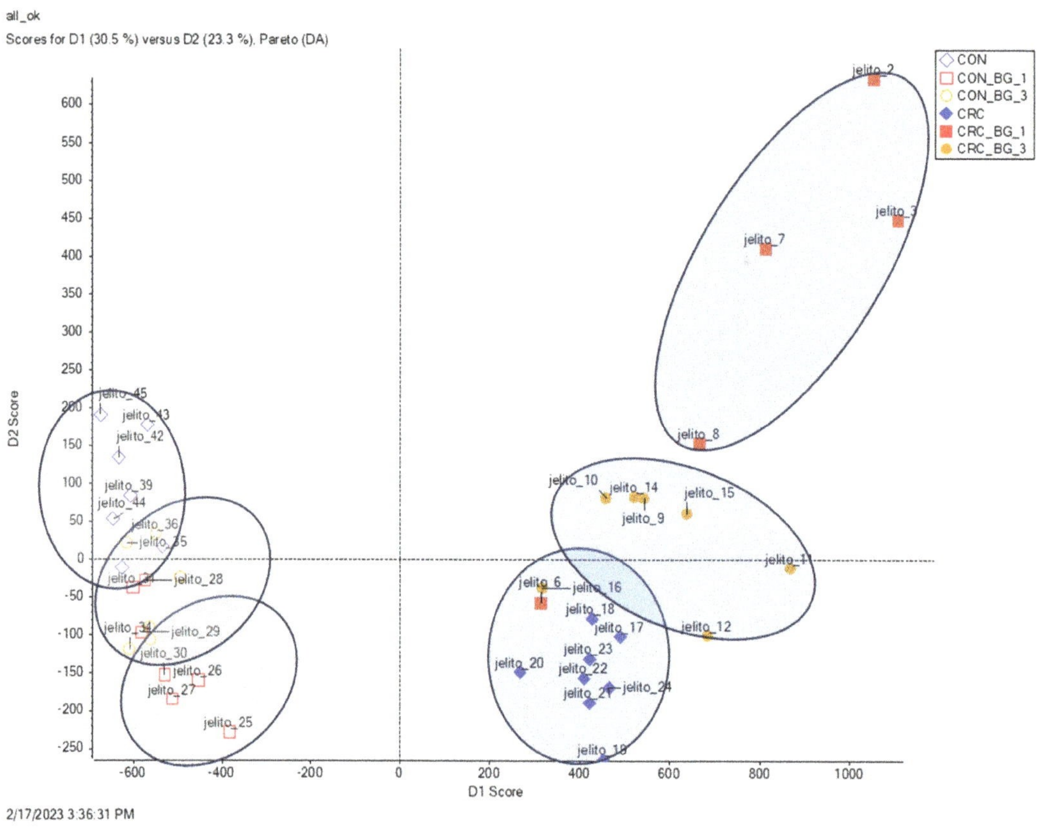

Figure 4. Principal components analysis (PCA) of colorectal tissue metabolome of rats from control (CON) and AOM-induced colorectal cancer (CRC) groups fed a diet with 1% or 3% of oat beta-glucan (BG_1 and BG_3, respectively) or control diet. Areas with a grey background represent metabolomes characteristic for control groups (CON), whereas areas with blue background represent metabolomes characteristic for CRC groups.

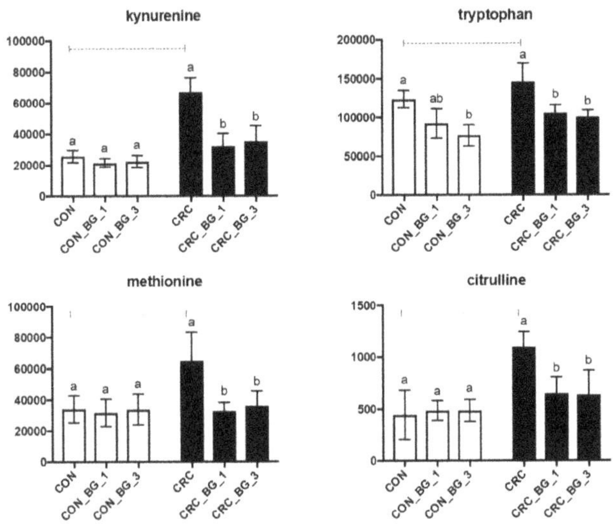

Figure 5. Sum of m/z metabolite counts of kynurenine, methionine, citrulline, tryptophan in colorectal tissue of rats from control (CON) and AOM-induced colorectal cancer (CRC) groups fed a diet with 1% or 3% of oat beta-glucan (BG_1 and BG_3, respectively) or control diet. Results are presented as means ± standard deviation. Statistically significant differences between groups and within groups were evaluated using two-way ANOVA with Tukey's post hoc test. Means followed by a common letter are not significantly different at the 5% level of significance ($p < 0.05$) when the comparison is made among dietary subgroups in the control/CRC group. Connecting dotted line is placed above bars that represent statistically different means between the control and CRC groups on the same feed at $p < 0.05$.

3. Discussion

The presented study aimed to determine the effect of low-molar-mass oat beta-glucan (OBG) on oxidative stress parameters, inflammation, and the metabolomic profile of the large intestine tissue in rats exposed to azoxymethane. The AOM model is a well-documented and widely used method for experimentally induced colorectal carcinogenesis [12]. This model, known for its resemblance to sporadic colorectal cancer in humans, not only shares similarities in response to certain promotional and preventive factors but also serves as an effective tool in evaluating the chemopreventive mechanisms of action of biologically active compounds in this specific type of cancer [13,14].

There are many types of beta-glucans, and the structure of these polysaccharides differs depending on the source of origin. Fungal beta-glucans are polymers of D-glucose residues joined in a β-1,3 configuration with additional β-(1,6)-linked branches that show a strong immunomodulatory effect [15]. In turn, branched or linear 1,4-β-glucans are believed to show limited biological activity [15]. Our research investigated the effect of OBG administered to rats for 8 weeks in the form of a feed supplement. The tested beta-glucan was isolated from oat bran and purified by a patented method described in detail elsewhere [16,17]. The structure of OBG is characterized by a mixture of β-D-glucose unbranched chains linked by β-(1,3) and β-(1,4) glycosidic bonds. Our previous reports demonstrated that the examined low-molar-mass OBG administered at a dose of 1% feed additive had a strong direct anti-inflammatory effect as well as an indirect antioxidant effect in models of colitis induced by 2,4,6-trinitrobenzenesulfonic acid (TNBS) and systemic inflammation induced by lipopolysaccharides (LPS) in rats [6,18,19]. To test whether the effectiveness of OBG is maintained or modified by an increased dose of OBG present in the diet, we also investigated the effect of 3% OBG.

One of the objectives of the present study was to investigate the antioxidative effect of dietary supplementation with OBG in animals with AOM-induced early-stage CRC. The role of oxidative stress in cancer, characterized by the overproduction of reactive oxygen species (ROS) and reactive nitrogen species (RNS) with simultaneous deficiency in endogenous antioxidant systems, has been extensively explored in many studies focused on carcinogenesis in the digestive tract, including the colon. Endogenous antioxidative systems include mainly superoxide dismutase (SOD), glutathione in reduced and oxidized forms, enzymes involved in glutathione metabolism (glutathione peroxidase and reductase) and others. Inhibition of the key enzymes involved in the synthesis of glutathione or enzymes capturing ROS causes prolonged oxidative stress, which may lead to induction and intensification of the process of carcinogenesis [20,21]. On the other hand, the inhibitory effect of biologically active compounds on redox processes in cancer cells suggests the possibility of their chemopreventive properties. In our study, AOM-induced CRC caused a statistically significant decrease in the GSH concentration (glutathione in reduced form), which also resulted in a decrease in the value of the GSH/GSSG ratio. Furthermore, we found increased activity of glutathione peroxidase (GPx) and decreased activity of glutathione reductase (GR) in all groups of animals with induced early-stage CRC. The 8-week dietary intervention with OBG, regardless of its dose (1% or 3%), improved the parameters characteristic for antioxidant protection related to glutathione oxidation status. Although the low-molar-mass OBG did not fully restore the activity of GPx and GR to the values observed in control groups, it modulated those changes towards physiological levels, particularly by decreasing GPx activity and increasing the activity of GR. The effect was especially pronounced in the case of 1% OBG supplementation. The mechanism of action of the low-molar-mass OBG in our experimental model does not seem to involve direct inhibition of AOM-induced carcinogenesis because no available data confirm the direct absorption of the beta-glucan into the intestinal cells. Oat beta-glucan exhibits high solubility in water and forms viscous gels due to long, poorly branched polysaccharide chains linked mainly by β-1,4-glycosidic bonds. The probable antioxidative mechanism of OBG action is based on limiting the promotion of the secondary oxidative factors that lead to the generation of oxidative stress. This causes the decreased concentration of free radicals, and thus, the diminished oxidation of GSH.

Glutathione peroxidase can also prevent the formation of toxic lipid peroxides by reducing ROS concentration in the cells [22]. Thus, we also analysed the concentration of lipid peroxidation products using the TBARS assay. The products of lipid peroxidation are used as markers of the degree of lipid damage, and the formed lipid peroxides show pro-mutagenic properties and induce mutations in oncogene/tumour suppressor genes in human cancers as well as in animal research models [23,24]. In our study, the colorectal tissue of rats with AOM-induced CRC showed a strong accumulation of lipid peroxidation products, expressed as the sum of TBARS. The lowest TBARS concentration was noted in control groups, irrespective of OBG supplementation, indicating that the protective effect of OBG was not observed in healthy animals.

Among the crucial parameters determining the ability of cells to resist oxidative stress is also superoxide dismutase (SOD). In the cancer cells microenvironment, there are several metabolites that activate enzymes involved in elimination of the effects of oxidative stress due to their pro-oxidative nature. The increase in SOD activity has been correlated with the increase in colorectal cancer severity, which has been demonstrated in neoplastic tissues compared to normal tissues [25]. The results from other studies have shown increased SOD activity in colon samples from patients with colon cancer at stages I and III when compared to the control group, whereas the lower SOD activity was demonstrated in patients with colon cancer at stage IV [26,27]. In our study, the SOD activity of animals in the control group (CON) was lower than that in the CRC group, confirming the initial stages of CRC development. This was also reflected by the results of the histopathological examination of colon sections from CRC and CON groups, revealing the initial stages of crypt aberration under the influence of AOM treatment, without noticeable changes in inflammatory cells infiltration. The consumption of feed supplemented with OBG, regardless of its concentration, significantly reduced SOD activity, confirming the potent strong chemopreventive effect of these polysaccharides in animals with early-stage CRC development. Taking together the results of the redox status of the colorectal tissue (GSH and GSSG concentrations, activity of GSH-related enzymes, activity of SOD, and TBARS concentration), it can be concluded that low-molar-mass OBG exhibits the potential to modulate the antioxidative balance in tissues altered by the early stage of CRC development, significantly diminishing the induced oxidative stress in the cancerous environment, but the effect did not depend on the doses selected in the present research.

In order to create a favourable tumour microenvironment, cancer cells primarily communicate with each other mainly through signalling molecules, such as cytokines, chemokines, or growth factors [28]. Nutritional interventions aim at modifying the secretory profile of the pro- and anti-inflammatory cytokines in order to minimize the risk of cancer development, growth, invasion, metastases, and resistance to therapy. Dietary bioactive compounds, such as beta-glucans, most probably are unable to reach further organs in an unchanged form. Their mechanism of action is limited to specific sections of the digestive tract, especially the large intestine. It is believed that oat beta glucans stimulate the immune cells to produce anti-inflammatory cytokines and inhibit the secretion of pro-inflammatory cytokines [29,30]. Our present study demonstrated that the concentration of pro-inflammatory interleukins IL-1α, IL-1β, IL-12, and C-reactive protein (CRP) was increased in the colorectal tissues of rats with AOM-induced CRC, whereas the concentration of the anti-inflammatory IL-10 was decreased in CRC. Supplementation of feed with the low-molar-mass OBG led to decreased concentrations of the pro-inflammatory cytokines (IL-1α, IL-1β, IL-12) in the colorectal tissue of rats with AOM-induced CRC to levels comparable with healthy control rats. Our previous research also showed that dietary supplementation with oat beta-glucans protected the large intestine from inflammation, in that case induced by LPS [6]. Since chronic inflammation may lead to epithelial–mesenchymal transition, dedifferentiation, and increased ROS production in cancer cells [31], our results support the hypothesis of the anti-cancer activity of the oat beta glucans, reflected among others by their potential for regulating the antioxidative balance and anti-inflammatory response.

One of the major objectives of the present study was to assess the difference in the metabolomic profile of colorectal tissue under physiological state (CON) and pathological conditions (CRC) induced by the AOM treatment. Additionally, the study aimed to identify metabolites that may exhibit differences after dietary intervention with the low-molar-mass oat beta-glucan. Therefore, for this purpose, metabolome comparisons of CRC vs. CON, CRC vs. CRC_1BG, and CRC vs. CRC_3BG were performed separately.

As a result of early-stage colorectal cancer induction by AOM treatment, changes in several key metabolic pathways were demonstrated. Some of the activated metabolic pathways have been previously described in the literature. Recently, a systemic review of studies reported from January 2012 to July 2021 described CRC biomarkers detected by metabolomics, pointing out metabolites that may be used as potential diagnostic markers in the future [32]. However, there are no data characterizing the metabolic profile of the large intestine under physiological conditions. Therefore, we compared the results of metabolomic analysis performed on control colorectal tissue samples with the CRC samples. Among the metabolic pathways that differed between the compared experimental conditions (CRC vs. CON), the crotonoyl-CoA and butanoyl-CoA pathway, involved in the beta-alanine pathway, biotin metabolism, butanoate and folate metabolic pathways, purine metabolism, and one-carbon metabolism, mediated by the folate cofactor, may be of special interest. Folate, acting as a donor and carrier of one-carbon units, is involved in various processes within the one-carbon metabolism, including the biosynthesis of guanine and adenine nucleotides necessary for nucleic acid synthesis [33]. In our study the pathways of purine metabolism were among the pathways differentially regulated in CRC group when compared with control group (CON). Moreover, carcinogenesis induction by AOM also induced changes in the sphingolipid pathway and a large part of amino acids metabolism.

Sphingolipid metabolites can mediate several signalling pathways, as well as affect cell growth, differentiation, autophagy, and apoptosis [34].

Our study also revealed a number of metabolic pathways regulated by the low-molar-mass OBG administered as a dietary supplement at 1% or 3% concentration. It seems that the most important metabolic pathways altered by 1% OBG are involved in the metabolism of polyunsaturated fatty acids (PUFAs), specifically arachidonic acid, linoleic acid, and their metabolites, as well as the metabolism of glycerophospholipids. Arachidonic acid metabolites are crucial for the function of cell membranes of cells stimulated by onco-carcinogens. Gholamalizadeh et al. [35] showed a relationship between the activity of enzymes involved in arachidonic acid metabolism and the type of fat in the diet, with highly inflammatory eicosanoids being products of those enzymatic reactions. Our study confirmed that the dietary oat beta-glucan reduced the concentration of pro-inflammatory cytokines, and given that PUFAs are precursors of prostanoids involved in inflammation, the modulation of arachidonic acid metabolism becomes particularly relevant. Prostanoids, which include leukotrienes, thromboxanes, and prostaglandins, are powerful signalling molecules synthesized by a diverse set of enzymes during inflammation. Among them, prostaglandin E2 (PGE2), the main product of cyclooxygenase (COX), plays a significant role in colorectal carcinogenesis. Many studies showed elevated levels of PGE2 in colon cancer [36]. An analysis using MetaboloAnalyst suggested that the change in the entire metabolic pathway of arachidonic acid was influenced to the greatest extent by the presence of the eicosanoids 16(R)-HETE, 9(S)-HETE, 11.12-EET, 5(S)-HETE, Hepoxilin A3-C, 15(S)-HETE, 19(S)-HETE, and 20(S)-HETE, which, as isomers, would require a thorough analysis in order to confirm their real impact on the metabolome of the tested samples. Despite the attempts of chromatographic separation and their analysis, we were unable to confirm the presence of these eicosanoids quantitatively. However, taking into account the significant changes in the arachidonic acid metabolic pathway and our previous observations regarding the decreased concentration of PGE2 in the colon tissue of rats with TNBS-induced colitis, receiving low-molar-mass OBG [19], it can be concluded that oat beta-glucans play an important role in the regulation of arachidonic acid metabolism.

The addition of OBG to the feed of rats with AOM-induced CRC, especially at the higher concentration (3%), also increased the concentration of metabolites involved in the pathway of methionine and cysteine metabolism. Cysteine, in particular, plays a key role in regulating the tumour microenvironment. This sulphur-containing amino acid takes part in cell biosynthesis processes, in enzymatic catalysis and, above all, in redox metabolism. Extracellular cysteine serves as the primary source for the intracellular cysteine pool that supports the cellular redox state [37]. During carcinogenesis, the demand for cysteine increases for glutathione production to manage oxidative stress. In our research, an increase in the GSH/GSSG ratio was noted in tissue samples from rats with AOM-induced CRC fed a diet supplemented with OBG (CRC_BG_1, CRC_BG_3) in comparison to colorectal tissue samples from the CRC group. These finding confirm the antioxidative properties of OBG, as discussed earlier, and show the relationship between the results of our metabolomic analysis with the investigated oxidative status.

Methionine, the second sulphur-containing amino acid, initiates protein synthesis during translation and serves as a source of methyl groups for most nucleotides, chromatin, and protein methylation. It also donates carbon skeletons for various aspects of the cellular antioxidant response and nucleotide biosynthesis. Cancer cells, unlike normal cells, exhibit high methionine cycle activity and depend on exogenous methionine for continued growth [38]. Methionine deficiency initiates extensive metabolic changes in cancer cells that enable them to survive despite the limited availability of this amino acid [39]. Furthermore, the combined effects of methionine and cysteine are important because of their direct involvement in the metabolism of the glutathione pathways and the antioxidant capacity of cells.

Our investigation also revealed the altered metabolism of glycine, serine, alanine, and threonine in the colorectal tissue of rats treated with AOM and fed a diet containing 3%

of OBG. Glycine, as a source of one-carbon units, together with increased expression of glycine decarboxylase, promotes the growth of cancer cells, e.g., for colorectal cancer [40]. Interestingly, an excess of glycine can be harmful to cancer cells, leading to a reduction in the rate of proliferation and inhibiting tumour growth. In colorectal cancer cells with an excess of glycine and a serine deficiency, the dominant reaction catalysed by methyltransferases is the conversion of glycine to serine. This reaction not only results in the depletion of one-carbon units but also inhibits nucleotide synthesis [41]. Studies have shown that the relationship between the breakdown of serine and the formation of glycine and the direction of these changes is crucial for colon cancer cell survival [41]. The observed changes in amino acid metabolism in our study emphasize the potential influence of OBG on the one-carbon metabolism and nucleotide synthesis in CRC.

Finally, four metabolites (kynurenine, methionine, citrulline, and tryptophan) were selected in the present study for further quantitative and qualitative analysis, thus allowing us to confirm the changes in their levels in the investigated experimental groups. Kynurenine plays a vital role in the nutritional modulation of immune cells and promotes the differentiation of Treg lymphocytes, which leads to the increased production of anti-inflammatory cytokines and inhibition of cytotoxic activity of T lymphocytes. However, an excessive activation of the kynurenine pathway in tumours has been linked to increased survival and invasion of tumour cells into surrounding tissues [42]. The kynurenine pathway strongly interacts with other molecular pathways involved in tumorigenesis, including the Wnt/β-catenin pathway, which is directly related to the AOM mechanism of action, and the COX2 and cyclin-dependent kinase pathways. The over-activation of the kynurenine pathway predicts a poor prognosis for several cancers, such as gastrointestinal cancers [43]. In our study, the quantitative analysis confirmed that in the group of rats with CRC and consuming feed supplemented with the oat beta-glucan (CRC_BG_1 and CRC_BG_3), the concentration of kynurenine decreased to the concentrations observed in animals from the control group (CON), while the highest concentration was observed in the large intestine of animals from the CRC group. Such modulation may contribute to a less favourable environment for tumour survival and invasion.

It is also important to link the metabolism of kynurenine with tryptophan (TRP)—one of the essential, exogenous amino acids, which plays an important role in both nutrition and a variety of physiological processes in organisms [44]. Two main metabolic pathways for tryptophan are known in mammals: the kynurenine metabolism pathway and the serotonin metabolism pathway. The former plays a significant role in the immune function and, therefore, in development [45]. Tryptophan is converted to kynurenine through the actions of enzymes such as tryptophan 2,3-dioxygenase (TDO) or indoleamine 2,3-dioxygenase (IDO). In the gastrointestinal tract and brain, residual TRP is transformed into kynurenine via IDO. Kynurenine then undergoes further metabolism leading to the production of various metabolites, including kynurenic acid (KA) and quinolinic acid, which is later transformed to NAD^+. Additionally, the gut microbiome can metabolize TRP to indole and its derivatives [46]. Tryptophan and its metabolites have multiple important functions in the gastrointestinal tract. Firstly, metabolites of tryptophan are believed to be ligands for the aryl hydrocarbon receptor (AhR), which is responsible for modulating immune response in the intestine [44]. Studies suggest that the number of T cells can be positively influenced by the high intake of tryptophan from the diet, and the indole acrylic acid improves the overall function of the epithelial barrier in the intestine [47,48]. Furthermore, TRP as well as kynurenine can inhibit the inflammatory response [49]. In the studies on CRC progression, the kynurenine pathway has been connected with an inhibitory effect on CRC cell proliferation. In the present research, the tryptophan level was reduced to control physiological values in the colorectal tissue of CRC rats fed the diet supplemented with OBG regardless of the dose used. This suggests a potential influence of oat beta-glucan on tryptophan metabolism and its associated pathways. The modulation of tryptophan metabolism, especially within the kynurenine pathway, appears to be a critical factor in the complex interplay between diet, immunity, and colorectal cancer. Understanding these intricate mechanisms provides valuable insights for potential therapeutic interventions and emphasizes the importance of dietary components in influencing metabolic responses.

4. Materials and Methods

4.1. Isolation and Characteristics of OBG

Beta-glucan was extracted from oat bran concentrate (Bestpharma, Warsaw, Poland) using a unique method described in details in our previous reports [16,17]. The detailed methodology for the isolation and purification of beta-glucan was described in our previous article [50]. Briefly, the oat bran was frozen and subjected to repeated milling, and was then extracted using an alkaline solution. The remaining substances were then separated by centrifuging and the supernatant was de-proteinised at its isoelectric point. The protein content, measured using the Lowry method, was found to be negligible, falling below the method's detection limit (<0.01 mg/mL). The purity of the beta-glucan was 99.3%, as determined by enzymatic methods (AACC Method 32-23.01). The molar mass of the beta-glucan used in the present study was determined by SEC-HPLC to be approximately $5.2 \times 10^4 \pm 0.6 \times 10^4$ g/mol.

4.2. In Vivo Experiment

The experiment was conducted on eight-week old male Sprague Dawley rats (CRL:CD(SD)), purchased from Charles River Laboratories (Sulzfeld, Germany). Detailed description of the in vitro experiment was presented in our recent research article [50]. Briefly, after a one-week adaptation period, the animals were divided into 2 main groups: a treatment group with chemically induced early-stage colorectal cancer (CRC, n = 24), and a control group, without pathological colorectum changes (CON, n = 21). The CRC was induced by peritoneal injection of azoxymethane (AOM) (Sigma-Aldrich, Saint Louis, MO, USA) at a dose of 15 mg/kg once a week for two weeks in accordance with the method described by Perše and Cerar [7]. Animals in control group (CON) received peritoneal injections of physiological saline using the same scheme of treatment. On the day of last AOM injection, both CRC and CON groups were subdivided into three sub-groups: (1) animals fed AIN-93M feed (ZooLab, Sędziszów, Poland) without beta-glucans (groups CRC and CON), (2) animals fed AIN-93M feed supplemented with 1% (*w*/*w*) of OBG (groups CRC_BG_1 and CON_BG_1), and (3) animals fed AIN-93M feed supplemented with 3% OBG (*w*/*w*) (groups CRC_BG_3 and CON_BG_3). Each CRC subgroup comprised 8 animals, whereas each CON subgroup comprised 7 animals. Detailed description of feed composition in all experimental groups and environmental conditions were provided in our previous article [50]. Each animal's body weight was recorded weekly, and feed consumption was monitored every day. At the end of the 56-day feeding period, the animals were subjected to general Isoflurane anaesthesia and blood was sampled by heart puncture, and after bleeding, the three parts of large intestine (cecum, colon, and rectum) were sampled. In order to minimize the impact of the ingesta components, all sampled parts were thoroughly rinsed in phosphate-buffered saline (PBS), then frozen in liquid nitrogen and stored at −80 °C for further analysis.

All experimental procedures were approved by the 2nd Local Ethical Committee in Warsaw, Poland (resolution no. WAW2/040/2019), in accordance with the EU Directive (2010/63/UE), the Polish law, and the 3R rules.

4.3. Histopathological Evaluation

Samples of the large intestine were fixed in 10% buffered formaldehyde, and dehydrated by successive immersion in a gradient of ethanol solutions. Subsequently, the samples were rinsed with xylene, embedded in paraffin blocks, and cut into 5 μm sections. The sections were stained with haematoxylin and eosin (H&E) for evaluation under a light microscope (Motic BA400, Olympus Corporation, Tokyo, Japan) by a veterinarian histopathologist.

4.4. The Immunological Parameters Evaluation

The large intestine samples were homogenized in PBS using electric homogenizer, vortexed for 15 min and centrifuged for 15 min at 500× *g*. Supernatants were used to analyse the concentrations of chosen cytokines interleukin IL-10 (cat.no.: RBMS629R), IL-12 (cat.no.: E90059Ra), IL-1α (cat.no.: E900233Ra), IL-1β (cat.no.: E900235Ra), and C reactive protein (CRP, cat.no.: E900878Ra) by competitive specific enzyme immunoassay (ELISA) commercial kits (USCN Life Science Inc., Wuhan, China). Analyses were performed according to the manufacturer's protocols.

4.5. The Colon Redox Balance Parameters Evaluation

Determination of total antioxidant status (TAS), glutathione reductase (GR), glutathione peroxidase (GPx), and superoxide dismutase (SOD) activity was conducted in tissue homogenates prepared according to the protocol described in Section 4.4. The parameters were analysed using the Randox assay kits (TAS, cat. no.: NX2332; GR, cat. no.: GR 2368; GPx, cat. no.: RS504/505/506; SOD, cat. no.: SD125) according to the protocols provided by the producer (Randox Laboratories Ltd., Crumlin, County Antrim, UK). Reduced (GSH) and oxidized (GSSG) gluthatione were detected using high-pressure liquid chromatography (HPLC) with a colourimetric electrochemical detector from ESA (Chelmsford, MA, USA) with a 4-channel electrochemical array for the simultaneous detection of both glutathione forms. The mobile phase for isocratic elution of the sulfhydryls was composed of 25 mM monobasic sodium phosphate, 0.5 mM 1-octane sulfonic acid, and 2.5% acetonitrile, adjusted to pH 2.7. All chemicals, including GSH and GSSG standards were purchased from Sigma-Aldrich (part of Merck KGaA, Darmstadt, Germany). The pH for the mobile phase was adjusted with 85% phosphoric acid. A flow rate of 1 mL min^{-1} was used with a C18 column (5 μM column, 4.6 × 250 mm). Acetonitrile was the key component in modulating the retention times of GSH and GSSG. With 2.5% of acetonitrile in the mobile phase, the retention times for GSH generally appeared at 5 min and GSSG at 20 min. The GSH/GSSG ratio was calculated on this basis. Thiobarbituric acid reactive substances (TBARSs) were analysed according to the method described by Aguilar Diaz De Leon and Borges [51].

4.6. Metabolomic Analysis

Metabolomic analysis was performed using a high-pressure liquid chromatography Symbiosis Pico UHPLC system. The detector used for this analysis was a SCIEX TripleTOF 5600+ DuoSpray Source for SCIEX TripleTOF 5600+ (TurboIonSpray and APCI). The acquired data were analysed using SCIEX MarkerView™ (ver. 1.3.1.), XCMSplus (on-line ver.), and MetaboAnalyst 5.0 software packages for comprehensive interpretation and extraction of metabolomic information.

The samples of colon tissue were homogenized in PBS and mixed with a mixture of acetonitrile and methanol in a 1:1 ratio. Samples were vortexed (2000 rpm for 15 min) and centrifuged for 15 min at 20,000× *g*. Supernatants were transferred to glass autosampler vials, and placed in an autosampler at 4 °C. Samples were injected directly into a Spark Holland Symbiosis™ Pico system. Chromatographic separation was performed on the Hypersil chromatographic column, BDS C18, (150 × 4.6 mm, 5 μm) with a Hypersil C18 guard column (10 × 2.1 mm, size 5 μm). The mobile phase consisted of methanol/formic acid (99:1, *v*/*v*) A and water/formic acid (99:1, *v*/*v*) B, and the flow rate was constant at 500 μL min^{-1}. The gradient elution of the mobile phase 100% A was started at 1.1–40 min linear gradient to 100% B, 40.1–55 min 100% B, and 55.1–60 min linear gradient to 100% A.

MS/MS parameters: the optimized detection conditions included curtain gas (N_2) 25 psi, nebulizer gas (N_2) 20 psi, heater gas (N_2) 50 psi, ion source voltage floating 5500 V, and source temperature 500 °C. Samples with a heated electrospray 3 ionization probe were measured in positive ionization (H-ESI+).

4.7. Statistical Analyses

Analyses of variance (ANOVA) were performed using GraphPad Prism ver.6.0 (GraphPad Software, Inc., La Jolla, CA, USA). To determine the effect of AOM treatment, as well as supplementation of the diet with 1% or 3% of OBG, two-way ANOVA was employed to assess differences between control and experimental groups. The significance of differences in results among the groups was determined by Tukey's post hoc test. A significance level of $p < 0.05$ was considered as a statistically significant difference. All data were expressed as mean $\pm$ standard deviation (mean $\pm$ SD).

The metabolite profiles, obtained within the 100–1100 Da range with a sensitivity of 5 cps, were analysed using SCIEX MarkerView™ software (ver.1.3.1.). Comparative assessments between groups were conducted using Student's t-tests and principal component analysis (PCA). The generated metabolomic profiling data sets were processed by the control software of the SCIEX Analyst® (ver.1.7.3) mass spectrometer and saved in a specific data format (*.raw). The first step was to convert data from Excalibur-specific raw files to open format files (*.mzXML) using MS Convertor software (MSConvert ver. 1.5.2.). Subsequently, the metabolomic data were processed using the XCMSplus platform. Additionally, PCA resulted in comparative profiles of metabolomes of specific groups, which were further analysed by SCIEX MarkerView™ software. The classification of identified metabolites within particular metabolic pathways was conducted using the XCMSplus on-line platform and MetaboAnalyst 5.0 software with a probability threshold of $p < 0.0001$. The identification of metabolites was carried out based on the ChemSpider database (accessed via SCIEX PeakView™ ver.2.2). Indications above an 80% probability of confirming a given structure were compared with the indications of metabolites from the MetaboAnalyst 5.0 database.

5. Conclusions

The obtained results indicate that dietary supplementation with the low-molar-mass oat beta glucan may play an important role in maintaining homeostasis of the intestinal microenvironment. The observed effects include exerting the anti-inflammatory and antioxidant activities, emphasizing the potential use of oat beta-glucan in dietary intervention. Additionally, low-molar-mass oat beta-glucan was found to modulate metabolic pathways, specifically those related to amino acids and fatty acids metabolism, which are crucial for maintaining the physiological functions of the colorectal tissue. The presented study is consistent with contemporary trends of research investigating the role of biologically active compounds in the prevention of colon cancer. However, only one type of beta-glucans was investigated, which does not allow for a direct comparison of the mechanism of action of beta-glucans of different origin and structure. Nevertheless, the present research, which refers only to oat beta-glucans, documents their strong effect in the early stages of carcinogenesis. Our results suggest potential therapeutic implications for dietary oat beta-glucan interventions aimed at supporting the intestinal homeostasis. The potential therapeutic use of beta-glucans isolated from oat grains should be further explored in the future preclinical and clinical studies.

Supplementary Materials: The following supporting information can be downloaded at: https://www.mdpi.com/article/10.3390/ijms25094635/s1.

Author Contributions: Conceptualization, J.W.; methodology, J.W. and A.P.; formal analysis, J.W., J.H. and A.P.; investigation, J.W., A.P., K.D., M.G., Ł.K., J.H. and M.O.; resources, J.W., A.P. and K.D.; writing—original draft preparation, J.W.; writing—review and editing, A.P., K.D., M.G., Ł.K., J.H., M.O. and J.G.-O.; visualization, J.W. and A.P.; supervision, J.G.-O.; project administration, K.D. and J.G.-O.; funding acquisition, J.G.-O. All authors have read and agreed to the published version of the manuscript.

Funding: This research was funded by the National Science Centre, Poland, through grant number 2018/29/B/NZ9/01060. The publication was financed by the Science development fund of the Warsaw University of Life Sciences—SGGW.

Institutional Review Board Statement: The animal study protocol was approved by the 2nd Local Ethical Committee in Warsaw (Resolution No. WAW2/040/2019) in accordance with the EU Directive 2010/63/EU for animal experiments and Polish law and 3R rules (Replacement, Reduction and Refinement).

Informed Consent Statement: Not applicable.

Data Availability Statement: The data that support the findings of this study are available on request from the corresponding author.

Acknowledgments: We would like to acknowledge Rafał Sapierzyński from Institute of Veterinary Medicine, Warsaw University of Life Sciences (IVM-WULS), Warsaw, Poland, for assistance in histology analysis.

Conflicts of Interest: The authors declare no conflicts of interest.

References

1. Sung, H.; Ferlay, J.; Siegel, R.L.; Laversanne, M.; Soerjomataram, I.; Jemal, A.; Bray, F. Global Cancer Statistics 2020: GLOBOCAN Estimates of Incidence and Mortality Worldwide for 36 Cancers in 185 Countries. *CA Cancer J. Clin.* **2021**, *71*, 209–249. [CrossRef]
2. Dekker, E.; Tanis, P.J.; Vleugels, J.L.A.; Kasi, P.M.; Wallace, M.B. Colorectal Cancer. *Lancet* **2019**, *394*, 1467–1480. [CrossRef]
3. Shivappa, N.; Godos, J.; Hébert, J.R.; Wirth, M.D.; Piuri, G.; Speciani, A.F.; Grosso, G. Dietary Inflammatory Index and Colorectal Cancer Risk—A Meta-Analysis. *Nutrients* **2017**, *9*, 1043. [CrossRef]
4. Thanikachalam, K.; Khan, G. Colorectal Cancer and Nutrition. *Nutrients* **2019**, *11*, 164. [CrossRef]
5. Suchecka, D.; Harasym, J.P.; Wilczak, J.; Gajewska, M.; Oczkowski, M.; Gudej, S.; Błaszczyk, K.; Kamola, D.; Filip, R.; Gromadzka-Ostrowska, J. Antioxidative and Anti-Inflammatory Effects of High Beta-Glucan Concentration Purified Aqueous Extract from Oat in Experimental Model of LPS-Induced Chronic Enteritis. *J. Funct. Foods* **2015**, *14*, 244–254. [CrossRef]
6. Wilczak, J.; Błaszczyk, K.; Kamola, D.; Gajewska, M.; Harasym, J.P.; Jałosińska, M.; Gudej, S.; Suchecka, D.; Oczkowski, M.; Gromadzka-Ostrowska, J. The Effect of Low or High Molecular Weight Oat Beta-Glucans on the Inflammatory and Oxidative Stress Status in the Colon of Rats with LPS-Induced Enteritis. *Food Funct.* **2015**, *6*, 590–603. [CrossRef]
7. Perše, M.; Cerar, A. Morphological and Molecular Alterations in 1,2 Dimethylhydrazine and Azoxymethane Induced Colon Carcinogenesis in Rats. *J. Biomed. Biotechnol.* **2011**, *2011*, 473964. [CrossRef]
8. Kobaek-Larsen, M.; Nielsen, D.S.; Kot, W.; Krych, Ł.; Christensen, L.P.; Baatrup, G. Effect of the Dietary Polyacetylenes Falcarinol and Falcarindiol on the Gut Microbiota Composition in a Rat Model of Colorectal Cancer. *BMC Res. Notes* **2018**, *11*, 41. [CrossRef]
9. Madka, V.; Kumar, G.; Pathuri, G.; Zhang, Y.; Lightfoot, S.; Asch, A.S.; Mohammed, A.; Steele, V.E.; Rao, C.V. Bisphosphonates Zometa and Fosamax Synergize with Metformin to Prevent AOM-Induced Colon Cancer in F344 Rat Model. *Cancer Prev. Res.* **2020**, *13*, 185–194. [CrossRef]
10. Yang, J.; Wei, H.; Zhou, Y.; Szeto, C.H.; Li, C.; Lin, Y.; Coker, O.O.; Lau, H.C.H.; Chan, A.W.H.; Sung, J.J.Y.; et al. High-Fat Diet Promotes Colorectal Tumorigenesis Through Modulating Gut Microbiota and Metabolites. *Gastroenterology* **2022**, *162*, 135–149. [CrossRef]
11. Harasym, J.; Dziendzikowska, K.; Kopiasz, Ł.; Wilczak, J.; Sapierzyński, R.; Gromadzka-Ostrowska, J. Consumption of Feed Supplemented with Oat Beta-Glucan as a Chemopreventive Agent against Colon Cancerogenesis in Rats. *Nutrients* **2024**, *16*, 1125. [CrossRef]
12. Raju, J. Azoxymethane-Induced Rat Aberrant Crypt Foci: Relevance in Studying Chemoprevention of Colon Cancer. *World J. Gastroenterol.* **2008**, *14*, 6632. [CrossRef]
13. Im, S.A.; Kim, J.W.; Kim, H.S.; Park, C.S.; Shin, E.; Do, S.G.; Park, Y.I.; Lee, C.K. Prevention of Azoxymethane/Dextran Sodium Sulfate-Induced Mouse Colon Carcinogenesis by Processed Aloe Vera Gel. *Int. Immunopharmacol.* **2016**, *40*, 428–435. [CrossRef]
14. Saki, E.; Saiful Yazan, L.; Mohd Ali, R.; Ahmad, Z. Chemopreventive Effects of Germinated Rough Rice Crude Extract in Inhibiting Azoxymethane-Induced Aberrant Crypt Foci Formation in Sprague-Dawley Rats. *Biomed Res. Int.* **2017**, *2017*, 9517287. [CrossRef]
15. Vetvicka, V.; Teplyakova, T.V.; Shintyapina, A.B.; Korolenko, T.A. Effects of Medicinal Fungi-derived Β-glucan on Tumor Progression. *J. Fungi* **2021**, *7*, 250. [CrossRef]
16. Harasym, J.; Suchecka, D.; Gromadzka-Ostrowska, J. Effect of Size Reduction by Freeze-Milling on Processing Properties of Beta-Glucan Oat Bran. *J. Cereal Sci.* **2015**, *61*, 119–125. [CrossRef]
17. Harasym, J.; Zyła, E.; Dziendzikowska, K.; Gromadzka-Ostrowska, J. Proteinaceous Residue Removal from Oat β-Glucan Extracts Obtained by Alkaline Water Extraction. *Molecules* **2019**, *24*, 1729. [CrossRef]
18. Kopiasz, Ł.; Dziendzikowska, K.; Gajewska, M.; Wilczak, J.; Harasym, J.; Żyła, E.; Kamola, D.; Oczkowski, M.; Królikowski, T.; Gromadzka-Ostrowska, J. Time-Dependent Indirect Antioxidative Effects of Oat Beta-Glucans on Peripheral Blood Parameters in the Animal Model of Colon Inflammation. *Antioxidants* **2020**, *9*, 375. [CrossRef]
19. Żyła, E.; Dziendzikowska, K.; Kamola, D.; Wilczak, J.; Sapierzyński, R.; Harasym, J.; Gromadzka-Ostrowska, J. Anti-Inflammatory Activity of Oat Beta-Glucans in a Crohn's Disease Model: Time-and Molar Mass-Dependent Effects. *Int. J. Mol. Sci.* **2021**, *22*, 4485. [CrossRef]

20. Jakubczyk, K.; Drużga, A.; Katarzyna, J.; Skonieczna-żydecka, K. Antioxidant Potential of Curcumin—A Meta-Analysis of Randomized Clinical Trials. *Antioxidants* **2020**, *9*, 1092. [CrossRef]
21. Sies, H.; Jones, D.P. Reactive Oxygen Species (ROS) as Pleiotropic Physiological Signalling Agents. *Nat. Rev. Mol. Cell Biol.* **2020**, *21*, 363–383. [CrossRef]
22. Zhang, X.; Ma, Y.; Ma, J.; Yang, L.; Song, Q.; Wang, H.; Lv, G. Glutathione Peroxidase 4 as a Therapeutic Target for Anti-Colorectal Cancer Drug-Tolerant Persister Cells. *Front. Oncol.* **2022**, *12*, 913669. [CrossRef]
23. Ayala, A.; Muñoz, M.F.; Argüelles, S. Lipid Peroxidation: Production, Metabolism, and Signaling Mechanisms of Malondialdehyde and 4-Hydroxy-2-Nonenal. *Oxid. Med. Cell. Longev.* **2014**, *2014*, 360438. [CrossRef]
24. Bardelčíková, A.; Šoltys, J.; Mojžiš, J. Oxidative Stress, Inflammation and Colorectal Cancer: An Overview. *Antioxidants* **2023**, *12*, 901. [CrossRef]
25. Skrzydlewska, E.; Sulkowski, S.; Koda, M.; Zalewski, B.; Kanczuga-Koda, L.; Sulkowska, M. Lipid Peroxidation and Antioxidant Status in Colorectal Cancer. *World J. Gastroenterol.* **2005**, *11*, 403. [CrossRef]
26. Skrzycki, M.; Majewska, M.; Podsiad, M.; Czeczot, H. Expression and Activity of Superoxide Dismutase Isoenzymes in Colorectal Cancer. *Acta Biochim. Pol.* **2009**, *56*, 663–670. [CrossRef]
27. Warsinggih; Irawan, B.; Labeda, I.; Lusikooy, R.E.; Sampetoding, S.; Kusuma, M.I.; Uwuratuw, J.A.; Syarifuddin, E.; Prihantono; Faruk, M. Association of Superoxide Dismutase Enzyme with Staging and Grade of Differentiation Colorectal Cancer: A Cross-Sectional Study. *Ann. Med. Surg.* **2020**, *58*, 194–199. [CrossRef]
28. Bhat, A.A.; Nisar, S.; Singh, M.; Ashraf, B.; Masoodi, T.; Prasad, C.P.; Sharma, A.; Maacha, S.; Karedath, T.; Hashem, S.; et al. Cytokine- and Chemokine-Induced Inflammatory Colorectal Tumor Microenvironment: Emerging Avenue for Targeted Therapy. *Cancer Commun.* **2022**, *42*, 689–715. [CrossRef]
29. Varnosfaderani, S.M.N.; Ebrahimzadeh, F.; Oryani, M.A.; Khalili, S.; Almasi, F.; Heris, R.M.; Payandeh, Z.; Li, C.; Afjadi, M.N.; Bahrami, A.A. Potential Promising Anticancer Applications of β-Glucans: A Review. *Biosci. Rep.* **2024**, *44*, BSR20231686. [CrossRef]
30. Volman, J.J.; Ramakers, J.D.; Plat, J. Dietary Modulation of Immune Function by β-Glucans. *Physiol. Behav.* **2008**, *94*, 276–284. [CrossRef]
31. Varga, J.; Greten, F.R. Cell Plasticity in Epithelial Homeostasis and Tumorigenesis. *Nat. Cell Biol.* **2017**, *19*, 1133–1141. [CrossRef] [PubMed]
32. Gold, A.; Choueiry, F.; Jin, N.; Mo, X.; Zhu, J. The Application of Metabolomics in Recent Colorectal Cancer Studies: A State-of-the-Art Review. *Cancers* **2022**, *14*, 725. [CrossRef] [PubMed]
33. Ducker, G.S.; Rabinowitz, J.D. One-Carbon Metabolism in Health and Disease. *Cell Metab.* **2017**, *25*, 27–42. [CrossRef] [PubMed]
34. Ségui, B.; Andrieu-Abadie, N.; Jaffrézou, J.P.; Benoist, H.; Levade, T. Sphingolipids as Modulators of Cancer Cell Death: Potential Therapeutic Targets. *Biochim. Biophys. Acta Biomembr.* **2006**, *1758*, 2104–2120. [CrossRef] [PubMed]
35. Gholamalizadeh, M.; Majidi, N.; Tajaddod, S.; Abdollahi, S.; Poorhosseini, S.M.; Ahmadzadeh, M.; Naimi Joubani, M.; Mirzaei Dahka, S.; Shafaei, H.; Hajiesmaeil, M.; et al. Interactions of Colorectal Cancer, Dietary Fats, and Polymorphisms of Arachidonate Lipoxygenase and Cyclooxygenase Genes: A Literature Review. *Front. Oncol.* **2022**, *12*, 865208. [CrossRef] [PubMed]
36. Lenihan-Geels, G.; Bishop, K.S.; Ferguson, L.R.; Brown, L.; Rauch, B.; Poudyal, H. Cancer Risk and Eicosanoid Production: Interaction between the Protective Effect of Long Chain Omega-3 Polyunsaturated Fatty Acid Intake and Genotype. *J. Clin. Med.* **2016**, *5*, 25. [CrossRef] [PubMed]
37. Colovic, M.B.; Vasic, V.M.; Djuric, D.M.; Krstic, D.Z. Sulphur-Containing Amino Acids: Protective Role Against Free Radicals and Heavy Metals. *Curr. Med. Chem.* **2018**, *25*, 324–335. [CrossRef] [PubMed]
38. Wanders, D.; Hobson, K.; Ji, X. Methionine Restriction and Cancer Biology. *Nutrients* **2020**, *12*, 684. [CrossRef] [PubMed]
39. Ji, M.; Xu, X.; Xu, Q.; Hsiao, Y.C.; Martin, C.; Ukraintseva, S.; Popov, V.; Arbeev, K.G.; Randall, T.A.; Wu, X.; et al. Methionine Restriction-Induced Sulfur Deficiency Impairs Antitumour Immunity Partially through Gut Microbiota. *Nat. Metab.* **2023**, *5*, 1526–1543. [CrossRef]
40. Kim, D.; Fiske, B.P.; Birsoy, K.; Freinkman, E.; Kami, K.; Possemato, R.L.; Chudnovsky, Y.; Pacold, M.E.; Chen, W.W.; Cantor, J.R.; et al. SHMT2 Drives Glioma Cell Survival in Ischaemia but Imposes a Dependence on Glycine Clearance. *Nature* **2015**, *520*, 363–367. [CrossRef]
41. Labuschagne, C.F.; van den Broek, N.J.F.; Mackay, G.M.; Vousden, K.H.; Maddocks, O.D.K. Serine, but Not Glycine, Supports One-Carbon Metabolism and Proliferation of Cancer Cells. *Cell Rep.* **2014**, *7*, 1248–1258. [CrossRef] [PubMed]
42. Gouasmi, R.; Ferraro-Peyret, C.; Nancey, S.; Coste, I.; Renno, T.; Chaveroux, C.; Aznar, N.; Ansieau, S. The Kynurenine Pathway and Cancer: Why Keep It Simple When You Can Make It Complicated. *Cancers* **2022**, *14*, 2793. [CrossRef] [PubMed]
43. Ala, M. The Footprint of Kynurenine Pathway in Every Cancer: A New Target for Chemotherapy. *Eur. J. Pharmacol.* **2021**, *896*, 173921. [CrossRef] [PubMed]
44. Zhang, H.L.; Zhang, A.H.; Miao, J.H.; Sun, H.; Yan, G.L.; Wu, F.F.; Wang, X.J. Targeting Regulation of Tryptophan Metabolism for Colorectal Cancer Therapy: A Systematic Review. *RSC Adv.* **2019**, *9*, 3072–3080. [CrossRef] [PubMed]
45. Wettersten, H.I.; Aboud, O.A.; Lara, P.N.; Weiss, R.H. Metabolic Reprogramming in Clear Cell Renal Cell Carcinoma. *Nat. Rev. Nephrol.* **2017**, *13*, 410–419. [CrossRef]

46. Breda, C.; Sathyasaikumar, K.V.; Idrissi, S.S.; Notarangelo, F.M.; Estranero, J.G.; Moore, G.G.L.; Green, E.W.; Kyriacou, C.P.; Schwarcz, R.; Giorgini, F. Tryptophan-2,3-Dioxygenase (TDO) Inhibition Ameliorates Neurodegeneration by Modulation of Kynurenine Pathway Metabolites. *Proc. Natl. Acad. Sci. USA* **2016**, *113*, 5435–5440. [CrossRef]
47. Cervantes-Barragan, L.; Chai, J.N.; Tianero, M.D.; Di Luccia, B.; Ahern, P.P.; Merriman, J.; Cortez, V.S.; Caparon, M.G.; Donia, M.S.; Gilfillan, S.; et al. Lactobacillus Reuteri Induces Gut Intraepithelial CD4+CD8αα+ T Cells. *Science* **2017**, *357*, 806–810. [CrossRef] [PubMed]
48. Wlodarska, M.; Luo, C.; Kolde, R.; D'Hennezel, E.; Annand, J.W.; Heim, C.E.; Krastel, P.; Schmitt, E.K.; Omar, A.S.; Creasey, E.A.; et al. Indoleacrylic Acid Produced by Commensal Peptostreptococcus Species Suppresses Inflammation. *Cell Host Microbe* **2017**, *22*, 25–37. [CrossRef] [PubMed]
49. Lee, K.; Kwak, J.H.; Pyo, S. Inhibition of LPS-Induced Inflammatory Mediators by 3-Hydroxyanthranilic Acid in Macrophages through Suppression of PI3K/NF-KB Signaling Pathways. *Food Funct.* **2016**, *7*, 3073–3082. [CrossRef]
50. Kopiasz, Ł.; Dziendzikowska, K.; Oczkowski, M.; Harasym, J.; Gromadzka-Ostrowska, J. Low-Molar-Mass Oat Beta-Glucan Impacts Autophagy and Apoptosis in Early Stages of Induced Colorectal Carcinogenesis in Rats. *Int. J. Biol. Macromol.* **2024**, *254*, 127832. [CrossRef]
51. Aguilar Diaz De Leon, J.; Borges, C.R. Evaluation of Oxidative Stress in Biological Samples Using the Thiobarbituric Acid Reactive Substances Assay. *JoVE J. Vis. Exp.* **2020**, *159*, e61122. [CrossRef] [PubMed]

International Journal of
Molecular Sciences

MDPI

Article

Chicoric Acid Effectively Mitigated Dextran Sulfate Sodium (DSS)-Induced Colitis in BALB/c Mice by Modulating the Gut Microbiota and Fecal Metabolites

Jiani Yang [1,2], Jie Lin [3], Ting Gu [1], Quancai Sun [3], Weidong Xu [4,*] and Ye Peng [2,*]

1 School of Food and Biological Engineering, Jiangsu University, Zhenjiang 212013, China; yangjiani0411@163.com (J.Y.); guting1029@163.com (T.G.)
2 Faculty of Medicine, Macau University of Science and Technology, Taipa, Macao SAR, China
3 Department of Nutrition and Integrative Physiology, Florida State University, Tallahassee, FL 32306, USA; lj23g@fsu.edu (J.L.); sqctp@hotmail.com (Q.S.)
4 School of Pharmacy, Jiangsu University, Zhenjiang 212013, China
* Correspondence: xuwd@ujs.edu.cn (W.X.); pengye@must.edu.mo (Y.P.)

Abstract: Chicoric acid (CA) has been reported to exhibit biological activities; it remains unclear, however, whether CA could regulate colitis via modulation of the gut microbiota and metabolites. This study aimed to assess CA's impact on dextran sulfate sodium (DSS)-induced colitis, the gut microbiota, and metabolites. Mice were induced with 2.5% DSS to develop colitis over a 7-day period. CA was administered intragastrically one week prior to DSS treatment and continued for 14 days. The microbial composition in the stool was determined using 16S rRNA sequencing, while non-targeted metabolomics was employed to analyze the metabolic profiles of each mouse group. The results show that CA effectively alleviated colitis, as evidenced by an increased colon length, lowered disease activity index (DAI) and histological scores, and decreased tumor necrosis factor-α (TNF-α) and interleukin-6 (IL-6) expression levels. CA intervention restored the structure of gut microbiota. Specifically, it decreased the abundance of Bacteroidetes and Cyanobacteria at the phylum level and *Bacteroides*, *Rosiarcus*, and unclassified *Xanthobacteraceae* at the genus level, and increased the abundance of unclassified *Lachnospiraceae* at the genus level. Metabolomic analysis revealed that CA supplementation reversed the up-regulation of asymmetric dimethylarginine, N-glycolylneuraminic acid, and N-acetylneuraminic acid, as well as the down-regulation of phloroglucinol, thiamine, 4-methyl-5-thiazoleethanol, lithocholic acid, and oxymatrine induced by DSS. Our current research provides scientific evidence for developing CA into an anti-colitis functional food ingredient. Further clinical trials are warranted to elucidate the efficacy and mechanism of CA in treating human inflammatory bowel disease (IBD).

Keywords: chicoric acid; colitis; DSS; gut microbiota; metabolites

1. Introduction

Inflammatory bowel disease (IBD) is a chronic inflammatory condition affecting the gastrointestinal mucosa, encompassing ulcerative colitis (UC) and Crohn's disease (CD), Behcet's syndrome and lymphocytic syndrome, and infectious or ischemic colitis [1,2]. The incidence and prevalence of UC and CD have shown a global increase over the past five decades, contributing to a significant health burden [3]. UC, characterized by idiopathic chronic relapsing–remitting inflammation in the colon that leads to symptoms like diarrhea and rectal bleeding, arises from a complex interplay of genetic, microbial, environmental, and other unidentified factors [4]. Standard treatments for ulcerative colitis primarily involve anti-inflammatory medications, including 5-aminosalicylates such as olsalazine, mesalazine, and balsalazide. For mild-to-moderate UC, immunosuppressants like thiopurines, 6-mercaptopurine, methotrexate, and cyclosporine are commonly prescribed [5]. In cases of moderate-to-severe colon inflammation, other drug classes, such as corticosteroids like prednisolone and anti-TNF-α antibodies, may also be utilized [6]. Surgery becomes a viable option for IBD patients in cases of refractory and fulminant disease stages [7]. Nonetheless, prolonged use of synthetic drugs can result in numerous complications and side-effects, including vomiting, nausea, headache, fatigue, liver and kidney toxicity, drug resistance, and allergic reactions [8,9].

In the gastrointestinal tract of mammals, gut microorganisms engage in intricate communication with intestinal epithelial cells, playing a pivotal role in the regulation of intestinal homeostasis [10]. Disruptions in the gut microbiota can adversely affect intestinal epithelial function and lead to heightened intestinal permeability [11]. Patients suffering from IBD frequently exhibit aberrations in the composition and activity of their gut microbiota, marked by reduced diversity, a decreased proportion of Firmicutes, and an increased proportion of Proteobacteria and Actinobacteria [11,12]. Moreover, the proliferation of pro-inflammatory microbiota, such as *Ruminococcus gnavus* and *Escherichia coli*, alongside the depletion of anti-inflammatory microbiota, including *Bacteroides*, *Lachnospiraceae*, and *Faecalibacterium prausnitzii*, has been associated with the progression of IBD [13]. Gender influences IBD through the gut flora, as the metabolism of androgens and estrogens is thought to be related to the gut microbiota [14,15]. One study reported that individuals, both men and women, with elevated levels of testosterone and estradiol exhibited more diverse bacterial communities. This suggests that sex hormones may contribute to the maintenance of gut health [16]. The majority of women encounter IBD symptoms during the premenstrual or menstrual cycle, pregnancy, and postpartum periods, as sex hormones may prolong the gastrointestinal transit time during the luteal phase [17]. Research has indicated that the dysregulation of bile acid metabolism and its psychological complications in CD patients may be linked to an overabundance of *Enterobacteriaceae* and *Lachnospiraceae*, and a diminished abundance of *Faecalibacterium prausnitzii* [18]. In a clinical trial involving 305 participants, the average number of *Lactobacillus*, *Bifidobacterium*, and *Bacteroides* increased after the consumption of probiotic yogurt, resulting in the restoration of intestinal function [19]. A recent experimental study demonstrated that a blend of four probiotic strains (including *L. acidophilus* LA1, *L. paracasei* 101/37, *B. animalis* spp. Lactis Bi1, and *B. breve* Bbr8) suppressed the production of IL-8, IL-23, and IL-1β cytokine in monocyte-derived dendritic cells obtained from UC patients [20]. The pronounced decrease in *Roseburia* in IBD may be associated with low levels of *R. intestinalis*, which is thought to maintain intestinal homeostasis and induce anti-inflammatory responses through IBD regulatory mechanisms [21]. Despite the robust connection between gut microbiota and IBD, the mechanisms of interaction remain incompletely elucidated. Metabolomics, frequently employed in clinical applications, serves to shed new light on physiological regulatory processes within complex mammalian systems, spanning disease etiology, diagnostic stratification, and potential mechanisms of therapeutic action. It is characterized by the exploration of intricate metabolic interactions between the host and its symbiotic microbiota [22]. Alterations in gut microbiota lead to modifications in bacterial metabolites, including the production of short-chain fatty acids (SCFA), tryptophan, and other small molecules within the gut [23]. Insufficient tryptophan levels can contribute to the development or worsening of IBD [24]. SCFAs, a category of fatty acid compounds with alkyl chains shorter than six carbons, are produced through the fermentation of anaerobic gut microbes abundant in dietary fiber substrates. They may modulate inflammation associated with IBD, a process crucial in the pathogenesis of IBD [25]. Recent research has demonstrated that butyrate can modify the gut microbiota in IBD patients, leading to a reduction in intestinal symptoms [26]. In vitro studies revealed that butyrate and propionate were more potent than acetate in suppressing lipopolysaccharide (LPS)-induced TNFα production by neutrophils and inhibiting TNFα-mediated nuclear factor kappa-B (NF-κB) activation in a human colon cancer cell line [27]. Within human colonic intestinal epithelial cells, butyrate enhances intestinal epithelial barrier function by inhibiting the activation of histone deacetylase, hypoxia-inducible factor-1, and signal transducer and activator of transcription-3. This modulation contributes to the integrity of epithelial tight junctions and hinders LPS-induced NF-κB activation [3]. Importantly, SCFAs can additionally impede the growth of potentially invasive pathogenic bacteria, such as *Escherichia coli*, by influencing intracellular pH and metabolic functions [28]. Consequently, it is crucial to understand the influence of intestinal microbiota and their metabolites on colitis development.

Chichoric acid (CA), also known as dicaffeoyl tartaric acid, was initially identified in 1958 [29,30]. CA is a natural phenolic compound, one of the important active ingredients in *Echinacea purpurea* and Cichorium intybus, reported to have anti-inflammatory, antioxidant, and antiviral activities [31]. CA has been demonstrated to attenuate lipopolysaccharide-induced inflammatory signaling nuclear factor kappa-B p65 (NF-κB p65) and cyclooxygenase-2 activity by inhibiting leucine-rich repeat (LRR) and pyrin domain (PYD) -containing protein 3 inflammasome activation, inhibiting caspase-1 activity, and inhibiting the inflammasome-mediated secretion of interleukin-10 and interleukin-18 [32]. In a model of LPS-induced acute lung injury, CA has been shown to reduce inflammatory cell infiltration, myeloperoxidase activity, and proinflammatory cytokine production in bronchoalveolar lavage fluid. Furthermore, in a rat model of collagen-induced arthritis, treatment with CA extracts from *Echinacea purpurea* (at doses of 8, 16, and 32 mg/kg) significantly reduced serum levels of TNF-α, interleukin-1β (IL-1β), and Prostaglandin E2 (PGE-2), leading to substantial relief in paw swelling [33]. However, the influence of CA on colitis, gut microbiota, and metabolite dysregulation induced by DSS remains to be determined.

Therefore, in this study, we aim to study the influence of CA on colitis induced by DSS in BALB/c mice. To further understand the potential role of gut microbiota and metabolites, 16S rRNA sequencing and fecal metabolomics were applied to elucidate the alterations in microbial community structure and crucial metabolites after DSS exposure as well as CA supplementation.

2. Results

2.1. Supplementation with CA Alleviated DSS-Induced Colitis in Mice

Weight loss during the modeling period is a characteristic feature indicative of colitis severity [34]. As shown in Figure 1a, mice in the DSS group experienced a continuous decline in body weight starting from the third day. In contrast, the CA group exhibited a slower rate of weight loss, which began on the fourth day. In comparison with the CON group, mice in the DSS group exhibited a significant decrease in body weight starting from the sixth day, and weight loss in the DSS group intensified with the duration of DSS administration. Interestingly, the weight of the CA group was significantly higher than that of the DSS group and CON group on the fourth day. This phenomenon may be attributed to the brief period of DSS administration before the mice started experiencing discomfort, and CA, improving the growth and development of the mice, leading to weight gain. Nevertheless, there is no other robust evidence to corroborate this, necessitating further investigation. After seven days of modeling, there was a significant decrease in the body weight of mice in the DSS group compared to the CON group ($p = 0.0009$). However, CA supplementation significantly mitigated the DSS-induced weight loss ($p = 0.0171$) (Figure 1b). Furthermore, visible bleeding in the stool was observed in the DSS group starting on the fifth day of modeling, while it occurred later in the CA group. By the seventh day post modeling, the severity of hematochezia in the CA group was notably lower than that in the DSS group (Figure 1c).

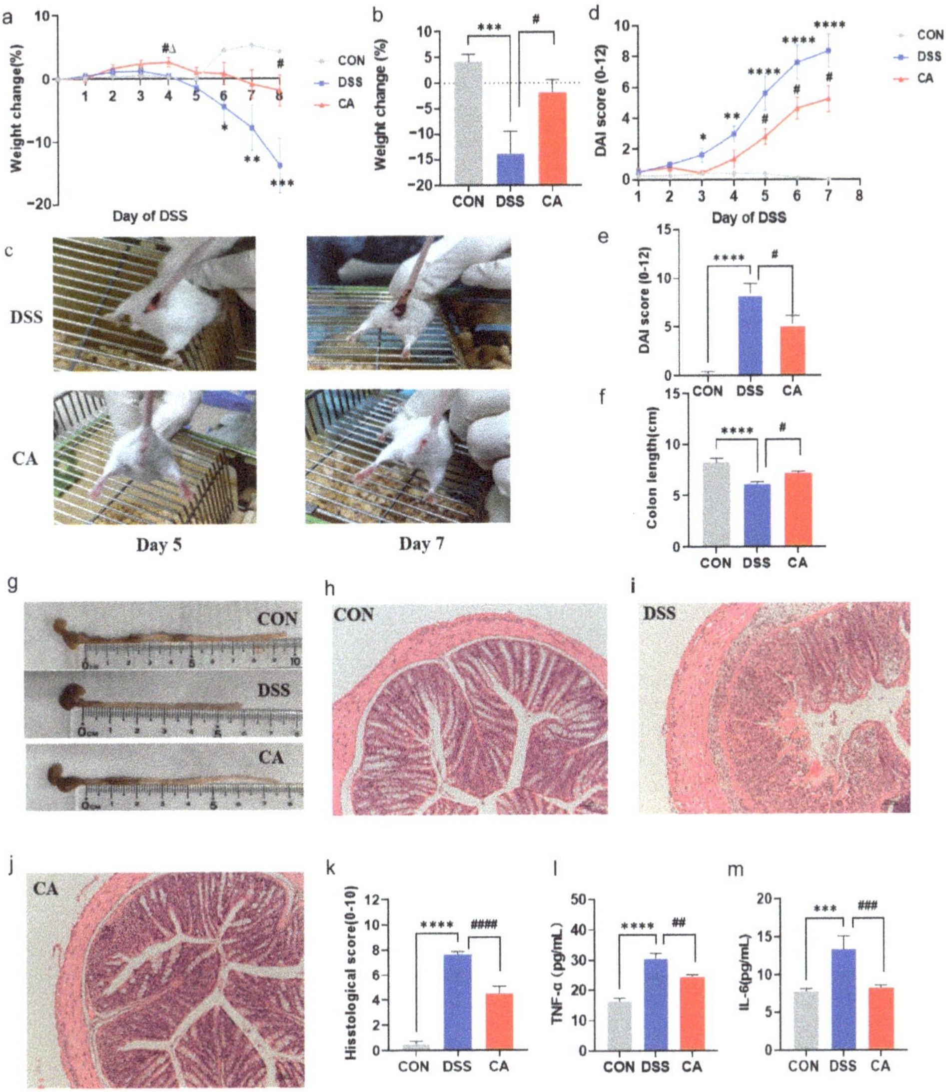

Figure 1. Supplementation with CA alleviated DSS-induced colitis. (**a**) Percent change in mouse body weight during DSS administration. (**b**) The percentage change in mouse body weight after seven days of DSS administration. (**c**) Presence of blood in the feces of mice on the fifth and seventh days of DSS modeling. (**d**) Disease activity index (DAI) scores of mice during DSS administration. (**e**) DAI score of mice after seven days of DSS modeling. (**f**) Colon length measurement in each group. (**g**) Representative colon from each group. (**h–j**) Representative images of H&E-stained colon (100× magnification). (**k**) Histological scores of colon in each group. (**l**,**m**) Expression levels of *TNF-α* and IL-6 levels in mouse serum. The data are presented as mean ± SEM. Statistical analysis was conducted using one-way ANOVA followed by the Tukey post hoc test between three groups and using T-test between two groups. * denotes comparisons between the CON and DSS groups, # indicates comparisons between the DSS and CA groups, and $^{\Delta}$ indicates comparisons between the CON and DSS groups. * $p < 0.05$; ** $p < 0.01$; *** $p < 0.001$; **** $p < 0.0001$; # $p < 0.05$; ## $p < 0.01$; ### $p < 0.001$; #### $p < 0.0001$; $^{\Delta}$ $p < 0.05$.

The disease activity index (DAI) was employed to assess the severity of colitis in mice, taking into account factors such as weight loss, stool consistency, and intestinal bleeding throughout the modeling period. As depicted in Figure 1d, the DAI score showed a progressive increase with the duration of DSS administration. At the conclusion of the experiment, the DAI score in the DSS group had significantly escalated, reaching a substantial value of 8.375, which was markedly higher than that in the CON group ($p < 0.0001$). Following CA treatment, the DAI score in the CA group was significantly lower than that in the DSS group ($p = 0.0369$) (Figure 1e). These observed phenotypic changes collectively indicate that CA effectively mitigates DSS-induced colitis.

2.2. Supplementation with CA Attenuated DSS-Induced Colon Shortening and Colonic Histological Damage

Colon length and colon histological assessment are effective disease indicators in DSS-induced colitis models. Disease severity is often correlated with a reduction in colon length due to intestinal inflammation [35]. Following seven days of DSS exposure, the colon length of mice in the DSS group was significantly shortened compared to the CON group ($p < 0.0001$), while CA supplementation significantly alleviated the DSS-induced colon length reduction ($p = 0.0162$) (Figure 1f,g). Histological examination of colon tissue stained with H&E revealed that the colon tissue in the CON group exhibited a normal histological morphology, with intact crypts and mucosal structures and no noticeable inflammatory cell infiltration. Conversely, the colonic mucosal tissue of mice in the DSS group displayed erosion and loss of epithelial integrity, accompanied by a substantial infiltration of inflammatory cells. Additionally, the crypt structure in the colon of DSS-exposed mice was disrupted. However, following CA supplementation, the mucosal structure of the colon improved, with only a minimal amount of inflammatory cell infiltration, and a reduction in crypt damage was observed (Figure 1h–j).

The histological scoring of the colon is determined by evaluating the extent of inflammatory cell infiltration, the degree of tissue damage, and the level of crypt damage. As depicted in Figure 1k, the histopathological score of the colon in the DSS group exhibited a significant increase ($p < 0.0001$) compared to the CON group. However, the supplementation of CA significantly prevented the elevated histopathological score induced by DSS ($p < 0.0001$). In summary, these findings strongly indicate the effectiveness of CA in preventing intestinal inflammation.

2.3. Supplementation with CA Reduced the Levels of Inflammatory Factors in Mouse Serum

Increased intestinal permeability and the production of pro-inflammatory cytokines, which play a pivotal role in the pathogenesis of IBD, contribute to the observed inflammatory cell infiltration [1]. To further validate the beneficial effects of CA on intestinal inflammation, the levels of TNF-α and IL-6 in the serum of mice from each group were assessed. As seen in Figure 1l,m, in comparison to the CON group, the levels of TNF-α and IL-6 in the serum of the mice in the DSS group were significantly elevated ($p < 0.0001$, $p = 0.0005$). After supplementing CA, the concentrations of TNF-α and IL-6 were markedly lower than those in the DSS group ($p = 0.0087$, $p = 0.0008$). These results collectively provide additional confirmation that CA effectively mitigates DSS-induced colitis.

2.4. Supplementation with CA Regulated Gut Microbiota

The balance of gut microbiota plays a crucial role in shaping host immunity and maintaining gut homeostasis. However, when this balance is disrupted, it can lead to intestinal immune dysfunction and trigger inflammation in IBD [36]. To investigate whether CA can modulate the gut microbiota, we assessed the composition and structure of the mouse gut microbiota using 16S rRNA sequencing. The Shannon index and Observed OTUs were employed to evaluate the α-diversity of the gut microbiota. As shown in Figure 2a,b, compared to the CON group, the α-diversity of the gut microbiota was significantly reduced in the DSS group, as indicated by a decrease in the Shannon index ($p = 0.009$) and Observed

OTUs (p = 0.008). To explore differences in the composition of the microbiota structure among samples, principal coordinate analysis (PCoA) and non-metric multidimensional scaling analysis (NMDS) were utilized to assess the β-diversity of the gut microbiota. The β-diversity analysis revealed a clear separation between the gut microbiota of the DSS and CON groups, indicating an overall change in the gut microbiota structure following DSS exposure. However, CA intervention did not alter the overall structure of the gut microbiota affected by DSS exposure (Figure 2c,d).

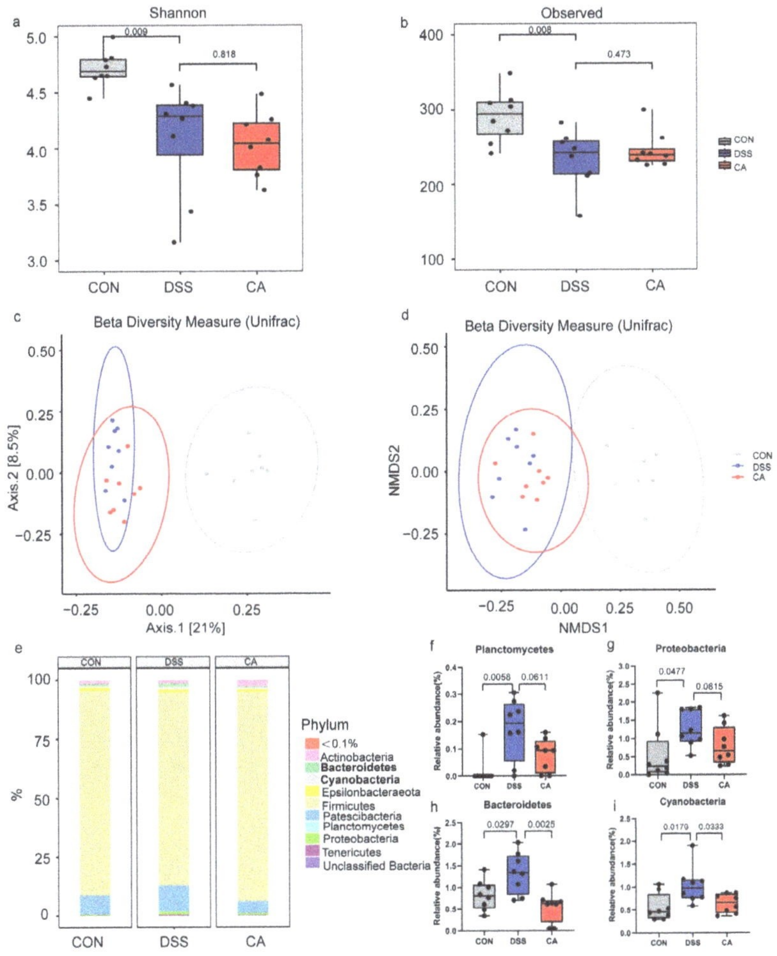

Figure 2. *Cont.*

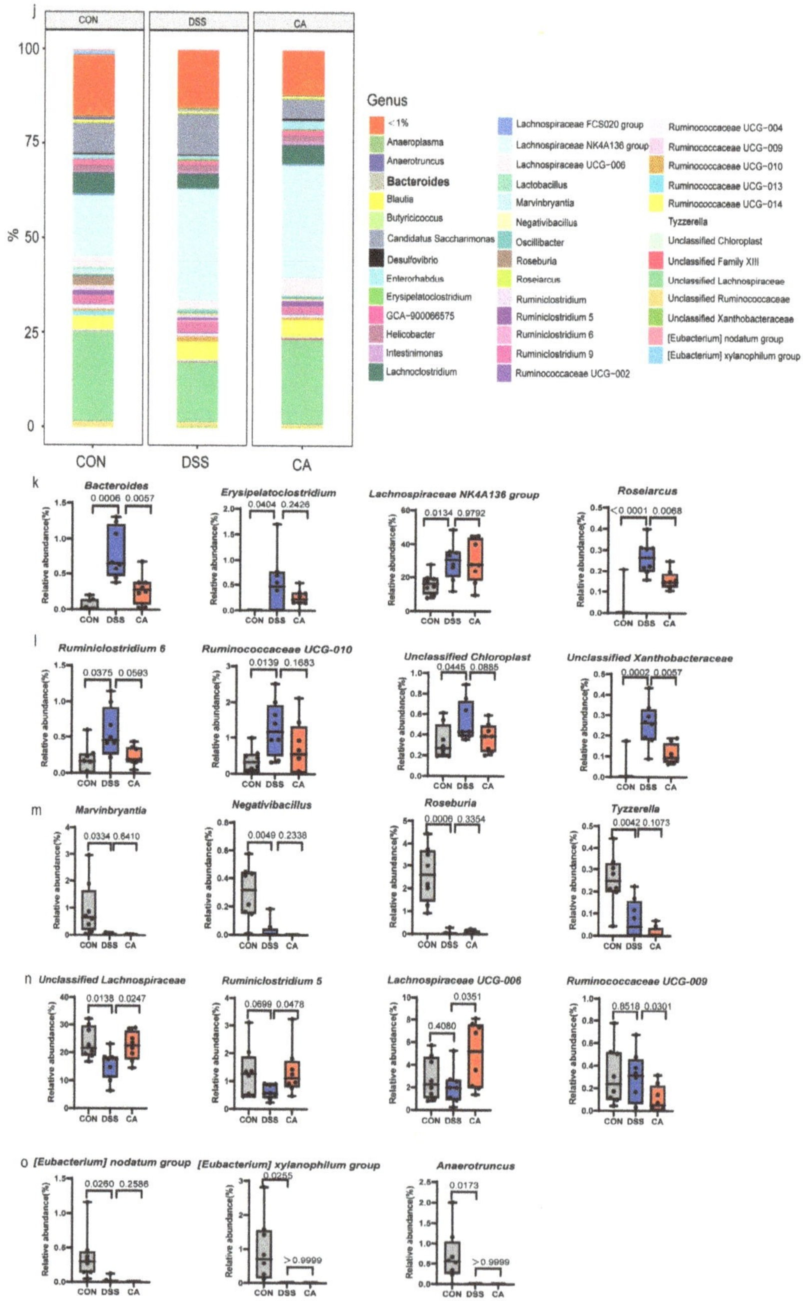

Figure 2. Supplementation with CA regulated gut microbiota. (**a**) Shannon index among three groups. (**b**) Observed OTUs among three groups. (**c**) Principal coordinate analysis (PCoA). (**d**) Non-metric multidimensional scaling analysis (NMDS). (**e**) Relative abundance of taxa at the phylum level. (**f–i**) Significant differences in the abundances of gut microbial community at the phylum level. (**j**) Relative abundance of taxa at the genus level. (**k–o**) Significant differences in the abundances of gut microbial community at the genus level.

Subsequently, we conducted an analysis of species composition and displayed the relative abundance of different bacteria in the three groups. At the phylum level, Firmicutes and Patescibacteria were the dominant bacteria (Figure 2e). Compared to the CON group, the DSS group exhibited a significant increase in the abundance of Planctomycetes (p = 0.0058) and Proteobacteria (p = 0.0477). Although there was no significant difference in the abundance of Planctomycetes and Proteobacteria between the CA group and the DSS group, there was a noticeable downward trend (p = 0.0611, p = 0.0815). Additionally, our results showed significant differences in Bacteroidetes and Cyanobacteria among the three groups. After DSS exposure, the abundance of Bacteroidetes (p = 0.0297) and Cyanobacteria (p = 0.0179) significantly increased compared to the CON group, while CA supplementation led to a significant decrease in the abundance of Bacteroidetes (p = 0.0025) and Cyanobacteria (p = 0.0333) (Figure 2f–i). At the genus level, the dominant gut microbiota included *Candidatus Saccharimonas*, *Lachnospiraceae NK4A136 group*, and unclassified *Lachnospiraceae* (Figure 2j). Figure 2k–o illustrate significant differences in gut microbiota at the genus level among the three groups. Specifically, in the DSS group, there was a notable increase in the abundance of *Bacteroides*, *Roseiarcus*, and unclassified *Xanthobacteraceae*, while the abundance of unclassified *Lachnospiraceae* decreased significantly when compared to the CON group. On the contrary, CA supplementation led to a significant increase in the abundance of unclassified *Lachnospiraceae*, whereas the abundances of *Bacteroides*, *Roseiarcus*, and unclassified *Xanthobacteraceae* were significantly reduced in the CA group. Results from the linear discriminant analysis effect size (LEfSe) revealed distinct species among the three groups. Genus-level discriminant species in the CON group included *Roseburia*, *Ruminococcaceae UCG-013*, *[Eubacterium] xylanophilum group*, *Marvinbryantia*, *Anaerotruncus*, *[Eubacterium] nodatum group*, *Negativibacillus*, and *Candidatus Arthromitus*. In the DSS group, discriminant species included *Bacteroides*, *Erysipelatoclostridium*, and *Roseiarcus*, whereas in the CA group, discriminant species comprised *[Ruminococcus] torques group*, *Ruminococcaceae UCG-014*, and *Family XIII AD3011 group* (Figure S1). Overall, the results indicated that DSS exposure disrupted the balance of gut microbiota in mice, and CA intervention had a beneficial effect on restoring gut microbiota homeostasis.

2.5. Supplementation with CA Altered Fecal Metabolites in Colitis Mice

Since 16S rRNA sequencing revealed significant changes in certain intestinal microbiota following CA supplementation, non-targeted liquid chromatograph mass spectrometry (LC-MS) was employed to further assess potential perturbations in the fecal metabolome after CA supplementation. To assess differences between groups (the dissimilarity among samples from the three groups) and within groups (the clustering degree of samples within each group), principal component analysis (PCA) was conducted. PCA results indicated that the first two principal component scores, PC1 and PC2, accounted for 36.86% and 18.18% of the variance, respectively (Figure 3a). There was a clear separation observed among the CON group, DSS group, and CA group, suggesting that DSS treatment induced alterations in mouse fecal metabolites, while CA intervention partially mitigated these changes. Differential metabolites were identified based on three criteria: VIP ≥ 1, Fold Change ≥ 1.2 or ≤ 0.83, and p-value < 0.05. In comparison to the CON group, the DSS group exhibited 359 up-regulated and 323 down-regulated differential metabolites, totaling 682 differential metabolites between the two groups. Conversely, the CA group, when compared to the DSS group, displayed 330 up-regulated and 340 down-regulated differential metabolites, amounting to 670 differential metabolites between these two groups. Upon clustering and analyzing the expression levels of these differential metabolites, distinct expression patterns of metabolites in the three sample groups became apparent (Figure 3b). Potential biomarker metabolites were defined as differential metabolites exhibiting an opposite trend following DSS exposure and CA supplementation. Table 1 lists potential biomarker metabolites in feces following DSS exposure and CA consumption. Asymmetric dimethylarginine, N-glycolylneuraminic acid, and N-acetylneuraminic acid showed significant increases in the DSS group compared to the CON group but were significantly down-regulated after CA supplementation. Additionally, following DSS exposure, metabolites such as phloroglucinol, thiamine, 4-methyl-5-thiazoleethanol, lithocholic acid, and oxymatrine were significantly down-regulated, with CA intervention leading to significant increases in these metabolites as well. While alpha-tocopherol, acetate, and ursolic acid did not exhibit significant differences between the CON group and DSS group, a decreasing trend was observed in the DSS group compared to the CON group. However, after CA supplementation, the levels of these metabolites were significantly increased (Figure 3c). Figure S2a showed key metabolites identified by LEfSe. To elucidate the metabolic pathways in which CA participates, we conducted enrichment analyses of altered metabolites using the KEGG database. Several significant metabolic pathways were identified (Figure S2b), including "Glycine, serine and threonine metabolism", "Arginine and proline metabolism", "Pyrimidine metabolism", "Tryptophan metabolism", and "Amino sugar and nucleotide sugar metabolism".

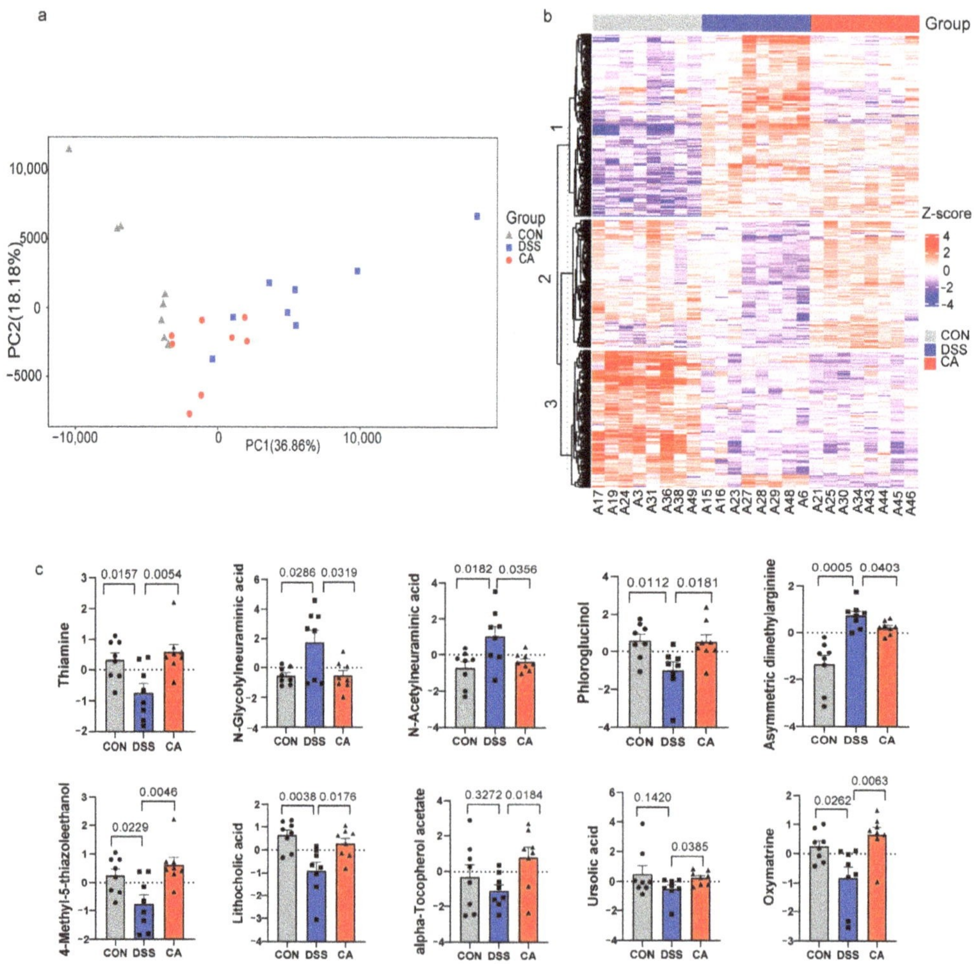

Figure 3. Supplementation with CA altered fecal metabolites. (**a**) Principal component analysis (PCA) score plot of metabolomic features between three groups. (**b**) Cluster analysis of differential metabolite expression level between three groups. (**c**) The metabolites with significant difference between three groups.

Table 1. Potential biomarkers after DSS exposure and CA supplementation.

Metabolite	Fold Change	*p*-Value	VIP	DSS vs. CON	CA vs. DSS	CA vs. CON
Asymmetric dimethylarginine	0.6509	0.0403	1.2106	↑	↓	↑
D-(+)-Raffinose	0.4744	0.0071	1.5651	↑	↓	↑
N-Acetyl-D-glucosamine	0.4189	0.0348	1.5544	↑	↓	n.s.
Quinone sulfate	0.1482	0.0248	3.6340	↑	↓	n.s.
Xanthurenic acid	0.245	0.0126	2.1214	↑	↓	n.s.
4-hydroxy-2-quinolinecarboxylic acid	1.8232	0.0039	1.5071	↓	↑	n.s.
1H-Indole-2,3-dione	0.5489	0.0261	1.4475	↑	↓	n.s.
Gamma-linolenic acid	0.6312	0.0408	1.1922	↑	↓	n.s.
Thiamine	2.0784	0.0054	1.4288	↓	↑	n.s.
4-Methyl-5-thiazoleethanol	2.1456	0.0046	1.5127	↓	↑	n.s.
N-Glycolylneuraminic acid	0.0199	0.0319	2.9523	↑	↓	n.s.
Allantoin	0.0886	0.0491	2.8324	↑	↓	n.s.
N-Acetylneuraminic acid	0.1702	0.0356	2.0773	↑	↓	n.s.
Phloroglucinol	3.5918	0.0181	2.0928	↓	↑	n.s.
Alpha -Aspartylphenylalanine	0.2381	0.0487	1.1902	↑	↓	n.s.
Lithocholic acid	1.9365	0.0176	1.4303	↓	↑	n.s.
Lipoic acid	0.5566	0.0216	1.2972	↑	↓	n.s.
D-(−)-Quinic acid	0.3136	0.0196	1.8506	↑	↓	n.s.
L-Alanine	0.3494	0.0040	2.0185	↑	↓	n.s.
Piperine	1.3955	0.0440	1.2827	↓	↑	n.s.
Oxymatrine	2.4038	0.0063	1.2889	↓	↑	n.s.
2-Hydroxy-4-(hydroxymethyl)-6-(1-hydroxy-3-methylbut-2-enyl)-3-[(E)-prop-1-enyl]-7-oxabicyclo[4.1.0]hept-3-en-5-one	1.8753	0.0037	1.4079	↓	↑	n.s.
Ibuprofen	0.3316	0.0118	1.4320	↑	↓	n.s.
15-Deoxy-delta 12,14-prostaglandins D2	2.74	0.0003	2.0166	↓	↑	n.s.
Thymidine 5′-monophosphate	0.1015	0.0162	2.6353	↑	↓	n.s.
Arachidonoyl ethanolamide phosphate	0.5875	0.0201	1.5616	↑	↓	n.s.
(S)-AL 8810	0.4339	0.0208	2.1068	↑	↓	n.s.
14-(Hydroxymethyl)-5,9-dimethyltetracyclo[11.2.1.0]	1.535	0.0295	1.2260	↓	↑	n.s.
11-Deoxy prostaglandin F2	0.4026	0.0184	1.2333	↑	↓	n.s.
Skatole	0.7237	0.0074	1.1165	n.s.	↓	n.s.
Alpha-tocopherol acetate	16.971	0.0184	2.6740	n.s.	↑	n.s.
Ursolic acid	1.5638	0.0385	1.1528	n.s.	↑	n.s.

The values of Fold Change, *p*-value, VIP are all from the comparison between CA group and DSS group. ↑ represents that the metabolite is up-regulated in the comparison of the two groups. ↓ represents that the metabolite is down-regulated in the comparison of the two groups. n.s. represents no significant difference between the two groups.

2.6. Correlation Analysis between Gut Microbiota, Fecal Differential Metabolites, and Host Phenotypes

Based on the aforementioned findings, it is evident that both DSS exposure and CA intervention induced changes in the abundance of certain intestinal microbiota species and the composition of fecal metabolites in mice. To explore the interplay between the gut microbiota, host metabolism, and different treatments, Spearman correlation analysis was performed on fecal metabolites, gut microbiota, and host phenotypic characteristics. The results unveiled significant correlations between the gut microbiota and host phenotypes. At the phylum level, Cyanobacteria, Planctomycetes, and Proteobacteria exhibited positive correlations with DAI and TNF-α levels. Tenericutes and unclassified bacteria were positively correlated with DAI. Bacteroidetes, Planctomycetes, and Proteobacteria demonstrated positive correlations with IL-6 levels. Planctomycetes, Tenericutes, and unclassified bacteria displayed negative correlations with colon length. Patescibacteria were positively associated with weight loss, while Tenericutes were negatively associated with weight loss (Figure 4a). At the genus level, *Bacteroides* and unclassified *Xanthobacteraceae* showed positive correlations with DAI, IL-6, and TNF-α levels, but exhibited negative correlations with colon length and weight loss. *Marvinbryantia, Roseburia, [Eubacterium] nodatum group, [Eubacterium] xylanophilum group, Anaerotruncus*, and unclassified *Lachnospiraceae* were negatively correlated with DAI and IL-6, while they were positively correlated with colon length. *Lachnospiraceae NK4A136 group, Roseiarcus*, and unclassified *Chloroplast* demonstrated positive correlations with DAI and TNF-α. *Erysipelatoclostridium, Rumiclostridium 6*, and *Ruminococcaceae UCG-010* were positively correlated with DAI, whereas *Tyzzerella* exhibited a negative correlation with DAI. *Lachnospiraceae NK4A136 group, Roseiarcus*, and *Rumiclostridium 6* were inversely associated with colon length and weight loss. *Marvinbryantia, Roseburia*, unclassified *Lachnospiraceae*, and *[Eubacterium] nodatum group* were positively correlated with weight loss, while *Ruminococcaceae UCG-010* displayed a positive correlation with IL-6 but a negative correlation with weight loss (Figure 4b). Importantly, the results also indicated that changes in fecal metabolites induced by DSS exposure and CA intervention are closely linked to gut microbiota. Cyanobacteria, Planctomycetes, Proteobacteria, and unclassified bacteria showed positive correlations with asymmetric dimethylarginine, while Bacteroidetes, Planctomycetes, and unclassified bacteria exhibited positive correlations with N-glycolylneuraminic acid. Bacteroidetes, Cyanobacteria, Planctomycetes, and unclassified bacteria displayed negative correlations with thiamine and 4-methyl-5-thiazoleethanol. In addition, Planctomycetes exhibited a positive correlation with N-acetylneuraminic acid and a negative correlation with lithocholic acid. Both Bacteroidetes and Tenericutes showed negative correlations with phloroglucinol (Figure 4c).

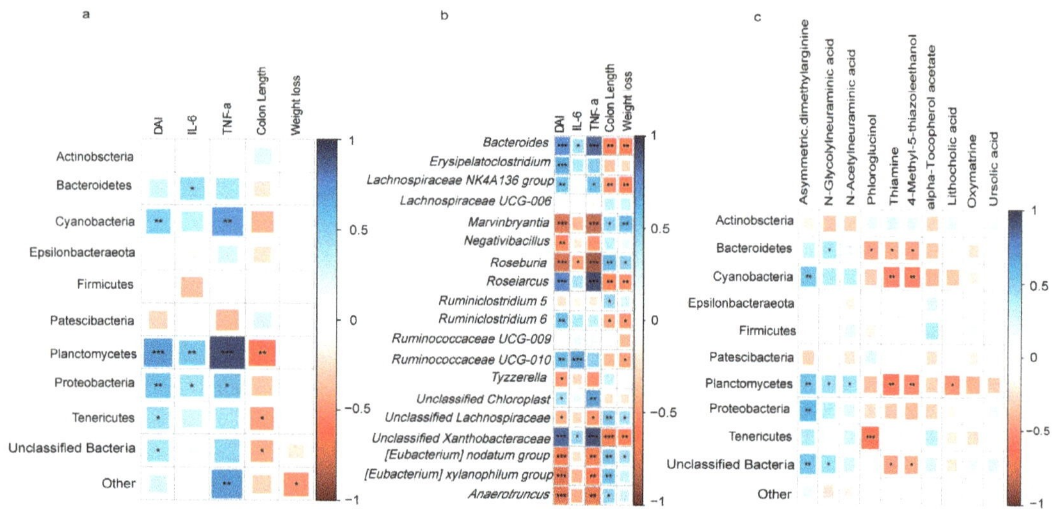

Figure 4. The heatmap of Spearman's correlation analysis between three groups. (**a**) Correlation among gut microbiota and host phenotypes at the phylum level. (**b**) Correlation among gut microbiota and host phenotypes at the genus level. (**c**) Correlation among candidate metabolites and gut microbiota. * $p < 0.05$; ** $p < 0.01$; *** $p < 0.001$.

3. Discussion

In our study, CA treatment demonstrated significant benefits in mitigating the effects of DSS-induced colitis. It effectively prevented weight loss and colon shortening, and it reduced the disease activity index (DAI) and histopathological scores that had been elevated due to DSS treatment. Additionally, CA supplementation led to the suppression of pro-inflammatory factors such as TNF-α and IL-6. Taken together, these findings provide compelling evidence of CA's effectiveness in ameliorating UC. Furthermore, our research results strongly suggest that CA exerts its anti-colitis effects through the regulation of gut microbiota and metabolites.

CA has garnered considerable attention for its well-documented anti-inflammatory properties. It has been reported that CA mitigated LPS-induced neuroinflammation, memory impairment, and amyloid production via inhibition of the NF-κB transcription pathway [37]. Landmann et al. reported that CA pretreatment significantly inhibited the mRNA expression of inducible nitric oxidesynthase (iNOS) and TNF-α in LPS-treated RAW264.7 macrophages. In the model of acute alcohol-induced hepatic steatosis, CA has demonstrated its ability to alleviate the condition in female C57BL/6J mice. This effect is achieved through the inhibition of oxidative stress and the exertion of anti-inflammatory effects [38]. *Echinacea purpurea* L. CA extract significantly reduced collagen-induced paw swelling in rats, which exerted anti-arthritic effects by reducing the levels of TNF-α, IL-1β, and PGE-2 in rat serum, and by reducing the levels of NF-κB, TNF-α, and PGE-2 in synovium tissues of the ankle joint [33]. In line with previous findings [33,37,38], our current research results demonstrate that CA treatment effectively improves colitis, reaffirming its well-established anti-inflammatory properties.

It is widely recognized that the gut harbors a vast number of bacteria that play pivotal roles in nutrition, energy metabolism, host defense, and immune system development [39]. Normally, a greater richness and diversity in the microbiota throughout life render the organism more resilient to external environmental challenges [40]. Indeed, alterations in the quantity and quality of the gut microbiota can result in dysbiosis, leading to intestinal inflammation [40]. Our results demonstrate that the mouse gut microbiota is altered following DSS exposure, revealing a reduction in diversity, which is consistent with previous reports of reduced gut microbiota diversity in IBD patients [41]. However, CA supplementation did not reverse this reduction in diversity, suggesting that CA may not exert its anti-colitis effect by increasing the overall diversity of the gut microbiota. In the current study, at the phylum level, the DSS group exhibited a significant increase in the abundance of Planctomycetes and Proteobacteria compared to the control group. The proliferation of Proteobacteria in the intestine signifies dysbiosis or an unstable microbial community structure and contributes to intestinal inflammation. Moreover, susceptibility to colitis positively correlates with the relative abundance of intestinal Proteobacteria [42]. Importantly, a study collecting biopsy samples from six UC patients through colonoscopy and utilizing Illumina high-throughput sequencing analysis has demonstrated that Proteobacteria significantly increases in UC patients during the severe inflammation stage compared to mild or moderate inflammation stages [43]. In the 1-methyl-4-phenyl-1,2,3,6-tetrahydropyridine-induced Parkinson's disease mouse model, CA significantly reduced the abundance of phylum Bacteroidota and genus *Parabacteroides*, as well as increased the abundance of phylum Firmicutes and genera *Lactobacillus* and *Ruminiclostridium* [44]. In a report exploring the effects of CA on non-alcoholic fatty liver disease (NAFLD) in C57BL/6 mice, CA was found to reduce Firmicutes levels and increase Bacteroidetes at the phylum level, while at the genus level, the levels of *Lactobacillus*, *Turicibacter*, and *Candidatus Saccharimonas* were reduced and the levels of *Ruminococcaceae UCG-014* and *Alloprevotella* were increased after CA treatment [45]. In our study, CA supplementation reduced the abundance of Bacteroidetes, which is consistent with previous reports. Interestingly, while CA has been shown in previous studies to have various effects on gut microbiota, our results indicate that CA supplementation reduced the abundance of Cyanobacteria at the phylum level and decreased the abundance of specific genera, such as *Bacteroides*, *Roseiarcus*, unclassified *Xanthobacteraceae*, and *Ruminococcaceae UCG-009*, while it increased the abundance of *Ruminiclostridium 5*, *Lachnospiraceae UCG-006*, and unclassified *Lachnospiraceae* at the genus level. These findings highlight the complex and context-dependent effects of CA on gut microbiota and suggest a unique impact of CA in the context of colitis. The observed inconsistencies in the effects of CA on gut microbiota could indeed be attributed to various factors, including differences in animal models, CA treatment concentrations, and treatment durations. It is reported that Cyanobacteria have been associated with potential neurotoxic or pro-inflammatory activity. A randomized controlled trial analyzed the fecal microbial profile of 50 amyotrophic lateral sclerosis (ALS) patients treated with probiotics and 50 healthy individuals treated with a placebo for 6 months. This study revealed a higher abundance of Cyanobacteria in ALS patients compared to the healthy individuals [46]. Furthermore, an analysis of microbial composition in the colorectal mucosa of 20 healthy volunteers and 31 patients with colorectal adenomatous polyps revealed a significantly higher abundance of Cyanobacteria in patients with adenomas [47]. Moreover, a clinical study collected fecal specimens from 18 infants with rotavirus (G9P8)-induced acute gastroenteritis (mean age 11.8 months) and 24 infants with human norovirus infections (GII)-induced acute gastroenteritis (mean age 8.8 months). This study demonstrated a higher abundance of Cyanobacteria in infants with viral diarrhea compared to healthy individuals [48]. In fecal samples from 38 children aged 7–17 years with pediatric Roman type III IBD who had previously completed a double-blind, randomized, placebo-controlled crossover trial (fructose versus maltodextrin), an elevated abundance of Cyanobacteria was observed after fructan ingestion compared to fructan-tolerant children [49]. These studies suggest a potential involvement of Cyanobacteria in the development of gastrointestinal and inflammatory diseases, though further research is needed to determine whether intestinal Cyanobacteria can cause diseases or whether changes in their abundance are a consequence of disease. In vivo studies have demonstrated a decrease in the abundance of *Rumiclostridium 5* in the feces of rats with necrotizing pancreatitis [50]. In a study that enrolled 13 patients with multiple kidney stones and 13 matched healthy controls, a diminished abundance of *Rumiclostridium 5* was observed in patients with kidney stones [51]. These studies suggest that a decreased abundance of *Rumiclostridium 5* could potentially contribute to the development of certain diseases. Overall, our current findings suggest that CA supplementation is beneficial in promoting the health of the gut microbial ecology to some extent.

Mouse feces encompass a diverse array of metabolites that can potentially serve as reflections of both the gut microbiota composition and the outcomes of nutrient uptake, digestion, and absorption within the gastrointestinal tract. The analysis of these fecal metabolites offers a window into the repercussions of interactions between the host and gut microbiota [52]. In our current investigation, we observed a noteworthy alteration in the fecal metabolic profile of mice following exposure to DSS and subsequent CA intervention. Our findings revealed a significant increase in the levels of asymmetric dimethylarginine, N-glycolneuraminic acid, and N-acetylneuraminic acid within the DSS-exposed group. These metabolites exhibited a significant down-regulation following CA supplementation. Notably, asymmetric dimethylarginine is a prominent inhibitor of endogenous nitric oxide synthase. Elevated levels of asymmetric dimethylarginine have been associated with oxidative stress and endothelial dysfunction [53,54]. Furthermore, a study gathered serum samples from 63 patients with colon cancer and 61 healthy individuals at the Fudan University Cancer Center. This study revealed heightened serum levels of asymmetric dimethylarginine in individuals with colon cancer [55]. Similarly, our study detected an elevation in asymmetric dimethylarginine levels in the feces of mice with colitis. On the other hand, N-glycolylneuraminic acid is a non-human glycan, and several studies have reported that the ingestion of N-glycolylneuraminic acid from red meat can trigger a heterophil-antibody-mediated inflammatory response, thus fostering the development of colitis and potentially colorectal cancer [56]. Furthermore, prior in vitro research has uncovered that N-glycolylneuraminic acid can promote the proliferation of colorectal cancer cells [56]. N-Acetylneuraminic acid, another non-human sialic acid, has been documented to induce cancer and inflammation [57].

Following exposure to DSS, we observed significant up-regulation of phloroglucinol, thiamine, 4-methyl-5-thiazoleethanol, lithocholic acid, and oxymatrine. However, upon CA intervention, these metabolites exhibited significant reductions. Phloroglucinol, categorized under phenols and derivatives, has demonstrated the ability to inhibit oxidative stress and decrease the expression levels of TNF-α, IL-1β, IL-6, and PGE2 in lipopolysaccharide-stimulated RAW264.7 cells [58]. In general, nutritional deficiencies in individuals with IBD stem from factors such as decreased appetite, malabsorption, and systemic inflammation triggered by active disease [59]. Notably, thiamine deficiency is relatively prevalent in individuals with IBD and can lead to symptoms like nausea, vomiting, and abdominal pain [60]. Furthermore, 4-methyl-5-thiazoleethanol, a natural sulfur-containing flavor compound, is a degradation product of thiamine [61]. In our study, we observed a similar pattern of change for 4-methyl-5-thiazoleethanol and thiamine. It is worth noting that lithocholic acid is a secondary bile acid produced only by microbiota, such as *Lachnospiraceae* and *Ruminococcaceae families*, which undergo a 7α-dehydroxylation reaction to generate lithocholic acid. These bile acids are synthesized in the colon and subsequently transported into the bloodstream [62–64]. More importantly, lithocholic acid, as a metabolite of cholesterol conversion, has been shown to promote the regeneration of the intestinal epithelium via activation of the Takeda G protein-coupled receptor 5 (TGR5) in intestinal stem cells to combat epithelial damage and subsequent barrier disruption [65]. Oxymatrine, classified as an alkaloid, has exhibited anti-inflammatory properties both in vitro and in vivo [66]. Additionally, it has been reported that oxymatrine ameliorated weight loss and histological damage induced by DSS in C57BL/6 wild-type mice [67].

Moreover, our findings revealed notable correlations between intestinal bacterial strains and fecal metabolites. Specifically, Bacteroidetes exhibited a significant positive correlation with N-glycolylneuraminic acid and a significant negative correlation with thiamine, 4-methyl-5-thiazoleethanol, and phloroglucinol. Cyanobacteria demonstrated a significant positive correlation with asymmetric dimethylarginine and a significant negative correlation with thiamine and 4-methyl-5-thiazoleethanol. Furthermore, Proteobacteria displayed a significant positive correlation with asymmetric dimethylarginine. Altogether, these comprehensive results collectively suggest that CA exerts its therapeutic effect in alleviating colitis by modulating the levels of key metabolites. CA appears to down-regulate the levels of asymmetric dimethylarginine, N-glycolylneuraminic acid, and N-acetylneuraminic acid, while simultaneously up-regulating the levels of phloroglucinol, thiamine, 4-methyl-5-thiazoleethanol, lithocholic acid, and oxymatrine.

Nevertheless, this study has certain limitations. While animal models can replicate human diseases, disparities in physiology and immune systems between animals and humans may exist. Therefore, crucial confirmation of the therapeutic efficacy of CA in human colitis necessitates future clinical trials. While we observed a correlation between microbiota and metabolites in mice, the causal relationship between these factors remains unclear. Therefore, further investigations are warranted to elucidate the intricate relationship between colonic microorganisms and metabolites. Such elucidation would provide deeper insights into the mechanisms underlying the anti-colitis effects of CA. Furthermore, in our experiments, DSS-induced enteritis was conducted for a duration of only one week. This limited timeframe may not comprehensively capture all stages of colitis onset and progression. Extending the experimental duration could contribute to a more thorough understanding of the disease evolution and therapeutic effects. Importantly, although CA may confer benefits in colitis, its dosage and safety require careful consideration. Varied dosages and treatment durations may yield divergent effects and potentially induce adverse reactions. Consequently, future research on the safety profile of CA is imperative.

4. Materials and Methods

4.1. Materials

Chicoric acid (≥98%) was procured from Chengdu Pufei De Biotech Co., Ltd. (Chengdu, China). Dextran sulfate sodium salt (DSS) with a molecular weight range of 36,000–50,000 was supplied by MP Biomedical (Irvine, CA, USA). Sodium carboxymethyl cellulose (CMC) was sourced from Sinopharm Chemical Reagent Co., Ltd. (Shanghai, China). The urine fecal occult blood test kit was obtained from Nanjing Jiancheng Bioengineering Institute (Nanjing, China). Enzyme-linked immunosorbent assay (ELISA) kits were purchased from Jiubang Biotechnology Co., Ltd. (Quanzhou, China).

4.2. Animals and Experimental Design

Seven-week-old male BALB/c mice purchased from Cavens Experimental Animal Co., Ltd. (Changzhou, China) were housed in the Experimental Animal Center of Jiangsu University under specific pathogen-free (SPF) conditions. Mice were kept in a room with a 12 h light–dark cycle at constant temperature (20–25 °C) and humidity (40–60%) with free access to water and food. After one week of adaptation, the mice were randomly assigned to three groups: control (CON) group, dextran sulfate sodium salt (DSS) treatment group, and chicoric acid (CA) treatment group. The CA group received oral administration of chicoric acid (50 mg/kg) once daily for 14 days, while the CON and DSS groups received an equivalent 0.5% CMC solution. Seven days after the initiation of oral pre-treatment, except for the CON group, which had unrestricted access to sterile water, the other two groups were given 2.5% DSS solution to induce colitis for the final 7 days of the experiment. Importantly, the DSS solution was freshly prepared every two days. Throughout this study, the mice were monitored daily, and their body weights were recorded. All animal experiments in this study were conducted in accordance with the Guide for the Care and Use of Laboratory Animals of Jiangsu University.

After the experiment, the mice were euthanized following standard protocols. Serum, colon tissue, and fecal samples were collected and stored in a −80 °C freezer until further analysis. A portion of the colon segment was also preserved in a 4% paraformaldehyde fixative solution for subsequent investigations.

4.3. Disease Activity Index (DAI) Assessment

The disease activity index (DAI) score was employed to assess the severity of colitis, primarily encompassing parameters such as weight loss, stool consistency, and intestinal bleeding. Weight loss was calculated as the percentage difference between initial body weight before modeling and body weight on any day after modeling (0: 0–1%; 1: 1–5%; 2: 5–10%; 3: 10–15%; 4: 15% or more). The consistency of stool was visually evaluated and classified based on the following criteria (0: normal feces; 1: soft but formed stool; 2: very soft and shapeless stool; 3: semi-diarrhea; 4: full diarrhea). Intestinal bleeding was detected with urine fecal occult blood test kit (0: the color of the stool does not change; 1: turn blue-green in 30 s; 2: turn blue-green in 10 s; 3: instantly turn blue-green; 4: instantly turn blue) [68]. The DAI score was calculated by summing the individual scores for these three parameters.

4.4. Colonic Hematoxylin and Eosin (H&E) Staining and Histopathological Analysis

The mouse colon was isolated and photographed, and its length was measured. A segment of the colon, obtained from the same region, was fixed in a 4% paraformaldehyde fixative solution. Following paraffin embedding, the colon tissue was sectioned into 5 μm-thin slices and stained with hematoxylin and eosin. Visualization was carried out using a Leica Microsystems microscope (Leica DM6000 B) for subsequent pathological analysis. Furthermore, histological score refers to the following criteria based on the severity of inflammatory cell infiltration, extent of damage, and degree of crypt damage. The severity of inflammatory cell infiltration was graded as follows: 0 for absent, 1 for slight and scattered cell infiltration, 2 for moderate cell infiltration with visible cell clusters, and 3 for extensive inflammatory cell infiltration resulting in loss of tissue structure. The extent of damage was scored as 0 for no damage, 1 for mucosal damage, 2 for mucosal and submucosal damage, and 3 for transmural damage. The degree of crypt damage was categorized as 0 for intact crypts, 1 for basal one-third damage, 2 for basal two-thirds damage, 3 for only superficial epithelial cells remaining intact, and 4 for complete loss of crypt and epithelial cells. The histological score was the sum of these three individual scores [69].

4.5. Inflammatory Cytokine Assay

The serum samples from the mice were retrieved from the −80 °C storage and allowed to thaw. Levels of TNF-α and IL-6 in mouse serum were assessed using an ELISA kit. In this assay, serum samples and horseradish peroxidase (HRP)-labeled detection antibodies were successively introduced into microwells that had been pre-coated with TNF-α or IL-6 capture antibodies. The mixture was incubated at 37 °C and subsequently subjected to thorough washing. Color development was initiated using the substrate 3,3′,5,5′-Tetramethylbenzidine (TMB), which, under the catalytic action of peroxidase, converted into a blue hue. Eventually, it transitioned into a final yellow color upon exposure to an acid. The absorbance at 450 nm was quantified using a microplate reader (Tecan Infinite® 200 PRO) within 15 min after the addition of the stop solution. Subsequently, the concentration of each sample was determined.

4.6. 16S rRNA DNA Sequencing

Feces samples of mice in each group were selected to extract deoxyribonucleic acid (DNA), and qualified genomic DNA samples were taken to perform polymerase chain reaction (PCR) amplification on the V3-V4 region (338F ACTCCTACGGGAGGCAGCAG, 806R GGACTACHVGGGTWTCTAAT) of the bacteria. Additionally, the PCR amplification products were purified using Agencourt AMPure XP magnetic beads and were then eluted in Elution Buffer to prepare the library. The fragment size distribution and concentration of the libraries were assessed using an Agilent 2100 Bioanalyzer (equipped with Agilent 2100 Expert software version B.02.08) Qualified libraries, based on insert size, were subsequently subjected to sequencing on the HiSeq platform. The raw 16S rRNA sequences were imported into Quantitative Insights Into Microbial Ecology version 2 (QIIME2) software for demultiplexing and quality filtering. Divisive Amplicon Denoising Algorithm (DADA2) was employed for denoising sequences and for the construction of a feature table. A custom Naïve Bayesian classifier was developed for taxonomic classification against the SILVA database release 132, utilizing the amplified 16S region. Subsequently, sequences were clustered into Operational Taxonomic Units (OTUs) based on their similarity. A sampling depth, defined as the number of sequences in the sample with the fewest sequences, was employed to generate Observed OTUs as an α diversity measure and Sim distances as a β diversity measure. The Kruskal–Wallis test was employed to assess differences in α diversity. Differences in β diversity were assessed using permutation-based analysis of variance, followed by Benjamini–Hochberg (BH) correction. Microbiome composition data were obtained by normalizing to the total number of reads in each sample (relative abundance). Analysis of compositions of microbiomes with bias correction was utilized for the differential abundance analysis at the phylum and genus levels, with significance set at $p < 0.05$. To determine the significantly important microbial taxa, the linear discriminant analysis effect size (LEfSe) analysis was carried out as previously reported [70].

4.7. Untargeted Metabolomics Analysis

Twenty-five milligrams of thawed fecal samples were placed in a 1.5 mL Eppendorf tube and mixed with 800 uL of pre-cooled extracting solution (−20 °C) composed of methanol/acetonitrile/water in a 2:2:1 (*v*/*v*/*v*) ratio, along with 10 uL of an internal standard. The mixture was then ground for 5 min. Afterward, the samples were sonicated for 10 min at 4 °C and allowed to stand at 20 °C for 1 h. Subsequently, the mixtures were centrifuged at 25,000× *g* for 15 min at 4 °C. The supernatant was collected, freeze-dried, and then redissolved in 600 uL of a mixed solution (methanol/water = 1:9, *v*/*v*). Following thorough mixing, the solution was centrifuged at 25,000× *g* for 15 min at 4 °C, and the resulting supernatant was retained.

The supernatant from each sample was collected and combined to create a quality control (QC) sample. Metabolite separation and detection were performed using a Waters UPLC I-Class Plus system (Waters, Milford, MA, USA) coupled to a Q Exactive high-resolution mass spectrometer (Thermo Fisher Scientific, Waltham, MA, USA). A BEH C18 column (1.7 μm, 2.1 × 100 mm, Waters, USA) was employed in this experiment. For the positive ion mode mobile phase, liquid A consisted of an aqueous solution with 0.1% formic acid, while liquid B was composed of methanol containing 0.1% formic acid. In contrast, for the negative ion mode mobile phase, liquid A contained an aqueous solution with 10 mM ammonium formate, and liquid B consisted of a solution containing 10 mM ammonium formate in 95% methanol. Gradient elution was conducted according to the following conditions: 0 to 1 min, 2% B solution; 1 to 9 min, a gradient from 2% to 98% B solution; 9 to 12 min, 98% B solution; 12 to 12.1 min, a transition from 98% B solution to 2% B solution; and 12.1 to 15 min, 2% B solution. The flow rate was set at 0.35 mL/min, the column temperature was maintained at 45 °C, and the injection volume was 5 μL.

The acquisition of primary and secondary mass spectrometry data was conducted using a Q Exactive mass spectrometer. The primary resolution was set at 70,000, with an automatic gain control (AGC) target of 3×10^6 and a maximum injection time of 100 ms. Fragmentation for secondary mass spectrometry was based on precursor ion intensity, selecting the top 3 for fragmentation. The secondary resolution differed from the primary, being set at 17,500, with an AGC target of 1×10^6 and a maximum injection time of 50 ms. Fragmentation energy levels were configured at 20, 40, and 60 eV. Additionally, the sheath gas had a flow rate of 40, and the auxiliary gas was set to a flow rate of 10. In the positive ion mode, the spray voltage was 3.80 |KV|, whereas in the negative ion mode, it was 3.20 |KV|. The ion transfer tube temperature was adjusted to 320 °C, and the temperature of the auxiliary gas heating was set to 350 °C.

The raw data were imported into Compound Discoverer 3.3 software, provided by Thermo Fisher Scientific, USA. Concurrently, mass spectrometry data analysis was carried out using the BMDB (BGI Metabolome Database), mzCloud database, and ChemSpider online database. In the R 4.0.2 software environment, the MetaX package was employed to conduct principal component analysis (PCA) and partial least squares discriminant analysis (PLS-DA) using the data derived from Compound Discoverer 3.3 software. The Variable Importance in Projection (VIP) obtained from the PLS-DA model, the Fold Change calculated as the ratio of the mean expression of each metabolite between two groups, and the P-value obtained from Student's T-test were utilized to assess statistical significance. The criteria for identifying differential metabolites were as follows: (1) VIP $\geq$ 1, (2) Fold Change $\geq$ 1.2 or $\leq$0.83, (3) p-value < 0.05 [71,72]. Enrichment analysis of differential metabolites was conducted based on the KEGG database (https://www.metaboanalyst.ca/MetaboAnalyst/ModuleView.xhtml (accessed on 28 December, 2023)).

4.8. Statistical Analysis

The results are presented as mean ± SEM (standard error of the mean), and data analysis was performed using GraphPad Prism 9.0 software. Statistical comparisons among the three groups were conducted using one-way ANOVA, with a p-value of <0.05 considered statistically significant. Phenotypic data, 16s rRNA sequencing results, and the metabolite dataset in each group were used for correlation analysis. The Spearman correlation coefficients were calculated between the data. Spearman correlation analysis was conducted using the R 4.0.2 software with the Corrplot and Psych package.

5. Conclusions

To conclude, this study establishes that the targeted administration of CA can alleviate DSS-induced colitis in mice. The anti-colitis effect of CA is, at the very least, partially mediated through the modulation of gut microbiota and metabolic processes. The structure of the gut microbiota was restored after CA supplementation, including reductions in levels of Bacteroidetes and Cyanobacteria at the phylum level and *Bacteroides*, *Roseiarcus*, and unclassified *Xanthobacteraceae* at the genus level, while the abundance of unclassified *Lachnospiraceae* at the genus level was increased. In addition, the metabolome results showed that CA prevents colitis by down-regulating asymmetric dimethylarginine, N-glycolylneuraminic acid, and N-acetylneuraminic acid, and up-regulating phloroglucinol, thiamine, 4-methyl-5-thiazoleethanol, lithocholic acid, and oxymatrine. In essence, this current study provides a critical foundation for considering CA as a promising functional food ingredient for combating colitis.

Supplementary Materials: The following supporting information can be downloaded at https://www.mdpi.com/article/10.3390/ijms25020841/s1.

Author Contributions: Methodology, J.Y. and T.G.; software, J.L.; formal analysis, J.Y.; investigation, J.Y., J.L. and T.G.; resources, Y.P.; data curation, J.Y.; writing—original draft, J.Y.; writing—review and editing, Q.S., W.X. and Y.P.; visualization, J.L. All authors have read and agreed to the published version of the manuscript.

Funding: This research received no external funding.

Institutional Review Board Statement: The animal study protocol was approved by the Animal Research Ethics Committee of Jiangsu University (UJS-IACUC-2022102802).

Informed Consent Statement: Not applicable.

Data Availability Statement: Data are contained within the article.

Acknowledgments: We thank Will Mumby (Department of Nutrition and Integrative Physiology, Florida State University, Tallahassee, FL 32306, USA) for critical revision. We also extend our heartfelt gratitude to Renhua Song (Epigenetics and RNA Biology Laboratory, The School of Medical Sciences, Faculty of Medicine and Health, The University of Sydney, Camperdown, NSW, 2006, Australia) for meticulously analyzing the data.

Conflicts of Interest: The authors affirm that the research was carried out without any commercial or financial associations that could be perceived as a potential conflict of interest.

Abbreviations

CA	Chicoric acid
DSS	Dextran sulfate sodium
DAI	Disease activity index
TNF-α	Tumor necrosis factor-α
IL-6	Interleukin- 6
IBD	Inflammatory bowel disease
UC	Ulcerative colitis
CD	Crohn's disease
SCFAs	Short-chain fatty acids
NF-κB p65	Nuclear factor kappa-B p65
NF-κB	Nuclear factor kappa-B
LRR	Leucine-rich repeat
PYD	Pyrin domain
LPS	Lipopolysaccharide
IL-1β	Interleukin -1β
PGE-2	Prostaglandin E2
CMC	Sodium carboxymethyl cellulose
ELISA	Enzyme-linked immunosorbent assay
SPF	Specific pathogen-free
CON	Control
HRP	Horseradish peroxidase
TMB	3,3′,5,5′-Tetramethylbenzidine
H&E	Colonic hematoxylin and eosin
QC	Quality control
AGC	Automatic gain control
DNA	Deoxyribonucleic acid
PCR	Polymerase chain reaction
QIIME2	Quantitative Insights Into Microbial Ecology version 2
PCoA	Principal coordinate analysis

NMDS	Non-metric multidimensional scaling analysis
DADA2	Divisive Amplicon Denoising Algorithm
LC-MS	Liquid chromatograph mass spectrometer
PCA	Principal component analysis
LEfSe	Linear discriminant analysis effective size
PLS-DA	Partial least squares discriminant analysis
VIP	Variable Importance in Projection
DA score	Differential abundance score
OTUs	Operational Taxonomic Units
iNOS	Inducible nitric oxidesynthase
ALS	Amyotrophic lateral sclerosis
NAFLD	Non-alcoholic fatty liver disease
TGR5	Takeda G protein-coupled receptor 5

References

1. Andlujar, I.; Recio, M.C.; Giner, R.M.; Cienfuegos-Jovellanos, E.; Laghi, S.; Muguerza, B.; Rios, J.L. Inhibition of Ulcerative Colitis in Mice after Oral Administration of a Polyphenol-Enriched Cocoa Extract Is Mediated by the Inhibition of STAT1 and STAT3 Phosphorylation in Colon Cells. *J. Agric. Food Chem.* **2011**, *59*, 6474–6483. [CrossRef] [PubMed]
2. Bahrami, G.; Malekshahi, H.; Miraghaee, S.; Madani, H.; Babaei, A. Improving Animal Model of Induced Colitis by Acetic Acid in Terms of Fibrosis and Inflammation Incidence in the Colon. *J. Investig. Surg.* **2022**, *35*, 214–222. [CrossRef] [PubMed]
3. Parada Venegas, D.; De la Fuente, M.K.; Landskron, G.; Julieta Gonzalez, M.; Quera, R.; Dijkstra, G.; Harmsen, H.J.M.; Faber, K.N.; Hermoso, M.A. Short Chain Fatty Acids (SCFAs)-Mediated Gut Epithelial and Immune Regulation and Its Relevance for Inflammatory Bowel Diseases. *Front. Immunol.* **2019**, *10*, 277; Erratum in *Front. Immunol.* **2019**, *10*, 1486. [CrossRef] [PubMed]
4. Low, D.; Nguyen, D.D.; Mizoguchi, E. Animal models of ulcerative colitis and their application in drug research. *Drug Des. Dev. Ther.* **2013**, *7*, 1341–1356. [CrossRef]
5. Isaacs, K.L.; Lewis, J.D.; Sandborn, W.J.; Sands, B.E.; Targan, S.R. State of the art: IBD therapy and clinical trials in IBD. *Inflamm. Bowel Dis.* **2005**, *11*, S3–S12. [CrossRef] [PubMed]
6. Taylor, K.M.; Irving, P.M. Optimization of conventional therapy in patients with IBD. *Nat. Rev. Gastroenterol. Hepatol.* **2011**, *8*, 646–656. [CrossRef]
7. Nidhi; Rashid, M.; Kaur, V.; Hallan, S.S.; Sharma, S.; Mishra, N. Microparticles as controlled drug delivery carrier for the treatment of ulcerative colitis: A brief review. *Saudi Pharm. J.* **2016**, *24*, 458–472. [CrossRef]
8. Cao, S.-Y.; Ye, S.-J.; Wang, W.-W.; Wang, B.; Zhang, T.; Pu, Y.-Q. Progress in active compounds effective on ulcerative colitis from Chinese medicines. *Chin. J. Nat. Med.* **2019**, *17*, 81–102. [CrossRef]
9. Li, S.; Wu, B.; Fu, W.; Reddivari, L. The Anti-inflammatory Effects of Dietary Anthocyanins against Ulcerative Colitis. *Int. J. Mol. Sci.* **2019**, *20*, 2588. [CrossRef]
10. Lozupone, C.A.; Stombaugh, J.I.; Gordon, J.I.; Jansson, J.K.; Knight, R. Diversity, stability and resilience of the human gut microbiota. *Nature* **2012**, *489*, 220–230. [CrossRef]
11. Sartor, R.B. Microbial influences in inflammatory bowel diseases. *Gastroenterology* **2008**, *134*, 577–594. [CrossRef] [PubMed]
12. Lee, D.; Albenberg, L.; Compher, C.; Baldassano, R.; Piccoli, D.; Lewis, J.D.; Wu, G.D. Diet in the Pathogenesis and Treatment of Inflammatory Bowel Diseases. *Gastroenterology* **2015**, *148*, 1087–1106. [CrossRef] [PubMed]
13. Ruiz, D.F.Z.; Kim, D.V.; Norwood, K.; Saldana-Morales, F.B.; Kim, M.; Ng, C.; Callaghan, R.; Uddin, M.; Chang, L.C.; Longman, R.S.; et al. Microbiota manipulation to increase macrophage IL-10 improves colitis and limits colitis-associated colorectal cancer. *Gut Microbes* **2022**, *14*, 2119054. [CrossRef]
14. Askari, H.; Shojaei-Zarghani, S.; Raeis-Abdollahi, E.; Jahromi, H.K.; Abdullahi, P.R.; Daliri, K.; Tajbakhsh, A.; Rahmati, L.; Safarpour, A.R. The Role of Gut Microbiota in Inflammatory Bowel Disease-Current State of the Art. *Mini-Rev. Med. Chem.* **2023**, *23*, 1376–1389. [CrossRef] [PubMed]
15. Kim, Y.S.; Unno, T.; Kim, B.-Y.; Park, M.-S. Sex Differences in Gut Microbiota. *World J. Mens Health* **2020**, *38*, 48–60. [CrossRef] [PubMed]
16. Shin, J.-H.; Park, Y.-H.; Sim, M.; Kim, S.-A.; Joung, H.; Shin, D.-M. Serum level of sex steroid hormone is associated with diversity and profiles of human gut microbiome. *Res. Microbiol.* **2019**, *170*, 192–201. [CrossRef] [PubMed]
17. Lovell, R.M.; Ford, A.C. Effect of Gender on Prevalence of Irritable Bowel Syndrome in the Community: Systematic Review and Meta-Analysis. *Am. J. Gastroenterol.* **2012**, *107*, 991–1000. [CrossRef]
18. Feng, L.; Zhou, N.; Li, Z.; Fu, D.; Guo, Y.; Gao, X.; Liu, X. Co-occurrence of gut microbiota dysbiosis and bile acid metabolism alteration is associated with psychological disorders in Crohn's disease. *Faseb J.* **2022**, *36*, e22100. [CrossRef]
19. Shadnoush, M.; Hosseini, R.S.; Navai, L.; Goudarzi, H.; Vaezjalali, M. Effects of Probiotics on Gut Microbiota in Patients with Inflammatory Bowel Disease: A Double-blind, Placebo-controlled Clinical Trial. *Korean J. Gastroenterol.* **2015**, *65*, 215–221. [CrossRef]

20. Leccese, G.; Bibi, A.; Mazza, S.; Facciotti, F.; Caprioli, F.; Landini, P.; Paroni, M. Probiotic Lactobacillus and Bifidobacterium Strains Counteract Adherent-Invasive *Escherichia coli* (AIEC) Virulence and Hamper IL-23/Th17 Axis in Ulcerative Colitis, but Not in Crohn's Disease. *Cells* **2020**, *9*, 1824. [CrossRef]
21. Nie, K.; Ma, K.; Luo, W.; Shen, Z.; Yang, Z.; Xiao, M.; Tong, T.; Yang, Y.; Wang, X. Roseburia intestinalis: A Beneficial Gut Organism From the Discoveries in Genus and Species. *Front. Cell. Infect. Microbiol.* **2021**, *11*, 757718. [CrossRef] [PubMed]
22. Collino, S.; Martin, F.P.J.; Rezzi, S. Clinical metabolomics paves the way towards future healthcare strategies. *Br. J. Clin. Pharmacol.* **2013**, *75*, 619–629. [CrossRef] [PubMed]
23. Dong, L.-N.; Wang, M.; Guo, J.; Wang, J.-P. Role of intestinal microbiota and metabolites in inflammatory bowel disease. *Chin. Med. J.* **2019**, *132*, 1610–1614. [CrossRef] [PubMed]
24. Nikolaus, S.; Schulte, B.; Al-Massad, N.; Thieme, F.; Schulte, D.M.; Bethge, J.; Rehman, A.; Tran, F.; Aden, K.; Haesler, R.; et al. Increased Tryptophan Metabolism Is Associated With Activity of Inflammatory Bowel Diseases. *Gastroenterology* **2017**, *153*, 1504–1516. [CrossRef] [PubMed]
25. Zhang, Z.; Zhang, H.; Chen, T.; Shi, L.; Wang, D.; Tang, D. Regulatory role of short-chain fatty acids in inflammatory bowel disease. *Cell Commun. Signal.* **2022**, *20*, 64. [CrossRef] [PubMed]
26. Facchin, S.; Vitulo, N.; Calgaro, M.; Buda, A.; Romualdi, C.; Pohl, D.; Perini, B.; Lorenzon, G.; Marinelli, C.; D'Inca, R.; et al. Microbiota changes induced by microencapsulated sodium butyrate in patients with inflammatory bowel disease. *Neurogastroenterol. Motil.* **2020**, *32*, e13914. [CrossRef] [PubMed]
27. Tedelind, S.; Westberg, F.; Kjerrulf, M.; Vidal, A. Anti-inflammatory properties of the short-chain fatty acids acetate and propionate: A study with relevance to inflammatory bowel disease. *World J. Gastroenterol.* **2007**, *13*, 2826–2832. [CrossRef]
28. Goncalves, P.; Araujo, J.R.; Di Santo, J.P. A Cross-Talk Between Microbiota-Derived Short-Chain Fatty Acids and the Host Mucosal Immune System Regulates Intestinal Homeostasis and Inflammatory Bowel Disease. *Inflamm. Bowel Dis.* **2018**, *24*, 558–572. [CrossRef]
29. Lee, J.; Scagel, C.F. Chicoric acid levels in commercial basil (*Ocimum basilicum*) and Echinacea purpurea products. *J. Funct. Foods* **2010**, *2*, 77–84. [CrossRef]
30. Lee, J.; Scagel, C.F. Chicoric acid: Chemistry, distribution, and production. *Front. Chem.* **2013**, *1*, 40. [CrossRef]
31. Lee, N.Y.; Chung, K.-S.; Jin, J.S.; Bang, K.S.; Eom, Y.-J.; Hong, C.-H.; Nugroho, A.; Park, H.-J.; An, H.-J. Effect of Chicoric Acid on Mast Cell-Mediated Allergic Inflammation in Vitro and in Vivo. *J. Nat. Prod.* **2015**, *78*, 2956–2962. [CrossRef] [PubMed]
32. Xue, Y.; Zhang, S.; Du, M.; Zhu, M.-J. Dandelion extract suppresses reactive oxidative species and inflammasome in intestinal epithelial cells. *J. Funct. Foods* **2017**, *29*, 10–18. [CrossRef]
33. Jiang, L.; Li, W.; Wang, Y.; Zhang, X.; Yu, D.; Yin, Y.; Xie, Z.; Yuan, Y. Effects of Cichoric Acid Extract from Echinacea purpurea on Collagen-Induced Arthritis in Rats. *American J. Chin. Med.* **2014**, *42*, 679–692. [CrossRef] [PubMed]
34. Chassaing, B.; Srinivasan, G.; Delgado, M.A.; Young, A.N.; Gewirtz, A.T.; Vijay-Kumar, M. Fecal Lipocalin 2, a Sensitive and Broadly Dynamic Non-Invasive Biomarker for Intestinal Inflammation. *PLoS ONE* **2012**, *7*, e44328. [CrossRef] [PubMed]
35. Klitgaard, M.; Kristensen, M.N.; Venkatasubramanian, R.; Guerra, P.; Jacobsen, J.; Berthelsen, R.; Rades, T.; Mullertz, A. Assessing acute colitis induced by dextran sulfate sodium in rats and its impact on gastrointestinal fluids. *Drug Deliv. Transl. Res.* **2023**, *13*, 1484–1499. [CrossRef] [PubMed]
36. Shi, N.; Li, N.; Duan, X.W.; Niu, H.T. Interaction between the gut microbiome and mucosal immune system. *Mil. Med. Res.* **2017**, *4*, 14. [CrossRef]
37. Liu, Q.; Chen, Y.; Shen, C.; Xiao, Y.; Wang, Y.; Liu, Z.; Liu, X. Chicoric acid supplementation prevents systemic inflammation-induced memory impairment and amyloidogenesis via inhibition of NF-kappa B. *Faseb J.* **2017**, *31*, 1494–1507. [CrossRef]
38. Landmann, M.; Kanuri, G.; Spruss, A.; Stahl, C.; Bergheim, I. Oral intake of chicoric acid reduces acute alcohol-induced hepatic steatosis in mice. *Nutrition* **2014**, *30*, 882–889. [CrossRef]
39. Michielan, A.; D'Inca, R. Intestinal Permeability in Inflammatory Bowel Disease: Pathogenesis, Clinical Evaluation, and Therapy of Leaky Gut. *Mediat. Inflamm.* **2015**, *2015*, 628157. [CrossRef]
40. Gancarcikova, S.; Lauko, S.; Hrckova, G.; Andrejcakova, Z.; Hajduckova, V.; Madar, M.; Fecskeova, L.K.; Mudronova, D.; Mravcova, K.; Strkolcova, G.; et al. Innovative Animal Model of DSS-Induced Ulcerative Colitis in Pseudo Germ-Free Mice. *Cells* **2020**, *9*, 2571. [CrossRef]
41. Sheehan, D.; Moran, C.; Shanahan, F. The microbiota in inflammatory bowel disease. *J. Gastroenterol.* **2015**, *50*, 495–507. [CrossRef] [PubMed]
42. Shin, N.-R.; Whon, T.W.; Bae, J.-W. Proteobacteria: Microbial signature of dysbiosis in gut microbiota. *Trends Biotechnol.* **2015**, *33*, 496–503. [CrossRef]
43. Walujkar, S.A.; Dhotre, D.P.; Marathe, N.P.; Lawate, P.S.; Bharadwaj, R.S.; Shouche, Y.S. Characterization of bacterial community shift in human Ulcerative Colitis patients revealed by Illumina based 16S rRNA gene amplicon sequencing. *Gut Pathog.* **2014**, *6*, 22. [CrossRef] [PubMed]
44. Wang, N.; Feng, B.-N.; Hu, B.; Cheng, Y.-L.; Guo, Y.-H.; Qian, H. Neuroprotection of chicoric acid in a mouse model of Parkinson's disease involves gut microbiota and TLR4 signaling pathway. *Food Funct.* **2022**, *13*, 2019–2032. [CrossRef] [PubMed]
45. Ding, X.Q.; Jian, T.Y.; Li, J.W.; Lv, H.; Tong, B.; Li, J.; Meng, X.H.; Ren, B.R.; Chen, J. Chicoric Acid Ameliorates Nonalcoholic Fatty Liver Disease via the AMPK/Nrf2/NF kappa B Signaling Pathway and Restores Gut Microbiota in High-Fat-Diet-Fed Mice. *Oxidative Med. Cell. Longev.* **2020**, *2020*, 9734560. [CrossRef] [PubMed]

46. Di Gioia, D.; Bozzi Cionci, N.; Baffoni, L.; Amoruso, A.; Pane, M.; Mogna, L.; Gaggia, F.; Lucenti, M.A.; Bersano, E.; Cantello, R.; et al. A prospective longitudinal study on the microbiota composition in amyotrophic lateral sclerosis. *Bmc Med.* **2020**, *18*, 153. [CrossRef]
47. Lu, Y.; Chen, J.; Zheng, J.; Hu, G.; Wang, J.; Huang, C.; Lou, L.; Wang, X.; Zeng, Y. Mucosal adherent bacterial dysbiosis in patients with colorectal adenomas. *Sci. Rep.* **2016**, *6*, 26337. [CrossRef] [PubMed]
48. Xiong, L.; Li, Y.; Li, J.; Yang, J.; Shang, L.; He, X.; Liu, L.; Luo, Y.; Xie, X. Intestinal microbiota profiles in infants with acute gastroenteritis caused by rotavirus and norovirus infection: A prospective cohort study. *Int. J. Infect. Dis.* **2021**, *111*, 76–84. [CrossRef]
49. Chumpitazi, B.P.; Hoffman, K.L.; Smith, D.P.; McMeans, A.R.; Musaad, S.; Versalovic, J.; Petrosino, J.F.; Shulman, R.J. Fructan-sensitive children with irritable bowel syndrome have distinct gut microbiome signatures. *Aliment. Pharmacol. Ther.* **2021**, *53*, 499–509. [CrossRef]
50. Huang, C.; Chen, J.; Wang, J.; Zhou, H.; Lu, Y.; Lou, L.; Zheng, J.; Tian, L.; Wang, X.; Cao, Z.; et al. Dysbiosis of Intestinal Microbiota and Decreased Antimicrobial Peptide Level in Paneth Cells during Hypertriglyceridemia-Related Acute Necrotizing Pancreatitis in Rats. *Front. Microbiol.* **2017**, *8*, 776. [CrossRef]
51. Tang, R.; Jiang, Y.; Tan, A.; Ye, J.; Xian, X.; Xie, Y.; Wang, Q.; Yao, Z.; Mo, Z. 16S rRNA gene sequencing reveals altered composition of gut microbiota in individuals with kidney stones. *Urolithiasis* **2018**, *46*, 503–514. [CrossRef] [PubMed]
52. Xu, J.; Zhang, Q.-F.; Zheng, J.; Yuan, B.-F.; Feng, Y.-Q. Mass spectrometry-based fecal metabolome analysis. *Trac-Trends Anal. Chem.* **2019**, *112*, 161–174. [CrossRef]
53. Owczarek, D.; Cibor, D.; Mach, T. Asymmetric Dimethylarginine (ADMA), Symmetric Dimethylarginine (SDMA), Arginine, and 8-Iso-Prostaglandin F2 alpha (8-iso-PGF2 alpha) Level in Patients with Inflammatory Bowel Diseases. *Inflamm. Bowel Dis.* **2010**, *16*, 52–57. [CrossRef] [PubMed]
54. Hosseinzadeh-Attar, M.J.; Sharifi, A.; Nedjat, S.; Mohamadkhani, A.; Vahedi, H. The Effect of Vitamin D on Serum Asymmetric Dimethylarginine in Patients with Mild to Moderate Ulcerative Colitis. *Int. J. Vitam. Nutr. Res.* **2020**, *90*, 17–22. [CrossRef] [PubMed]
55. Li, H.; Zhou, Y.; Zhao, A.; Qiu, Y.; Xie, G.; Jiang, Q.; Zheng, X.; Zhong, W.; Sun, X.; Zhou, Z.; et al. Asymmetric dimethylarginine attenuates serum starvation-induced apoptosis via suppression of the Fas (APO-1/CD95)/JNK (SAPK) pathway. *Cell Death Dis.* **2013**, *4*, e830. [CrossRef]
56. Samraj, A.N.; Bertrand, K.A.; Luben, R.; Khedri, Z.; Yu, H.; Nguyen, D.; Gregg, C.J.; Diaz, S.L.; Sawyer, S.; Chen, X.; et al. Polyclonal human antibodies against glycans bearing red meat-derived non-human sialic acid N-glycolylneuraminic acid are stable, reproducible, complex and vary between individuals: Total antibody levels are associated with colorectal cancer risk. *PLoS ONE* **2018**, *13*, e0197464. [CrossRef]
57. Jahan, M.; Thomson, P.C.; Wynn, P.C.; Wang, B. Red Meat Derived Glycan, N-acetylneuraminic Acid (Neu5Ac) Is a Major Sialic Acid in Different Skeletal Muscles and Organs of Nine Animal Species-A Guideline for Human Consumers. *Foods* **2023**, *12*, 337. [CrossRef]
58. Kim, M.-M.; Kim, S.-K. Effect of phloroglucinol on oxidative stress and inflammation. *Food Chem. Toxicol.* **2010**, *48*, 2925–2933. [CrossRef]
59. Owczarek, D.; Rodacki, T.; Domagala-Rodacka, R.; Cibor, D.; Mach, T. Diet and nutritional factors in inflammatory bowel diseases. *World J. Gastroenterol.* **2016**, *22*, 895–905. [CrossRef]
60. Vitali, R.; Prioreschi, C.; Rebenaque, L.L.; Colantoni, E.; Giovannini, D.; Frusciante, S.; Diretto, G.; Marco-Jimenez, F.; Mancuso, M.; Casciati, A.; et al. Gut-Brain Axis: Insights from Hippocampal Neurogenesis and Brain Tumor Development in a Mouse Model of Experimental Colitis Induced by Dextran Sodium Sulfate. *Int. J. Mol. Sci.* **2022**, *23*, 11495. [CrossRef]
61. Lau, S.K.P.; Lam, C.W.; Curreem, S.O.T.; Lee, K.C.; Chow, W.N.; Lau, C.C.Y.; Sridhar, S.; Wong, S.C.Y.; Martelli, P.; Hui, S.W.; et al. Metabolomic profiling of Burkholderia pseudomallei using UHPLC-ESI-Q-TOF-MS reveals specific biomarkers including 4-methyl-5-thiazoleethanol and unique thiamine degradation pathway. *Cell Biosci.* **2015**, *5*, 26. [CrossRef] [PubMed]
62. Wahlstrom, A.; Sayin, S.I.; Marschall, H.-U.; Backhed, F. Intestinal Crosstalk between Bile Acids and Microbiota and Its Impact on Host Metabolism. *Cell Metab.* **2016**, *24*, 41–50. [CrossRef] [PubMed]
63. Sinha, S.R.; Haileselassie, Y.; Nguyen, L.P.; Tropini, C.; Wang, M.; Becker, L.S.; Sim, D.; Jarr, K.; Spear, E.T.; Singh, G.; et al. Dysbiosis-Induced Secondary Bile Acid Deficiency Promotes Intestinal Inflammation. *Cell Host Microbe* **2020**, *27*, 659–670.e5. [CrossRef] [PubMed]
64. Perino, A.; Demagny, H.; Velazquez-Villegas, L.; Schoonjans, K. MOLECULAR PHYSIOLOGY OF BILE ACID SIGNALING IN HEALTH, DISEASE, AND AGING. *Physiol. Rev.* **2021**, *101*, 683–731. [CrossRef] [PubMed]
65. Sorrentino, G.; Perino, A.; Yildiz, E.; El Alam, G.; Sleiman, M.B.; Gioiello, A.; Pellicciari, R.; Schoonjans, K. Bile Acids Signal via TGR5 to Activate Intestinal Stem Cells and Epithelial Regeneration. *Gastroenterology* **2020**, *159*, 956–968.e8. [CrossRef]
66. Liang, J.; Chang, B.; Huang, M.; Huang, W.; Ma, W.; Liu, Y.; Tai, W.; Long, Y.; Lu, Y. Oxymatrine prevents synovial inflammation and migration via blocking NF-kappa B activation in rheumatoid fibroblast-like synoviocytes. *Int. Immunopharmacol.* **2018**, *55*, 105–111. [CrossRef] [PubMed]
67. Guzman, J.R.; Koo, J.S.; Goldsmith, J.R.; Muehlbauer, M.; Narula, A.; Jobin, C. Oxymatrine Prevents NF-kappa B Nuclear Translocation And Ameliorates Acute Intestinal Inflammation. *Sci. Rep.* **2013**, *3*, 1629. [CrossRef] [PubMed]

68. Tang, S.; Liu, W.; Zhao, Q.; Li, K.; Zhu, J.; Yao, W.; Gao, X. Combination of polysaccharides from Astragalus membranaceus and Codonopsis pilosula ameliorated mice colitis and underlying mechanisms. *J. Ethnopharmacol.* **2021**, *264*, 113280. [CrossRef]
69. Backer, V.; Cheung, F.Y.; Siveke, J.T.; Fandrey, J.; Winning, S. Knockdown of myeloid cell hypoxia-inducible factor-1 alpha ameliorates the acute pathology in DSS-induced colitis. *PLoS ONE* **2017**, *12*, e0190074. [CrossRef]
70. Segata, N.; Izard, J.; Waldron, L.; Gevers, D.; Miropolsky, L.; Garrett, W.S.; Huttenhower, C. Metagenomic biomarker discovery and explanation. *Genome Biol.* **2011**, *12*, R60. [CrossRef]
71. Zhou, J.; Hong, H.; Zhao, J.; Fang, R.; Chen, S.; Tang, C. Metabolome analysis to investigate the effect of heavy metal exposure and chemoprevention agents on toxic injury caused by a multi-heavy metal mixture in rats. *Sci. Total Environ.* **2024**, *906*, 167513. [CrossRef] [PubMed]
72. Yue, T.; Tan, H.; Shi, Y.; Xu, M.; Luo, S.; Weng, J.; Xu, S. Serum Metabolomic Profiling in Aging Mice Using Liquid Chromatography-Mass Spectrometry. *Biomolecules* **2022**, *12*, 1594. [CrossRef] [PubMed]

International Journal of
Molecular Sciences

MDPI

Review

Dietary Flavonoids Vitexin and Isovitexin: New Insights into Their Functional Roles in Human Health and Disease Prevention

Weiqi Yan [1,2], Junying Cheng [1] and Baojun Xu [1,*]

[1] Food Science and Technology Program, Department of Life Sciences, Beijing Normal-Hong Kong Baptist University, Zhuhai 519087, China; yanweiqi@uic.edu.cn (W.Y.); r130013004@mail.uic.edu.cn (J.C.)

[2] Centre for Cancer and Inflammation Research, School of Chinese Medicine, Hong Kong Baptist University, Hong Kong SAR, China

[*] Correspondence: baojunxu@uic.edu.cn; Tel.: +86-0756-3620636

Abstract: Vitexin and isovitexin are dietary flavonoids widely distributed in food and medicinal plants. They have attracted increasing attention owing to their diverse pharmacological activities and favorable safety profiles. These compounds exhibit therapeutic potential across multiple biological systems, including the immune, nervous, respiratory, cardiovascular, and endocrine systems, through antioxidant, anti-inflammatory, anticancer, antibacterial, and neuroprotective mechanisms. Although previous reviews have addressed the pharmacological effects of vitexin and isovitexin, most are limited in scope—either focusing solely on vitexin or restricted to specific disease models such as cancer or diabetes. Moreover, some studies are outdated and do not reflect the recent advances in synthetic modification, green extraction technologies, and systems pharmacology. This review aims to provide a comprehensive evaluation of the pharmacological properties, pharmacokinetics, and clinical relevance of vitexin and isovitexin, highlighting their potential in disease prevention and treatment. A literature search was conducted using Web of Science, PubMed, and Google Scholar, with keywords including "vitexin", "isovitexin", "disease", and "mechanism". Here, we summarize the current research on the pharmacological effects of vitexin and isovitexin in metabolic disorders, inflammatory diseases, cancer, and neurodegenerative conditions, focusing on their molecular mechanisms and therapeutic targets. Furthermore, we discussed their toxicity, bioavailability, pharmacokinetics, and clinical research findings. Vitexin and isovitexin hold promise as therapeutic agents or adjuncts for multiple diseases with potential applications in modern medicine and healthcare. However, their pharmacological mechanisms, clinical efficacy, and potential synergistic effects with other therapeutic agents remain unclear. Further systematic research is needed to clarify molecular targets and optimize their therapeutic applications.

Keywords: vitexin; isovitexin; dietary flavonoid; chronic disease prevention; bioavailability; health promotion

1. Introduction

Vitexin (5, 7, 4-trihydroxyflavone-8-glucoside, Figure 1) is a C-glycosylated flavonoid compound widely present in various kinds of plants and is an active ingredient in many traditional Chinese medicines and foods. Flavonoids often have multiple pharmacological activities; therefore, vitexin has received increasing attention because of its wide range of pharmacological effects, including anticancer, antioxidant, anti-inflammatory, anti-Alzheimer's disease (AD), blood pressure-lowering, and anti-hypoxia/ischemic injury [1–4]. These pharmacological effects are related to multiple systems, such as the central nervous, cardiovascular, intestinal system, and endocrine systems [5].

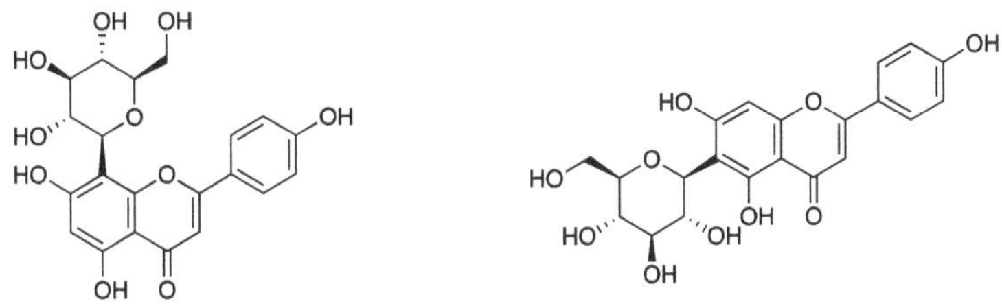

Vitexin **Isovitexin**

Figure 1. Structure of vitexin and isovitexin.

Isovitexin (apigenin-6-C-glucoside, Figure 1), an isomer of vitexin, also exists in plants containing vitexin, such as pigeon peas, passion flowers, bamboo, mimosa, and wheat leaves, and has been screened as a bioactive ingredient [6,7] because of its similar chemical structure. Isovitexin has also been shown to exert pharmacological effects similar to those of vitexin, including antioxidant, anti-inflammatory, and anti-AD effects.

However, a review of the literature reveals several limitations in previous studies. Some works have focused exclusively on vitexin and overlooked isovitexin, while others have addressed both compounds but restricted discussion to individual diseases such as cancer or diabetes, or emphasized isolated domains like extraction processes. Many existing reviews have yet to incorporate recent findings from multiple disciplines, including medicinal chemistry, systems biology, and clinical pharmacokinetics [1–3]. These limitations hinder a comprehensive understanding of their therapeutic potential.

To bridge this gap, the present review systematically integrates the recent research on vitexin and isovitexin, spanning from basic pharmacological studies and molecular mechanisms to structural optimization and potential clinical translation. We highlight their roles in treating metabolic diseases, cardiovascular conditions, neurodegenerative disorders, and cancer. This review also emphasizes their complementary pharmacological profiles and structural similarities, aiming to elucidate their synergistic mechanisms and inform future drug development strategies.

This review provides a comprehensive summary of the pharmacological and molecular mechanisms of action of vitexin and isovitexin as well as a brief overview of their pharmacokinetic studies, highlights the potential of vitexin and isovitexin in health prevention and treatment, and proposes future research directions and recommendations, providing a reference for subsequent scientific exploration and clinical applications.

2. Methodology

The literature for this study was searched on 10 December 2024, in three electronic research databases: Web of Science (Clarivate Analytics, Philadelphia, PA, USA), Google Scholar (Google LLC, Mountain View, CA, USA), and PubMed (National Library of Medicine, Bethesda, MD, USA). The selection was based on a series of clear inclusion criteria: only research published in English journals must be included, and the keywords "vitexin" and "isovitexin" must be included in the article. The system review excluded review articles, abstracts, editorials/letters, conferences, and conference papers. Articles unrelated to "vitexin" and "isovitexin" in terms of antioxidant, anti-inflammatory, safety, and toxicological applications were also excluded. To further screen the articles, the research results were manually screened to exclude articles that contained duplicate data or were duplicated in the database, as well as some articles on the phytochemical separation of "vitexin" and "isovitexin" derivatives. The remaining articles constituted the main source of information for writing this article. Figure 2 shows the study selection process, through which the researchers ensured the quality and relevance of the included studies to more accurately evaluate the therapeutic potential and health protection of "vitexin" and "isovitexin" in diseases. This systematic approach helps extract the most valuable data from a large amount of work in the literature, providing a scientific basis for further research and clinical applications. Through this process, researchers can provide deeper insights into the use of "vitexin" and "isovitexin" as disease treatment methods and provide guidance for future research.

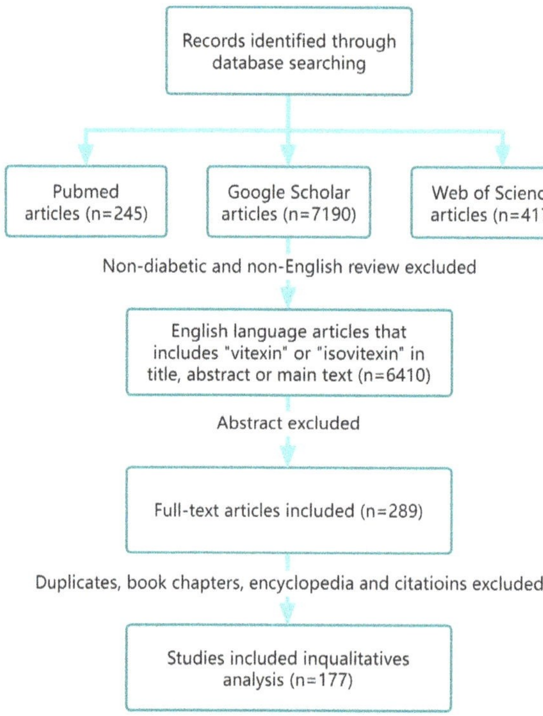

Figure 2. Flowchart of the literature retrieval and screening.

Chemical Structure and Sources

Flavonoids are a class of compounds characterized by a 2-phenyl-1-benzopyran-4-one skeleton. Among them, apigenin (4′,5,7-trihydroxyflavone) is a typical natural flavonoid. Vitexin and isovitexin are C-glycosylated derivatives of apigenin, with glucose units attached at different positions—C-8 in vitexin and C-6 in isovitexin. Both compounds share similar chemical structures and properties. Vitexin, also known as 8-D-glucosyl-4′,5,7-trihydroxyflavone or apigenin 8-C-glucoside, has a molecular formula of $C_{21}H_{20}O_{10}$ and a molecular weight of 432.38 g/mol. Isovitexin, the 6-C-glucoside isomer of vitexin, has the same molecular formula ($C_{21}H_{20}O_{10}$) and molecular weight (432.38 g/mol), but differs in glycosylation position. Both contain seven hydroxyl groups, and their polyhydroxylated structures contribute significantly to their biological activity. In vitexin, the order of radical scavenging ability among hydroxyl groups is 4′-OH > 7-OH > 5-OH. Additionally, the C-glycoside bond in vitexin is more stable than O-glycosides, which enhances the antioxidant capacity by reducing the negative charge at the C-3 oxygen. C-glycosyl flavonoids generally show stronger antioxidant and antidiabetic properties than their O-glycoside or aglycone counterparts [8].

Vitexin exists in various plants, among which hawthorn leaves are the main source. In recent years, technology for extracting the biological activities of vitexin and isovitexin from plants has been optimized. Compared to the traditional Soxhlet extraction method, there are limitations in solvent selectivity, extraction rate, and toxicity. Microwave-assisted extraction (MAE) is the optimal technique for achieving maximum total polyphenol and flavonoid content and high yield, obtaining high concentrations of antioxidants and anti-inflammatory and antibacterial compounds in a short period of time, with reasonable yield and good selectivity. The preparation and separation of vitexin and isovitexin can be easily and effectively achieved through adsorption and desorption of the ADS-5 resin. This method can also be used for the separation of flavonoid C-glucosides from other Chinese herbal medicines.

3. Health Promotion Effects of Vitexin and Isovitexin

This section highlights the therapeutic effects of vitexin and isovitexin, including cardiovascular protection, blood sugar regulation, anti-obesity, anticancer, antioxidant, anti-inflammatory, and neuroprotective properties. Additionally, they exhibit antimicrobial and antibacterial activities. Table 1 and Figures 3 and 4 illustrate the diseases and treatment types that vitexin and isovitexin can improve, along with their key mechanisms. The yellow inner circle of Figure 3 illustrates the key molecular mechanisms through which vitexin and isovitexin exert their pharmacological effects, while the outer ring shows the corresponding disease conditions or therapeutic areas.

Table 1. Typical pathways and target cases of vitexin and isovitexin in the treatment of diseases.

Vitexin/ Isovitexin	Biological Activities	Cell/Animal	Dosage	Mechanisms and Pathways	Targets	References
Vitexin	Myocardial ischemia/reperfusion (I/R) injury	Isolated *Sprague-Dawley* rat hearts, H9c2 cells	10 μM	Reducing ROS levels; improving mitochondrial activity, mitochondrial membrane potential, and ATP content; markedly increasing *MFN2* expression and reducing the recruitment of Drp1 in mitochondria.	*MFN2*, Drp1, Epac1-Rap1	[4]
				Inhibiting ischemic myocardial mitochondrial dysfunction and reducing cardiomyocyte apoptosis by regulating Epac1-Rap1 signaling.		[9]
Vitexin	Protect against DOX-induced acute cardiotoxicity	Rats	30 mg/kg	Vitexin induced elevated FOXO3a protein expression levels, by suppressing oxidative stress, inflammation, and apoptotic signals.	FOXO3a	[10]
Vitexin	Pre-eclampsia	Pregnant rats	45–60 mg/kg	Decreased sFlt-1, increased PlGF, and alleviated oxidative stress	HIF-1α, VEGF	[11]
Vitexin/ Isovitexin	Diabetes, overweight	Diabetic rats	1 mg/kg	Inhibits the negative regulator of insulin signaling, protein tyrosine phosphatase (PTP)-1B, inhibits α-amylase and α-glucosidase.	PTP-1B	[12]
Vitexin/ Isovitexin		HepG2				[13]

Table 1. *Cont.*

Vitexin/ Isovitexin	Biological Activities	Cell/Animal	Dosage	Mechanisms and Pathways	Targets	References
Vitexin		HUVECs	5–100 μM	Vitexin disrupted Wnt/β-catenin signaling pathway, vitexin activated nuclear factor-erythroid 2-related factor 2 (Nrf2) in HUVEC under high glucose.	Nrf2	[14]
Vitexin/ isovitexin	Diabetes, overweight	HepG2 cells within an insulin-resistant system		Regulating glycemia, through changes in anti-hyperglycemic activity and in the gut microbiota in overweight individuals.		[15]
Vitexin	Diabetic nephropathy	HK-2 cells/DN rat	0–40 μM	Vitexin could alleviate diabetic nephropathy by attenuated ferroptosis via activating GPX4.	GPX4	[16]
Vitexin	Non-alcoholic fatty liver disease	NAFLD mice	6 mg/kg	Vitexin degraded lipids in HFD-induced NAFLD mice liver by inducing autophagy and restoring both ER and mitochondrial biological proteins.		[17]
		CSHFD mice	40 mg/kg	Inhibit TLR4/NF-κB signaling and reduce fatty acid synthesis proteins.		[2]
Vitexin	Obesity	*C57BL/6J*, 3T3-L1 adipocytes	30 mg/kg	Vitexin may prevent HFD-induced obesity/adipogenesis via the AMPKα mediated pathway.	AMPKα	[18]
Isovitexin	Liver fibrosis	Hepatic stellate cell models		Regulation of *miR-21*, targeting PTEN-Akt signaling and the GSH metabolic pathway.	PTEN	[19]
Vitexin	Triple-negative breast cancer			Vitexin promoted M1 polarization and suppressed M2 polarization, affecting EGFR phosphorylation and downstream signaling.	EGFR	[20]
Vitexin	Gastric cancer	Nude mice/GC cells	2 mg/kg 10–160 μM	Vitexin inhibited the malignant progression of GC in vitro and in vivo by suppressing HMGB1-mediated activation of PI3K/Akt/HIF-1α signaling pathway.	HMGB-1	[21]
Vitexin	Colon cancer	HCT-116 cells	1–300 μM	Inhibit colon cancer HCT-116 cell proliferation by suppressing CDK1/cyclin B expression, leading to cell cycle arrest in the G2/M phase.		[22]
Isovitexin	Colon cancer			Promoted apoptosis and suppressed cell proliferation by activating the p53 signaling pathway.		[23]
Vitexin/ isovitexin	Non-small cell lung cancer cells	A549/ H1299 cells, nude mice	1–120 μM	Suppressed NF-κB, AKT and ERK activation.		[24]
Vitexin		A549 cells, nude mice	0–40 μM	Reduced the levels of p-PI3K, p-Akt, and p-mTOR.		[25]
Isovitexin	Hepatocarcinoma	SK-Hep-1 cells		Mediated *miR-34a* upregulation induces apoptosis and suppresses the stemness of SK-SC.	*miR-34a*	[26]
Vitexin	Vitiligo	human melanocyte PIG1	0–40 μM	Protected melanocytes from oxidative stress by activating MAPK-Nrf2/ARE signaling pathway.	Nrf2	[27]

Table 1. *Cont.*

Vitexin/Isovitexin	Biological Activities	Cell/Animal	Dosage	Mechanisms and Pathways	Targets	References
Isovitexin	Acute lung injury	RAW 264.7 cells	0–50 μM	Inhibiting MAPK and NF-κB and activating HO-1/Nrf2 pathways.	Nrf2	[28]
Vitexin	Autoimmune hepatitis	EAH mice	5 mg/kg	Vitexin ameliorated hepatic injury in EAH mice through activation of the AMPK/AKT/GSK-3β pathway and upregulation of the *Nrf2* gene.	*Nrf2*	[29]
Vitexin	Mastitis	MAC-T cells	20 μM	Vitexin inhibited the production of ROS through promoting PPARγ, increased the activity of antioxidant enzymes, and reduced inflammatory cytokines and apoptosis by alleviating ER stress and inactivation MAPKs and NF-κB signaling pathway.	PPARγ	[30]
Isovitexin	Acute gouty arthritis	*Sprague-Dawley* rats	100 mg/kg	Isovitexin ameliorates joint inflammation in acute GA via the TLR4/MyD88/NF-κB pathway.	TLR4	[31]
Isovitexin	Chronic kidney disease	SV40-MES-13 cells/*C57BL/6* mice	0–50 μM, 5 mg/kg	Ameliorated renal injury, inflammation, and increased protected autophagy by anti-ROS production, anti-inflammation, and anti-pyroptosis.		[32]
Vitexin/isovitexin	Alzheimer's disease	Microglial cells	0.1–100 μg/mL	Mediating the nuclear factor-kappa B (NF-κB) signaling pathway.	NF-κB	[33]
Vitexin	Cerebral ischemia/reperfusion	Rat	50 mg/kg	Protect the neuron cells and brain related with the Keap1/Nrf2/HO-1 signaling pathway.	Keap1, Nrf2	[34]

3.1. Regulating Cardiovascular Protection

Vitexin has a wide range of pharmacological effects, particularly protective effects on the cardiovascular system. The research has shown that vitexin plays an important role in the cardiovascular system by affecting the myocardium, blood vessels, and platelets through multiple signaling pathways. Vitexin has been shown to have a protective effect against myocardial ischemia–reperfusion injury (MIRI). Experimental data showed that vitexin inhibited the expression of *Epac1* and regulated the Epac1–Rap1 signaling pathway, alleviating MIRI-induced mitochondrial dysfunction and inhibiting the activation of mitochondria-mediated apoptosis pathways, thus exerting a protective effect against MIRI [4]. This discovery provided new targets and a theoretical basis for the potential application of vitexin in the treatment of ischemic myocardial injury. However, this study also has limitations in exploring the specific regulatory mechanism of Epac in ischemia/reperfusion (I/R), and further research is needed to determine the therapeutic potential of vitexin through *Epac1* [9].

Vitexin can also improve the pathological score of the myocardial tissue and reduce the apoptosis index, which is related to its antioxidant activity, inhibition of inflammatory cytokine release, and myocardial cell apoptosis. Experiments have confirmed that doxorubicin (DOX) is an effective chemotherapeutic drug; however, its clinical application is limited by the development of cardiotoxicity. Vitexin exerts a cardioprotective effect against DOX-induced cardiac toxicity by reducing oxidative stress, lowering cardiac inflammatory cytokines, increasing FOXO3a, and inhibiting caspase-3 activation. Therefore, vitexin may serve as an effective therapeutic agent for preventing DOX-induced cardiomyopathy [10,20].

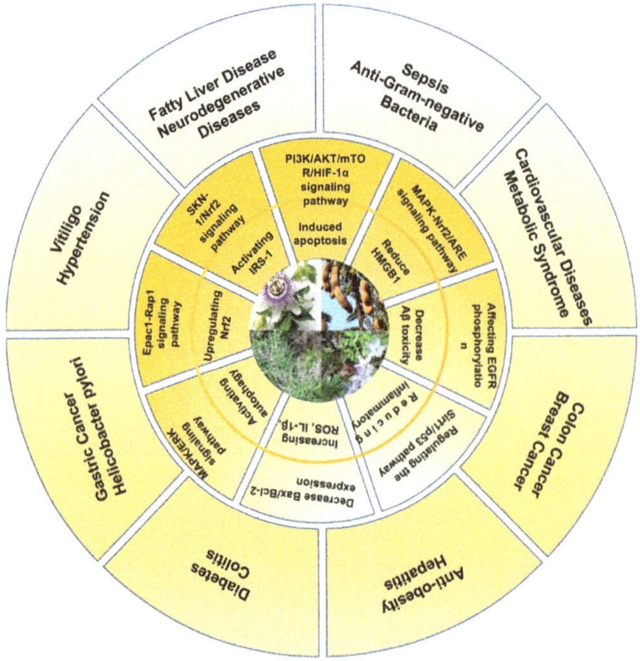

Figure 3. The main mechanisms and treatment types of vitexin and isovitexin.

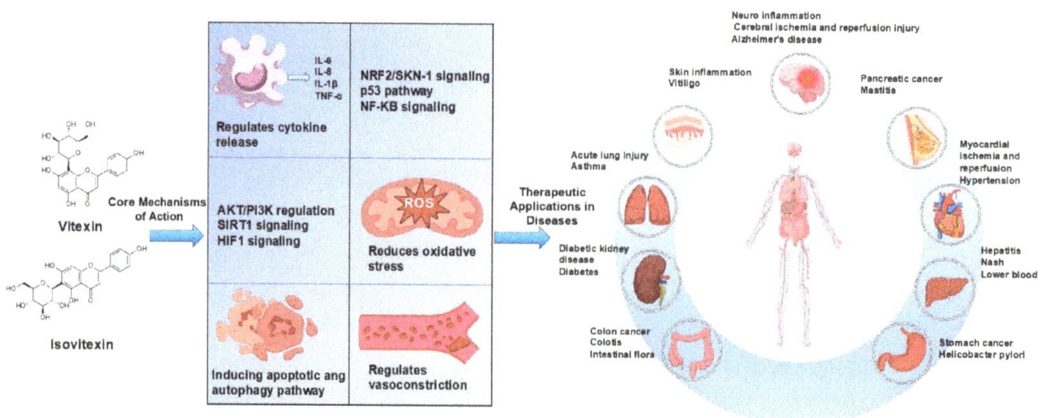

Figure 4. The mechanisms of vitexin and isovitexin and the types of diseases in various systems.

Research has shown that flavonoids, such as vitexin, can regulate vasoconstriction in an agonist-dependent manner, not only through endothelial dependence, but also by inhibiting the MAPK/ERK pathway and partially inhibiting Rho kinase. *Lagenaria siceraria* fruits containing vitexin exhibit anti-hypertensive and anti-cardioprotective effects induced by L-NAME (a NOS inhibitor) [11]. Although further research is needed to determine whether vitexin has anti-hypertensive effects associated with NOS, the existing studies have shown its importance in vascular health.

In summary, vitexin exerts multiple protective effects on the cardiovascular system by affecting the myocardium and blood vessels. These effects include anti-myocardial ischemia/reperfusion injury and anti-vasoconstriction effects, thus playing an important role in the prevention and treatment of cardiovascular diseases. With further research, the cardiovascular protective mechanism of vitexin will be further elucidated, providing a more theoretical basis for clinical applications.

3.2. Regulating Blood Sugar and Combating Obesity

Vitexin and isovitexin have shown good therapeutic potential in diabetes as natural compounds with protective effects. The research has indicated that pancreatic injury or pancreatic tissue damage often leads to insufficient insulin secretion or insulin resistance, which in turn affects blood glucose processing and increases blood glucose levels [35,36]. Vitexin has been shown to have a protective effect on pancreatic and islet tissues by upregulating *Nrf2* and antioxidant enzymes, as well as inhibiting MAPK pathways, including c-Jun N-terminal kinase (JNK) and p38, to alleviate the toxicity caused by streptozotocin or lipopolysaccharide (LPS) [37,38].

Oral administration of vitexin and isovitexin can significantly reduce postprandial blood glucose levels in normal blood glucose mice and sucrose-induced diabetic rats. They can also provide potential therapeutic approaches for the prevention of diabetes or other pathological complications through free radical scavenging and metal ion-trapping activities. Simultaneously, α-glucosidase and α-amylase, the key enzymes for carbohydrate digestion, are located on the facial mask on the brush edge of intestinal cells. Vitexin and isovitexin have significant inhibitory effects on α-glucosidase [12,39]. Vitexin and isovitexin were found to exhibit inhibitory effects on AGE formation. For the first time, vitexin and isovitexin were confirmed to inhibit PTP-1B activity, improve insulin signaling, and increase glucose uptake [13]. Research has suggested that acute and long-term hyperglycemia can activate resting endothelial cells and impaired endothelial function of vascular cells. High blood sugar levels induced endothelial damage and apoptosis by reducing the bioavailability of NO and increasing reactive oxygen species (ROS). Vitexin inhibited endothelial activation caused by high glucose levels and reduced vascular inflammation. It also increased superoxide dismutase (SOD) and glutathione peroxidase (GPx) levels and decreased GPX4-mediated ferroptosis. Furthermore, vitexin promoted GLUT4's translocation from the cytoplasm to the cell membrane, potentially facilitating the uptake of glucose [16,40].

High-dose glucose was added to an in vitro model to create a human umbilical vein endothelial cell (HUVEC) model and examine the protective effect of vitexin against high glucose-induced HUVEC damage. It was demonstrated that vitexin reduced the proliferation and death of HUVECs driven by high glucose, activated Nrf2 in HUVECs, and controlled the Wnt/β-catenin signaling pathway. Vitexin administration simultaneously increased SOD activity and decreased ROS and malondialdehyde (MDA) content in high glucose-induced HUVECs [14].

Obesity-related insulin resistance and glucose metabolism disorders cause cells' metabolism of fats and carbohydrates to become unbalanced, which eventually results in diabetes. According to the research, instead of depending only on one active ingredient, vitexin and isovitexin are more likely to work in concert to prevent disease. Synergistic gut microbiota regulation may be a component of the pharmacological effects. Insulin resistance in obesity is characterized by decreased insulin-triggered glucose transport and processing in skeletal muscles and adipocytes, as well as inefficient hepatic regulation of glucose synthesis. Vitexin and isovitexin flavonoids not only affected the absorption of peripheral glucose in insulin and non-insulin sensitive tissues but also showed the potential to restore insulin resistance in HepG2 cells by enhancing cellular uptake of glucose. Furthermore, *proteobacteria* are the predominant gut microbiota in obesity and type 2 diabetes, and vitexin and isovitexin have the ability to increase the number of beneficial bacteria while suppressing the levels of harmful bacteria [15].

Through a variety of mechanisms, such as lowering postprandial blood glucose, blocking important digestive enzymes, and controlling intestinal flora, vitexin and isovitexin have demonstrated both preventative and therapeutic effects in diabetes. In addition to being possible medications for atherosclerosis and cardiovascular problems in diabetes, vitexin and isovitexin may offer novel approaches for the prevention and management of vascular diseases. Vitexin and isovitexin can be used as functional ingredients in the industrial food sector. To clarify the metabolic pathways of bacteria and understand the specific microbial glucose metabolism process, more research is required. Additional investigation of the distinct routes of action of these chemicals will offer a scientific foundation for the development of novel treatment approaches.

Adipose tissue abnormalities and excessive accumulation are intimately linked to inflammation, as evidenced by elevated TNF-α and IL6 release. Vitexin dramatically decreased serum levels of IL6 and TNF-α in *C57BL/6J* mice fed a high-fat diet. Fat formation is inhibited when AMPK α, the primary regulator of energy metabolism, is phosphorylated [17,18,27]. Previous studies have shown that vitexin ameliorated weight gain and obesity induced by a high-fat diet (HFD) in *C57BL/6J* male mice, inhibited adipogenesis in adipocytes, dramatically increased the expression of phosphorylated AMPK α, and decreased the expression of C/EBP α and fatty acid synthase (FAS) in white adipose tissue and 3T3-L1 adipocytes [41].

Vitexin-induced decrease in adipogenesis may involve additional signaling pathways besides AMPK alpha. Vitexin has been shown to reduce the inflammation and apoptosis induced by ER stress by blocking adipogenesis through ER stress-induced pathways. Conversely, vitexin prevents HFD-induced vasculitis by blocking TMAO-mediated RNA N6-methyladenosine (m6A) alteration, which paves the way for the creation of functional foods [42].

In conclusion, vitexin decreased the expression of adipogenic genes in white adipose tissue and relieved obesity induced by a high-fat diet [41]. Further research is necessary to determine the possible impacts of vitexin on the upstream regulators of the AMPK α and ER pathways. Vitexin could therefore be another dietary supplement that can be utilized in conjunction with other methods for treating and preventing obesity and its associated problems.

3.3. Lowering Blood Cholesterol and Treating Fatty Liver

Although liver damage from chronic alcohol use is inevitable, few effective liver-protective medications are currently available to treat ethanol-induced liver damage. The development of alcoholic liver disease (ALD) is significantly influenced by apoptosis associated with liver injury. Hepatitis can result from the discharge of apoptotic cells into the extracellular space, which can cause inflammation if they are not quickly removed. Furthermore, the phagocytic action of apoptotic cells has a pro-fibrotic effect, and they also release pro-fibrotic mediators, which cause fibrosis of liver cells. Thus, ALD can be slowed or reversed by blocking apoptosis [43]. Studies have demonstrated that vitexin possesses antiapoptotic properties, preserves the structural integrity of liver cells, controls lipid metabolism, and dramatically reduces ethanol-induced LO2 cell damage. The protective effect of vitexin against ethanol-induced liver damage was closely linked to the Sirt1/p53-mediated apoptosis pathway, as evidenced by the fact that vitexin inhibited serum TG and TC levels and significantly improved ethanol-induced ALD [44], and its protective effect against ethanol-induced liver injury was closely linked to the Sirt1/p53-mediated apoptosis pathway, as Sirt1 could remove the acetyl group of p53, reduce the transfer of p53 from cytoplasm to mitochondria, inhibit the release of mitochondrial pro-apoptotic proteins, and ultimately restrict apoptosis [45].

Currently, there is no approved treatment for non-alcoholic fatty liver disease (NAFLD), a prevalent liver illness characterized by the accumulation of liver fat. Since diabetes and obesity are the primary causes of NAFLD, substances with anti-obesity properties and the ability to lower insulin resistance are regarded as promising candidates for the treatment of NAFLD [46]. Studies have demonstrated that vitexin can decrease the body and liver weight of *C57BL/6J* mice fed an HFD, as well as the serum and liver levels of cholesterol and triglycerides. In the HFD group, vitexin markedly decreased the serum levels of aspartate aminotransferase (AST) and alanine aminotransferase (ALT). Activating receptor gamma (PPAR), FAS, steroid regulatory element binding protein-1c (SREBP-1c), CCAAT/enhancer binding protein alpha (C/EBPα), peroxisome proliferators, and acetyl CoA carboxylase (ACC) were also downregulated by vitexin, which inhibited lipogenesis. By activating insulin receptor base-1 (IRS-1) and its downstream target AKT, vitexin enhances insulin signaling [17,47]. NAFLD may worsen because of chronic stress (CS). Research has demonstrated that vitexin helps treat NAFLD caused by a high-fat diet and chronic stress. This mechanism is directly linked to lowering the accumulation of fatty acids and suppressing the TLR4/NF-κ B signaling pathway. Vitexin addresses lipid metabolic issues, reduces liver inflammation, and decreases the accumulation of fat in the liver. In addition, vitexin suppresses the expression of proteins and genes linked to the production of fat in the liver [2].

The initial hypothesis was that isovitexin prevents liver fibrosis. Isovitexin could alleviate liver fibrosis in two hepatic stellate cell models induced by platelet-derived growth factor BB (PDGF-BB) and in mouse liver fibrosis models induced by carbon tetrachloride (CCl_4) by reducing collagen deposition and hepatic stellate cell activation. According to transcriptomic and metabolomic analyses, the primary regulatory processes of isovitexin in liver fibrosis may be the phosphatidylinositol 3-kinase (PI3K)/protein kinase B (Akt) signaling pathway and the glutathione (GSH) metabolic pathway, also confirming that *miR-21* is a crucial node in the isovitexin regulation of the PI3K/Akt signaling pathway and GSH metabolic pathway [19].

The biological roles of vitexin and isovitexin, which are possible medications for the treatment of fatty liver disease, may include immune control, lipid metabolism regulation, and anti-inflammatory activities. These results place vitexin and isovitexin as a solid scientific foundation for novel medications for fatty liver and offer evidence in favor of further clinical studies and development.

3.4. Anticancer Effect

As the second greatest cause of mortality and one of the main global public health concerns, cancer not only endangers lives, but also has a significant financial impact. Thus, creating successful plans for long-term cancer prevention and control is essential [48]. Plant-derived secondary metabolites have shown beneficial anticancer properties. Studies conducted both in vitro and in vivo have demonstrated that the chemopreventive chemicals vitexin and isovitexin exhibit pharmacological efficacy against a variety of malignancies by boosting autophagy or apoptotic processes and preventing migration and proliferation [1]. According to the preliminary research, vitexin and isovitexin demonstrated antitumor activity through targeted cell apoptosis in HepG2 hepatocytes, HeLa cervical cells, MCF-7 human breast cancer cells, U937 human leukemia cells, and HCT116 colorectal cells [49–52].

Changes in the pathological milieu from an inflammatory state to a pro-tumor microenvironment are key risk factors for the development of colitis-associated cancer (CAC), which is caused by chronic colitis. The deadly effects of inflammatory cytokines generated by M1 macrophages were maintained by vitexin's ability to bind to the vitamin D receptor (VDR) protein, which inhibits macrophage development towards the M2 phenotype. In vitro co-culture studies have shown that vitexin enhances the anticancer activity of M1 macrophages by sustaining their pro-inflammatory phenotype and inhibiting the M2 polarization typically associated with tumor promotion. The targeting effect of vitexin on VDR was further validated using myeloid-specific VDR gene knockout mice. The shift from colitis to colorectal cancer is a major pathogenic role that VDR inhibits [53,54]. The important role of VDR as a nuclear receptor transcription factor in colitis and colorectal cancer has been confirmed by clinical and basic research. Isovitexin has been found to activate the p53 signaling pathway in a dose-dependent manner, thereby exerting inhibitory effects on colorectal cancer cells. The possible use of isovitexin in the treatment of colorectal cancer now has a scientific foundation for this discovery [23].

Gastric cancer (GC) is a serious threat to public health and its high recurrence rate after surgery is the main reason for the short survival of patients. The primary cause of stomach cancer recurrence is metastasis, and epithelial–mesenchymal transition (EMT) is a crucial step in the metastatic process. Two distinct gastric cancer cell lines (SGC-7901 and AGS) and a normal, non-transformed gastric cell line (GES-1) were used to assess the antitumor potential of vitexin, a medication that is thought to be a contender for the treatment of gastric cancer. These findings showed that vitexin had a dose-dependent effect on gastric cancer cells' production of HIF-1 α. Additionally, by deactivating the PI3K/AKT/mTOR/HIF-1 α pathway, vitexin therapy slowed the growth of gastric cancer cells. In particular, it was discovered for the first time that vitexin inhibits the proliferation, migration, invasion, and EMT of gastric cancer cells by downregulating the expression of HMGB1 in these cells. More significantly, vitexin inhibited the activation of the PI3K/AKT/mTOR pathway by promoting the nuclear translocation of HMGB1. According to the study results, vitexin might be a novel and potent medication for the treatment of stomach cancer. Nevertheless, it is still unknown how vitexin inhibits HMGB1 in gastric cancer cells [21].

According to the research, vitexin can cause cell death by triggering apolipoprotein L1 (ApoL1)-mediated autophagy and JNK and suppresses the activity of heat shock factor 1 (HSF-1) to prevent the growth of colorectal cancer cells [53]. Many cancers overexpress cyclin-dependent kinases (CDKs), which are important targets of new anticancer medications. According to experimental findings, vitexin's IC_{50} value for HCT116 colorectal cancer cells was 203.27 ± 9.85 μ mol/L. According to Zhao et al. (2024) [22], vitexin may have an inhibitory effect on colorectal cancer cell proliferation, G2/M conversion, cell cycle progression, and CDK1/cyclin B protein production in HCT116 cells.

Triple-negative breast cancer (TNBC) is a type of breast cancer with a dismal prognosis and is insensitive to targeted and endocrine therapies. Following chemotherapy, it has a poor prognosis and significant recurrence rate. Tumor-associated macrophages (TAMs) are essential for the development of TNBC. According to previous studies, vitexin can stop TNBC cells from proliferation and invading, and can also prevent the migration of cancer cells. Furthermore, vitexin can influence the downstream signaling and phosphorylation of the epidermal growth factor receptor (EGFR), block M2 polarization, enhance M1 macrophage polarization, and exert anticancer effects by targeting the EGFR/PI3K/AKT/mTOR pathway [20,55]. When combined with doxorubicin (Dox), vitexin can reduce tumor growth and show synergistic effects in in vivo tests, increasing antitumor efficacy. However, more research is needed to determine how vitexin controls this pathway and how it affects macrophage polarization. Furthermore, there is potential for the therapeutic application of vitexin's effect on other immunological targets in breast cancer, which has yet to be investigated.

Chemotherapy remains the principal treatment for patients with non-small cell lung cancer (NSCLC), although anticancer medications frequently have negative side effects. Therefore, finding novel, low-toxicity anti-NSCLC medications is crucial. The possible antitumor action of vitexin in NSCLC has been verified in previous studies. Treatment with vitexin and isovitexin decreased the A549 cells in vitro survival rate and lessened the increase in lactate dehydrogenase (LDH) release caused by cell membrane disruption. Further evidence of the anti-NSCLC potential of vitexin came from the administration of both vitexin and isovitexin, which both suppressed the growth of NSCLC tumors in vivo. Based on the findings of this study, vitexin and isovitexin lowered the Bcl-2/Bax ratio and caused cytochrome c to be released from the mitochondria into the cytoplasm, which cleaved caspase-3 in A549 cells. It was hypothesized that vitexin uses a mitochondrial-dependent pathway to partially trigger apoptosis in A549 cells. Through in vitro and in vivo experiments, the survival rate of A549 cells damaged by vitexin treatment was revealed for the first time, and apoptosis was partially induced through the mitochondrial pathway and the PI3K/Akt/mTOR signaling pathway. Vitexin and isovitexin are expected to be innovative and effective drugs for the treatment of NSCLC [24,25].

Through controlling the expression of *miR-34a*, isovitexin was shown to suppress the stem cell properties of SK-Hep-1 cells. Isovitexin decreased the number of CD44+ cells and prevented the development of spheroids and colonies. Furthermore, isovitexin showed promise in treating liver cancer by increasing the amount of Bax protein in SK-SC cells and lowering the levels of Bcl-2 and Mcl-1 proteins, which are linked to apoptosis. This provides isovitexin's new approach to liver cancer treatment as a scientific foundation, particularly its regulatory effect on the traits of liver cancer stem cells, which could serve as crucial targets for the development of novel therapeutic approaches [26].

Vitexin and isovitexin have antitumor properties through a number of mechanisms, such as triggering cell death, preventing tumor growth, inhibiting tumor cell invasion and migration, regulating immune responses and tumor cell autophagy, and directly suppressing tumor development in specific cancers. These mechanisms provide a theoretical basis for exploring vitexin and isovitexin as promising candidates for anticancer drug development.

3.5. Antioxidant Effect

Oxidative stress is a harmful process at the cellular or subcellular level and may become an important mediator of damage to cellular structures and various disease states. Reactive oxygen species-induced lipid peroxidation produces lipid peroxides in the cell membrane, causing extensive damage to the membrane structure and membrane-mediated chromosomal damage [56]. In recent years, vitexin and isovitexin have shown good antioxidant properties in both in vitro and in vivo studies.

Excessive nitrosation and oxidative stress can trigger inflammatory responses, leading to an increase in inflammatory mediators, which subsequently activate apoptotic pathways, are crucial for neuronal cell death, and play an important role in the development of hyperactivity disorder (OD) [38]. In vitro experiments confirmed for the first time that the flavonoid compound vitexin protects against neuro-inflammation, nitrosation, oxidative damage, mitochondrial dysfunction, apoptotic activation through the Nrf2 pathway, and haloperidol-induced hyperactivity disorder (HPD-OD). According to the study, the antioxidant qualities of vitexin were ascribed to its capacity to directly scavenge oxygen free radicals and shield antioxidant enzymes. Vitexin is a potent free-radical scavenger because it can supply electrons via the nearby dihydroxy structure of the A ring. Vitexin also increased *Nrf2* expression and boosted GSH and antioxidant enzymes such as SOD, CAT, GPx, and GST. Additionally, vitexin can exert anti-inflammatory effects by inhibiting NF kB transcription factors, downregulating pro-inflammatory mediators such as TNF-α and IL-1 β, and upregulating anti-inflammatory mediators such as IL-4 and IL-10 [29,57].

Vitiligo is a common skin depigmentation disease, and its pathogenesis is related to cellular oxidative stress caused by reactive oxygen species (ROS). Vitexin has been shown to inhibit the apoptosis of melanocytes caused by H_2O_2, promote cell proliferation, reduce the expression of inflammatory factors and ROS, and upregulate the expression of antioxidant genes HO-1 and SOD by activating the MAPK-Nrf2/ARE signaling pathway, thereby exerting its antioxidant effects. When Nrf2 was knocked down, the protective effect of vitexin was reversed, indicating that Nrf2 plays a key role in the antioxidant effect of vitexin. This study not only emphasizes the importance of the Nrf2/ARE axis in the antioxidant defense of melanocytes but also provides new strategies for the treatment of vitiligo [27].

Numerous pathophysiological processes such as tumor development, aging, inflammation, diabetes, and neurological illnesses are significantly affected by DNA damage. DNA damage results from highly active endogenous aldehydes produced by cellular processes such as oxidative stress, lipid peroxidation, and glycosylation. These aldehydes directly react with DNA to form DNA aldehyde-derived adducts. Mice that drink diluted alcohol suffer harm to their hematopoietic stem cell DNA as a result of the metabolism of ethanol into acetaldehyde. Cellular homeostasis problems caused by unrepaired DNA damage can accelerate the development and accumulation of the disease phenotypes. Different doses of vitexin have been shown to have antioxidant effects that can repair ethanol-induced DNA damage, suggesting its possible use in the treatment of liver injury [58].

In *Caenorhabditis* elegans, studies have shown that vitexin and isovitexin, as putative SKN-1/Nrf2 activators, increase lifespan and support a healthy lifespan. To eliminate ROS generated under stress, vitexin and isovitexin increased the expression of antioxidant genes and proteins, decreased the accumulation of ROS, and increased the accumulation of SKN-1 in the nucleus. Vitexin and isovitexin have shown promise as prospective compounds for novel approaches in the chemoprevention of aging and oxidative-related disorders, and they offer fresh perspectives on activating the SKN-1/Nrf2 pathway [59].

By lowering intracellular ROS levels, suppressing the expression of heat shock proteins (Hsp70 and Hsp90), and shielding cells from the synthesis of apoptotic proteins, vitexin can effectively prevent heat stress-induced cell death. In addition to offering possible application opportunities for the management of people working in hot environments and patients suffering from heat-related illnesses, this discovery provides a fresh perspective for the prevention and treatment of heat stress [60]. Research has found that vitexin inhibits ROS production by promoting PPARγ activity, enhancing antioxidant enzyme activity [30].

Chronic kidney disease (CKD) is a systemic inflammatory syndrome characterized by tubulointerstitial inflammation. LPS, an outer membrane component of Gram-negative bacteria, increases the production of ROS and triggers cellular inflammation. Isovitexin has been demonstrated in vitro to improve cell viability in SV40-MES-13 cells and inhibit ROS generation generated by LPS. Isovitexin also decreased inflammatory and apoptotic markers and enhanced the mitochondrial membrane potential. Isovitexin decreased kidney damage and inflammation in a *C57BL/6* mouse kidney injury model by enhancing protective autophagy, reducing ROS generation, and exhibiting anti-inflammatory and anti-pyroptotic properties. Isovitexin also regulates the NF-κB, P53, PI3K/Akt, and MAPK signaling pathways, which can cause autophagy in HepG2 human liver cancer cells. These findings suggest that isovitexin may exert a protective effect against kidney injury through multiple mechanisms [22,32].

Both vitexin and isovitexin demonstrated antioxidant activity both in vitro and in vivo. They achieve this by directly inhibiting oxidative stress markers, reducing inflammation, encouraging autophagy, and shielding endothelial cells. These processes complement each other, making them promising candidates for antioxidant medications. These findings offer a solid scientific foundation for investigating the antioxidant mechanisms and the possible uses of isovitexin and vitexin.

3.6. Anti-Inflammatory Effect

Vitexin has demonstrated possible protective effects against autoimmune hepatitis (AIH), a chronic progressive liver disease [3]. According to previous studies, taking vitexin dramatically lowered serum ALT and AST levels, reduced oxidative stress-induced liver damage, and decreased the infiltration of inflammatory cells and CD4 + T cells into the liver. Furthermore, vitexin ameliorated liver damage in EAH animals by upregulating the *Nrf2* gene and activating the AMPK/AKT/GSK-3 β pathway, offering scientific support for vitexin as a novel and potent medication for AIH treatment [29].

Several inflammatory cytokines, including IL-6, TNF-α, and IL-8, are associated with skin inflammation, aging, and photoaging. Following UV exposure, these cytokines increase and affect the epidermal barrier function. The excessive generation of ROS and reactive nitrogen species (RNS) is also linked to inflammation, which accelerates the development of wrinkles and skin dryness. According to previous research, vitexin has anti-inflammatory properties and helps treat skin irritation; however, it has no discernible effect on TNF-α levels and may instead control keratinocyte differentiation rather than death. Through free radical scavenging and anti-inflammatory mechanisms in primary human skin cells, vitexin was shown to eradicate UV-induced ROS/RNS for the first time [61].

Since Nrf2 is thought to be a desirable biological target for vascular inflammation, it is possible to create small-molecule medicines that specifically target it to treat metabolic disorders. Non-covalent inhibitors may be less hazardous than covalent Nrf2 activators for Nrf2 activation [62]. Vitexin is absorbed by HUVECs in a dose-dependent manner. Vitexin inhibits vascular inflammation by directly binding to the Kelch domain of Keap1, breaking the Keap1-Nrf2 connection, and releasing Nrf2 for transport to the nucleus, highlights a new function for Nrf2 in vascular inflammation and offers intriguing molecular pathways for innovative approaches to treating metabolic and cardiovascular disorders [28,63].

Lipopolysaccharide (LPS) is a microbial toxin that is one of the main causes of sepsis. During inflammation, LPS is known to impair the function of P-glycoprotein (P-gp), and molecular docking studies have shown that vitexin is an effective substrate for P-gp, while verapamil is an effective inhibitor of P-gp. Mice were intraperitoneally injected with LPS (10 mg/kg), and macrophages showed increased levels of H_2O_2, superoxide, and NO, whereas antioxidant activity decreased. Mice treated with vitexin (5 mg/kg body weight) and verapamil (5 mg/kg body weight) showed higher antioxidant enzyme activities (SOD, CAT, and GRx), thereby reducing oxidative stress. This combination therapy simultaneously downregulated the expression of TLR4, NF-κB, and P-gp in mouse peritoneal macrophages, leading to a polarization transition from M1 to M2 macrophages and reduced inflammatory responses. Therefore, combined therapy with vitexin and verapamil may be a potential therapeutic strategy to shield the body from inflammatory damage and sepsis caused by LPS [64].

A study using an LPS/D-GalN-induced acute liver injury (ALI) animal model found that isovitexin has potential therapeutic effects in ALI. Treatment with isovitexin dramatically decreased the total m6A and m6A modification levels of phosphatase and tensin homolog (PTEN) and binding immunoglobulin protein (BiP) in the liver, while upregulating the expression of PI3K, Akt, and mTOR. PTEN is a negative regulator of the PI3K/Akt pathway and plays a crucial role in controlling immune cell activation and pro-inflammatory signaling. BiP, an endoplasmic reticulum (ER) chaperone, is a key indicator of ER stress, which is often linked to inflammation and cell apoptosis. These findings suggest that isovitexin may alleviate liver inflammation by modulating PTEN and BiP expression through m6A modification, thereby influencing key inflammatory signaling cascades. To further validate these mechanistic insights, another study demonstrated that isovitexin treatment reduced the inflammatory response induced by lipopolysaccharide (LPS) in both in vitro macrophage models and an in vivo LPS-induced mouse ALI model, highlighting its potential as a novel therapeutic agent for acute liver injury [65].

Ulcerative colitis (UC) is an idiopathic inflammatory disease of the colonic mucosa. Studies have shown that dextran sulfate sodium (DSS) administration can cause chemical damage to the intestinal epithelium of mice, leading to colon atrophy and inflammation. Vitexin treatment significantly inhibited colonic atrophy induced by DSS in colitis mice and reduced colonic mucosal inflammation, ulceration, and necrosis [66,67]. Vitexin reduced inflammation and DSS-induced colon damage by concurrently blocking the phosphorylation of the p65, I κ B, and STAT1 proteins in colon tissue. In addition to its anti-inflammatory properties, the ability of vitexin to alter the colonic mucosal barrier is another crucial molecular mechanism in the fight. The gut microbiota is vital for controlling the function of the intestinal mucosal barrier and the inflammatory response, and there is mounting evidence that it plays a significant role in the onset and progression of colitis. Increased intestinal permeability and inflammatory activation brought on by gut microbiota dysfunction can hasten the onset of colitis [68,69]. Therefore, regulation of the gut microbiota is considered an important strategy for the treatment of UC. Research has found that vitexin regulated the gut microbiota of DSS-induced colitis mice and significantly increased the abundance of protective symbiotic bacteria, revealing that vitexin can alleviate colitis by reversing dysbiosis of the gut microbiota in DSS-induced colitis mice. In summary, vitexin exerts significant therapeutic effects in mouse models of DSS-induced colitis by suppressing intestinal mucosal inflammation, preserving intestinal barrier integrity, and reversing gut microbiota dysbiosis associated with colitis. In addition, vitexin has been shown to significantly improve chronic colitis symptoms in CAC mouse models induced by azoxymethane and dextran sulfate sodium (AOM/DSS), alleviate colon damage, and regulate inflammatory cytokine levels in the colon [53]. Therefore, vitexin may be a potential drug for treating or preventing colitis and provides a scientific basis for potential therapeutic drugs to prevent the progression of colitis to colorectal cancer [15,57].

Mastitis is caused by several pathogenic microorganisms. *Staphylococcus aureus*-induced mastitis produces a large number of DEGs, including immune signaling pathways, apoptosis, and endoplasmic reticulum stress. Vitexin may also reduce inflammatory factors and apoptosis by alleviating endoplasmic reticulum stress and inactivating the MAPK and NF-κB signaling pathways. Vitexin may have great potential as a preventive and therapeutic agent for mastitis [30].

Allergic contact dermatitis (ACD) is a skin disease caused by environmental and occupational allergens. During the sensitization stage of ACD, allergens activate innate immunity by releasing pro-inflammatory cytokines and chemokines through keratinocytes, recruiting T cells, and producing cytokines, such as TNF-α, IFN-γ, and IL-17A. Appropriate immune regulation is of great significance in preventing ACD. A study optimized an ACD mouse model induced by *Ginkgo biloba* acid (GA). This study showed that isovitexin reduced MAPK phosphorylation, decreased NF-κB activation, and promoted the M2 polarization of macrophages. It was also found that isovitexin blocked SHP2 with immunomodulatory effects, dose-dependently upregulated apoptosis, and inhibited cytokines, such as TNF-α, IFN-γ, IL-2, and IL-17A [70].

When the serum uric acid level in patients with gouty arthritis (GA) increases, supersaturated urate salts precipitate to form crystals and deposit in the synovial tissue. These urate crystals are recognized by TLR4 and activate the TLR4 signaling pathway, leading to the transcription, synthesis, and release of downstream pro-inflammatory factors, causing acute inflammatory reactions in joints. Research has found that isovitexin attenuated the recognition and internalization of TLR4 in inflammatory responses, blocked intracellular transmission of MyD88 connections, and inhibited the activation of NF-κB. In addition, treatment with isovitexin + TLR4 inhibitors could inhibit the release of related inflammatory factors [31].

Vitexin and isovitexin exert their anti-inflammatory activity through various mechanisms, including regulating immune responses and inflammatory cytokines, activating the Nrf2 pathway, affecting LPS-induced inflammatory responses, and exhibiting good anti-inflammatory effects on colitis, skin inflammation, and mastitis. The discovery of these anti-inflammatory mechanisms provides a scientific basis for the application of vitexin and isovitexin in the treatment of metabolic diseases.

3.7. *Neuroprotective Effect*

Vitexin and isovitexin have various pharmacological properties, including neuroprotective effects. By reducing the levels of pro-inflammatory and cytotoxic factors in LPS-induced microglia and mediating the NF-κB signaling pathway, neuroprotective effects have been demonstrated [33,71]. Research has shown that vitexin can regulate cell activity through the AKT/mTOR, p53, or Bcl-2/Bax pathways. In the treatment of cerebral ischemia–reperfusion injury (CIRI), vitexin can improve the behavior of CIRI mice and inhibit oxidative damage in the mouse brain, thereby showing its potential as a drug candidate for stroke treatment [72].

The short-term and long-term neuroprotective effects of vitexin were assessed by using a rat neonatal model. Vitexin, as an HIF1-α inhibitor, has demonstrated good protective effects by decreasing neuronal damage and reducing infarct volume, particularly when administered early after hypoxic–ischemic (HI) injury. Moreover, vitexin reduced cerebral edema, which is linked to the downregulation of VEGF and HIF-1-α, and helped preserve the integrity of the blood–brain barrier (BBB). It has been demonstrated that vitexin protects against a cerebral ischemia/reperfusion (I/R)-induced increase in the permeability of brain endothelial cells [34]. Vitexin decreased inflammation and LDH and caspase 3 levels in the human brain microvascular endothelial cells (HBMEc) I/R damage model. Additionally, vitexin suppressed matrix metalloproteinase (MMP) activity and increased the expression of tight junction proteins [46].

Research has shown that vitexin has a protective effect on cerebral ischemia–reperfusion injury by upregulating p-ERK1/2, downregulating p-JNK and p-P38, increasing Bcl-2, and decreasing Bax expression in the cortex and hippocampus. The data suggested that vitexin may exert its effect by regulating the mitogen-activated protein kinase (MAPK) signaling pathway [73].

In summary, vitexin, as a neuroprotective agent, may regulate multiple signaling pathways, including inhibition of HIF-1-α, modulation of the MAPK signaling pathway, and impact on the expression of apoptosis-related proteins. In addition, the neuroprotective effect of vitexin may be achieved through other pathways and targets, such as reducing free radical levels, combating neuronal apoptosis, regulating inflammatory factors and their pathways, and modulating neurotransmitters and related receptors. Vitexin has a significant protective effect on the increased permeability of brain endothelial cells caused by cerebral ischemia/reperfusion through various mechanisms, including the reduction in cell damage markers, alleviation of inflammatory responses, maintenance of the integrity of the blood–brain barrier, and regulation of nitric oxide synthase activity. These findings provide scientific evidence that vitexin is a potential neuroprotective agent. This may also provide new strategies for the treatment of cerebral ischemia–reperfusion injury in the future.

Another important neuroprotective role of vitexin has been demonstrated through its potential role in Alzheimer's disease (AD). AD is a chronic neurodegenerative disease characterized by the deposition of the beta amyloid protein (Aβ) to form senile plaques. Research using a *Caenorhabditis* elegans model of AD has shown that vitexin treatment can significantly prolong the lifespan of nematodes and delay Aβ-induced paralysis, which may be related to its antioxidant activity. In nematodes, *acetylcholinesterase gene 1* (*ace-1*) and *acetylcholinesterase gene 2* (*ace-2*) together account for approximately 95% of total acetylcholinesterase activity and are critically involved in the termination of cholinergic neurotransmission. Vitexin treatment significantly inhibited acetylcholinesterase activity and markedly downregulated the expression of *ace-1* and *ace-2*. In addition, it conferred neuroprotection against Aβ-induced toxicity in nematodes and upregulated the expression of the acr-8 gene. These findings suggest that vitexin, as a multifunctional compound, holds promise as a potential therapeutic agent for the treatment of Alzheimer's disease [74,75].

3.8. Antimicrobial and Antibacterial Effect

Vitexin and isovitexin have also shown promising potential in modulating intestinal microbiota and in turn play a significant role in regulating various diseases such as overweight [15], acute colitis [67], neuro-inflammation [57], and lipid metabolism disorders [76]. The gut microbiota is a key mediator of health, influencing a wide range of physiological processes such as metabolism, immune response, and inflammation [68,69]. Recent studies have suggested that vitexin and isovitexin exert therapeutic effects by altering the composition of gut microbiota, thereby mitigating disease symptoms.

Research has shown that vitexin has inhibitory effects on Gram-negative bacteria, including *Proteus mirabilis*, *Escherichia coli*, and *Enterobacter cloacae*, and even exhibits antibacterial effects against *Pseudomonas aeruginosa*, which has the potential to resist biofilms [77,78]). Vitexin can also resist *Helicobacter pylori* infection, which may be related to its anti-myeloperoxidase (MPO) enzyme activity and inhibition of H- and K-ATPase activity [79,80].

In terms of antiviral activity, vitexin is resistant to rotavirus and parainfluenza type 3 virus but has relatively weak resistance to human immunodeficiency virus (HIV). Although the molecular mechanism of its antiviral activity is still unclear, vitexin has shown resistance to plant viruses such as *Tobacco mosaic* virus (TMV) [81].

Vitexin's anti-inflammatory and antioxidant properties, as well as its inhibitory effect on a variety of pathogens, have potential applications in the treatment of mastitis and other infectious diseases [30], indicating that vitexin may become a new anti-infective drug. In a human gut simulation model, both vitexin and isovitexin were shown to affect the abundance of several gut bacterial genera, including *Adlercreutzia*, *Akkermansia*, and *Streptococcus*. These changes were associated with a reduction in the risk of T2DM in overweight individuals. In an acute colitis mice model, vitexin significantly regulated the abundance of beneficial gut bacteria such as *Veillonella*, *Parabacteroides*, and *Flavonifractor*, which are associated with reduced gut inflammation. This modulation of gut microbiota was linked to the alleviation of colitis symptoms, supporting the potential of vitexin as a therapeutic agent for inflammatory bowel diseases. Vitexin has been shown to exert neuroprotective effects by modulating the gut microbiota in a mice model of neuro-inflammation. This research suggests that vitexin's ability to regulate gut bacteria contributes to its neuroprotective effects, potentially offering a novel approach for treating neurodegenerative diseases. In a study of lipid metabolism disorders, vitexin was found to influence the levels of *Rombousia* and *Faecalibaculum* bacteria, both of which play a role in regulating lipid metabolism. This highlights the importance of gut microbiota modulation in addressing conditions related to dyslipidemia and obesity.

These findings demonstrate that vitexin and isovitexin can influence a broad range of diseases through their ability to modulate the intestinal microbiota. This emerging area of research suggests that these flavonoid compounds not only exert direct pharmacological effects but also influence disease processes indirectly via the gut microbiota, offering potential therapeutic benefits for a variety of conditions.

4. Mechanisms

To address the potential redundancy suggested in Section 3 and to enhance clarity, we have summarized the core mechanisms of vitexin and isovitexin in Table 2 and Figure 5. This table consolidates the major pharmacological effects alongside their corresponding molecular pathways and target molecules, providing a clear and systematic overview. Rather than repeating detailed mechanistic descriptions, this section highlights the key molecular targets and regulatory axes that have been consistently observed across different disease contexts, thereby facilitating a more integrated understanding of the bioactivities of vitexin and isovitexin.

Table 2. Mechanistic overview of vitexin and isovitexin.

Pharmacological Effect	Compound	Mechanism/Pathway	Target Molecules	Reference
Anti-inflammatory	Vitexin and Isovitexin	Inhibits NF-κB, activates Nrf2/HO-1, AMPK/AKT/GSK-3β	TNF-α, IL-6, IL-1β, Nrf2, ICAM-1, VCAM	[28,29,63]
Antioxidant	Isovitexin	Activates HO-1/Nrf2, reduces ROS, inhibits MAPK	ROS, GPx, SOD, HO-1, CAT	[28,32]
Anti-cancer	Vitexin	Inhibits PI3K/Akt, promotes apoptosis, suppresses HMGB1, modulates Bcl-2/Bax ratio	HMGB1, caspase-3, CDK1, Bcl-2, Bax	[21,22,25]
Hepatoprotective	Vitexin	Modulates Sirt1/p53, reduces apoptosis and lipid accumulation, activates AMPK, enhances IRS-1/AKT signaling	Sirt1, p53, TG, ALT, AST, PPARγ, SREBP-1c	[44,47]
Neuroprotective	Vitexin	Regulates HIF-1-α, MAPK, Keap1/Nrf2, AKT/mTOR, reduces inflammation	VEGF, MMPs, Bcl-2, Bax, caspase-3	[34,71,73]
Cardioprotective	Vitexin	Regulates Epac1-Rap1 pathway, FOXO3a, MAPK/ERK	FOXO3a, *Epac1*, Drp1, *MFN2*	[4,9,10]
Antidiabetic	Vitexin and Isovitexin	Inhibits α-glucosidase/α-amylase, promotes GLUT4, modulates gut microbiota	PTP-1B, GLUT4, GPx4, Nrf2	[12,15,16,40]
Anti-obesity	Vitexin	Activates AMPKα, inhibits C/EBPα, FAS, activates Hedgehog signaling	AMPKα, C/EBPα, FAS	[17,18,41]

Table 2. *Cont.*

Pharmacological Effect	Compound	Mechanism/Pathway	Target Molecules	Reference
Anti-fibrotic	Isovitexin	Suppresses PI3K/Akt, modulates *miR-21* and GSH pathway	*miR-21*, PI3K, GSH, PTEN	[19]
Antimicrobial	Vitexin	Inhibits H+/K+-ATPase, suppresses MPO and biofilm formation, interferes with bacterial efflux pumps	MPO, H+/K+-ATPase	[77,79,80]

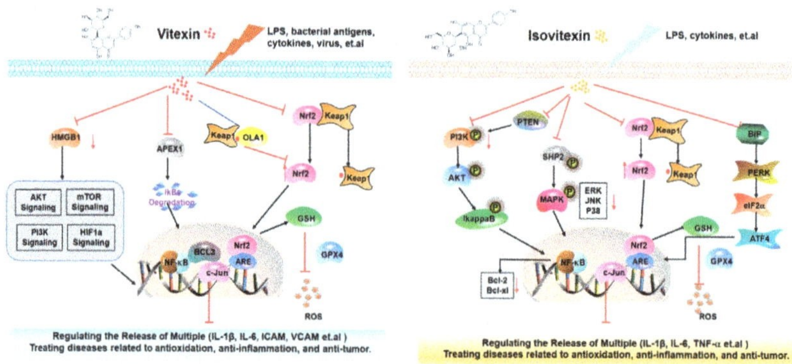

Figure 5. Target mechanism diagram of vitexin and isovitexin.

Building upon the summarized evidence, we further elaborate on the biological activities and molecular functions of vitexin and isovitexin in this section. In Table 2, these compounds have demonstrated antioxidant, anti-inflammatory, anticancer, neuroprotective, and cardioprotective properties. Additionally, they have demonstrated lipid-lowering effects, regulation of glucose metabolism, and hepatoprotective activity. Both in vitro and in vivo studies have validated their anticancer potential, primarily through the induction of apoptosis and autophagy [25,26]. Consequently, the molecular targets and therapeutic relevance of vitexin and isovitexin in disease treatment have attracted significant research interest.

Nrf2 is widely regarded as an attractive biological target, and small-molecule compounds that activate Nrf2 can be developed for the treatment of metabolic diseases. Research has confirmed that vitexin significantly enhances the expression of the Nrf2 protein. Using tissue adenosine triphosphate (TPP), molecular docking, siRNA-mediated RNA interference, and in vitro cell-based studies, Obg-like ATPase 1 (OLA1) was reported for the first time as a novel protein target of vitexin that activates Nrf2 and inhibits colitis through non-covalent interactions with Keap1. Vitexin significantly reduced the elevated levels of TNF-α and IL-1 β in the serum of DSS-induced colitis. In addition, after treatment with vitexin, the levels of pro-inflammatory cytokines (IL-1 β, IL-6, ICAM, and VCAM) in the colon of DSS-induced colitis mice were significantly reduced at both the gene and protein levels [40,63,82]. OLA1 is a novel target protein that contributes to the anti-inflammatory effects by increasing the Nrf2 protein expression. OLA1 provides a molecular mechanism for a new therapeutic strategy for colitis and related systemic inflammations [82].

Vitexin has therapeutic effects on chronic kidney disease (CKD), a progressive and destructive disease that may ultimately lead to irreversible renal failure, end-stage renal disease (ESRD), and premature death. Previous studies have confirmed that vitexin protects the kidneys and prevents the formation of kidney stones by inhibiting pyroptosis, apoptosis, epithelial–mesenchymal transition (EMT), and macrophage activation. Vitexin also has a protective effect against lipopolysaccharide (LPS)-induced apoptosis of rat renal tubular epithelial cells and reduces cadmium-induced nephrotoxicity. In addition, vitexin inhibited the development of diabetes nephropathy by regulating the NF-κB and AMPK signaling pathways. Data showed that vitexin inhibited ferroptosis and reduced GPX4-related lipid peroxidation by regulating the KEAP1/NRF2/HO-1 pathway, thereby exerting a renoprotective effect [27,29]. In addition, vitexin is considered a candidate drug for cancer treatment because of its ability to inhibit EMT in various cancers. Mechanistic studies have found that vitexin exerts its effects by inhibiting the expression of HMGB1 and activating downstream PI3K/AKT/mTOR/HIF-1 α signaling pathways, providing evidence for potential drugs for the clinical treatment of CKD [21].

APEX1 plays a role in promoting the atheromatous phenotype of the vascular endothelium. The expression of pro-inflammatory cytokines (SELE, VCAM1, and ICAM1) induced by oxidative stress is mediated by APEX1 via the NF-κB pathway. In vivo vitexin administration alleviated neointimal formation and atherosclerosis induced by disturbed blood flow. Through the use of target participation analysis (CETSA) and biophysical properties (SPR), vitexin, as a direct inhibitor of APEX1 in endothelial cells, plays a key role in the anti-inflammatory treatment of atherosclerosis, providing a theoretical basis for the clinical use of vitexin [83].

5. Safety Profile and Toxicological Assessment of Vitexin and Isovitexin

Given the potential of vitexin and isovitexin as drug candidates, it is essential to evaluate their safety profiles for human application. At present, a large number of in vitro and in vivo studies have confirmed the safety of vitexin, whereas research on isovitexin is relatively scarce. In vitro experiments have shown that vitexin had no cytotoxicity (IC_{50} > 200 μg/mL). In in vivo studies, high concentrations of vitexin and isovitexin did not show significant toxicity in terms of acute and sub-chronic toxicity, as well as genetic toxicity. Furthermore, in terms of liver and gastric mucosal injury, the long-term repeated use of high-dose vitexin (10 mg/kg, i. p.) was shown to be safe. Vitexin did not cause significant toxic reactions, even at high doses in in vivo experiments [84,85]. These results indicate that vitexin has potential advantages in terms of safety. However, further clinical research on these compounds is necessary to comprehensively evaluate their safety and efficacy, especially for their application in humans.

6. Absorption and Metabolism of Vitexin and Isovitexin in Human

Vitexin and isovitexin have poor absorption in the gastrointestinal tract, with significant first-pass effects in the intestine (approximately 94%), stomach (30%), and liver (50%), resulting in a lower bioavailability (F) (approximately 5%). Vitexin and isovitexin directly reach the colon and are hydrolyzed by the gut microbiota through deglycosylation and heterocyclic C-ring opening. The intestinal bacteria *Eubacterium cellulosolvens* can completely convert isovitexin into apigenin by decomposing it into small molecule phenols and various aromatic acids. However, vitexin cannot undergo this complete conversion. This difference may be attributed to the presence of two adjacent hydroxyl groups at the C6 position in isovitexin, which are absent in vitexin [86].

Vitexin is rapidly and widely distributed in various tissues after intravenous and oral administration in rats or mice and has the highest content in the stomach and intestines within 0.5 h, which may be related to residual drugs and enterohepatic circulation. The highest accumulation was observed in the liver. In contrast, the enterohepatic circulation of vitexin also affects its accumulation in the body and it is mainly excreted through the urine. Through intravenous administration in rats, vitexin is mainly distributed in the liver and kidneys, with the lowest concentration in the brain and adipose tissue, and most of it is excreted through the urine and bile. In contrast, after intravenous administration of isovitexin in rats, the levels were highest in the kidneys, liver, and lungs and lowest in the brain. The isovitexin content in the ovary was much higher than that in other organs [86].

The absorption and metabolism processes of vitexin and isovitexin in the human body are complex, and their bioavailability is influenced by multiple factors, including the action of the gut microbiota, interactions with dietary components, first-pass effects, and plasma protein-binding rates. As shown in Table 3, different nano-delivery systems have been developed to enhance the bioavailability of vitexin by improving its solubility, stability, and sustained release properties. The vitexin-loaded bilayer nanoparticles are designed by assembling soybean peptides and coating them with a goblet cell-targeting peptide. They significantly increase the bioaccessibility and bioavailability of vitexin while providing better antioxidant activity through sustained release in the intestine [87]. A *zein*–pectin nanoparticle system exhibits slow-release properties and enhanced absorption in the duodenum, providing a robust delivery system that improves the bioavailability of vitexin, with potential applications in sustained-release formulations [88]. Researchers have also developed D-α-tocopherol polyethylene glycol succinate, polyvinylpyrrolidone K30, and sodium cholate-mixed micelles to enhance the bioavailability and demonstrated the anti-osteoporotic effect of vitexin. The oral bioavailability of vitexin increased by 5.6-fold compared to free vitexin [89]. Another delivery system of mPEG-g-CTS/ALG polyelectrolyte complex nanoparticles enhances vitexin's gastrointestinal digestion and is suitable for oral, intestinal-specific delivery, offering a new approach for improving vitexin's absorption and bioavailability [90]. Liposomal encapsulation using the 'thin-film hydration' method provides an effective strategy for treating liver cirrhosis by enhancing the bioavailability and therapeutic effectiveness of vitexin through oral delivery [91]. Moreover, the vitexin-rhamnoside (VR) and *zein*-VR-pectin nanoparticles were found to improve the bioavailability of vitexin while alleviating chronic inflammation and hepatic injury in HFD mice [76]. Collectively, these factors determine the distribution and excretion of these two compounds in the body, thereby affecting their efficacy and safety [86–88]. Therefore, in clinical applications, it is necessary to optimize administration strategies based on pharmacokinetic characteristics.

Table 3. Design types and mechanisms of vitexin-loaded nanocarriers.

Nano Types	Efficacy Tested	Model Type	Results	References
Vitexin-loaded bilayer nanoparticles by the assembly of soybean peptides and the coating of goblet cell targeting peptide CSKSSDYQC (CSK) coupled N-trimethyl chitosan (TMC)	The bilayer nanoparticles could protect vitexin from being released in stomach and promote sustained release in intestine	In vitro	Bioaccessibility and bioavailability of vitexin was significantly increased by the bilayer nanoparticles and vitexin exhibited better antioxidant activity after encapsulation.	[87]
Encapsulated by the *zein*-pectin nanoparticles system	Nanoparticles exhibited significant slow-release properties and the highest absorption rate in the duodenal segment of rats	In vivo/ In vitro	It provides a theoretical and technical approach for the construction of vitexin delivery system with sustained-release properties and higher bioavailability	[88]

Table 3. *Cont.*

Nano Types	Efficacy Tested	Model Type	Results	References
Vitexin (Vi)-loaded D-α-tocopherol polyethylene glycol succinate, polyvinylpyrrolidone K30, and sodium cholate-mixed micelles	Vi-MMs exhibited enhanced bioavailability and anti-osteoporotic effect	In vivo	The oral bioavailability of Vi-MMs was increased by 5.6-fold compared to free vitexin.	[89]
Vitexin into poly(ethylene glycol) methyl ether-grafted chitosan (mPEG-g-CTS)/alginate (ALG) polyelectrolyte complex nanoparticles.	The gastrointestinal digestion of vitexin increased by encapsulating into mPEG-g-CTS/ALG nanoparticles.	In vitro	Nanoparticles are suitable for oral intestinal-specific delivery systems.	[90]
Vitexin-encapsulated liposomes were synthesized by the 'thin-film hydration' method	VLP-treated group also showed better results up to some extent.	In vivo	Liposomal encapsulation of vitexin and subsequent PEG coating to be a substantial strategy for treating liver cirrhosis through oral drug delivery.	[91]
Vitexin-rhamnoside (VR) and *zein*-VR-pectin nanoparticles (VRN)	Alleviating chronic inflammation and hepatic injury in HFD mice.	In vivo	Provided new evidence that nanoparticles enhance the bioavailability of vitexin bioactive ingredients.	[76]

7. Bioavailability

Vitexin and isovitexin have poor water solubility, resulting in low oral bioavailability and poor gastrointestinal absorption, which limit their effectiveness in vivo. Studies have confirmed that loading it into a nano-delivery system can improve its bioavailability and biological activity. A novel pH-stable targeted oral drug delivery nanosystem was constructed by assembling soy peptides to form a hydrophobic inner layer for loading vitexin and crosslinking the peptide CSKSSDYQC (CSK) targeting goblet cells coupled with N-trimethyl chitosan (TMC) with TPP ions to form an outer layer [87]. By encapsulating vitexin in a corn-*zein*-pectin nanoparticle system, the solubility, stability, and bioavailability of vitexin, as well as the sustained-release properties of the nanoparticles and their high absorption rate in the small intestine, can be improved [88]. D-α-tocopherol polyethylene glycol succinate, polyvinylpyrrolidone K30, and sodium cholate-mixed micelles (Vi MMs) loaded with vitexin improved the oral bioavailability of vitexin and enhanced its anti-osteoporosis effect [89]. Researchers have successfully loaded vitexin into polyethylene glycol methyl ether-grafted chitosan/alginate (mPEG-g-CTS/ALG) nanoparticles, which could enhance their digestion [76,90,91].

The results showed that nanoparticles can protect vitexin from release in the stomach and promote its sustained release in the intestine. In vitro and in vivo experiments showed that nanoparticles significantly improved the bioavailability of vitexin, and vitexin exhibited better antioxidant activity after encapsulation, providing a new strategy for improving the bioavailability of vitexin [76,87–91].

8. Prospective

Flavonoids are a class of polyphenols widely present in nature. They have a variety of biological activities, including antioxidant, anti-inflammatory, antitumor, cognitive improvement, cardiovascular protection, and anti-diabetic effects [81,92,93]. These biological activities may be related to the phenolic hydroxyl groups in their molecular structure, especially the adjacent trihydroxy or dihydroxy groups on the A or B ring, which may serve as active functional groups. In addition, the skeletal structure of flavonoids may have good compatibility with certain proteins, although the current evidence is insufficient and requires further validation [1–3].

As known flavonoid compounds, the apigenin 8/6-C-glucoside derivatives vitexin and isovitexin not only possess the common biological activities of flavonoids but also exhibit specificity due to their unique chemical structures. For instance, their shared adjacent dihydroxy structure on the A ring facilitates effective free radical scavenging, while the presence of C-8 glucoside enhances the antioxidant capacity of vitexin by reducing the negative charge of the oxygen atom at C-3 [1,2,7]. Recent studies have suggested that these compounds possess a broad spectrum of biological activities, which may contribute to the treatment of cancer, cognitive impairment, depression, Alzheimer's disease, brain injury, ischemia/reperfusion injury, pain, cardiac hypertrophy, hypertension, diabetes, obesity, infections, and metabolic disorders [81].

These compounds influence numerous physiological processes, including antioxidant, anti-inflammatory [28,29,63], and pro-apoptotic effects [45]. Additionally, they exhibit protective effects on the nervous [33] and cardiovascular systems [4,9,10,20], thyroid gland, liver, and other organs [81] Several biological mechanisms have been proposed to explain their pharmacological actions, such as the apoptotic signaling pathway, inflammatory cytokine network, the MAPK pathway, and key molecular targets including *HIF-1α*, *P53*, and related signaling cascades [28,29,44].

Despite its structural similarity to vitexin, isovitexin has not been extensively studied. Although research on isovitexin has been relatively recent, its potential in antioxidation, anti-inflammation, blood glucose regulation, and liver fibrosis improvement has been demonstrated [81]. As research progresses, isovitexin has emerged as a promising drug candidate for the treatment of various diseases. Further investigations into isovitexin will not only deepen our understanding of flavonoids but also provide essential scientific insights for the development of novel therapeutic agents.

The gut microbiota plays a crucial role in host health by regulating immune responses, preventing pathogen invasion, and influencing the progression of intestinal diseases. Given their anti-inflammatory and antioxidant properties, vitexin and isovitexin may exert beneficial effects by modulating the gut microbiota composition, as shown in Table 4 [15,57]. Future studies should explore the mechanisms through which vitexin and isovitexin interact with the gut microbiota, including their effects on specific microbial populations and their influence on host immune responses and gut barrier function. Additionally, investigating their potential synergy with probiotics and their role in treatment strategies, such as fecal microbiota transplantation, could provide valuable insights.

Thus far, experimental approaches for studying vitexin and isovitexin have included screening assays for medicinal plant extracts, protein-level analyses, observations of cell death and tissue damage, and behavioral tests in animal models. However, most studies have used doses exceeding those typically used for the oral administration of vitexin and isovitexin as plant-derived compounds. Determining the most effective and physiologically relevant dose remains a significant challenge, necessitating further research to facilitate future clinical trials.

Table 4. The regulation and related mechanisms of vitexin and isovitexin on intestinal flora.

Vitexin/ Isovitexin	Disease	Model	Change at Genus Levels	Results	References
Vitexin/ Isovitexin	Overweight	Simulation of Human Gut Model	*Adlercreutzia, Terrisporobactel, Promicromonospor, Pseudonocardia, Anaerostipes, Akkermansia, Alistipes, Parabacteroides, Enterocloster, Peptacetobacter, Collinsella, Paraclostridium, Duncaniella, Streptococcus, Gillisia*	Industry can use this optimal ratio to formulate more effective functional ingredients for functional foods and create nutraceuticals designed to reduce the risk of T2DM in overweight individuals.	[15]
Vitexin	Acute colitis	An acute colitis mice model	*Veillonella, Terrisporobacter, Klebsiella, Paeniclostridium, Parabacteroides, Flavonifractor, Blautia*	Vitexin could alleviate colitis by regulating gut microbiota and attenuating gut inflammation.	[67]
Vitexin	Neuro-inflammation	Mice model	*Akkermansia, Lachnospiraceae*	Vitexin exerted neural protective effects via anti-oxidant, anti-inflammatory, and gut microbiota modulating properties.	[57]
Vitexin	Lipid metabolism disorders		*Rombousia* and *Faecalibaculum*	Vitexin can regulate the gut microbiota and thus improve lipid metabolism.	[76]

9. Conclusions

The current understanding of the pharmacological and biological activities, as well as the underlying mechanisms of vitexin and isovitexin, remains limited and warrants systematic investigation. Future research should consider the following:

1. Identification of molecular targets involved in neuroprotection using animal models.
2. Investigation of potential synergistic effects and structure–activity relationships between vitexin, isovitexin, and other therapeutic agents.

Ultimately, in-depth studies encompassing both in vitro and in vivo approaches would enhance our understanding of the pharmacological effects of vitexin and isovitexin. These compounds may serve as promising alternative therapeutic agents for various diseases or as treatments that target specific molecular pathways. Furthermore, vitexin and isovitexin may hold potential as adjunct therapies or preventative health supplements for multiple diseases, paving the way for their broader applications in medicine and healthcare.

Author Contributions: All authors have made substantial contributions to the study, including data collection, manuscript drafting, revision, and final approval of the version to be published. Their specific contributions are as follows: W.Y.: contributed to the conception and design of the study, data collection and analysis, drafting of the manuscript, and critical revision of the manuscript; J.C.: assisted in manuscript revision, provided input on data organization, and helped with the literature review and reference management; B.X.: responsible for the overall structure of the paper, supervision of the research, manuscript review, and final approval. All authors have read and agreed to the published version of the manuscript.

Funding: This work was jointly supported by the Guangdong Higher Education Upgrading Plan (2021–2025) (No. of UICR0400015-24B and UICR0400016-24B) at Beijing Normal Hong Kong Baptist University, Zhuhai, China.

Conflicts of Interest: The authors declare no conflicts of interest.

References

1. Ganesan, K.; Xu, B. Molecular targets of vitexin and isovitexin in cancer therapy: A critical review. *Ann. N. Y. Acad. Sci.* **2017**, *1401*, 102–113. [CrossRef] [PubMed]
2. Lin, J.; Ran, H.; Feng, Q.; Shen, Q.; Zhou, S.; Sun, Y.; Hou, D. Unveiling the differences between vitexin and isovitexin: From the perspective of sources, green advanced extraction technologies, biological activities, and safety. *Food Chem.* **2025**, *415*, 144600. [CrossRef] [PubMed]
3. Abdulai, I.L.; Kwofie, S.K.; Gbewonyo, W.S.; Boison, D.; Puplampu, J.B.; Adinortey, M.B. Multitargeted effects of vitexin and isovitexin on diabetes mellitus and its complications. *Sci. World J.* **2021**, *2021*, 6641128. [CrossRef] [PubMed]
4. Xue, W.; Wang, X.; Tang, H.; Sun, F.; Zhu, H.; Huang, D.; Dong, L. Vitexin attenuates myocardial ischemia/reperfusion injury in rats by regulating mitochondrial dysfunction induced by mitochondrial dynamics imbalance. *Biomed. Pharmacother.* **2020**, *124*, 109849. [CrossRef] [PubMed]
5. Zhang, D.; Ning, T.; Wang, H. Vitexin alleviates inflammation and enhances apoptosis through the regulation of the JAK/STAT/SOCS signaling pathway in the arthritis rat model. *J. Biochem. Mol. Toxicol.* **2022**, *36*, e23201. [CrossRef] [PubMed]
6. Cao, D.; Li, H.; Yi, J.; Zhang, J.; Che, H.; Cao, J.; Yang, L.; Zhu, C.; Jiang, W. Antioxidant properties of the mung bean flavonoids on alleviating heat stress. *PLoS ONE* **2011**, *6*, e21071. [CrossRef] [PubMed]
7. Pereira, C.A.M.; Yariwake, J.H.; McCullagh, M. Distinction of the C-glycosylflavone isomer pairs orientin/isoorientin and vitexin/isovitexin using HPLC-MS exact mass measurement and in-source CID. *Phytochem. Anal.* **2005**, *16*, 295–301. [CrossRef] [PubMed]
8. Praveena, R.; Sadasivam, K.; Kumaresan, R.; Deepha, V.; Sivakumar, R. Experimental and DFT studies on the antioxidant activity of a C-glycoside from *Rhynchosia capitata*. *Spectrochim. Acta A Mol. Biomol. Spectrosc.* **2013**, *103*, 442–452. [CrossRef] [PubMed]
9. Yang, H.; Xue, W.; Ding, C.; Wang, C.; Xu, B.; Chen, S.; Zha, B.; Sun, Y.; Zhu, H.; Zhang, J.; et al. Vitexin mitigates myocardial ischemia/reperfusion injury in rats by regulating mitochondrial dysfunction via Epac1-Rap1 signaling. *Oxid. Med. Cell. Longev.* **2021**, *2021*, 9921982. [CrossRef] [PubMed]
10. Sun, Z.; Yan, B.; Yu, W.Y.; Yao, X.; Ma, X.; Sheng, G.; Ma, Q. Vitexin attenuates acute doxorubicin cardiotoxicity in rats via the suppression of oxidative stress, inflammation and apoptosis and the activation of FOXO3a. *Exp. Ther. Med.* **2016**, *12*, 1879–1884. [CrossRef] [PubMed]
11. Zheng, L.; Huang, J.; Su, Y.; Wang, F.; Kong, H.; Xin, H. Vitexin ameliorates preeclampsia phenotypes by inhibiting *TFPI-2* and *HIF-1α/VEGF* in a l-NAME induced rat model. *Drug Dev. Res.* **2019**, *80*, 1120–1127. [CrossRef] [PubMed]
12. Choo, C.Y.; Sulong, N.Y.; Man, F.; Wong, T.W. Vitexin and isovitexin from the leaves of *Ficus deltoidea* with in-vivo α-glucosidase inhibition. *J. Ethnopharmacol.* **2012**, *142*, 776–781. [CrossRef] [PubMed]
13. Saeting, O.; Chandarajoti, K.; Phongphisutthinan, A.; Hongsprabhas, P.; Sae-Tan, S. Water extract of mungbean (*Vigna radiata* L.) inhibits protein tyrosine phosphatase-1B in insulin-resistant HepG2 cells. *Molecules* **2021**, *26*, 1452. [CrossRef] [PubMed]
14. Zhang, S.; Jin, S.; Zhang, S.; Li, Y.Y.; Wang, H.; Chen, Y.; Lu, H. Vitexin protects against high glucose-induced endothelial cell apoptosis and oxidative stress via Wnt/β-catenin and Nrf2 signalling pathway. *Arch. Physiol. Biochem.* **2024**, *130*, 275–284. [CrossRef] [PubMed]
15. Yutharaksanukul, P.; Tangpromphan, P.; Tunsagool, P.; Sae-Tan, S.; Nitisinprasert, S.; Somnuk, S.; Nakphaichit, M.; Pusuntisumpun, N.; Wanikorn, B. Effects of purified vitexin and iso-vitexin from mung bean seed coat on antihyperglycemic activity and gut microbiota in overweight individuals' modulation. *Nutrients* **2024**, *16*, 3017. [CrossRef] [PubMed]
16. Zhang, S.; Zhang, S.; Wang, H.; Chen, Y. Vitexin ameliorated diabetic nephropathy via suppressing GPX4-mediated ferroptosis. *Eur. J. Pharmacol.* **2023**, *951*, 175787. [CrossRef] [PubMed]
17. Jiang, Y.; Gong, Q.; Gong, Y.; Zhuo, C.; Huang, J.; Tang, Q. Vitexin attenuates non-alcoholic fatty liver disease lipid accumulation in high fat-diet fed mice by activating autophagy and reducing endoplasmic reticulum stress in liver. *Biol. Pharm. Bull.* **2022**, *45*, 260–267. [CrossRef] [PubMed]

18. Peng, Y.; Sun, Q.; Xu, W.; He, Y.; Jin, W.; Yuan, L.; Gao, R. Vitexin ameliorates high fat diet-induced obesity in male *C57BL/6J* mice via the AMPKα-mediated pathway. *Food Funct.* **2019**, *10*, 1940–1947. [CrossRef] [PubMed]
19. Huang, Y.; Luo, W.; Chen, S.; Su, H.; Zhu, W.; Wei, Y.; Qiu, Y.; Long, Y.; Shi, Y.; Wei, J. Isovitexin alleviates hepatic fibrosis by regulating *miR-21*-mediated PI3K/Akt signaling and glutathione metabolic pathway: Based on transcriptomics and metabolomics. *Phytomedicine* **2023**, *121*, 155117. [CrossRef] [PubMed]
20. Lin, Y.; Li, L.; Huang, H.; Wen, X.; Zhang, Y.; Zhang, R.; Huang, W. Vitexin inhibits TNBC progression and metastasis by modulating macrophage polarization through EGFR signaling. *J. Immunother.* **2024**, *47*, 303–312. [CrossRef] [PubMed]
21. Zhou, P.; Zheng, Z.H.; Wan, T.; Wu, J.; Liao, C.W.; Sun, X.J. Vitexin inhibits gastric cancer growth and metastasis through HMGB1-mediated inactivation of the PI3K/AKT/mTOR/*HIF-1α* signaling pathway. *J. Gastric Cancer* **2021**, *21*, 439–456. [CrossRef] [PubMed]
22. Zhao, C.; He, Y.; Shi, H.; Han, C.; Zhu, X.; Wang, C.; Wang, B.; Liu, J.; Shi, Y.; Hua, D. Investigating the molecular mechanism of vitexin targeting *CDK1* to inhibit colon cancer cell proliferation via GEO chip data mining, computer simulation, and biological activity verification. *Naunyn Schmiedebergs Arch. Pharmacol.* **2024**, *398*, 1637–1652. [CrossRef] [PubMed]
23. Li, J.; Shang, L.; Zhou, F.; Wang, S.; Liu, N.; Zhou, M.; Lin, Q.; Zhang, M.; Cai, Y.; Chen, G.; et al. *Herba Patriniae* and its component isovitexin show anti-colorectal cancer effects by inducing apoptosis and cell-cycle arrest via p53 activation. *Biomed. Pharmacother.* **2023**, *168*, 115690. [CrossRef] [PubMed]
24. Chen, M.; Wang, X.; Zha, D.; Cai, F.; Zhang, W.; He, Y.; Huang, Q.; Zhuang, H.; Hua, Z. Apigenin potentiates TRAIL therapy of non-small cell lung cancer via upregulating *DR4/DR5* expression in a *p53*-dependent manner. *Sci. Rep.* **2016**, *6*, 35468. [CrossRef] [PubMed]
25. Liu, X.; Jiang, Q.; Liu, H.; Luo, S. Vitexin induces apoptosis through mitochondrial pathway and PI3K/Akt/mTOR signaling in human non-small cell lung cancer A549 cells. *Biol. Res.* **2019**, *52*, 7. [CrossRef] [PubMed]
26. Xu, C.; Cao, X.; Cao, X.; Liu, L.; Qiu, Y.; Li, X.; Zhou, L.; Ning, Y.; Ren, K.; Cao, J. Isovitexin inhibits stemness and induces apoptosis in hepatocellular carcinoma SK-Hep-1 spheroids by upregulating *miR-34a* expression. *Anticancer Agents Med. Chem.* **2020**, *20*, 1654–1663. [CrossRef] [PubMed]
27. Li, X.-S.; Tang, X.-Y.; Su, W.; Li, X. Vitexin protects melanocytes from oxidative stress via activating MAPK-*Nrf2*/ARE pathway. *Immunopharmacol. Immunotoxicol.* **2020**, *42*, 594–603. [CrossRef] [PubMed]
28. Lv, H.; Yu, Z.; Zheng, Y.; Wang, L.; Qin, X.; Cheng, G.; Ci, X. Isovitexin exerts anti-inflammatory and anti-oxidant activities on lipopolysaccharide-induced acute lung injury by inhibiting MAPK and NF-κB and activating HO-1/Nrf2 pathways. *Int. J. Biol. Sci.* **2016**, *12*, 72–86. [CrossRef] [PubMed]
29. Zhang, L.; Chen, D.; Tu, Y.; Sang, T.; Pan, T.; Lin, H.; Cai, C.; Jin, X.; Wu, F.; Xu, L.; et al. Vitexin attenuates autoimmune hepatitis in mouse induced by syngeneic liver cytosolic proteins via activation of AMPK/AKT/GSK-3β/Nrf2 pathway. *Eur. J. Pharmacol.* **2022**, *917*, 174720. [CrossRef] [PubMed]
30. Chen, Y.; Yang, J.; Huang, Z.; Yin, B.; Umar, T.; Yang, C.; Zhang, X.; Jing, H.; Guo, S.; Guo, M.; et al. Vitexin mitigates *Staphylococcus* aureus-induced mastitis via regulation of ROS/ER stress/NF-κB/MAPK pathway. *Oxid. Med. Cell Longev.* **2022**, *2022*, 7977433. [CrossRef] [PubMed]
31. Sun, X.; Li, P.; Qu, X.; Liu, W. Isovitexin alleviates acute gouty arthritis in rats by inhibiting inflammation via the TLR4/MyD88/NF-κB pathway. *Pharm. Biol.* **2021**, *59*, 1326–1333. [CrossRef] [PubMed]
32. Tseng, C.Y.; Yu, P.R.; Hsu, C.C.; Lin, H.H.; Chen, J.H. The effect of isovitexin on lipopolysaccharide-induced renal injury and inflammation by induction of protective autophagy. *Food Chem. Toxicol.* **2023**, *172*, 113581. [CrossRef] [PubMed]
33. Zolkiffly, S.Z.I.; Stanslas, J.; Abdul Hamid, H.; Mehat, M.Z. *Ficus deltoidea*: Potential inhibitor of pro-inflammatory mediators in lipopolysaccharide-induced activation of microglial cells. *J. Ethnopharmacol.* **2021**, *279*, 114309. [CrossRef] [PubMed]
34. Guo, L.; Shi, L. Vitexin improves cerebral ischemia-reperfusion injury by attenuating oxidative injury and ferroptosis via Keap1/Nrf2/HO-1 signaling. *Neurochem. Res.* **2023**, *48*, 980–995. [CrossRef] [PubMed]
35. Li, Z.M.; Liu, N.; Jiang, Y.P.; Yang, J.M.; Zheng, J.; Sun, M.; Li, Y.X.; Sun, T.; Wu, J.; Yu, J.Q. Vitexin alleviates streptozotocin-induced sexual dysfunction and fertility impairments in male mice via modulating the hypothalamus-pituitary-gonadal axis. *Chem. Biol. Interact.* **2019**, *297*, 119–129. [CrossRef] [PubMed]
36. Nurdiana, S.; Goh, Y.M.; Ahmad, H.; Dom, S.M.; Syimal'ain Azmi, N.; Noor Mohamad Zin, N.S.; Ebrahimi, M. Changes in pancreatic histology, insulin secretion and oxidative status in diabetic rats following treatment with *Ficus deltoidea* and vitexin. *BMC Complement. Altern. Med.* **2017**, *17*, 290. [CrossRef]
37. Nurdiana, S.; Goh, Y.M.; Hafandi, A.; Dom, S.M.; Nur Syimal'ain, A.; Noor Syaffinaz, N.M.; Ebrahimi, M. Improvement of spatial learning and memory, cortical gyrification patterns and brain oxidative stress markers in diabetic rats treated with *Ficus deltoidea* leaf extract and vitexin. *J. Tradit. Complement. Med.* **2017**, *8*, 190–202. [CrossRef] [PubMed]
38. Chen, S.; Wang, M.; Chang, K.; Fang, C.; Lin, Y.; Tseng, H. Vitexin mitigates haloperidol-induced orofacial dyskinesia in rats through activation of the Nrf2 pathway. *Int. J. Mol. Sci.* **2024**, *25*, 10206. [CrossRef] [PubMed]

39. Chen, Y.G.; Li, P.; Li, P.; Yan, R.; Zhang, X.Q.; Wang, Y.; Zhang, X.T.; Ye, W.C.; Zhang, Q.W. α-Glucosidase inhibitory effect and simultaneous quantification of three major flavonoid glycosides in *Microctis folium*. *Molecules* **2013**, *18*, 4221–4232. [CrossRef] [PubMed]
40. Song, J.; Wang, H.; Sheng, J.; Zhang, W.; Lei, J.; Gan, W.; Cai, F.; Yang, Y. Vitexin attenuates chronic kidney disease by inhibiting renal tubular epithelial cell ferroptosis via *NRF2* activation. *Mol. Med.* **2023**, *29*, 147. [CrossRef] [PubMed]
41. Qiu, S.; Chen, J.; Kim, J.T.; Zhou, Y.; Moon, J.H.; Lee, S.B.; Park, H.J.; Lee, H.J. Suppression of adipogenesis and fat accumulation by vitexin through activation of hedgehog signaling in 3T3-L1 adipocytes. *J. Med. Food* **2022**, *25*, 313–323. [CrossRef] [PubMed]
42. Li, W.; Deng, Z.; Xiao, S.; Du, Q.; Zhang, M.; Song, H.; Zhao, C.; Zheng, L. Protective effect of vitexin against high fat-induced vascular endothelial inflammation through inhibiting trimethylamine N-oxide-mediated RNA m6A modification. *Food Funct.* **2024**, *15*, 6988–7002. [CrossRef] [PubMed]
43. Mackowiak, B.; Fu, Y.; Maccioni, L.; Gao, B. Alcohol-associated liver disease. *J. Clin. Investig.* **2024**, *134*, e176345. [CrossRef] [PubMed]
44. Yuan, H.; Duan, S.; Guan, T.; Yuan, X.; Lin, J.; Hou, S.; Lai, X.; Huang, S.; Du, X.; Chen, S. Vitexin protects against ethanol-induced liver injury through *Sirt1/p53* signaling pathway. *Eur. J. Pharmacol.* **2020**, *873*, 173007. [CrossRef] [PubMed]
45. Cui, Z.; Zhao, X.; Amevor, F.K.; Du, X.; Wang, Y.; Li, D.; Shu, G.; Tian, Y.; Zhao, X. Therapeutic application of quercetin in aging-related diseases: SIRT1 as a potential mechanism. *Front. Immunol.* **2022**, *13*, 943321. [CrossRef] [PubMed]
46. Younossi, Z.M.; Koenig, A.B.; Abdelatif, D.; Fazel, Y.; Henry, L.; Wymer, M. Global epidemiology of nonalcoholic fatty liver disease-Meta-analytic assessment of prevalence, incidence, and outcomes. *Hepatology* **2016**, *64*, 73–84. [CrossRef] [PubMed]
47. Inamdar, S.; Joshi, A.; Malik, S.; Boppana, R.; Ghaskadbi, S. Vitexin alleviates non-alcoholic fatty liver disease by activating *AMPK* in high fat diet fed mice. *Biochem. Biophys. Res. Commun.* **2019**, *519*, 106–112. [CrossRef] [PubMed]
48. Ejiyi, C.J.; Qin, Z.; Agbesi, V.K.; Yi, D.; Atwereboannah, A.A.; Chikwendu, I.A.; Bamisile, O.F.; Bakanina Kissanga, G.-M.; Bamisile, O.O. Advancing cancer diagnosis and prognostication through deep learning mastery in breast, colon, and lung histopathology with ResoMergeNet. *Comput. Biol. Med.* **2024**, *185*, 109494. [CrossRef] [PubMed]
49. Najafipour, R.; Momeni, A.M.; Mirmazloomi, Y.; Moghbelinejad, S. Vitexin induces apoptosis in MCF-7 breast cancer cells through the regulation of specific miRNAs expression. *Int. J. Mol. Cell Med.* **2022**, *11*, 197–206. [CrossRef] [PubMed]
50. Scarpa, E.S.; Antonini, E.; Palma, F.; Mari, M.; Ninfali, P. Antiproliferative activity of vitexin-2-O-xyloside and avenanthramides on CaCo-2 and HepG2 cancer cells occurs through apoptosis induction and reduction of pro-survival mechanisms. *Eur. J. Nutr.* **2018**, *57*, 1381–1395. [CrossRef] [PubMed]
51. Sowa, I.; Mołdoch, J.; Paduch, R.; Strzemski, M.; Szkutnik, J.; Tyszczuk-Rotko, K.; Dresler, S.; Szczepanek, D.; Wójciak, M. Polyphenolic composition of *Carlina acaulis* L. extract and cytotoxic potential against colorectal adenocarcinoma and cervical cancer cells. *Molecules* **2023**, *28*, 6148. [CrossRef] [PubMed]
52. Bhardwaj, M.; Paul, S.; Jakhar, R.; Khan, I.; Kang, J.I.; Kim, H.M.; Yun, J.W.; Lee, S.J.; Cho, H.J.; Lee, H.G.; et al. Vitexin confers *HSF-1* mediated autophagic cell death by activating *JNK and ApoL1* in colorectal carcinoma cells. *Oncotarget* **2017**, *8*, 112426–112441. [CrossRef] [PubMed]
53. Chen, Y.; Wang, B.; Yuan, X.; Lu, Y.; Hu, J.; Gao, J.; Lin, J.; Liang, J.; Hou, S.; Chen, S. Vitexin prevents colitis-associated carcinogenesis in mice through regulating macrophage polarization. *Phytomedicine* **2021**, *83*, 153489. [CrossRef] [PubMed]
54. Chen, Y.; Liang, J.; Chen, S.; Lin, N.; Xu, S.; Miao, J.; Zhang, J.; Chen, C.; Yuan, X.; Xie, Z.; et al. Discovery of vitexin as a novel VDR agonist that mitigates the transition from chronic intestinal inflammation to colorectal cancer. *Mol. Cancer* **2024**, *23*, 196. [CrossRef] [PubMed]
55. Ganesan, K.; Xu, C.; Wu, J.; Du, B.; Liu, Q.; Sui, Y.; Song, C.; Zhang, J.; Tang, H.; Chen, J. Ononin inhibits triple-negative breast cancer lung metastasis by targeting the EGFR-mediated PI3K/Akt/mTOR pathway. *Sci. China Life Sci.* **2024**, *67*, 1849–1866. [CrossRef] [PubMed]
56. Chen, P.; Chang, Y.; Tsai, K.; Shen, C.; Lee, S. Vitexin suppresses high-glucose-upregulated adhesion molecule expression in endothelial cells through Inhibiting NF-κB signaling pathway. *ACS Omega* **2024**, *9*, 32727–32734. [CrossRef] [PubMed]
57. Li, S.; Liang, T.; Zhang, Y.; Huang, K.; Yang, S.; Lv, H.; Chen, Y.; Zhang, C.; Guan, X. Vitexin alleviates high-fat diet induced brain oxidative stress and inflammation via anti-oxidant 2734, anti-inflammatory and gut microbiota modulating properties. *Free Radic. Biol. Med.* **2021**, *171*, 332–344. [CrossRef] [PubMed]
58. Ding, C.; Shen, H.; Tian, Z.; Kang, M.; Ma, J.; He, Q.; Wang, J.; Zhang, Y.; Deng, Y.; Wang, D. Protective effect of hawthorn vitexin on the ethanol-injured DNA of BRL-3A hepatocytes. *Medicine* **2021**, *100*, e28228. [CrossRef] [PubMed]
59. Tao, M.; Li, R.; Xu, T.; Zhang, Z.; Zheng, D.; Xia, Z.; Wu, T.; Pan, S.; Xu, X. Vitexin and isovitexin delayed ageing and enhanced stress-resistance through the activation of the SKN-1/Nrf2 signaling pathway. *Int. J. Food Sci. Nutr.* **2023**, *74*, 685–694. [CrossRef] [PubMed]
60. Wu, T.; Sheng, Y.; Tian, Y.; Wang, C. Vitexin regulates heat shock protein expression by modulating ROS levels thereby protecting against heat-stress-induced apoptosis. *Molecules* **2023**, *28*, 7639. [CrossRef] [PubMed]

61. Hu, S.; Simon, J.E.; Wang, M.; Wu, Y.; Huang, Y.; Wu, Q. Kinkéliba (*Combretum micranthum*) leaf extract alleviates skin inflammation: In vitro and in vivo study. *Molecules* **2023**, *28*, 1791. [CrossRef] [PubMed]
62. Hu, X.; Mu, L.; Zhu, L.; Chang, X.; Nie, L.; Wang, L.; Li, G. *Lycium barbarum* polysaccharides attenuate cardiovascular oxidative stress injury by enhancing the Keap1/Nrf2 signaling pathway in exhaustive exercise rats. *Mol. Med. Rep.* **2021**, *24*, 643. [CrossRef] [PubMed]
63. Gan, T.; Xing, Q.; Li, N.; Deng, Z.; Pan, C.; Liu, X.; Zheng, L. Protective effect of vitexin against IL-17-induced vascular endothelial inflammation through Keap1/Nrf2-dependent signaling pathway. *Mol. Nutr. Food Res.* **2024**, *68*, e2300331. [CrossRef] [PubMed]
64. Kundu, A.; Ghosh, P.; Bishayi, B. Vitexin along with verapamil downregulates efflux pump P-glycoprotein in macrophages and potentiate M1 to M2 switching via TLR4-NF-κB-TNFR2 pathway in lipopolysaccharide treated mice. *Immunobiology* **2024**, *229*, 152767. [CrossRef] [PubMed]
65. Huang, Y.; Chen, S.; Pang, L.; Feng, Z.; Su, H.; Zhu, W.; Wei, J. Isovitexin protects against acute liver injury by targeting PTEN, PI3K and BiP via modification of m6A. *Eur. J. Pharmacol.* **2022**, *917*, 174749. [CrossRef] [PubMed]
66. Zhang, J.; Liang, F.; Chen, Z.; Chen, Y.; Yuan, J.; Xiong, Q.; Hou, S.; Huang, S.; Liu, C.; Liang, J. Vitexin protects against dextran sodium sulfate-induced colitis in mice and its potential mechanisms. *J. Agric. Food Chem.* **2022**, *70*, 12041–12054. [CrossRef] [PubMed]
67. Li, S.; Luo, L.; Wang, S.; Sun, Q.; Zhang, Y.; Huang, K.; Guan, X. Regulation of gut microbiota and alleviation of DSS-induced colitis by vitexin. *Eur. J. Nutr.* **2023**, *62*, 3433–3445. [CrossRef] [PubMed]
68. Duan, S.; Du, X.; Chen, S.; Liang, J.; Huang, S.; Hou, S.; Gao, J.; Ding, P. Effect of vitexin on alleviating liver inflammation in a dextran sulfate sodium (DSS)-induced colitis model. *Biomed. Pharmacother.* **2020**, *121*, 109683. [CrossRef] [PubMed]
69. Matar, A.; Damianos, J.A.; Jencks, K.J.; Camilleri, M. Intestinal barrier impairment, preservation, and repair: An update. *Nutrients* **2024**, *16*, 3494. [CrossRef] [PubMed]
70. Zhang, Y.; Qi, Z.; Wang, W.; Wang, L.; Cao, F.; Zhao, L.; Fang, X. Isovitexin inhibits ginkgolic acids-induced inflammation through downregulating SHP2 activation. *Front. Pharmacol.* **2021**, *12*, 630320. [CrossRef] [PubMed]
71. Cui, Y.H.; Zhang, X.Q.; Wang, N.D.; Zheng, M.D.; Yan, J. Vitexin protects against ischemia/reperfusion-induced brain endothelial permeability. *Eur. J. Pharmacol.* **2019**, *853*, 210–219. [CrossRef] [PubMed]
72. Long, Y.; Yang, Q.; Xiang, Y.; Zhang, Y.; Wan, J.; Liu, S.; Li, N.; Peng, W. Nose to brain drug delivery—A promising strategy for active components from herbal medicine for treating cerebral ischemia reperfusion. *Pharmacol. Res.* **2020**, *159*, 104795. [CrossRef] [PubMed]
73. Wang, Y.; Zhen, Y.; Wu, X.; Jiang, Q.; Li, X.; Chen, Z.; Zhang, G.; Dong, L. Vitexin protects brain against ischemia/reperfusion injury via modulating mitogen-activated protein kinase and apoptosis signaling in mice. *Phytomedicine* **2015**, *22*, 379–384. [CrossRef] [PubMed]
74. Malar, D.S.; Prasanth, M.I.; Jeyakumar, M.; Balamurugan, K.; Devi, K.P. Vitexin prevents Aβ proteotoxicity in transgenic *Caenorhabditis elegans* model of Alzheimer's disease by modulating unfolded protein response. *J. Biochem. Mol. Toxicol.* **2021**, *35*, e22632. [CrossRef] [PubMed]
75. de Oliveira, N.K.S.; Almeida, M.R.S.; Pontes, F.M.M.; Barcelos, M.P.; Silva, G.M.; de Paula da Silva, C.H.T.; Cruz, R.A.S.; da Silva Hage-Melim, L.I. Molecular docking, physicochemical properties, pharmacokinetics and toxicity of flavonoids present in *Euterpe oleracea* Martius. *Curr. Comput. Aided Drug Des.* **2021**, *17*, 589–617. [CrossRef] [PubMed]
76. Huang, X.; Hu, X.; Li, S.; Li, T. Vitexin-rhamnoside encapsulated with *zein*-pectin nanoparticles relieved high-fat diet induced lipid metabolism disorders in mice by altering the gut microbiota. *Int. J. Biol. Macromol.* **2024**, *264*, 130704. [CrossRef] [PubMed]
77. Das, M.C.; Sandhu, P.; Gupta, P.; Rudrapaul, P.; De, U.C.; Tribedi, P.; Akhter, Y.; Bhattacharjee, S. Attenuation of *Pseudomonas aeruginosa* biofilm formation by vitexin: A combinatorial study with azithromycin and gentamicin. *Sci. Rep.* **2016**, *6*, 23347. [CrossRef] [PubMed]
78. Mohammed, H.S.; Abdel-Aziz, M.M.; Abu-Baker, M.S.; Saad, A.M.; Mohamed, M.A.; Ghareeb, M.A. Antibacterial and potential antidiabetic activities of flavone C-glycosides isolated from *Beta vulgaris* subspecies *cicla* L. var. Flavescens (Amaranthaceae) cultivated in Egypt. *Curr. Pharm. Biotechnol.* **2019**, *20*, 595–604. [CrossRef] [PubMed]
79. Chiavaroli, A.; Sinan, K.I.; Zengin, G.; Mahomoodally, M.F.; Sadeer, N.B.; Etienne, O.K.; Cziáky, Z.; Jekő, J.; Glamocilja, J.; Sokovic, M.; et al. Identification of chemical profiles and biological properties of *Rhizophora racemosa* G. Mey. extracts obtained by different methods and solvents. *Antioxidants* **2020**, *9*, 533. [CrossRef] [PubMed]
80. Król-Kogus, B.; Głód, D.; Hałasa, R.; Krauze-Baranowska, M.; Pobłocka-Olech, L. 2D LC as a tool for standardization of *Foenugraeci semen* extracts containing compounds with anti-*Helicobacter pylori* activity. *Food Funct.* **2021**, *12*, 2686–2692. [CrossRef] [PubMed]
81. He, M.; Min, J.-W.; Kong, W.-L.; He, X.-H.; Li, J.-X.; Peng, B.-W. A review on the pharmacological effects of vitexin and isovitexin. *Fitoterapia* **2016**, *115*, 74–85. [CrossRef] [PubMed]
82. Li, N.; Wang, R.; Li, W.; Du, Q.; Deng, Z.; Fan, Y.; Zheng, L. Identification of OLA1 as a novel protein target of vitexin to ameliorate dextran sulfate sodium-induced colitis with tissue thermal proteome profiling. *J. Agric. Food Chem.* **2023**, *71*, 16057–16066. [CrossRef] [PubMed]

83. Zhao, C.R.; Yang, F.F.; Cui, Q.; Wang, D.; Zhou, Y.; Li, Y.S.; Zhang, Y.P.; Tang, R.Z.; Yao, W.J.; Tang, R.Z. Vitexin inhibits *APEX1* to counteract the flow-induced endothelial inflammation. *Proc. Natl. Acad. Sci. USA* **2021**, *118*, e2115158118. [CrossRef] [PubMed]
84. Rosa, S.I.G.; Rios-Santos, F.; Balogun, S.O.; Martins, D.T.d.O. Vitexin reduces neutrophil migration to inflammatory focus by down-regulating pro-inflammatory mediators via inhibition of p38, ERK1/2 and JNK pathway. *Phytomedicine* **2016**, *23*, 9–17. [CrossRef] [PubMed]
85. Liu, Y.; Zhu, S.; Liu, J.; Chen, Y.; Zhong, S.; Lian, D.; Liang, J.; Huang, S.; Hou, S. Vitexin regulates angiogenesis and osteogenesis in ovariectomy-induced osteoporosis of rats via the VDR/PI3K/AKT/eNOS signaling pathway. *J. Agric. Food Chem.* **2023**, *71*, 546–556. [CrossRef] [PubMed]
86. Bai, Y.; Zhang, Q.; Wang, B.; Zhang, M.; Xu, Y.; Li, S.; Zhao, Y.; Yu, Z. Plasma pharmacokinetics, bioavailability, and tissue distribution of four C-glycosyl flavones from mung bean (*Vigna radiata* L.) seed extracts in rat by ultrahigh-performance liquid chromatography-tandem mass spectrometry. *J. Agric. Food Chem.* **2017**, *65*, 5570–5580. [CrossRef] [PubMed]
87. Li, S.; Lv, H.; Chen, Y.; Song, H.; Zhang, Y.; Wang, S.; Luo, L.; Guan, X. N-trimethyl chitosan coated targeting nanoparticles improve the oral bioavailability and antioxidant activity of vitexin. *Carbohydr. Polym.* **2022**, *286*, 119273. [CrossRef] [PubMed]
88. Huang, X.; Li, T.; Li, S. Encapsulation of vitexin-rhamnoside based on *zein*/pectin nanoparticles improved its stability and bioavailability. *Curr. Res. Food Sci.* **2022**, *6*, 100419. [CrossRef] [PubMed]
89. Zhang, J.; Li, X.; Xia, X.; Adu-Frimpong, M.; Shen, X.; He, Q.; Rong, W.; Shi, F.; Cao, X.; Ji, H.; et al. Vitexin loaded mixed polymeric micelles: Preparation, optimization, evaluation and anti-osteoporotic effect. *Biomed. Mater.* **2023**, *18*, 045005. [CrossRef] [PubMed]
90. Yoksan, R.; Towongphaichayonte, P. Vitexin-loaded poly(ethylene glycol) methyl ether-grafted chitosan/alginate nanoparticles: Preparation, physicochemical properties and in vitro release behaviors. *J. Sci. Food Agric.* **2024**, *104*, 956–966. [CrossRef] [PubMed]
91. Farooq, A.; Iqbal, A.; Rana, N.F.; Fatima, M.; Maryam, T.; Batool, F.; Rehman, Z.; Menaa, F.; Azhar, S.; Nawaz, A.; et al. A novel Sprague-Dawley rat model presents improved NASH/NAFLD symptoms with PEG coated vitexin liposomes. *Int. J. Mol. Sci.* **2022**, *23*, 3131. [CrossRef] [PubMed]
92. Ramanan, M.; Sinha, S.; Sudarshan, K.; Aidhen, I.S.; Doble, M. Inhibition of the enzymes in the leukotriene and prostaglandin pathways in inflammation by 3-aryl isocoumarins. *Eur. J. Med. Chem.* **2016**, *124*, 428–434. [CrossRef] [PubMed]
93. Sudarshan, K.; Aidhen, I.S. Convenient synthesis of 3-glycosylated isocoumarins. *Eur. J. Org. Chem.* **2017**, *2017*, 34–38. [CrossRef]

International Journal of
Molecular Sciences

MDPI

Review

The Potential of Superoxide Dismutase-Rich *Tetraselmis chuii* as a Promoter of Cellular Health

Stuart P. Cocksedge [1,2], Lalia Mantecón [3], Enrique Castaño [3], Carlos Infante [3] and Stephen J. Bailey [1,*]

1 School of Sport, Exercise and Health Sciences, Loughborough University, Loughborough LE11 3TU, UK; stuart.cocksedge@coventry.ac.uk
2 Centre for Physical Activity, Sport and Exercise Sciences, Coventry University, Coventry CV1 5FB, UK
3 Fitoplancton Marino, S.L., Dársena Comercial s/n, 11500 El Puerto de Santa María, Cádiz, Spain; emg@easyalgae.com (L.M.); ecp@easyalgae.com (E.C.); carlos.infante@easyalgae.com (C.I.)
* Correspondence: s.bailey2@lboro.ac.uk

Abstract: *Tetraselmis chuii* (*T. chuii*) is a green, marine, eukaryotic, microalgae that was authorized in the European Union (EU) as a novel food for human consumption in 2014, and as a food supplement in 2017. This narrative review will provide an overview of preclinical and clinical trials assessing the efficacy of a *T. chuii*-derived ingredient, characterized by a high superoxide dismutase (SOD) activity (SOD-rich *T. chuii*), to improve various aspects of cellular health. Collectively, results from in vitro, and more importantly in vivo research, support SOD-rich *T. chuii* as a potential promoter of cellular health. Principally, the ingredient appears to function as an indirect antioxidant by boosting intracellular antioxidant systems. Moreover, it can positively modulate inflammatory status by up-regulating anti-inflammatory and down-regulating pro-inflammatory cytokines and factors. In addition, SOD-rich *T. chuii* appears to promote cellular health though protecting from DNA damage, boosting immune function, strengthening cell structure and integrity, and positively modulating cell signaling pathways. There is also some evidence to suggest that SOD-rich *T. chuii* may improve aspects of mitochondrial function through the up-regulation of genes linked to mitochondrial biogenesis and ATP synthesis. From the trials conducted to date, transcriptional activation of nuclear factor erythroid 2-related factor 2 (NRF2) and sirtuin 1 (SIRT1) appear to be important in mediating the effects of SOD-rich *T. chuii* on cellular health. These exciting preliminary observations suggest that SOD-rich *T. chuii* may represent a natural blue food supplement with the potential to enhance various aspects of cellular health.

Keywords: *Tetraselmis chuii*; SOD; food supplement; cellular health; antioxidant; anti-inflammatory

1. Introduction

Microalgae are a wide group of single-celled photosynthetic organisms that represent the basis of aquatic food chains in both freshwater and seawater environments. From a phylogenetic standpoint, the term microalgae comprises both prokaryotic (cyanobacteria) and eukaryotic organisms, and it is considered to include 200,000–800,000 potential species [1]. One noteworthy feature of microalgae is their immense physiological plasticity, a characteristic that allows these organisms to survive in the most hostile environments, ranging from hypersaline waters to soils and rocks [2,3]. The capacity of microalgae to colonize such a diverse range of environmental conditions has been linked to their ability to produce primary and secondary metabolites with a highly diverse chemical nature (peptides and proteins, polyunsaturated fatty acids (PUFAs), polysaccharides, pigments, phytosterols, and/or phenolic compounds). These chemicals not only aid survival in microalgae, but are recognized to confer potential benefits for human health by exerting antidiabetic, antihypertensive, anti-obesity, anti-inflammatory, antimicrobial, or antiviral effects [4–6]. Consequently, there is increasing interest in the potential of consuming microalgae-containing supplements to enhance human health outcomes.

The marine eukaryotic microalgae *Tetraselmis chuii* (*T. chuii*) is a Chlorophyta included in the family *Chlorodendraceae*. A freeze-dried ingredient derived from this species is produced using a patent-protected technology under culture conditions that ensure a superoxide dismutase (SOD) activity above 30,000 U/g of the end product, which is significantly higher than that exhibited by other microalgae species [7,8]. In addition to high SOD activity, this SOD-rich *T. chuii*-derived ingredient (SOD-rich *T. chuii*) contains other compounds with potential bioactivity including PUFAs, vitamins, carotenoids, polyphenols, or phytosterols [9,10]. Initial work using pre-clinical experimental models has indicated that SOD-rich *T. chuii* can promote antioxidant and metal chelating responses, and inhibit acetylcholinesterase activities in in vitro tests, which could have positive implications for the prevention and/or management of neurological disorders, such as Alzheimer's disease [11]. Moreover, immunostimulant effects of *T. chuii* were revealed in vivo in the gilthead seabream (*Sparus aurata*), as evidenced by increased hemolytic complement activity, phagocytic activity, serum levels of immunoglobulin M (IgM), expression of β-defensin, major histocompatibility complex II alpha chain (MHCIIα), and colony-stimulating factor-1 receptor (CSF1-R) in head kidney, as well as expression of occludin (a marker of the integrity of tight junctions) in the intestine [12–14]. Subsequent studies in preclinical mammalian animal models (rat) have reported indirect antioxidant properties of SOD-rich *T. chuii*, as well as its anti-inflammatory and immunostimulatory effects [15,16]. However, whilst these observations from preclinical studies were encouraging, clinical trials testing the bioactive effects of *T. chuii* and/or SOD-rich *T. chuii* in humans were lacking as such products had not yet been approved for human consumption.

Although *T. chuii* has been widely administered for fish, crustacean, and mollusk larvae feeding in the aquaculture industry [17], its use for humans has been limited by regulatory constraints. In 2014, *T. chuii* was authorized as a novel food for human consumption in the European Union (EU) according to Regulation (EC) No 258/97. Subsequently, in 2017, freeze-dried *T. chuii* was authorized as a food supplement in the EU [18]. These legislative changes opened new avenues for the application of SOD-rich *T. chuii*, including the possibility of assessing for potential health benefits of ingesting SOD-rich *T. chuii* in humans in vivo. This is important because, whilst SOD-rich *T. chuii* administration has been reported to enhance various aspects of cellular health in in vitro settings [19], the translation of such effects to humans could not be studied until recently and it is important to verify or refute the potential efficacy of SOD-rich *T. chuii* to promote human health. In this regard, clinical trials conducted in recent years have shown SOD-rich *T. chuii* to enhance physical performance and endogenous antioxidant responses in human skeletal muscle and to promote anti-inflammatory and immunostimulatory effects [16,20–24]. Importantly, these recent studies have shed light on the candidate molecular mechanisms underlying such physiological effects. Therefore, this narrative review aims to provide an overview of studies assessing the potential of SOD-rich *T. chuii* as a promoter of cellular health.

Factors Contributing to Cellular Health

Cellular health broadly refers to the well-being and optimal functioning of individual cells. As cells aggregate to form tissues, which in turn form organs, and then organ systems that comprise the human body, good cellular health is crucial and foundational for overall health. Cellular health can be compromised by various endogenous and exogenous stressors (e.g., pollution, radiation, smoke, and/or oxidative stress). A myriad of factors influence cellular health, but this review will focus on redox balance, inflammation, immune and mitochondrial function, and cell structural integrity and signaling, which have the potential to be modulated by SOD-rich *T. chuii* supplementation.

Historically, redox state has been considered one of the most important factors affecting cellular health [25]. This redox state is determined by the balance between oxidants, such as reactive oxygen species (ROS) and antioxidants. Cells produce ROS as a consequence of metabolic processes. Organelles such as peroxisomes and mitochondria generate ROS as by-products, and different enzymes (xanthine oxidase, p450 cytochromes, and NADPH oxidase family) are well known to contribute to ROS production. While ROS have conventionally been considered cytotoxic derivatives of cellular metabolic processes, more recent work indicates that ROS production within a 'physiological' range is important for cellular redox signaling and health [26]. In healthy conditions, production of ROS is balanced by an elaborate defense system that includes both enzymes such as SOD, glutathione peroxidase (GPx), catalase (CAT), peroxiredoxins, and thioredoxins, and also non-enzymatic antioxidant scavengers such as vitamins C and E, and glutathione. However, when ROS production becomes excessive and 'pathophysiological', this can overwhelm the antioxidant defense systems, leading to a condition of oxidative stress, which reflects disruptive redox signaling and leads to the damaging of cellular constituents such as proteins, lipids, and DNA [27,28]. As such, maintaining an appropriate balance between ROS production and antioxidant response mechanisms is crucial to maintain cellular redox balance and health, and to prevent ill health and disease morbidity [25,29,30].

Inflammation is another important factor affecting cellular health. It is a fundamental part of immune function, being activated in response to microbial infections, tissue injury, or toxic compounds that disrupt tissue homeostasis [31]. Inducers of inflammation can be exogenous (pathogen-associated molecular patterns or virulence factors) or endogenous (signals derived from cells, tissues, plasma or extracellular matrix). These inflammatory responses are recognized by macrophages and mast cells, which initiate the production of inflammatory mediators, like chemokines and cytokines, with interleukin-1β (IL-1β), interleukin-6 (IL-6), and tumor necrosis factor-α (TNFα) being among the most important pro-inflammatory cytokines. Subsequent interaction of pro-inflammatory stimuli with their specific receptors triggers signaling pathways such as the mitogen-activated protein kinases (MAPK), and the nuclear factor kappa-B (NF-κB) pathway, which are known to play a key role in the immune response, survival, and apoptosis [32]. Although the inflammatory signaling cascade facilitates the elimination of the noxious stimuli and infection, it can also be harmful and damage healthy cells and tissues when the process is dysregulated [33]. This happens, for instance, during the so-called low-grade chronic inflammation, which has been associated with several age-related conditions, including the metabolic syndrome and cardiovascular disease [34]. To control inflammation, some immune cells produce anti-inflammatory cytokines such as interleukin-10 (IL-10), which inhibits the secretion of pro-inflammatory cytokines and promotes tissue repair [31,32].

The immune system is also recognized as being integral to the maintenance of optimal cellular health. In addition to providing defense against microorganisms, the immune system promotes elimination of damaged or abnormal cells that might degenerate in malignancies [35]. The effective execution of these functions is underpinned by an intricate network comprising physical barriers, like the skin; protein systems, like the complement and immunoglobulins; cellular elements, such as macrophages, lymphocytes, neutrophils, basophils, or eosinophils; surface receptors, like T- and B-cell receptors; and recognition molecules, like major histocompatibility complex I and II. Collectively, these act in host defense to restore tissue homeostasis and to protect and maintain health [36].

An additional factor contributing to cellular health is DNA integrity. Damage to DNA, which can be caused by ROS or radiation, can lead to mutations and contribute to premature aging and to several diseases, including cancer, if not properly repaired [25]. Moreover, it has been suggested that DNA mutations may alter transcriptional activity and alternative splicing of genes, which in turn might alter cellular homeostasis due to inadequate responses to endogenous or exogenous stressors [37,38].

Various aspects of mitochondrial function are key factors in determining cellular health. Principally, mitochondrial respiration is the most recognized aspect of mitochondrial function, during which the chemical energy-carrying molecule, ATP, is resynthesized via oxidative phosphorylation, and provides the necessary energy to fuel various essential cellular processes to preserve health and function. However, it is now clear that mitochondria are key orchestrators in a variety of additional key cellular processes, such as intracellular calcium signaling, proliferation, differentiation, and death or survival. Moreover, mitochondria contribute to the production of lipids, proteins, nucleic acids, and also neurotransmitters and hormones, and have an important role in iron homeostasis and the folate cycle [39,40]. This central role of mitochondrial function in cellular health is revealed by the number of diseases either underpinned or accompanied by dysfunction in one or more mitochondrial processes [41,42].

Two further factors with a significant effect on cellular health and are worthwhile mentioning within the context of the current review are cellular structural integrity and cell signaling. Regarding the structural integrity of the cell, eukaryotic cells exhibit a plasma membrane, acting as a barrier that separates the inside of the cell from the environment, and well-defined membrane-bound intracellular compartments. The membrane-bound intracellular compartments include a nucleus, endoplasmic reticulum, Golgi apparatus, and mitochondria, and whilst each compartment can orchestrate discrete cellular functions, mechanisms for inter-compartment communication pathways are recognized. Moreover, the dynamic structures belonging to the cytoskeleton (microtubules, actin filaments and intermediate filaments) are necessary to maintain cellular shape and internal organization, and are also important for supporting cellular processes such as cell division, cell movement, gene expression, or signal transduction. Maintaining this overall organization is crucial for cellular homeostasis and function [43–45]. Cellular health is also influenced by cell signaling, which allows cells to receive information from both the intracellular and extracellular stimuli and to initiate various effectors in response to such stimuli to preserve cellular and tissue homeostasis [46]. Amongst the effectors, intracellular signaling pathways that operate to convert intracellular or extracellular stimuli into cellular responses, MAPK cascades are one of the most conserved transduction pathways and play essential roles in key cellular processes such as division, motility, metabolism, apoptosis, differentiation, and stress responses [47].

There has been great interest in the health benefits that may be conferred by nutritional supplements with the potential to positively modulate cellular redox balance, inflammation, immune, and mitochondrial function, and cell structural integrity and signaling responses. The following sections will provide an overview of pre-clinical and clinical studies assessing the effects of SOD-rich *T. chuii* on these aspects of cellular health followed by a discussion of the results and a summary of the potential of SOD-rich *T. chuii* to improve cellular health.

2. Reported Studies Conducted with SOD-Rich *T. chuii* in Preclinical Models

In human skeletal muscle myoblasts, SOD-rich *T. chuii* has been reported to increase activity of the primary antioxidant enzymes (SOD, GPx, CAT) after 24 h of treatment, together with the transcriptional up-regulation of genes encoding such enzymes (SOD1, SOD2, GPx1, CAT) [19]. Moreover, this report was the first to demonstrate transcriptional activation of nuclear factor erythroid 2-related factor 2 (NRF2), a key transcription factor considered as the master regulator of the cellular antioxidant response, by SOD-rich *T. chuii* [48].

Results obtained in studies conducted with SOD-rich *T. chuii* using rodent models (rat) are summarized in Table 1. These studies can also be split into two completely different experimental designs. In three of these studies, animals consumed SOD-rich *T. chuii* (2.55 or 5.1 mg/Kg/day) for 6 weeks, and were subjected to a usual procedure for animal exercise training, including an initial week of adaptation to a treadmill system for the novel and stress impacts. Thereafter, animals ran on the treadmill five days per week at 25 m/min, 45 min/day, a model known to allow physical remodeling of the heart to increased oxygen consumption, and to improve contractile function and calcium handling of cardiac muscle [49]. These experiments were conducted in parallel with human trials (see Table 2), and were designed to serve as mechanistic models to investigate and help to understand the effects of SOD-rich *T. chuii* in physical performance and recovery, muscle strength and damage, and inflammatory and immunoregulatory responses [16,20,21]. Different markers were analyzed in rat muscle and serum samples, including antioxidant and oxidative stress-related parameters, markers of muscle tissue damage, myogenic factors, and pro-inflammatory cytokines.

Table 1. Published studies reporting the effects of SOD-rich *T. chuii* in rodent models.

Strain	N	Age	Dose (mg/Kg/day)	Duration	Main Outcomes	Reference
Wistar (male)	28	8 weeks old	2.55/5.1	6 weeks	↑ SOD, GPx and CAT intramuscular activity ↓ Serum MDA and myoglobin ↓ Serum CK activity	[16]
Wistar (male)	28	8 weeks old	2.55/5.1	6 weeks	↑ Muscle MyoD, NCAM ↓ Muscle myostatin, MAFbx, MuRF-1 ↓ Serum CK activity	[20]
Wistar (male)	28	8 weeks old	2.55/5.1	6 weeks	↓ Muscle IL-1β and TNFα ↓ Serum IL-1β and TNFα ↓ Muscle MDA	[21]
Sprague-Dawley (male)	50	7 weeks old	0.17/1.7/17	8 weeks	↓ Serum oxLDL ↑ Serum IL-10 ↑ Liver GPx activity ↑ Liver GSH content Gene expression in liver: ↑ GPx1, GSR, GSH-S, SOD1, SOD2, GCLM ↓ TGF-β1, NF-κB1 Gene expression in MWAT: ↑ ACDC, IL-10 ↓ IL-1β, TNFα, IFNγ Gene expression in thymus: ↑ IL-10 ↓ IL-1β, TNFα, IFNγ, NF-κB1 Gene expression in spleen: ↑ IL-10 ↓ IL-1β, IFNγ, NF-κB1	[15]

ACDC: adiponectin; CAT: catalase; CK: creatine kinase; GCLM: glutamate-cysteine ligase modifier subunit; GPx: glutathione peroxidase; GPx1: glutathione peroxidase 1; GR: glutathione-disulfide reductase; GSH: glutathione; GSH-S: glutathione synthetase; IFNγ: interferon γ; IL-1β: interleukin-1β; IL-10: interleukin-10; MAFbx: muscle atrophy F-box; MDA: malondialdehyde; MuRF-1: muscle RING-finger protein-1; MWAT: mesenteric white adipose tissue; MyoD: myogenic differentiation factor; N: number of animals; NCAM: neural cell adhesion molecules; NF-κB1: nuclear factor kappa B subunit 1; oxLDL: oxidized low-density lipoprotein; SOD: superoxide dismutase; SOD1: superoxide dismutase 1; SOD2: superoxide dismutase 2; TGF-β1: transforming growth factor-β1; TNFα: tumor necrosis factor-α; *T. chuii*: *Tetraselmis chuii*. ↑ represents an increase in parameters or in expression of target genes, whereas ↓ represents a decrease in parameters or down-regulation of target genes.

The last of the trials was conducted to study the potential therapeutic effect of SOD-rich *T. chuii* in an animal model of metabolic syndrome, which is a cluster of metabolic disturbances such as abdominal obesity, hypertension, and hyperglycemia, and is characterized by low-grade chronic inflammation. Such dysregulations are known to represent a serious risk for developing cardiovascular disease and type 2 diabetes [50]. Metabolic syndrome was induced in animals by a diet rich in carbohydrates and fat, and low in fiber (cafeteria diet). Three different doses of SOD-rich *T. chuii* were evaluated (0.17, 1.7, and 17 mg/Kg/day), and a range of antioxidant, anti-inflammatory, and immune-modulatory markers were measured, including gene expression analysis in the liver, adipose tissue, thymus, and spleen [15].

3. An Overview of the Clinical Trials Conducted with SOD-Rich *T. chuii*

A summary of the most relevant data from clinical trials conducted with SOD-rich *T. chuii* in humans is presented in Table 2. In six of the reported studies, the effects of SOD-rich *T. chuii* supplementation on physical performance and related physiological parameters of healthy individuals were evaluated [16,20–23,51]. It is well known that the production of ROS increases in an intensity- and duration-dependent manner during skeletal muscle contractions, and although redox balance may be preserved by cellular antioxidant mechanisms, continued high rates of ROS production during intense or prolonged exercise can lead to exercise-induced oxidative stress [52]. Importantly, this exercise-induced oxidative stress has been linked to impairments in muscle perfusion and contractility, culminating in neuromuscular fatigue and impaired exercise performance [53]. In this scenario, it has been reported that dietary supplementation with some antioxidants has the potential to mitigate exercise-induced oxidative stress and to improve exercise performance, but such effects are inconsistent and require further investigation [54]. Given the potential for SOD-rich *T. chuii* supplementation to promote endogenous antioxidant enzyme responses, trials have been conducted in healthy human participants to examine its effects on various physiological and functional responses, as well as exercise performance and recovery [16,20–24]. In the only trial conducted to date to assess the effects in human skeletal muscle tissue directly via muscle biopsy sampling, gene expression changes in more than 100 key genes involved in antioxidant and inflammatory response were assessed using OpenArray™ technology (Thermo Fisher Scientific, Waltham, MA, USA). This work aimed to unravel some of the molecular basis underlying the potential ergogenic effects of SOD-rich *T. chuii* [24].

In addition to effects on exercise performance and recovery, two clinical trials have been conducted which investigated the effects of SOD-rich *T. chuii* supplementation in seminal parameters of idiopathic infertile men [51,55]. In this regard, oxidative stress is thought to be related to impairments in spermatogenesis, epididymal maturation, or sperm capacitation [56], and it has been negatively correlated with sperm count, motility, and morphology [57]. Whilst dietary consumption of antioxidants has been linked to improvements in sperm quality, outcomes are variable and appear to be depending on the ingredient investigated [58]. Thus, the potential effects of SOD-rich *T. chuii* as an indirect antioxidant on sperm quality were assessed in idiopathic infertile men. Participants consumed SOD-rich *T. chuii* for 90 days to cover a complete spermatogenesis cycle. Semen samples were collected for the determination of semen volume, as well as concentration, total number, progressive motility, and normal forms of sperm [51]. Subsequently, a second clinical trial has been conducted to confirm the previous outcomes in which additional relevant parameters related to sperm DNA integrity and seminal redox state have been investigated, and preliminary results have been already reported [55].

Table 2. Published studies reporting the effects of SOD-rich *T. chuii* in humans.

Trial Model	N	Dose (mg/day)	Duration	Main Outcomes	Reference
R, DB, PC, PM	18	25	14 days	↑ IMTP strength ↑ SJ power ↓ Serum CK activity	[16]
R, DB, PC, PM	32	25	30 days	↓ Heart rate ↑ $\dot{V}O_{2max}$ ↑ Hemoglobin and MCH	[22]
R, DB, PC, PM	22	25	6 weeks	↑ PRS ↑ CMJ strength	[20]
R, DB, PC, PM	19	25	21 days	↑ IMTP force ↑ sIgA	[21]
R, DB, PC, PM	46	25/200	60 days	↑ Muscle percentage ↓ Fat percentage ↑ Basophils, monocytes, lymphocytes	[23]
R, DB, PM	40	25/250	90 days	↑ Semen volume ↑ Sperm concentration ↑ Total sperm number ↑ Sperm progressive motility	[51]
R, DB, PC, CO	13	25	14 days	↑ $\dot{V}O_2$ peak Gene expression in muscle: ↑ NRF2, SIRT1, GPx7, PRDX6, PRDX3, c-JUN, MAPK14, GSR, GCLM, GSTM3, SOD2, CAT, CAPN3, AIFM1, CCL2, CASP8, IL-18, CUL3, BACH1 ↓ MAPK7, JUND	[24]
R, DB, PC, PM	80	250	90 days	↑ Total sperm number ↓ sORP ↓ DSBs	[55]

AIFM1: apoptosis inducing factor mitochondria associated 1; BACH1: BTB and CNC homology 1, basic leucine zipper transcription factor 1; BTBD1: BTB domain containing 1; CAPN3: calpain 3; CASP8: caspase-8; CAT: catalase; CCL2: C-C motif chemokine ligand 2; CK: creatine kinase; CMJ: counter movement jump; CO: cross-over; CUL3: cullin 3; DB: double blind; DSBs: double-strand DNA breaks; GCLM: glutamate-cysteine ligase modifier subunit; GPx7: glutathione peroxidase 7; GSR: glutathione-disulfide reductase; GSTM3: glutathione S-transferase mu 3; IMTP: isometric mid-thigh pull; IL-18: interleukin-18; MAPKP7: mitogen-activated protein kinase 7; MAPKP14: mitogen-activated protein kinase 14; MCH: mean corpuscular hemoglobin; N: number of participants; NRF2: nuclear factor erythroid 2-related factor 2; PC: placebo controlled; PM: parallel manner; PRDX3: peroxiredoxin 3; PRDX6: peroxiredoxin 6; PRS: perceived recovery status; R: randomized; sIgA: salivary immunoglobulin A; SIRT1: sirtuin 1; SJ: squat jump; SOD2: superoxide dismutase 2; sORP: static oxidation–reduction potential; *T. chuii*: *Tetraselmis chuii*. ↑ represents an increase in parameters or in expression of target genes, whereas ↓ represents a decrease in parameters or down-regulation of target genes.

4. Data Supporting SOD-Rich *T. chuii* as a Promoter of Cellular Health

When interpreted together, the outcomes from both preclinical and clinical trials previously shown collectively appear to support a role for SOD-rich *T. chuii* to positively promote some aspects of cellular health (Figure 1). Each of these points is substantiated in the sub-sections below.

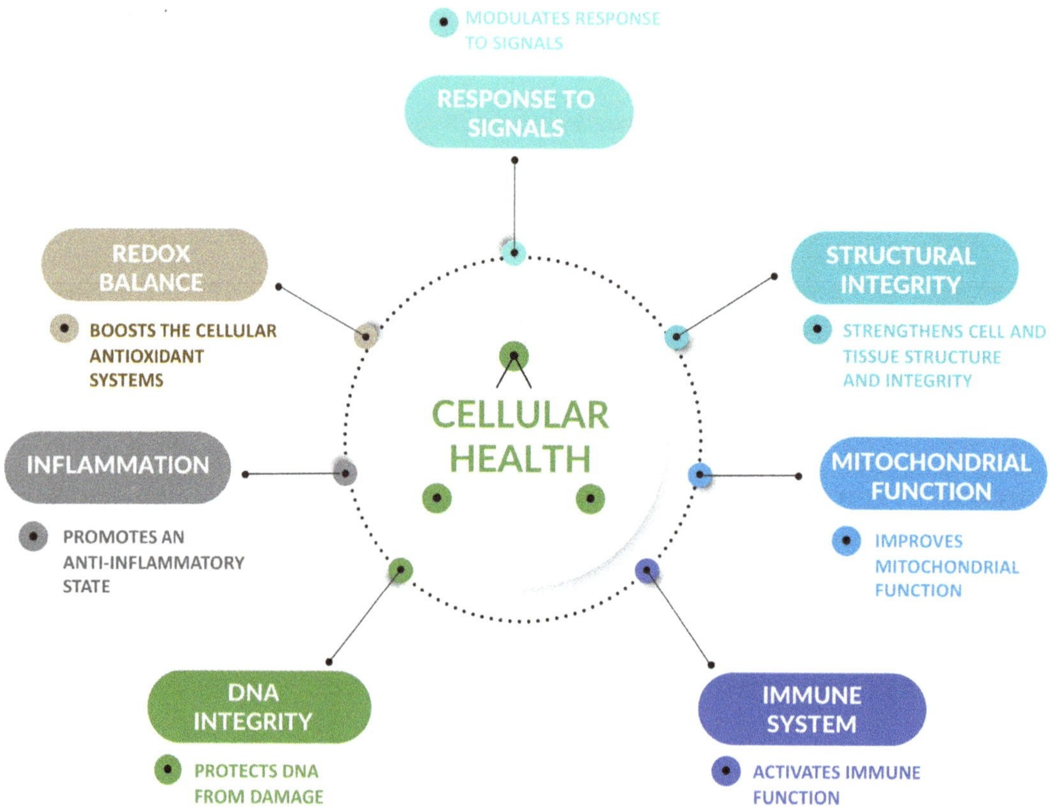

Figure 1. The effects of SOD-rich *T. chuii* on promoting cellular health.

4.1. SOD-Rich T. chuii Boosts the Cellular Antioxidant Systems

In a rodent model of metabolic syndrome, in which oxidative stress and low-grade chronic inflammation were induced by diet [15], supplementation with SOD-rich *T. chuii* reduced circulating levels of oxidized low-density lipoprotein (oxLDL). Serum oxLDL is derived from the reaction of LDL with peroxides or products generated from their degradation [59], and is considered a representative parameter of oxidative stress. As such, changes in plasma oxLDL in response to antioxidant-rich foods or dietary supplements are used to interpret their efficacy in improving redox balance [60–62]. Malondialdehyde (MDA) is another physiological parameter frequently measured as an indicator of oxidative stress [63–65]. MDA is probably the principal and most analyzed compound derived from lipid peroxidation, particularly from PUFAs, as they contain multiple carbon–carbon double bonds. In a physiological scenario of oxidative stress induced by physical exercise, SOD-rich *T. chuii* has demonstrated reduced MDA concentration, not only in serum but also in muscle tissue in rodent models, which supports the potential antioxidant effect afforded by SOD-rich *T. chuii* supplementation [16,21]. Another potential oxidative stress-related biomarker that has been reported to be attenuated after SOD-rich *T. chuii* consumption is the static oxidation–reduction potential (sORP) in seminal samples of idiopathic infertile men [55]. sORP is considered a direct measurement of oxidative stress or abnormal redox state in biological samples [66]. Particularly in semen samples, sORP has been demonstrated to provide very useful information about sperm function [67,68]. In aggregate, the improvement of these three oxidative stress parameters supports enhanced redox balance in cells after SOD-rich *T. chuii* administration.

Activation of the primary antioxidant enzymes (SOD, GPx, CAT) in response to SOD-rich *T. chuii* supports an improvement in the capacity of cells to restore and preserve redox homeostasis when encountering pro-oxidative insults. This concomitant activation of the three primary antioxidant enzymes has been observed in vitro in human skeletal muscle cells [19] and rat skeletal muscle tissue [16]. Moreover, an increase in the hepatic GPx activity after SOD-rich *T. chuii* consumption has been reported in a rodent model of metabolic syndrome [15]. These enzymes represent the first line of antioxidant defenses in the scavenging of free radicals in cells. SOD catalyzes the dismutation of superoxide radicals generated by different enzymatic systems in cells to hydrogen peroxide. Subsequently, GPx and CAT catalyze the reduction of hydrogen peroxide to oxygen and water, thereby converting a potentially pernicious molecule into stable byproducts [69]. In this context, activation of these primary antioxidant enzymes by dietary supplements with purported antioxidant effects is considered a much more effective therapeutic or complementary approach to combat cellular redox imbalance than supplementation with direct antioxidants [70–72]. The increase in the activities of these enzymes following SOD-rich *T. chuii* administration might be related to the transcriptional up-regulation of genes encoding such enzymes. In this regard, SOD-rich *T. chuii* increased the expression of SOD1 (the cytosolic Cu,Zn-SOD), SOD2 (the mitochondrial Mn-SOD), GPx1 (the most abundant selenoperoxidase, being virtually present in all cells) and CAT (mainly located at peroxisomes) genes in human muscle cells [19]. Similarly, an increase in transcripts of SOD1, SOD2, and GPx1 was detected in the liver of obese rats after SOD-rich *T. chuii* supplementation [15]. In human skeletal muscle, SOD2 and CAT genes were up-regulated by SOD-rich *T. chuii*, together with GPx7 [24]. Of note, GPx7 is a particular GPx that cannot bind glutathione, and hence, it has no capacity to participate in redox reactions directly. However, it is considered an essential sensor for oxidative stress and endoplasmic reticulum stress as it promotes signal transduction through interaction with additional factors [73].

SOD-rich *T. chuii* administration has been reported to increase the hepatic glutathione levels in a rodent model of metabolic syndrome [15]. Glutathione is a tripeptide comprising cysteine, glycine, and glutamic acid that can exist as a reduced form (GSH, or g-glutamyl–cysteinyl–glycine), which contains a thiol group on the cysteine residue and an oxidized form (GSSG), which corresponds to two glutathione molecules bound at the sulfur atoms via a disulfide bridge. Glutathione plays a crucial role in cellular antioxidant protection through direct chemical neutralization of free radicals (superoxide anion, hydroxyl radical, or nitric oxide), but it is also involved in the regeneration of oxidized vitamins C and E, and in the neutralization of reactive compounds such as peroxides or xenobiotics produced by detoxification enzymes. Moreover, it facilitates the plasma membrane transport of metabolites from toxins via formation of glutathione S-conjugates [74,75]. Glutathione biosynthesis occurs in the cytosol via two steps, both of which require ATP hydrolysis. The first one is the rate-limiting step and is catalyzed by the glutamate-cysteine ligase (GCL), in which g-glutamylcysteine is formed. The GCL comprises two subunits, a heavy catalytic subunit (GCLC), and a light regulatory or modifier subunit (GCLM) [76]. The second step is catalyzed by the glutathione synthetase (GSH-S), which adds glycine to form the final tripeptide. Another important enzyme is the glutathione-disulfide reductase (GSR), which catalyzes the reduction of GSSG to GSH [74,75]. The increase in liver glutathione content after SOD-rich *T. chuii* supplementation in obese rats might, therefore, be related to the up-regulated gene expression of GCLM, GSH-S, and GSR [15]. Moreover, SOD-rich *T. chuii* has been reported to increase the transcript levels of GCLM and GSR in human skeletal muscle of healthy subjects [24]. Collectively, these data support a role for SOD-rich *T. chuii* in the activation of de novo synthesis and recycling of GSH. This represents a valuable finding as dietary interventions that can directly stimulate GSH synthesis or prevent GSH depletion are of therapeutic interest [75].

Additional genes with an antioxidant function that are up-regulated after SOD-rich *T. chuii* supplementation in human skeletal muscle include peroxiredoxin 3 (PRDX3) and 6 (PRDX6) [24]. Peroxiredoxins are small non-selenoperoxidases found in all organisms that decompose hydrogen peroxide, lipid hydroperoxides, and peroxynitrite to form water and alcohols, thus protecting against oxidative damage. PRDX3 is restricted to the mitochondria, whereas PRDX6 is mainly located in the cytosol, and seems to play an important role in repairing oxidized cell membranes [77]. In addition, up-regulation of the glutathione-transferase mu 3 gene (GSTM3) in human skeletal muscle after SOD-rich *T. chuii* supplementation also is of value as GSTM3 serves as a phase II detoxification enzyme involved in maintaining redox homeostasis [78].

Regarding the underlying cell signaling that may contribute to the up-regulation of the aforementioned transcriptionally activated antioxidant genes by SOD-rich *T. chuii*, the Kelch-like ECH associated protein 1 (KEAP1)-NRF2-Antioxidant Response Element (ARE) is likely a key candidate signaling cascade. Indeed, KEAP1 interacts with NRF2 and acts as a sensor of cellular stress. It functions as an adaptor for the ubiquitin ligase complex, targeting NRF2 for ubiquitination and further degradation by the proteasome. Under oxidative conditions, KEAP1 undergoes a conformational change, which renders it unable to bind NRF2, and allows transport of NRF2 to the nucleus to activate transcription of target genes [79]. NRF2 belongs to the family of cap 'n' collar basic region leucine-zipper transcription factors, and binds to the AREs in the promoter region of target genes through heterodimerization with small musculoaponeurotic fibrosarcoma (Maf) proteins. Consequently, the NRF2 transcription factor is considered the master regulator of the antioxidant response in cells, and also controls the adaptive response to various environmental stressors. It is now recognized that NRF2 regulates the expression of more than 200 genes, most of them with a cytoprotective role, including genes involved in drug detoxification, lipid, and carbohydrate metabolism, as well as additional transcription factors [79,80]. In particular, NRF2 controls the expression of a fundamental set of genes involved in redox metabolism, like genes of the GSH-based system (e.g., GCLM, GCLC, or GSR), thioredoxin-based system (such as PRDX6), and other antioxidant systems such as SOD [79–83]. The pivotal role of NRF2 in cellular function is revealed by the complex regulatory mechanisms controlling its activity not only at transcriptional level but also at protein level, with polyubiquitination and further proteasomal degradation being one of the most important mechanisms [84]. Up-regulation of NRF2 by SOD-rich *T. chuii* has been demonstrated not only in vitro in human muscle cells [19], but, more importantly, in vivo in human skeletal muscle after a two-week supplementation period [24]. This represents a potential key finding in explaining the molecular mechanisms underlying the widespread stimulation of different cellular antioxidant systems by SOD-rich *T. chuii*, as outlined above, including up-regulation of both GCLM and GSR in rat liver and human skeletal muscle [15,24], SOD2 in human muscle cells and human skeletal muscle [19,24], or PRDX6 in human skeletal muscle [24]. Interestingly, another one of the known NRF2 target genes is GSTM3 [80], which is up-regulated by SOD-rich *T. chuii* in human skeletal muscle [24]. It has been reported that GSTM3 can prevent NFR2 polyubiquitination and further degradation by the proteasome, thus leading to activation and enhancement of NRF2 function [78].

In addition to NRF2, transcriptional activation of sirtuin 1 (SIRT1) by SOD-rich *T. chuii* in human skeletal muscle has also been observed. Sirtuins mediate the deacetylation of both histones and non-histone proteins in an NAD^+-dependent manner, comprising a total of seven members in mammals, referred to as SIRT1 to SIRT7. Among them, SIRT1 is well known for its role in multiple biological processes, including cellular senescence, cell death, sugar and lipid metabolism, maintenance of genomic stability, inflammation, and also oxidative stress responses [85]. In this sense, it has been reported that SIRT1 can deacetylate NRF2, and this modification has been related to an increase in stability, nuclear localization, and transcriptional activity of NRF2 [86]. Thus, SIRT1 might improve cell resistance to oxidative stress-induced damage by increasing the expression of NRF2 and the downstream genes it activates with ARE [86–88]. Moreover, a positive regulation of SIRT1 by NRF2 at both protein expression and deacetylase activity has been reported, representing a positive feedback pathway in the cellular antioxidative response mediated by NRF2 [89]. As such, this interaction of SIRT1 with NRF2 could be considered as a mechanism to enhance the key antioxidant function of NRF2 and SOD-rich *T. chuii* might be mediating NRF2 signaling through SIRT1 up-regulation.

4.2. SOD-Rich *T. chuii* Promotes an Anti-Inflammatory State

Supplementation with SOD-rich *T. chuii* promoted an anti-inflammatory state in a rodent model of metabolic syndrome [22]. Indeed, one of the key findings was the transcriptional down-regulation of NF-κB1/p50 in the liver, thymus, and spleen of obesity-induced animals, restoring the transcript amounts to levels similar to those found in healthy animals. NF-κB1/p50 belongs to the family of inducible transcription factors involved in the inflammatory response. This family comprises four additional members, all of them being structurally related: NF-κB2/p52, RelA/p65, RelB, and c-Rel. All the members bind to the promoter region of the target genes in the form of homo- or heterodimers [90]. In normal conditions, family members are sequestered in the cytoplasm by inhibitory proteins like the IkB family members, with IkBa considered most important. Under several stimuli, a multi-subunit IkB kinase complex, or IKK, phosphorylates IkBa, triggering its degradation by the proteasome. This then allows the nuclear translocation of NF-κB members to the nucleus to mediate the transcriptional induction of pro-inflammatory cytokines in innate immune cells such as IL-1β or TNFα [91]. Thus, down-regulation of NF-κB1 after dietary supplementation with SOD-rich *T. chuii* might be related to the parallel reductions in IL-1β and TNFα transcripts in adipose tissue and thymus, or IL-1β in spleen [15]. Interestingly, SOD-rich *T. chuii* has also been demonstrated to reduce both serum and muscle levels of IL-1β and TNFα in response to intense exercise in a murine model [21], which supports more general anti-inflammatory protective properties of SOD-rich *T. chuii*.

Significant down-regulation of the interferon γ (IFNγ) gene in adipose tissue, thymus, and spleen of dietary-induced obese animals has also been observed after supplementation with SOD-rich *T. chuii* [15]. IFNγ is the only characterized Type II interferon to date, and it is considered one of the most important cytokines mediating systemic and pathogenic inflammation in obesity [92,93]. In parallel to the IFNγ gene, SOD-rich *T. chuii* also elicited a significant down-regulation of the transforming growth factor-β1 (TGF-β1) gene in the liver. Significant and concomitant up-regulation of the IL-1β, TNFα, IFNγ, and TGF-β1 gene has been observed in the liver of high-fat fed mice, which were linked to the development of inflammation [94]. Additionally, transcriptional up-regulation of proinflammatory cytokines has been shown in primary human brain pericytes in response to TGF-β1 treatment [95]. Thus, down-regulation of the IFNγ and TGF-β1 genes in response to SOD-rich *T. chuii* might contribute to the anti-inflammatory effect of the ingredient.

Interestingly, dietary supplementation with SOD-rich *T. chuii* positively regulated the pleiotropic anti-inflammatory cytokine IL-10 in a rodent model of metabolic syndrome [15]. In this regard, serum levels of IL-10 in supplemented animals were restored to similar values exhibited by healthy animals. In parallel, the IL-10 gene was up-regulated in adipose tissue, and to a higher degree in immune organs such as the thymus and spleen. It has been shown that IL-10 inhibits, at the transcriptional level, the expression of the pro-inflammatory cytokines IL-1β and TNFα via the blockade of NF-κB nuclear localization [96,97], which agrees with the down-regulation of both cytokines induced by SOD-rich *T. chuii*, as previously mentioned.

In adipose tissue of obese rats, SOD-rich *T. chuii* significantly increased adiponectin (ACDC) transcripts [15]. ACDC is an adipokine with anti-diabetic, anti-atherogenic, and anti-inflammatory effects, which is expressed mainly in adipose tissue but also in a variety of different tissues like myocytes or epithelial cells [98]. In obese mice, ACDC has been shown a negative effect on IFNγ production by CD4+ T cells [99]. Moreover, it has been reported that ACDC can induce the production of anti-inflammatory IL-10 in macrophages and dendritic cells, while concomitantly suppressing the production of proinflammatory TNFα and IFNγ in stimulated macrophages [100]. Therefore, the existing data suggest that SOD-rich *T. chuii* can modulate the complex and interconnected network of cytokines and factors controlling the inflammatory response to promote anti-inflammatory effects in a variety of tissues and organs.

The transcriptional activation of both NRF2 and SIRT1 observed in human skeletal muscle following SOD-rich *T. chuii* supplementation might also play a key role in its anti-inflammatory effects. In this regard, it has been shown that a fine-tuning regulatory mechanism operates between NRF2 and NF-κB, with NRF2 negatively regulating the pro-inflammatory NF-κB signaling pathway via different pathways [101,102]. For instance, as a consequence of NRF2 activation, cellular antioxidant defenses are increased, which reduces ROS bioavailability and thus, activation of the IKK complex and further phosphorylation of IkBa, thereby inhibiting nuclear translocation of NF-κB members to the nucleus. In this way, transcriptional induction of pro-inflammatory cytokines is down-regulated. Moreover, it has been shown that NRF2 activation can prevent transcriptional up-regulation of pro-inflammatory cytokines, including IL-1β or IL-6, by directly binding to the promoter region of these genes, thus inhibiting recruitment of RNA polymerase II and transcription initiation [103]. Considering SIRT1 expression is increased following SOD-rich *T. chuii* supplementation, alternative mechanisms might contribute to the anti-inflammatory effects elicited by SOD-rich *T. chuii* [104]. For instance, SIRT1 is known to directly deacetylate RelA/p65, one of the NF-κB subunits, thus inhibiting pro-inflammatory cytokine expression mediated by NF-κB [105]. Moreover, this deacetylation has been related to an enhancement of RelA/p65 methylation, which leads to an increase in RelA/p65 ubiquitination and further degradation by the proteasome, and hence, inhibition of NF-κB transcriptional activity [106]. SIRT1 can also negatively affect DNA binding of NF-κB subunits through accumulation in the promoter region of pro-inflammatory cytokines [107]. Additionally, it has been reported that SIRT1 can inhibit IkB degradation, thus avoiding nuclear accumulation of NF-κB components and down-regulating NF-κB function [108].

4.3. SOD-Rich T. chuii Protects DNA from Damage

In a clinical trial (Table 2), preliminary data analysis has demonstrated that SOD-rich *T. chuii* lowered double-strand DNA breaks (DSBs) in the sperm of idiopathic infertile men after 90 days of supplementation [55]. This may have important implications for sperm quality as high levels of DSBs have been related to a diminished chance of conceiving and to a higher incidence of miscarriage when fathering a pregnancy [109]. Although less frequent than single-strand DNA breaks (SSBs), DSBs are considered particularly harmful lesions as they provoke genome instability and chromosomal rearrangements. They can originate because of under replicated DNA during cell division, and also by the action of transposable elements or ionizing radiation. This ionizing radiation breaks water molecules to create hydroxyl free radicals, which can react with DNA to produce SSBs and these SSBs can be spontaneously converted in DSBs directly [110]. Thus, and as previously addressed, the induction of the cellular antioxidant mechanisms and the corresponding increase in the scavenging capacity of free radicals mediated by SOD-rich *T. chuii* might be related to the decrease in DSBs. Indeed, antioxidants have been demonstrated to protect against DSBs [111–114].

Reduction of DSBs might also be related to an improvement in the activity of DNA repair mechanisms. The two main methods involved in DSB repair are homologous recombination (HR), in which sister chromatids serve as templates during the process of repair, and nonhomologous end joining (NHEJ). In both instances, a complex molecular machinery involved first in the recognition of DSBs and then in DNA synthesis and repair is known to operate [115]. In this scenario, up-regulation of SIRT1 mediated by SOD-rich *T. chuii* might have a role in decreasing DSBs as SIRT1 activity has been strongly related to repair of DSBs [116]. For instance, SIRT1 deacetylates the repair factor Ku70, enhancing DNA repair capacity [117], and maintains the acetylation level of nibrin (NBS1), which is a component of a conserved nuclease complex that acts as a critical sensor in regulating cellular responses to DSBs for efficient DNA damage repair [118]. Moreover, SIRT1 is also involved in the selection of the DSB repair pathway via deacetylation of the KRAB-associated protein 1 (KAP1) to promote NHEJ, suppressing the HR repair pathway [119].

The potential contribution of NRF2 activation by SOD-rich *T. chuii* in the reduction of DSBs may also be of value regarding DNA repair. In this sense, several DNA repair genes involved in the HR pathway are likely to be regulated by NRF2 as they exhibit AREs in their promoter regions [120]. Moreover, NRF2 can activate the ataxia telangiectasia mutated (ATM), which is a master regulator of the DNA damage response, leading to G2 cell cycle arrest and promoting the HR repair of DSBs to preserve genome stability [121]. In addition, NRF2 seems to be also involved in NHEJ [122].

4.4. SOD-Rich T. chuii Activates Immune Function

Supplementation with SOD-rich *T. chuii* for three weeks has been shown to sustain immune function in individuals after an intensified resistance training protocol, as revealed by better maintenance of salivary immunoglobulin A (sIgA), compared to a placebo group [21]. The sIgA molecule as a secretory IgA is formed by a dimeric IgA and a glycoprotein known as the secretory component, stabilizing and protecting the molecule from degradation by bacterial and digestive enzymes. It is known that sIgA can prevent bacterial colony formation on mucosal surfaces, and can also neutralize toxins and enzymes produced by bacteria, and also pathogenic viruses inhibiting penetration into epithelial cells [123]. Regarding athlete illness, sIgA has been used as a valuable biomarker to evaluate the risk of developing respiratory tract infections [124]. Indeed, a decrease in sIgA levels after a long duration and high-intensity exercise seems to be associated with increased upper respiratory symptoms [124,125]. For instance, college football players exhibited a reduction in sIgA concentration and an increase in the incidence of upper respiratory tract infections over 12 months [126]. Conversely, a decrease in respiratory symptoms was observed in individuals after 12 weeks of moderate exercise training, with a parallel increase in the concentration of sIgA [127]. Thus, a possible contribution of SOD-rich *T. chuii* to improving immune function is inferred from better sIgA during intensified exercise training.

An increase in counts of different white cells such as basophils, monocytes, and particularly, lymphocytes, has also been observed after supplementation with SOD-rich *T. chuii* [23]. Other dietary interventions have also been shown to increase lymphocyte count. Indeed, an association between an increase in lymphocyte numbers and the improvement in the immune status of individuals has been reported after supplementation with different antioxidants, such as vitamins, b-carotene, and selenium [128–130], and also with the microalgae Spirulina [131]. The presence of acidic and sulfated polysaccharides in the cell wall of the microalgae *Tetraselmis* has been suggested as a potential inductor of immune cell proliferation through the increase in cytokine and chemokine production by macrophages [23].

Transcriptionally activated genes by SOD-rich *T. chuii* in human skeletal muscle are known to play a role in immune function [24], specifically, the monocyte chemoattractant protein-1 (MCP-1)/CC chemokine ligand-2 (CCL2). Chemokines are small (8–14 kDa) signaling proteins d, which are secreted by different immune cells. Chemokines comprise four families with two main subgroups (CXC and CC) and two small subgroups (CX3C and C), with MCP-1/CCL2 belonging to the CC family. MCP-1/CCL2 protein is mainly produced by epithelial cells, endothelial cells, smooth muscle cells, monocytes/macrophages, fibroblasts, astrocytes, and microglial cells, which are regulated by several other cytokines and factors. Importantly, it is known that MCP-1/CCL2 can direct the migration and infiltration of monocytes at the site of injury and infection, and are also involved in proliferation of T cells, thus important to the immune response [132]. Interleukin-18 (IL-18) is another of the genes up-regulated by SOD-rich *T. chuii*. It belongs to the IL-1 family of cytokines, and is a potent pro-inflammatory cytokine involved in host defense against infections through innate and acquired immune stimulation responses. IL-18 is produced by hematopoietic cells (such as monocytes and macrophages) and non-hematopoietic cells (for instance, keratinocytes and mesenchymal cells). Together with IL-12, IL-18 triggers the innate immune system, stimulating NK cells to respond to cancer and infections, as well as to activate macrophages. In the adaptive immune system, IL-18 promotes the activation and differentiation of T cells and is essential for the development of natural killer (NK) cells, and also up-regulates the cytotoxic activities of NK and CD8+ T cells [133]. Interestingly, IL-18, as occurs with other interleukins of the same family, is synthesized in the cytoplasm as an inactive precursor referred to as pro-IL-18. This precursor is further transformed into the active IL-18 in multiprotein cytosolic complexes named inflammasomes in a caspase-1 (CASP1)-mediated process. However, an additional activated gene by SOD-rich *T. chuii* is caspase-8 (CASP8), which has also been involved in the processing of inactive IL-18 into the active form in a CASP1-independent process [134].

Transcriptional up-regulation of NRF2 by SOD-rich *T. chuii* [24] might also have a direct influence in the regulation of immune function based on the known roles of this transcription factor in immunity [79,135]. For instance, NRF2 is involved in bacterial clearance via up-regulation of the macrophage receptor with collagenous structure (MARCO) gene, which encodes a scavenger receptor necessary for bacterial phagocytosis [136]. An antitumor and antiviral role has also been observed for NRF2 via activation of the cytokine IL-17D, which exhibits tumor rejection activity mediated by NK cells [137]. Moreover, a role of NRF2 in T-cell differentiation has been observed, favoring Th2 but decreasing Th17 [138,139]. On the other hand, SIRT1 has also been implicated in the immune response, and hence up-regulation of SIRT1 gene expression could contribute to improved immune responses after SOD-rich *T. chuii* supplementation [24]. For instance, regarding innate immunity, SIRT1 influences myeloid-derived suppressor cells (MDSCs) differentiation, and regulates the generation of cytokines by dendritic cells, subsequently modulating their function. In the adaptive immune response, SIRT1 can influence the differentiation of inflammatory T cells, and plays an essential role in Th17 formation and in the activation of B cells, facilitating immune function [140,141].

4.5. SOD-Rich *T. chuii* Potentially Improves Mitochondrial Function

Although there is currently no direct experimental evidence of SOD-rich *T. chuii* modifying key parameters of mitochondrial physiology, data from clinical trials suggest a potential positive effect of SOD-rich *T. chuii* on aspects of mitochondrial function. Positive regulation of SIRT1 by SOD-rich *T. chuii* [24] might be involved in increasing mitochondrial content as SIRT1 is a well-known promoter of mitochondrial biogenesis through the activation, by deacetylation, of the peroxisome proliferator-activated receptor γ-coactivator-1α (PGC-1α). In turn, activated PGC-1α activates the mitochondrial transcription factor A (TFAM) in the cytoplasm, eliciting the import of both SIRT1 and PGC-1α into the mitochondria and the recruitment of TFAM to the D-loop region of mitochondrial DNA, where it forms a multiprotein complex with SIRT1 and PGC-1α. Finally, this complex drives the replication and transcription of mitochondrial DNA to improve mitochondrial biogenesis [142]. In addition, up-regulation of NRF2 induced by SOD-rich *T. chuii* [24] may also have contributed to mitochondrial biogenesis. Indeed, NRF2 controls the expression of the nuclear respiratory factor 1 (Nrf-1), which in turn activates TFAM, leading to mitochondrial DNA replication. NRF2 can also contribute to mitochondrial function through the induction of mitophagy, a process in which damaged mitochondria are removed from the cell, and for its key role in maintaining mitochondrial membrane potential, which increases the efficiency of oxidative phosphorylation and ATP production [143,144]. Another gene known to play a key homeostatic role in mitochondrial function is the apoptosis-inducing factor mitochondria associated 1 (AIFM1), which was also found to be up-regulated by SOD-rich *T. chuii* in skeletal muscle [24]. This gene encodes a mitochondrial oxidoreductase that takes part in the electron chain assembly, and thus regulates oxidative phosphorylation and ATP production [145].

Up-regulation of genes encoding the key mitochondrial antioxidant proteins SOD2 and PRDX3 [24] by SOD-rich *T. chuii* might also contribute to improved mitochondrial function as they act to protect the organelle from damage caused by ROS. In this sense, a role for the NRF2 pathway in maintaining mitochondrial homeostasis through the activation of antioxidant and quality control genes has been stated [146].

The MAPK signaling pathway has also been involved in mitochondrial physiology. In this regard, activation of MAPK/p38 signaling has been shown to enhance PGC-1α levels and activity, thus promoting mitochondrial biogenesis [147,148]. Indeed, MAPK/p38 has been shown to phosphorylate PGC-1α in muscle cells directly, enhancing its activity via the increase in protein stability and through the inhibition of the interaction with its co-repressor [149]. Moreover, MAPK/p38 signaling is currently known to regulate the activity of key mitochondrial proteins involved in oxidative phosphorylation and iron homeostasis [150]. In this scenario, up-regulation of MAPK14/p38α by SOD-rich *T. chuii* might be an additional pathway contributing to the improvement of mitochondrial function.

4.6. SOD-Rich *T. chuii* Strengthens Cell and Tissue Structure and Integrity

Supplementation with SOD-rich *T. chuii* has been related to a decrease in markers of cellular and tissue damage, which suggests a strengthening effect on cellular structure. In this regard, serum creatine kinase (CK) and myoglobin were significantly reduced after a cross-training event in endurance-trained individuals and mechanistic rodent models after SOD-rich *T. chuii* supplementation [16,20]. Serum content of skeletal muscle enzymes (such as CK) and proteins (such as myoglobin) are markers of the functional status of muscle tissue, and can vary widely depending on physiological conditions. An increase in CK is considered an index of tissue damage following acute and chronic muscle injuries, as it is known that strenuous exercise, particularly incorporating eccentric contractions, can damage skeletal muscle cell structure, which increases CK efflux into the systemic circulation [151,152]. Myoglobin, a monomer protein involved in oxygen storage, is released to serum as a result of degradation of muscle cells following strenuous exercise [151,153]. These results might be related, at least in part due to the indirect antioxidant effects mediated by the ingredient helping to mitigate the harmful effects caused by exercise-induced oxidative stress, thus reducing protein damage (e.g., carbonylation and cross-linking) and further degradation by the proteasome, and also damage to cellular membranes related to lipid peroxidation [154]. Moreover, in a mechanistic rodent model, SOD-rich *T. chuii* has been shown to increase the muscle protein content of the muscle atrophy F-box (MAFbx) and muscle RING-finger protein-1 (MuRF-1) [20], two ubiquitin ligases involved in protein degradation by the proteasome [155], which might be related to an increase in protein turnover to counteract protein breakdown linked to exercise and oxidative stress [20].

Some well-known markers linked to tissue repair and regeneration have been shown to be activated by SOD-rich *T. chuii*. Indeed, an increase in the muscle protein content of the myogenic differentiation factor, MyoD, and neural cell adhesion molecule, NCAM, was detected in response to SOD-rich *T. chuii* in a rodent model [20]. Both proteins are key positive regulators of satellite cell progression [156,157], which are mononuclear dormant cells that are activated to promote regeneration upon fiber damage. After muscle injury, these satellite cells are activated and differentiate to myoblasts, which exit the cell cycle and become myocytes after several rounds of proliferation. Ultimately, myocytes undergo a fusion process to form multinucleated myotubes that eventually mature into myofibers [158]. In contrast and in the same mechanistic model, SOD-rich *T. chuii* reduced levels of myostatin [20], a paracrine signaling molecule mainly secreted by skeletal myocytes that behaves as a negative regulator of muscle growth and differentiation [159].

In human skeletal muscle, SOD-rich *T. chuii* transcriptionally activated different genes with myogenic functions [24], thereby representing an additional support role for SOD-rich *T. chuii* in the processes of repair and regeneration of tissues. For instance, BTBD1, a gene that encodes a BTB (broad-complex, tramtrack, and bric-a-brac) domain and facilitates interaction with other proteins lacking this domain, is known to interact with DNA topoisomerase I, and is expressed in many human tissues, with higher levels in heart and skeletal muscle [160]. Interestingly, BTBD1 has been demonstrated to be essential for myogenic differentiation [161]. In addition, calpain 3 (CAPN3) gene expression was also up-regulated by SOD-rich *T. chuii*. Calpains are a family of Ca^{2+}-dependent cysteine proteases that participate in various cellular processes. In particular, CAPN3 is the only muscle-specific calpain, and has important roles in the promotion of calcium release from skeletal muscle fibers, calcium uptake of sarcoplasmic reticulum, and also in muscle formation and remodeling [162]. Therefore, as a potential contribution of CASP8 and/or caspase-10 (CASP10), both genes also up-regulated in response to SOD-rich *T. chuii* treatment, and might also be considered to aid tissue structure and integrity. Caspases are a family of cysteine-aspartate proteases involved in cell death processes [163], but it is currently known that some of these proteins like caspase-2 (CASP2) and caspase-3 (CASP3) can play a direct role in skeletal muscle differentiation and myogenesis [164,165], allowing the muscle remodeling required for differentiation but being inhibited before their activation can lead to cell death. Finally, given the role of NRF2 in the stimulation of muscle repair, growth, and differentiation [166], up-regulation of this key transcription factor by SOD-rich *T. chuii* might also contribute to improved muscle regeneration and integrity.

4.7. SOD-Rich *T. chuii* Modulates Cellular Signaling

A role for SOD-rich *T. chuii* in the modulation of cellular signaling is supported by the modification, in human skeletal muscle tissue, of transcript amounts of up to four different genes of the MAPK signaling pathway: MAPK1/ERK2, MAPK6/ERK3, MAPK7/ERK5, and MAPK14/p38α [24]. The three conventional MAPKs (MAPK1/ERK2, MAPK7/ERK5, and MAP14/p38a) were up-regulated, and the atypical MAPK6/ERK3 was down-regulated by SOD-rich *T. chuii*. In eukaryotic cells, involvement of the MAPK pathway in key physiological processes (mitosis, metabolism, motility, survival, apoptosis, differentiation, and response to stress) is well established and works in four different signaling cascades: the extracellular signal-regulated kinases 1/2 (ERK1/2), c-Jun amino (N)-terminal kinases 1/2/3 (JNK1/2/3), p38 isoforms (α, β, γ, and δ), and ERK5 [167]. In a general view, different studies suggest that the MAPK/ERK signaling pathway is closely related to processes such as cellular proliferation and differentiation, whereas the JNK and p38 pathways seem to be more related to response to stress and apoptosis [47].

At the cellular level, the actual physiological significance of the modified expression of MAPK genes is currently unknown, but the cross-talk between MAPKs and key factors and metabolic processes are worthy of discussion. For instance, the MAPKs signaling cascade can be activated by ROS, which has been associated with NRF2 function [168]. In this regard, it is known that MAPK1/ERK2 can directly phosphorylate NRF2, which positively regulates NRF2 activity. Moreover, it has been demonstrated that MAPK14/p38α can phosphorylate NRF2, although both positive and negative regulation of NRF2 activity has been reported. Irrespective, it seems that phosphorylation of NRF2 by MAPKs only slightly affects NRF2 transactivation and further expression of NRF2 target genes, suggesting that MAPKs might regulate NRF2 activity mainly in an indirect manner [169]. In this regard, an additional point of relevance is the fact that phosphorylation and nuclear accumulation of MAP14/p38a can be stimulated by SIRT1, which has been related to the promotion of cell proliferation [170].

The activator protein-1 (AP-1) is also relevant regarding MAPK cross-talk. AP-1 is a transcription factor that consists of different components, such as the Jun family (which includes c-JUN and JUND), Fos family, Jun-dimerizing partners (JDP), musculoaponeurotic fibrosarcomas (Maf) family, and activating transcription factor (ATF) family. Several stimuli, including oxidative stress and intracellular MAPK signaling, can activate AP-1, mediating functions such as cell growth and differentiation [171]. In human skeletal muscle, SOD-rich *T. chuii* has been shown to up-regulate both c-JUN and ATF1 and to down-regulate JUND [24]. Understanding the physiological relevance of such modifications in gene expression will require further research, but the existence of a complex fine-tuned regulatory network modulated by SOD-rich *T. chuii* to maintain an adequate physiological balance in the cell is suggested. Indeed, JUND can activate NRF2-induced transcription of downstream genes, with MAPK signaling playing a central role in this process [172], with c-JUN, a known target of SIRT1, inhibiting AP-1 transcriptional activity [173]. It should also be mentioned that AP-1 mediates the expression of inflammatory mediators, such as cyclooxygenase 2 and prostaglandin E2 [141], and hence the inhibition of AP-1 signaling by SIRT1 potentially represents an additional anti-inflammatory mechanism of SOD-rich *T. chuii* which may be mediated through SIRT1 up-regulation.

4.8. SOD-Rich *T. chuii* Induces a Homeostatic Response

In human skeletal muscle, a deeper gene expression analysis revealed a potentially interesting series of observations. In more than half of the genes analyzed using the OpenArray™ technology, a negative correlation was found between the baseline gene expression and the magnitude of increase in transcript amounts after supplementation with SOD-rich *T. chuii* for two weeks [24]. This suggests that individuals with the lowest expression levels exhibited the highest responsiveness to SOD-rich *T. chuii* supplementation. As such, SOD-rich *T. chuii* seems to act as a modulator of cellular responses that need enhancement. This finding is of potential value as both direct and indirect antioxidant supplements have been used to offset the potentially harmful effects of excess free radicals. However, it has more recently been considered that physiological levels of such free radicals control fundamental biological processes [174], and excess antioxidants could interfere with beneficial redox-mediated cellular signaling [26]. Consistent with this interpretation, it has been shown that over-stimulation of the NRF2/SIRT1 pathway, which regulates the activity of the cellular antioxidant and anti-inflammatory mechanisms, can lead to deleterious outcomes, even causing reductive stress and adverse health effects [175–177]. Instead, the existing data suggest that SOD-rich *T. chuii* may behave as a transcriptional activator of this pathway in a controlled manner, as supported by concomitant improvement in various physiological responses, in both preclinical models and clinical trials.

As observed in human skeletal muscle, transcriptional activation of additional genes, such as cullin 3 (CUL3) and BTB and CNC homology 1, basic leucine zipper transcription factor 1 (BACH1) by SOD-rich *T. chuii* might somehow elicit a balanced and controlled cellular response to stress mediated by the NRF2/SIRT1 pathway [24]. Indeed, CUL3 is a protein known to behave as a negative regulator of NRF2 function through mediating its ubiquitination and proteasomal degradation [84]. BACH1 is a transcription repressor that is conserved and ubiquitously expressed in tissues. In the absence of cellular stress, BACH1 forms heterodimers in the nucleus with small Maf proteins that bind to the AREs and repress defensive gene expression mediated by NRF2. Thus, BACH1 acts as a negative regulator of NRF2 nuclear levels and function [178]. Therefore, increased CUL3 and BACH1 gene expression after SOD-rich *T. chuii* might modulate and fine-tune NRF2 expression and its subsequent effects to promote a healthy adaptive response, as opposed to excessive and maladaptive responses.

4.9. Summary of Data Supporting SOD-Rich T. chuii as a Promoter of Cellular Health

Considering the existing empirical data, SOD-rich *T. chuii* may be considered as a potential cellular health promoter. This statement is based on the ability of SOD-rich *T. chuii* to (i) boost the cellular antioxidant systems to help combat the harmful effects of oxidative stress; (ii) promote an anti-inflammatory state through activating anti-inflammatory and down-regulating pro-inflammatory factors; (iii) protect DNA damage by reducing double-strand DNA breaks and up-regulating key genes involved in maintaining genome stability; (iv) improve immune function by up-regulating genes known to be involved in the host defense against pathogens, increasing levels of immunoglobulins and counts of different immune cells; (v) potentially improve mitochondrial function by up-regulating pivotal genes involved in mitochondrial biogenesis and ATP production, as well as genes coding for mitochondrial markers involved in the protection against free radicals; (vi) strengthen cell and tissue structure and integrity through reducing markers of cellular and tissue damage, and up-regulating genes involved in the protection and repair of membranes, and promoting cellular differentiation and tissue regeneration; (vii) modulate cellular signaling by enhancing the expression of key genes that participate in signaling pathways. Two key factors transcriptionally up-regulated by SOD-rich *T. chuii*, NRF2 and SIRT1, appear as central players that might mediate the cellular health effects of the ingredient as they are themselves considered as promoters of cellular homeostasis and health [179,180].

5. Prospects

Despite encouraging initial evidence to support improved cellular health and homeostasis after SOD-rich *T. chuii* supplementation, it is recognized that further research is necessary to unravel the full molecular mechanisms and pathways that support the health effects SOD-rich *T. chuii*. In this regard, some key mediators, like NRF2 and SIRT1, which are well-known pivotal factors for cellular health and homeostasis, have been identified and are worthy of further empirical investigation. Indeed, NRF2 and SIRT1 genes are up-regulated after SOD-rich *T. chuii* supplementation [24]. However, transcriptional control is not the only regulatory level of gene function acting on cells. Particularly for NRF2, its activity is affected by post-translational modifications such as ubiquitination, sumoylation, acetylation, or phosphorylation, which control protein degradation by the proteasome and its nuclear accumulation and stability [84]. SIRT1 activity is also modulated by post-translational modifications such as ubiquitination, sumoylation, phosphorylation, and methylation, which regulate protein degradation, stability, and affinity toward its target proteins [142]. Thus, future attempts to identify all the regulatory mechanisms modulated by SOD-rich *T. chuii*, particularly concerning NRF2 and SIRT1, will undoubtedly help to better understand the molecular bases of the physiological benefits it may confer.

As presented in this review, the clinical trials conducted to assess the potential beneficial health effects of SOD-rich *T. chuii* have been performed in two particular and quite different physiological scenarios, including sports nutrition and male infertility. By taking into account the molecular mechanisms supporting SOD-rich *T. chuii* as a promoter of cellular homeostasis and health, particularly with the transcriptional activation of NRF2 and SIRT1, additional health applications become of potential interest to explore with the supplement. For example, a novel therapeutic use of SOD-rich *T. chuii* could be useful to combat age-related health declines. Aging is characterized by a decline in various functional responses owing to the accumulation of cellular damage. Several factors have been identified as key contributors to the aging process, including, among others, genome instability, loss of proteostasis, mitochondrial dysfunction, and chronic inflammation [181,182]. It has been shown that NRF2 transcriptional activity decreases with age [183], and this gradual reduction of NRF2 is considered to drive aging owing to increased oxidative stress, which contributes to various hallmarks of aging [184]. In this scenario, the induction and activation of the NRF2 pathway to maintain cellular antioxidant function and redox balance has been suggested as a targeted therapeutic strategy to reduce cell and tissue damage, which is known to occur in age-related ocular [185], joint [186], skin [187], kidney [188], liver [189], cardiovascular [190], and neurodegenerative [191] disorders, and could, therefore, help ameliorate symptoms and offset disease morbidity and progression. In addition to NRF2, it is known that SIRT1 exhibits an age-dependent decrease in expression both at protein and transcription levels [192], such that activation of SIRT1 has also been proposed as an effective means to improve age-related disorders [193–196]. Hence, in this scenario, the potential contribution of SOD-rich *T. chuii* to the improvement of age-related disorders and diseases through the activation of NRF2 and SIRT1 becomes of interest, although specific clinical trials need to be conducted to evaluate its efficacy in this regard.

A potential therapeutic use of SOD-rich *T. chuii* for women's health through the improvement of cellular function is also of potential interest. It is known that oxidative stress and inflammation are important negative factors in several aspects of women's reproductive physiology, and hence, SOD-rich *T. chuii* might help to restore the redox balance and to promote an anti-inflammatory state in target cells. For instance, it is known that ROS negatively affect the maturation and development of the oocyte, implantation and luteolysis, and hence, a favorable redox balance appears to be crucial for the oocyte maturation and quality, and also for placentation, fetal growth, and organ development. Moreover, oxidative stress and inflammation mediates the acceleration of pathology in the female reproductive tract, including primary ovarian insufficiency, polycystic ovary syndrome, endometriosis, endometrial hyperplasia, and preeclampsia [197]. In this scenario, interventions to reduce the impact of ROS, and hence, the quality of embryos and implantation, are considered an adequate strategy for a successful pregnancy. Thus, it has been stated that the use of supplements to activate the NRF2/SIRT1 pathway and hence to improve antioxidant and anti-inflammatory activities in female reproductive organs represents a fruitful approach in treating female reproductive disorders [198–201]. In addition, it has been reported that postmenopausal women are at a high risk for oxidative stress due to a marked reduction in estrogen production. The increase in the serum concentration of inflammatory cytokines, together with the increase in pro-oxidant biomarkers and the decrease in antioxidant biomarkers are among the main features related to the menopausal transition [202,203]. Thus, the elevation of cytokines and pro-oxidant markers suggests that there is a high degree of oxidative stress in the postmenopausal state. All these features are mainly due to estrogen deprivation, and hence, it might be linked with the development of postmenopause-associated increased cardiovascular risk, bone density loss in osteoporosis, hot flushes, oral dryness, loss of muscle mass and strength, atherosclerosis, weight gain, or disorders such as depression and anxiety [204–208]. In this scenario, a large body of evidence suggests that the NRF2/SIRT1 pathway is involved in protection against the physiological impairments that develop observed in postmenopausal women. Indeed, it is known that estrogens activate NRF2 and SIRT1, and the decrease in estrogen production during menopause has been related to a range of pathologies related to oxidative stress and inflammation including, among others, osteoporosis, cardiovascular diseases, dyslipidemia, changes in corporal composition, or immunological disorders [209,210]. Thus, oxidative stress and inflammation observed during menopause and its stages might be effectively modulated by dietary supplementation with SOD-rich *T. chuii* via activation of the NRF2/SIRT1 pathway, which could promote health benefits through achieving a better redox balance. However, further research and clinical trials are required to evaluate the potential benefits of SOD-rich *T. chuii* for women's health applications.

Although SOD-rich *T. chuii* supplementation may hold particular promise for individuals with a more pro-oxidative or pro-inflammatory state, potential benefits in already healthy individuals wishing to use the ingredient as part of a healthy and active lifestyle should not be discounted. Indeed, SOD-rich *T. chuii* has already been reported to elicit beneficial effects in such individuals (Table 2). A compromised redox state, which can result from poor lifestyle choices (e.g., smoking, imbalanced diet, and/or sedentary behavior) can lead to cell damage as a consequence of oxidative stress and inflammation [211–213]. Therefore, dietary interventions with SOD-rich *T. chuii* might help to prevent health challenges in many individuals who do not meet physical activity and dietary intake guidelines.

Author Contributions: Conceptualization, S.J.B.; writing—original draft preparation, S.J.B.; writing—review and editing, S.P.C., L.M., E.C. and C.I.; visualization, S.P.C. All authors have read and agreed to the published version of the manuscript.

Funding: The APC was funded by Fitoplancton Marino, S.L.

Conflicts of Interest: Authors Lalia Mantecón, Enrique Castaño, and Carlos Infante were employed by the company Fitoplancton Marino, S.L. The remaining authors declare that the research was conducted in the absence of any commercial or financial relationships that could be construed as a potential conflict of interest.

References

1. Ramos-Romero, S.; Torrella, J.R.; Pagès, T.; Viscor, G.; Torres, J.L. Edible Microalgae and Their Bioactive Compounds in the Prevention and Treatment of Metabolic Alterations. *Nutrients* **2021**, *13*, 563. [CrossRef] [PubMed]
2. Camacho, F.; Macedo, A.; Malcata, F. Potential Industrial Applications and Commercialization of Microalgae in the Functional Food and Feed Industries: A Short Review. *Mar. Drugs* **2019**, *17*, 312. [CrossRef] [PubMed]
3. Dittami, S.M.; Heesch, S.; Olsen, J.L.; Collén, J. Transitions between Marine and Freshwater Environments Provide New Clues about the Origins of Multicellular Plants and Algae. *J. Phycol.* **2017**, *53*, 731–745. [CrossRef]
4. Barkia, I.; Saari, N.; Manning, S.R. Microalgae for High-Value Products towards Human Health and Nutrition. *Mar. Drugs* **2019**, *17*, 304. [CrossRef]
5. Ampofo, J.; Abbey, L. Microalgae: Bioactive Composition, Health Benefits, Safety and Prospects as Potential High-Value Ingredients for the Functional Food Industry. *Foods* **2022**, *11*, 1744. [CrossRef] [PubMed]
6. Borowitzka, M.A. Microalgae in Medicine and Human Health: A Historical Perspective. In *Microalgae in Health and Disease Prevention*; Academic Press: Cambridge, MA, USA, 2018; pp. 195–210; ISBN 9780128114056.
7. Qiu, C.; Wang, W.; Zhang, Y.; Zhou, G.J.; Bi, Y. Response of Antioxidant Enzyme Activities of the Green Microalga *Chlorococcum* sp. AZHB to Cu^{2+} and Cd^{2+} Stress. *Sustainability* **2022**, *14*, 10320. [CrossRef]
8. Zhao, Z.; Rasool, M.A.; Chen, C.; Ma, S.; Wang, L.; Huang, G. Identification and Screening of Multiple Tropical Microalgal Strains for Antioxidant Activity in Vitro. *Food Biosci.* **2020**, *36*, 100649. [CrossRef]
9. Cokdinleyen, M.; Alvarez-Rivera, G.; Tejera, J.L.G.; Mendiola, J.A.; Valdés, A.; Kara, H.; Ibáñez, E.; Cifuentes, A. *Tetraselmis chuii* Edible Microalga as a New Source of Neuroprotective Compounds Obtained Using Fast Biosolvent Extraction. *Int. J. Mol. Sci.* **2024**, *25*, 3897. [CrossRef]
10. Mantecón, L.; Moyano, R.; Cameán, A.M.; Jos, A. Safety Assessment of a Lyophilized Biomass of *Tetraselmis chuii* (TetraSOD®) in a 90 Day Feeding Study. *Food Chem. Toxicol.* **2019**, *133*, 110810. [CrossRef]
11. Custódio, L.; Justo, T.; Silvestre, L.; Barradas, A.; Duarte, C.V.; Pereira, H.; Barreira, L.; Rauter, A.P.; Alberício, F.; Varela, J. Microalgae of Different Phyla Display Antioxidant, Metal Chelating and Acetylcholinesterase Inhibitory Activities. *Food Chem.* **2012**, *131*, 134–140. [CrossRef]
12. Cerezuela, R.; Guardiola, F.A.; Meseguer, J.; Esteban, M.Á. Enrichment of Gilthead Seabream (*Sparus aurata* L.) Diet with Microalgae: Effects on the Immune System. *Fish Physiol. Biochem.* **2012**, *38*, 1729–1739. [CrossRef]
13. Cerezuela, R.; Meseguer, J.; Esteban, M.Á. Effects of Dietary Inulin, *Bacillus subtilis* and Microalgae on Intestinal Gene Expression in Gilthead Seabream (*Sparus aurata* L.). *Fish Shellfish Immunol.* **2013**, *34*, 843–848. [CrossRef] [PubMed]
14. Cerezuela, R.; Guardiola, F.A.; González, P.; Meseguer, J.; Esteban, M.Á. Effects of Dietary *Bacillus subtilis*, *Tetraselmis chuii*, and *Phaeodactylum tricornutum*, Singularly or in Combination, on the Immune Response and Disease Resistance of Sea Bream (*Sparus aurata* L.). *Fish Shellfish Immunol.* **2012**, *33*, 342–349. [CrossRef] [PubMed]
15. Gil-Cardoso, K.; Del Bas, J.M.; Caimari, A.; Lama, C.; Torres, S.; Mantecón, L.; Infante, C. TetraSOD®, a Unique Marine Microalgae Ingredient, Promotes an Antioxidant and Anti-Inflammatory Status in a Metabolic Syndrome-Induced Model in Rats. *Nutrients* **2022**, *14*, 4028. [CrossRef]
16. Sharp, M.; Sahin, K.; Stefan, M.; Orhan, C.; Gheith, R.; Reber, D.; Sahin, N.; Tuzcu, M.; Lowery, R.; Durkee, S.; et al. Phytoplankton Supplementation Lowers Muscle Damage and Sustains Performance across Repeated Exercise Bouts in Humans and Improves Antioxidant Capacity in a Mechanistic Animal. *Nutrients* **2020**, *12*, 1990. [CrossRef] [PubMed]
17. Kaparapu, J. Application of Microalgae in Aquaculture. *Phykos* **2018**, *48*, 21–26.
18. Hurtado, M.C.; Pilar, M.; Moreno, C.; Daschner, Á.; Riba, R.E.; María, R.; Pons, G.; Fandos, E.G.; Arnau, S.G.; Gallego, J.; et al. Report of the Scientific Committee of the Spanish Agency for Consumer Affairs, Food Safety and Nutrition (AECOSAN) on a Request for Initial Assessment for Marketing of the Dried Marine Microalgae *Tetraselmis chuii* in Food Supplements under Regulation (EC). *Rev. Com. Cient. AECOSAN* **2017**, *25*, 11–21.
19. Ramírez, P.; Torres, S.; Lama, C.; Mantecón, L.; Unamunzaga, C.; Infante, C. TetraSOD® Activates the Antioxidant Response Pathway in Human Cells: An in Vitro Approach. *Afr. J. Biotechnol.* **2020**, *19*, 367–373. [CrossRef]
20. Sharp, M.H.; Sahin, K.; Stefan, M.W.; Gheith, R.H.; Reber, D.D.; Ottinger, C.R.; Orhan, C.; Tuzcu, M.; Sahin, N.; Lowery, R.P.; et al. Marine Phytoplankton Improves Exercise Recovery in Humans and Activates Repair Mechanisms in Rats. *Int. J. Sports Med.* **2021**, *42*, 1070–1082. [CrossRef]
21. Sharp, M.; Wilson, J.; Stefan, M.; Gheith, R.; Lowery, R.; Ottinger, C.; Reber, D.; Orhan, C.; Sahin, N.; Tuzcu, M.; et al. Marine Phytoplankton Improves Recovery and Sustains Immune Function in Humans and Lowers Proinflammatory Immunoregulatory Cytokines in a Rat Model. *Phys. Act. Nutr.* **2021**, *25*, 42–55. [CrossRef]
22. Toro, V.; Siquier-Coll, J.; Bartolomé, I.; Robles-Gil, M.C.; Rodrigo, J.; Maynar-Mariño, M. Effects of *Tetraselmis chuii* Microalgae Supplementation on Ergospirometric, Haematological and Biochemical Parameters in Amateur Soccer Players. *Int. J. Environ. Res. Public Health* **2020**, *17*, 6885. [CrossRef] [PubMed]

23. García, Á.; Toro-Román, V.; Siquier-Coll, J.; Bartolomé, I.; Muñoz, D.; Maynar-Mariño, M. Effects of *Tetraselmis chuii* Microalgae Supplementation on Anthropometric, Hormonal and Hematological Parameters in Healthy Young Men: A Double-Blind Study. *Int. J. Environ. Res. Public Health* **2022**, *19*, 6060. [CrossRef]
24. Cocksedge, S.P.; Infante, C.; Torres, S.; Lama, C.; Mantecón, L.; Manchado, M.; Acton, J.P.; Alsharif, N.S.; Clifford, T.; Wadley, A.J.; et al. *Tetraselmis chuii* Supplementation Increases Skeletal Muscle Nuclear Factor Erythroid 2-Related Factor 2 and Antioxidant Enzyme Gene Expression, and Peak Oxygen Uptake in Healthy Adults. *Antioxidants* **2025**, Submitted.
25. Kotur-Stevuljević, J.; Savić, J.; Simić, M.; Ivanišević, J. Redox Homeostasis, Oxidative Stress and Antioxidant System in Health and Disease: The Possibility of Modulation by Antioxidants. *Arh. Farm.* **2023**, *73*, 251–263. [CrossRef]
26. Nikolaidis, M.G.; Margaritelis, N. V Free Radicals and Antioxidants: Appealing to Magic. *Trends Endocrinol. Metab.* **2023**, *34*, 503–504. [CrossRef] [PubMed]
27. Brieger, K.; Schiavone, S.; Miller, F.J.; Krause, K.H. Reactive Oxygen Species: From Health to Disease. *Swiss Med. Wkly.* **2012**, *142*, w13659. [CrossRef]
28. Tebay, L.E.; Robertson, H.; Durant, S.T.; Vitale, S.R.; Penning, T.M.; Dinkova-Kostova, A.T.; Hayes, J.D. Mechanisms of Activation of the Transcription Factor Nrf2 by Redox Stressors, Nutrient Cues, and Energy Status and the Pathways through Which It Attenuates Degenerative Disease. *Free Radic. Biol. Med.* **2015**, *88*, 108–146. [CrossRef] [PubMed]
29. Hajam, Y.A.; Rani, R.; Ganie, S.Y.; Sheikh, T.A.; Javaid, D.; Qadri, S.S.; Pramodh, S.; Alsulimani, A.; Alkhanani, M.F.; Harakeh, S.; et al. Oxidative Stress in Human Pathology and Aging: Molecular Mechanisms and Perspectives. *Cells* **2022**, *11*, 552. [CrossRef] [PubMed]
30. Jena, A.B.; Samal, R.R.; Bhol, N.K.; Duttaroy, A.K. Cellular Red-Ox System in Health and Disease: The Latest Update. *Biomed. Pharmacother.* **2023**, *162*, 114606. [CrossRef]
31. Ashley, N.T.; Weil, Z.M.; Nelson, R.J. Inflammation: Mechanisms, Costs, and Natural Variation. *Annu. Rev. Ecol. Evol. Syst.* **2012**, *43*, 385–406. [CrossRef]
32. Chen, L.; Deng, H.; Cui, H.; Fang, J.; Zuo, Z.; Deng, J.; Li, Y.; Wang, X.; Zhao, L. Inflammatory Responses and Inflammation-Associated Diseases in Organs. *Oncotarget* **2018**, *9*, 7204–7218. [CrossRef] [PubMed]
33. Herold, K.; Mrowka, R. Inflammation—Dysregulated Inflammatory Response and Strategies for Treatment. *Acta Physiol.* **2019**, *226*, e13284. [CrossRef] [PubMed]
34. Minihane, A.M.; Vinoy, S.; Russell, W.R.; Baka, A.; Roche, H.M.; Tuohy, K.M.; Teeling, J.L.; Blaak, E.E.; Fenech, M.; Vauzour, D.; et al. Low-Grade Inflammation, Diet Composition and Health: Current Research Evidence and Its Translation. *Br. J. Nutr.* **2015**, *114*, 999–1012. [CrossRef]
35. Paludan, S.R.; Pradeu, T.; Masters, S.L.; Mogensen, T.H. Constitutive Immune Mechanisms: Mediators of Host Defence and Immune Regulation. *Nat. Rev. Immunol.* **2021**, *21*, 137–150. [CrossRef] [PubMed]
36. Chaplin, D.D. Overview of the Immune Response. *J. Allergy Clin. Immunol.* **2010**, *125*, S3–S23. [CrossRef] [PubMed]
37. Snyman, M.; Xu, S. The Effects of Mutations on Gene Expression and Alternative Splicing. *Proc. R. Soc. B Biol. Sci.* **2023**, *290*, 20230565. [CrossRef] [PubMed]
38. Landry, C.R.; Lemos, B.; Rifkin, S.A.; Dickinson, W.J.; Hartl, D.L. Genetic Properties Influencing the Evolvability of Gene Expression. *Science* **2007**, *317*, 118–121. [CrossRef]
39. Suomalainen, A.; Nunnari, J. Mitochondria at the Crossroads of Health and Disease. *Cell* **2024**, *187*, 2601–2627. [CrossRef] [PubMed]
40. Casanova, A.; Wevers, A.; Navarro-Ledesma, S.; Pruimboom, L. Mitochondria: It Is All about Energy. *Front. Physiol.* **2023**, *14*, 1114231. [CrossRef]
41. Gorman, G.S.; Chinnery, P.F.; DiMauro, S.; Hirano, M.; Koga, Y.; McFarland, R.; Suomalainen, A.; Thorburn, D.R.; Zeviani, M.; Turnbull, D.M. Mitochondrial Diseases. *Nat. Rev. Dis. Prim.* **2016**, *2*, 16080. [CrossRef]
42. Morris, A.A.M.; Leonard, J.V.; Brown, G.K.; Bidouki, S.K.; Bindoff, L.A.; Woodward, C.E.; Harding, A.E.; Lake, B.D.; Harding, B.N.; Farrell, M.A.; et al. Deficiency of Respiratory Chain Complex I Is a Common Cause of Leigh Disease. *Ann. Neurol.* **1996**, *40*, 25–30. [CrossRef] [PubMed]
43. Zhen, Y.; Radulovic, M.; Vietri, M.; Stenmark, H. Sealing Holes in Cellular Membranes. *EMBO J.* **2021**, *40*, e106922. [CrossRef] [PubMed]
44. Kim, Y.J.; Cho, M.J.; Yu, W.D.; Kim, M.J.; Kim, S.Y.; Lee, J.H. Links of Cytoskeletal Integrity with Disease and Aging. *Cells* **2022**, *11*, 2896. [CrossRef] [PubMed]
45. Dias, C.; Nylandsted, J. Plasma Membrane Integrity in Health and Disease: Significance and Therapeutic Potential. *Cell Discov.* **2021**, *7*, 4. [CrossRef]
46. Valls, P.O.; Esposito, A. Signalling Dynamics, Cell Decisions, and Homeostatic Control in Health and Disease. *Curr. Opin. Cell Biol.* **2022**, *75*, 102066. [CrossRef] [PubMed]
47. Guo, Y.; Pan, W.; Liu, S.; Shen, Z.; Xu, Y.; Hu, L. ERK/MAPK Signalling Pathway and Tumorigenesis (Review). *Exp. Ther. Med.* **2020**, *19*, 1997–2007. [CrossRef] [PubMed]

48. Tkachev, V.O.; Menshchikova, E.B.; Zenkov, N.K. Mechanism of the Nrf2/Keap1/ARE Signaling System. *Biochemistry* **2011**, *76*, 407–422. [CrossRef] [PubMed]
49. Rivas-Estany, E.; Sixto-Fernández, S.; Barrera-Sarduy, J.; Hernández-García, S.; González-Guerra, R.; Stusser-Beltranena, R. Efectos Del Entrenamiento Físico de Larga Duración Sobre La Función y Remodelación Del Ventrículo Izquierdo En Pacientes Con Infarto Miocárdico de Pared Anterior. *Arch. Cardiol. Mex.* **2013**, *83*, 167–173. [CrossRef]
50. Fahed, G.; Aoun, L.; Zerdan, M.B.; Allam, S.; Zerdan, M.B.; Bouferraa, Y.; Assi, H.I. Metabolic Syndrome: Updates on Pathophysiology and Management in 2021. *Int. J. Mol. Sci.* **2022**, *23*, 786. [CrossRef] [PubMed]
51. Corral, J.M.; Jodar, M.; Matheu, R.; González, E.; Castaño, E.; Mantecón, L.; Infante, C.; Juan, C.; Corral, M.; Matheu, R.; et al. Improvement of Semen Quality in Idiopathic Infertile Men by Dietary Supplementation with TetraSOD®: A Pilot Clinical Trial. *Bimed. J. Sci. Tech. Res.* **2023**, *50*, 41971–41975. [CrossRef]
52. Powers, S.K.; Deminice, R.; Ozdemir, M.; Yoshihara, T.; Bomkamp, M.P.; Hyatt, H. Exercise-Induced Oxidative Stress: Friend or Foe? *J. Sport Health Sci.* **2020**, *9*, 415–425. [CrossRef] [PubMed]
53. D'Angelo, S.; Rosa, R. Oxidative Stress and Sport Performance. *Sport Sci.* **2020**, *13*, 18–22.
54. Clemente-Suárez, V.J.; Bustamante-Sanchez, Á.; Mielgo-Ayuso, J.; Martínez-Guardado, I.; Martín-Rodríguez, A.; Tornero-Aguilera, J.F. Antioxidants and Sports Performance. *Nutrients* **2023**, *15*, 2371. [CrossRef] [PubMed]
55. González, E.; Leiva, M.; Domingo, M.; Corral, J.M.; Jodar, M.; Unamunzaga, C.; Mantecón, L.; Infante, C. Use of TetraSOD® in Male Infertility: Double-Blind Randomized Placebo-Controlled Study. In Proceedings of the Abstracts 40th Hybrid Annual Meeting of the European Society of Human Reproduction and Embriology, Amsterdam, The Netherlands, 7–10 July 2024; Oxford University Press: Amsterdam, The Nederlands, 2024; p. 227.
56. O'Flaherty, C. Reactive Oxygen Species and Male Fertility. *Antioxidants* **2020**, *9*, 287. [CrossRef] [PubMed]
57. Hussain, T.; Kandeel, M.; Metwally, E.; Murtaza, G.; Kalhoro, D.H.; Yin, Y.; Tan, B.; Chughtai, M.I.; Yaseen, A.; Afzal, A.; et al. Unraveling the Harmful Effect of Oxidative Stress on Male Fertility: A Mechanistic Insight. *Front. Endocrinol.* **2023**, *14*, 1070692. [CrossRef] [PubMed]
58. Li, K.P.; Yang, X.S.; Wu, T. The Effect of Antioxidants on Sperm Quality Parameters and Pregnancy Rates for Idiopathic Male Infertility: A Network Meta-Analysis of Randomized Controlled Trials. *Front. Endocrinol.* **2022**, *13*, 810242. [CrossRef] [PubMed]
59. Parthasarathy, S.; Raghavamenon, A.; Garelnabi, M.O.; Santanam, N. Oxidized Low-Density Lipoprotein. In *Methods in Molecular Biology*; Uppu, R., Murthy, S., Pryor, W., Parinandi, N., Eds.; Humana Press: Totowa, NJ, USA, 2010; Volume 610, pp. 403–417; ISBN 978-1-58829-710-5.
60. Bender, C.; Candi, I.; Rogel, E. Efficacy of Hydroxytyrosol-Rich Food Supplements on Reducing Lipid Oxidation in Humans. *Int. J. Mol. Sci.* **2023**, *24*, 5521. [CrossRef] [PubMed]
61. Boushehri, S.N.; Yusof, R.M.; Taib, M.N.M.; Mirzaei, K.; Yazdekhasti, N.; Akbarzadeh, S. Effect of Vitamin Supplementation on Serum Oxidized Low-Density Lipoprotein Levels in Male Subjects with Cardiovascular Disease Risk Factors. *Iran. J. Basic Med. Sci.* **2012**, *15*, 958–964.
62. Covas, M.I.; De La Torre, K.; Farré-Albaladejo, M.; Kaikkonen, J.; Fitó, M.; López-Sabater, C.; Pujadas-Bastardes, M.A.; Joglar, J.; Weinbrenner, T.; Lamuela-Raventós, R.M.; et al. Postprandial LDL Phenolic Content and LDL Oxidation Are Modulated by Olive Oil Phenolic Compounds in Humans. *Free Radic. Biol. Med.* **2006**, *40*, 608–616. [CrossRef]
63. Nielsen, F.; Mikkelsen, B.B.; Nielsen, J.B.; Andersen, H.R.; Grandjean, P. Plasma Malondialdehyde as Biomarker for Oxidative Stress: Reference Interval and Effects of Life-Style Factors. *Clin. Chem.* **1997**, *43*, 1209–1214. [CrossRef] [PubMed]
64. Yilgor, A.; Demir, C. Determination of Oxidative Stress Level and Some Antioxidant Activities in Refractory Epilepsy Patients. *Sci. Rep.* **2024**, *14*, 6688. [CrossRef] [PubMed]
65. Cordiano, R.; Di Gioacchino, M.; Mangifesta, R.; Panzera, C.; Gangemi, S.; Minciullo, P.L. Malondialdehyde as a Potential Oxidative Stress Marker for Allergy-Oriented Diseases: An Update. *Molecules* **2023**, *28*, 5979. [CrossRef]
66. Agarwal, A.; Sharma, R.; Roychoudhury, S.; Du Plessis, S.; Sabanegh, E. MiOXSYS: A Novel Method of Measuring Oxidation Reduction Potential in Semen and Seminal Plasma. *Fertil. Steril.* **2016**, *106*, 566–573.e10. [CrossRef] [PubMed]
67. Agarwal, A.; Bui, A.D. Oxidation-Reduction Potential as a New Marker for Oxidative Stress: Correlation to Male Infertility. *Investig. Clin. Urol.* **2017**, *58*, 385–399. [CrossRef] [PubMed]
68. Agarwal, A.; Panner Selvam, M.; Arafa, M.; Okada, H.; Homa, S.; Killeen, A.; Balaban, B.; Saleh, R.; Armagan, A.; Roychoudhury, S.; et al. Multi-Center Evaluation of Oxidation-Reduction Potential by the MiOXSYS in Males with Abnormal Semen. *Asian J. Androl.* **2019**, *21*, 565–569. [CrossRef]
69. Ighodaro, O.M.; Akinloye, O.A. First Line Defence Antioxidants-Superoxide Dismutase (SOD), Catalase (CAT) and Glutathione Peroxidase (GPX): Their Fundamental Role in the Entire Antioxidant Defence Grid. *Alex. J. Med.* **2018**, *54*, 287–293. [CrossRef]
70. Romao, S. Therapeutic Value of Oral Supplementation with Melon Superoxide Dismutase and Wheat Gliadin Combination. *Nutrition* **2015**, *31*, 430–436. [CrossRef] [PubMed]
71. Nelson, S.K.; Bose, S.K.; Grunwald, G.K.; Myhill, P.; McCord, J.M. The Induction of Human Superoxide Dismutase and Catalase In Vivo: A Fundamentally New Approach to Antioxidant Therapy. *Free Radic. Biol. Med.* **2006**, *40*, 341–347. [CrossRef]

72. Décordé, K.; Ventura, E.; Lacan, D.; Ramos, J.; Cristol, J.P.; Rouanet, J.M. An SOD Rich Melon Extract Extramel® Prevents Aortic Lipids and Liver Steatosis in Diet-Induced Model of Atherosclerosis. *Nutr. Metab. Cardiovasc. Dis.* **2010**, *20*, 301–307. [CrossRef] [PubMed]
73. Pei, J.; Pan, X.; Wei, G.; Hua, Y. Research Progress of Glutathione Peroxidase Family (GPX) in Redoxidation. *Front. Pharmacol.* **2023**, *14*, 1147414. [CrossRef] [PubMed]
74. Pizzorno, J. Glutathione! *Integr. Med.* **2014**, *13*, 8–12.
75. Di Giacomo, C.; Malfa, G.A.; Tomasello, B.; Bianchi, S.; Acquaviva, R. Natural Compounds and Glutathione: Beyond Mere Antioxidants. *Antioxidants* **2023**, *12*, 1445. [CrossRef]
76. Lu, S.C. Regulation of Glutathione Synthesis. *Mol. Asp. Med.* **2009**, *30*, 42–59. [CrossRef]
77. Wu, M.; Deng, C.; Lo, T.; Chan, K.; Li, X.; Wong, C. Peroxiredoxin, Senescence, and Cancer. *Cells* **2022**, *11*, 1772. [CrossRef] [PubMed]
78. Chen, T.; Jinlin, D.; Wang, F.; Yuan, Z.; Xue, J.; Lu, T.; Huang, W.; Liu, Y.; Zhang, Y. GSTM3 Deficiency Impedes DNA Mismatch Repair to Promote Gastric Tumorigenesis via CAND1/NRF2-KEAP1 Signaling. *Cancer Lett.* **2022**, *538*, 215692. [CrossRef] [PubMed]
79. He, F.; Ru, X.; Wen, T. NRF2, a Transcription Factor for Stress Response and Beyond. *Int. J. Mol. Sci.* **2020**, *21*, 4777. [CrossRef] [PubMed]
80. Hayes, J.D.; Dinkova-Kostova, A.T. The Nrf2 Regulatory Network Provides an Interface between Redox and Intermediary Metabolism. *Trends Biochem. Sci.* **2014**, *39*, 199–218. [CrossRef] [PubMed]
81. Eleutherio, E.C.A.; Silva Magalhães, R.S.; de Araújo Brasil, A.; Monteiro Neto, J.R.; de Holanda Paranhos, L. SOD1, More than Just an Antioxidant. *Arch. Biochem. Biophys.* **2021**, *697*, 108701. [CrossRef]
82. Vahidinia, Z.; Azami Tameh, A.; Barati, S.; Izadpanah, M.; Seyed Hosseini, E. Nrf2 Activation: A Key Mechanism in Stem Cell Exosomes-Mediated Therapies. *Cell Mol. Biol. Lett.* **2024**, *29*, 30. [CrossRef] [PubMed]
83. Singh, B.; Bhat, H.K. Superoxide Dismutase 3 Is Induced by Antioxidants, Inhibits Oxidative DNA Damage and Is Associated with Inhibition of Estrogen-Induced Breast Cancer. *Carcinogenesis* **2012**, *33*, 2601–2610. [CrossRef] [PubMed]
84. McCord, J.M.; Gao, B.; Hybertson, B.M. The Complex Genetic and Epigenetic Regulation of the Nrf2 Pathways: A Review. *Antioxidants* **2023**, *12*, 366. [CrossRef] [PubMed]
85. Rahman, S.; Islam, R. Mammalian Sirt1: Insights on Its Biological Functions. *Cell Commun. Signal.* **2011**, *9*, 11. [CrossRef] [PubMed]
86. Kratz, E.M.; Sołkiewicz, K.; Kubis-Kubiak, A.; Piwowar, A. Sirtuins as Important Factors in Pathological States and the Role of Their Molecular Activity Modulators. *Int. J. Mol. Sci.* **2021**, *22*, 630. [CrossRef] [PubMed]
87. Ma, F.; Wu, J.; Jiang, Z.; Huang, W.; Jia, Y.; Sun, W.; Wu, H. P53/NRF2 Mediates SIRT1's Protective Effect on Diabetic Nephropathy. *Biochim. Biophys. Acta—Mol. Cell Res.* **2019**, *1866*, 1272–1281. [CrossRef]
88. Huang, K.; Huang, J.; Xie, X.; Wang, S.; Chen, C.; Shen, X.; Liu, P.; Huang, H. Sirt1 Resists Advanced Glycation End Products-Induced Expressions of Fibronectin and TGF-B1 by Activating the Nrf2/ARE Pathway in Glomerular Mesangial Cells. *Free Radic. Biol. Med.* **2013**, *65*, 528–540. [CrossRef] [PubMed]
89. Huang, K.; Gao, X.; Wei, W. The Crosstalk between Sirt1 and Keap1/Nrf2/ARE Anti-Oxidative Pathway Forms a Positive Feedback Loop to Inhibit FN and TGF-B1 Expressions in Rat Glomerular Mesangial Cells. *Exp. Cell Res.* **2017**, *361*, 63–72. [CrossRef] [PubMed]
90. Biswas, R.; Bagchi, A. NFkB Pathway and Inhibition: An Overview. *Comput. Mol. Biol.* **2016**, *6*, 1–20. [CrossRef]
91. Liu, T.; Zhang, L.; Joo, D.; Sun, S.C. NF-KB Signaling in Inflammation. *Signal Transduct. Target. Ther.* **2017**, *2*, e17023. [CrossRef] [PubMed]
92. Lee, H.; Lee, I.S.; Choue, R. Obesity, Inflammation and Diet. *Pediatr. Gastroenterol. Hepatol. Nutr.* **2013**, *16*, 143–152. [CrossRef]
93. Bae, H.R.; Choi, M.-S.; Kim, S.; Young, H.A.; Gershwin, M.E.; Jeon, S.-M.; Kwon, E.-Y. IFNγ Is a Key Link between Obesity and Th1-Mediated AutoImmune Diseases. *Int. J. Mol. Sci.* **2021**, *22*, 208. [CrossRef] [PubMed]
94. Stanton, M.C.; Chen, S.-C.; Jackson, J.V.; Rojas-Triana, A.; Kinsley, D.; Cui, L.; Fine, J.S.; Greenfeder, S.; Bober, L.A.; Jenh, C.-H. Inflammatory Signals Shift from Adipose to Liver during High Fat Feeding and Influence the Development of Steatohepatitis in Mice. *J. Inflamm.* **2011**, *8*, 8. [CrossRef] [PubMed]
95. Rustenhoven, J.; Aalderink, M.; Scotter, E.L.; Oldfield, R.L.; Bergin, P.S.; Mee, E.W.; Graham, E.S.; Faull, R.L.M.; Curtis, M.A.; Park, T.I.H.; et al. TGF-Beta1 Regulates Human Brain Pericyte Inflammatory Processes Involved in Neurovasculature Function. *J. Neuroinflamm.* **2016**, *13*, 37. [CrossRef]
96. de Waal Malefyt, R.; Abrams, J.; Bennett, B.; Figdor, C.G.; de Vries, J.E. Interleukin 10 (IL-10) Inhibits Cytokine Synthesis by Human Monocytes: An Autoregulatory Role of IL-10 Produced by Monocytes. *J. Exp. Med.* **1991**, *174*, 1209–1220. [CrossRef] [PubMed]
97. Fiorentino, D.F.; Zlotnik, A.; Mosmann, T.R.; Howard, M.; O'Garra, A. IL-10 Inhibits Cytokine Production by Activated Macrophages. *J. Immunol.* **1991**, *147*, 3815–3822. [CrossRef]

98. Achari, A.E.; Jain, S.K. Adiponectin, a Therapeutic Target for Obesity, Diabetes, and Endothelial Dysfunction. *Int. J. Mol. Sci.* **2017**, *18*, 1321. [CrossRef] [PubMed]
99. Surendar, J.; Frohberger, S.J.; Karunakaran, I.; Schmitt, V.; Stamminger, W.; Neumann, A.-L.; Wilhelm, C.; Hoerauf, A.; Hübner, M.P. Adiponectin Limits IFN-γ and IL-17 Producing CD4 T Cells in Obesity by Restraining Cell Intrinsic Glycolysis. *Front. Immunol.* **2019**, *10*, 2555. [CrossRef] [PubMed]
100. Tilg, H.; Moschen, A.R. Adipocytokines: Mediators Linking Adipose Tissue, Inflammation and Immunity. *Nat. Rev. Immunol.* **2006**, *6*, 772–783. [CrossRef] [PubMed]
101. Ahmed, S.M.U.; Luo, L.; Namani, A.; Wang, X.J.; Tang, X. Nrf2 Signaling Pathway: Pivotal Roles in Inflammation. *Biochim. Biophys. Acta—Mol. Basis Dis.* **2017**, *1863*, 585–597. [CrossRef] [PubMed]
102. Saha, S.; Buttari, B.; Panieri, E.; Profumo, E.; Saso, L. An Overview of Nrf2 Signaling Pathway and Its Role in Inflammation. *Molecules* **2020**, *25*, 5474. [CrossRef] [PubMed]
103. Kobayashi, E.H.; Suzuki, T.; Funayama, R.; Nagashima, T.; Hayashi, M.; Sekine, H.; Tanaka, N.; Moriguchi, T.; Motohashi, H.; Nakayama, K.; et al. Nrf2 Suppresses Macrophage Inflammatory Response by Blocking Proinflammatory Cytokine Transcription. *Nat. Commun.* **2016**, *7*, 11624. [CrossRef] [PubMed]
104. Yang, Y.; Liu, Y.; Wang, Y.; Chao, Y.; Zhang, J.; Jia, Y.; Tie, J.; Hu, D. Regulation of SIRT1 and Its Roles in Inflammation. *Front. Immunol.* **2022**, *13*, 831168. [CrossRef] [PubMed]
105. Yeung, F.; Hoberg, J.E.; Ramsey, C.S.; Keller, M.D.; Jones, D.R.; Frye, R.A.; Mayo, M.W. Modulation of NF-KB-Dependent Transcription and Cell Survival by the SIRT1 Deacetylase. *EMBO J.* **2004**, *23*, 2369–2380. [CrossRef]
106. Yang, X.-D.; Tajkhorshid, E.; Chen, L.-F. Functional Interplay between Acetylation and Methylation of the RelA Subunit of NF-KB. *Mol. Cell Biol.* **2010**, *30*, 2170–2180. [CrossRef] [PubMed]
107. Liu, T.F.; Yoza, B.K.; El Gazzar, M.; Vachharajani, V.T.; McCall, C.E. NAD+-Dependent SIRT1 Deacetylase Participates in Epigenetic Reprogramming during Endotoxin Tolerance. *J. Biol. Chem.* **2011**, *286*, 9856–9864. [CrossRef]
108. Yoshizaki, T.; Schenk, S.; Imamura, T.; Babendure, J.L.; Sonoda, N.; Bae, E.J.; Oh, D.Y.; Lu, M.; Milne, J.C.; Westphal, C.; et al. SIRT1 Inhibits Inflammatory Pathways in Macrophages and Modulates Insulin Sensitivity. *Am. J. Physiol.—Endocrinol. Metab.* **2010**, *298*, E419–E428. [CrossRef]
109. Alvarez, J.G.; García-Peiró, A.; Barros, A.; Ferraz, L.; Sousa, M.; Sakkas, D. Double Strand DNA Breaks in Sperm: The Bad Guy in the Crowd. *J. Assist. Reprod. Genet.* **2023**, *40*, 745–751. [CrossRef] [PubMed]
110. Cannan, W.J.; Pederson, D.S. Mechanisms and Consequences of Double-Strand DNA Break Formation in Chromatin. *J. Cell Physiol.* **2016**, *231*, 3–14. [CrossRef] [PubMed]
111. Stehli, J.; Fuchs, T.A.; Ghadri, J.R.; Gaemperli, O.; Fiechter, M.; Kaufmann, P.A. Antioxidants Prevent DNA Double-Strand Breaks from X-Ray-Based Cardiac Examinations: A Randomized, Double-Blinded, Placebo-Controlled Trial. *J. Am. Coll. Cardiol.* **2014**, *64*, 117–118. [CrossRef]
112. Lottner, S.; Shehata, M.; Hickel, R.; Reichl, F.X.; Durner, J. Effects of Antioxidants on DNA-Double Strand Breaks in Human Gingival Fibroblasts Exposed to Methacrylate Based Monomers. *Dent. Mater.* **2013**, *29*, 991–998. [CrossRef]
113. Bicheru, N.S.; Haidoiu, C.; Călborean, O.; Popa, A.; Porosnicu, I.; Hertzog, R. Effect of Different Antioxidants on X-Ray Induced DNA Double-Strand Breaks Using γ-H2AX in Human Blood Lymphocytes. *Health Phys.* **2020**, *119*, 101–108. [CrossRef]
114. Brand, M.; Sommer, M.; Ellmann, S.; Wuest, W.; May, M.S.; Eller, A.; Vogt, S.; Lell, M.M.; Kuefner, M.A.; Uder, M. Influence of Different Antioxidants on X-Ray Induced DNA Double-Strand Breaks (DSBs) Using γ-H2AX Immunofluorescence Microscopy in a Preliminary Study. *PLoS ONE* **2015**, *10*, e0127142. [CrossRef] [PubMed]
115. Tan, J.; Sun, X.; Zhao, H.; Guan, H.; Gao, S.; Zhou, P.K. Double-Strand DNA Break Repair: Molecular Mechanisms and Therapeutic Targets. *MedComm* **2023**, *4*, e388. [CrossRef] [PubMed]
116. Lagunas-Rangel, F.A. Current Role of Mammalian Sirtuins in DNA Repair. *DNA Repair* **2019**, *80*, 85–92. [CrossRef]
117. Jeong, J.; Juhn, K.; Lee, H.; Kim, S.; Min, B.; Lee, K.-M.; Cho, M.; Park, G.; Lee, K. SIRT1 Promotes DNA Repair Activity and Deacetylation of Ku70. *Exp. Mol. Med.* **2007**, *39*, 8–13. [CrossRef]
118. Yuan, Z.; Zhang, X.; Sengupta, N.; Lane, W.S.; Seto, E. SIRT1 Regulates the Function of the Nijmegen Breakage Syndrome Protein. *Mol. Cell* **2007**, *27*, 149–162. [CrossRef] [PubMed]
119. Lin, Y.H.; Yuan, J.; Pei, H.; Liu, T.; Ann, D.K.; Lou, Z. KAP1 Deacetylation by SIRT1 Promotes Non-Homologous End-Joining Repair. *PLoS ONE* **2015**, *10*, e0123935. [CrossRef]
120. Jayakumar, S.; Pal, D.; Sandur, S.K. Nrf2 Facilitates Repair of Radiation Induced DNA Damage through Homologous Recombination Repair Pathway in a ROS Independent Manner in Cancer Cells. *Mutat. Res.—Fundam. Mol. Mech. Mutagen.* **2015**, *779*, 33–45. [CrossRef]
121. Sun, X.; Wang, Y.; Ji, K.; Liu, Y.; Kong, Y.; Nie, S.; Li, N.; Hao, J.; Xie, Y.; Xu, C.; et al. NRF2 Preserves Genomic Integrity by Facilitating ATR Activation and G2 Cell Cycle Arrest. *Nucleic Acids Res.* **2020**, *48*, 9109–9123. [CrossRef] [PubMed]
122. Li, J.; Xu, C.; Liu, Q. Roles of NRF2 in DNA Damage Repair. *Cell Oncol.* **2023**, *46*, 1577–1593. [CrossRef] [PubMed]

123. Tsujita, S.; Morimoto, K. Secretory IgA in Saliva Can Be a Useful Stress Marker. *Environ. Health Prev. Med.* **1999**, *4*, 1–8. [CrossRef] [PubMed]
124. Turner, S.E.G.; Loosemore, M.; Shah, A.; Kelleher, P.; Hull, J.H. Salivary IgA as a Potential Biomarker in the Evaluation of Respiratory Tract Infection Risk in Athletes. *J. Allergy Clin. Immunol. Pract.* **2021**, *9*, 151–159. [CrossRef] [PubMed]
125. Martin, S.A.; Pence, B.D.; Woods, J.A. Exercise and Respiratory Tract Viral Infections. *Exerc. Sport Sci. Rev.* **2009**, *37*, 157–164. [CrossRef]
126. Fahlman, M.M.; Engels, H.J. Mucosal IgA and URTI in American College Football Players: A Year Longitudinal Study. *Med. Sci. Sports Exerc.* **2005**, *37*, 374–380. [CrossRef] [PubMed]
127. Klentrou, P.; Cieslak, T.; MacNeil, M.; Vintinner, A.; Plyley, M. Effect of Moderate Exercise on Salivary Immunoglobulin A and Infection Risk in Humans. *Eur. J. Appl. Physiol.* **2002**, *87*, 153–158. [CrossRef] [PubMed]
128. Gombart, A.F.; Pierre, A.; Maggini, S. A Review of Micronutrients and the Immune System–Working in Harmony to Reduce the Risk of Infection. *Nutrients* **2020**, *12*, 236. [CrossRef] [PubMed]
129. Haryanto, B.; Suksmasari, T.; Wintergerst, E.; Maggini, S. Multivitamin Supplementation Supports Immune Function and Ameliorates Conditions Triggered by Reduced Air Quality. *Vitam. Miner.* **2015**, *4*, 1000128. [CrossRef]
130. Wood, S.M.; Beckham, C.; Yosioka, A.; Darban, H.; Watson, R.R. β-Carotene and Selenium Supplementation Enhances Immune Response in Aged Humans. *Integr. Med.* **2000**, *2*, 85–92. [CrossRef] [PubMed]
131. Simpore, J.; Pignatelli, S.; Musumeci, S. The Effects of Spiruline on the Immune Functions of HIV-Infected Undernourished Children. *J. Infect. Dev. Ctries.* **2007**, *1*, 112–117.
132. Singh, S.; Anshita, D.; Ravichandiran, V. MCP-1: Function, Regulation, and Involvement in Disease. *Int. Immunopharmacol.* **2021**, *101*, 107598. [CrossRef] [PubMed]
133. Ihim, S.A.; Abubakar, S.D.; Zian, Z.; Sasaki, T.; Saffarioun, M.; Maleknia, S.; Azizi, G. Interleukin-18 Cytokine in Immunity, Inflammation, and Autoimmunity: Biological Role in Induction, Regulation, and Treatment. *Front. Immunol.* **2022**, *13*, 919973. [CrossRef]
134. Mandal, R.; Barrón, J.C.; Kostova, I.; Becker, S.; Strebhardt, K. Caspase-8: The Double-Edged Sword. *Biochim. Biophys. Acta—Rev. Cancer* **2020**, *1873*, 188357. [CrossRef] [PubMed]
135. van der Horst, D.; Carter-Timofte, M.E.; van Grevenynghe, J.; Laguette, N.; Dinkova-Kostova, A.T.; Olagnier, D. Regulation of Innate Immunity by Nrf2. *Curr. Opin. Immunol.* **2022**, *78*, 102247. [CrossRef] [PubMed]
136. Harvey, C.J.; Thimmulappa, R.K.; Sethi, S.; Kong, X.; Yarmus, L.; Brown, R.H.; Feller-Kopman, D.; Wise, R.; Biswal, S. Targeting Nrf2 Signaling Improves Bacterial Clearance by Alveolar Macrophages in Patients with COPD and in a Mouse Model. *Sci. Transl. Med.* **2011**, *3*, 78ra32. [CrossRef] [PubMed]
137. Saddawi-Konefka, R.; Seelige, R.; Gross, E.T.E.; Levy, E.; Searles, S.C.; Washington, A.; Santosa, E.K.; Liu, B.; O'Sullivan, T.E.; Harismendy, O.; et al. Nrf2 Induces IL-17D to Mediate Tumor and Virus Surveillance. *Cell Rep.* **2016**, *16*, 2348–2358. [CrossRef]
138. Chen, X.; Su, W.; Wan, T.; Yu, J.; Zhu, W.; Tang, F.; Liu, G.; Olsen, N.; Liang, D.; Zheng, S.G. Sodium Butyrate Regulates Th17/Treg Cell Balance to Ameliorate Uveitis via the Nrf2/HO-1 Pathway. *Biochem. Pharmacol.* **2017**, *142*, 111–119. [CrossRef] [PubMed]
139. Rockwell, C.E.; Zhang, M.; Fields, P.E.; Klaassen, C.D. Th2 Skewing by Activation of Nrf2 in CD^{4+} T Cells. *J. Immunol.* **2012**, *188*, 1630–1637. [CrossRef]
140. Yu, Q.; Dong, L.; Li, Y.; Liu, G. SIRT1 and HIF1α Signaling in Metabolism and Immune Responses. *Cancer Lett.* **2018**, *418*, 20–26. [CrossRef]
141. Shen, P.; Deng, X.; Chen, Z.; Ba, X.; Qin, K.; Huang, Y.; Huang, Y.; Li, T.; Yan, J.; Tu, S. SIRT1: A Potential Therapeutic Target in Autoimmune Diseases. *Front. Immunol.* **2021**, *12*, 779177. [CrossRef] [PubMed]
142. Yuan, Y.; Cruzat, V.F.; Newshome, P.; Cheng, J.; Chen, Y.; Lu, Y. Regulation of SIRT1 in Aging: Roles in Mitochondrial Function and Biogenesis. *Mech. Ageing Dev.* **2016**, *155*, 10–21. [CrossRef] [PubMed]
143. Gumeni, S.; Papanagnou, E.D.; Manola, M.S.; Trougakos, I.P. Nrf2 Activation Induces Mitophagy and Reverses Parkin/Pink1 Knock down-Mediated Neuronal and Muscle Degeneration Phenotypes. *Cell Death Dis.* **2021**, *12*, 671. [CrossRef]
144. Dinkova-Kostova, A.T.; Abramov, A.Y. The Emerging Role of Nrf2 in Mitochondrial Function. *Free Radic. Biol. Med.* **2015**, *88*, 179–188. [CrossRef] [PubMed]
145. Wischhof, L.; Scifo, E.; Ehninger, D.; Bano, D. AIFM1 beyond Cell Death: An Overview of This OXPHOS-Inducing Factor in Mitochondrial Diseases. *eBioMedicine* **2022**, *83*, 104231. [CrossRef] [PubMed]
146. Tjahjono, E.; Kirienko, D.R.; Kirienko, N.V. The Emergent Role of Mitochondrial Surveillance in Cellular Health. *Aging Cell* **2022**, *21*, e13710. [CrossRef] [PubMed]
147. Puigserver, P.; Rhee, J.; Lin, J.; Wu, Z.; Yoon, J.C.; Zhang, C.Y.; Krauss, S.; Mootha, V.K.; Lowell, B.B.; Spiegelman, B.M. Cytokine Stimulation of Energy Expenditure through P38 MAP Kinase Activation of PPARγ Coactivator-1. *Mol. Cell* **2001**, *8*, 971–982. [CrossRef] [PubMed]
148. Akimoto, T.; Pohnert, S.C.; Li, P.; Zhang, M.; Gumbs, C.; Rosenberg, P.B.; Williams, R.S.; Yan, Z. Exercise Stimulates Pgc-1α Transcription in Skeletal Muscle through Activation of the P38 MAPK Pathway. *J. Biol. Chem.* **2005**, *280*, 19587–19593. [CrossRef]

149. Fernandez-Marcos, P.J.; Auwerx, J. Regulation of PGC-1α, a Nodal Regulator of Mitochondrial Biogenesis. *Am. J. Clin. Nutr.* **2011**, *93*, 884S–890S. [CrossRef] [PubMed]
150. Sarg, N.H.; Zaher, D.M.; Abu Jayab, N.N.; Mostafa, S.H.; Ismail, H.H.; Omar, H.A. The Interplay of P38 MAPK Signaling and Mitochondrial Metabolism, a Dynamic Target in Cancer and Pathological Contexts. *Biochem. Pharmacol.* **2024**, *225*, 116307. [CrossRef]
151. Brancaccio, P.; Lippi, G.; Maffulli, N. Biochemical Markers of Muscular Damage. *Clin. Chem. Lab. Med.* **2010**, *48*, 757–767. [CrossRef]
152. Yeom, D.C.; Hwang, D.J.; Lee, W.B.; Cho, J.Y.; Koo, J.H. Effects of Low-Load, High-Repetition Resistance Training on Maximum Muscle Strength and Muscle Damage in Elite Weightlifters: A Preliminary Study. *Int. J. Mol. Sci.* **2023**, *24*, 17079. [CrossRef]
153. Saita, Y.; Hattori, K.; Hokari, A.; Ohyama, T.; Inoue, J.; Nishimura, T.; Nemoto, S.; Aoyagi, S. Plasma Myoglobin Indicates Muscle Damage Associated with Acceleration/Deceleration during Football. *J. Sports Med. Phys. Fitness* **2023**, *63*, 1337–1342. [CrossRef] [PubMed]
154. Agrawal, S.; Chakole, S.; Shetty, N.; Prasad, R.; Lohakare, T.; Wanjari, M. Exploring the Role of Oxidative Stress in Skeletal Muscle Atrophy: Mechanisms and Implications. *Cureus* **2023**, *15*, e42178. [CrossRef]
155. Bodine, S.C.; Latres, E.; Baumhueter, S.; Lai, V.K.M.; Nunez, L.; Clarke, B.A.; Poueymirou, W.T.; Panaro, F.J.; Na, E.; Dharmarajan, K.; et al. Identification of Ubiquitin Ligases Required for Skeletal Muscle Atrophy. *Science* **2001**, *294*, 1704–1708. [CrossRef] [PubMed]
156. Zammit, P.S. Function of the Myogenic Regulatory Factors Myf5, MyoD, Myogenin and MRF4 in Skeletal Muscle, Satellite Cells and Regenerative Myogenesis. *Semin. Cell Dev. Biol.* **2017**, *72*, 19–32. [CrossRef] [PubMed]
157. Capkovic, K.L.; Stevenson, S.; Johnson, M.C.; Thelen, J.J.; Cornelison, D.D.W. Neural Cell Adhesion Molecule (NCAM) Marks Adult Myogenic Cells Committed to Differentiation. *Exp. Cell Res.* **2008**, *314*, 1553–1565. [CrossRef] [PubMed]
158. Dumont, N.A.; Bentzinger, C.F.; Sincennes, M.C.; Rudnicki, M.A. Satellite Cells and Skeletal Muscle Regeneration. *Compr. Physiol.* **2015**, *5*, 1027–1059. [CrossRef] [PubMed]
159. Abati, E.; Manini, A.; Comi, G.P.; Corti, S. Inhibition of Myostatin and Related Signaling Pathways for the Treatment of Muscle Atrophy in Motor Neuron Diseases. *Cell Mol. Life Sci.* **2022**, *79*, 374. [CrossRef]
160. Xu, L.; Yang, L.; Hashimoto, K.; Anderson, M.; Kohlhagen, G.; Pommier, Y.; D'Arpa, P. Characterization of BTBD1 and BTBD2, Two Similar BTB-Domain-Containing Kelch-like Proteins That Interact with Topoisomerase I. *BMC Genom.* **2002**, *3*, 1. [CrossRef]
161. Pisani, D.F.; Cabane, C.; Derijard, B.; Dechesne, C.A. The Topoisomerase 1-Interacting Protein BTBD1 Is Essential for Muscle Cell Differentiation. *Cell Death Differ.* **2004**, *11*, 1157–1165. [CrossRef]
162. Chen, L.; Tang, F.; Gao, H.; Zhang, X.; Li, X.; Xiao, D. CAPN3: A Muscle-specific Calpain with an Important Role in the Pathogenesis of Diseases (Review). *Int. J. Mol. Med.* **2021**, *48*, 203. [CrossRef] [PubMed]
163. Kesavardhana, S.; Malireddi, R.K.S.; Kanneganti, T.D. Caspases in Cell Death, Inflammation, and Pyroptosis. *Annu. Rev. Immunol.* **2020**, *38*, 567–595. [CrossRef] [PubMed]
164. Boonstra, K.; Bloemberg, D.; Quadrilatero, J. Caspase-2 Is Required for Skeletal Muscle Differentiation and Myogenesis. *Biochim. Biophys. Acta—Mol. Cell Res.* **2018**, *1865*, 95–104. [CrossRef]
165. Fernando, P.; Kelly, J.F.; Balazsi, K.; Slack, R.S.; Megeney, L.A. Caspase 3 Activity Is Required for Skeletal Muscle Differentiation. *Proc. Natl. Acad. Sci. USA* **2002**, *99*, 11025–11030. [CrossRef]
166. Łoboda, A.; Dulak, J. Nuclear Factor Erythroid 2-Related Factor 2 and Its Targets in Skeletal Muscle Repair and Regeneration. *Antioxid. Redox Signal.* **2023**, *38*, 619–642. [CrossRef] [PubMed]
167. Cargnello, M.; Roux, P.P. Activation and Function of the MAPKs and Their Substrates, the MAPK-Activated Protein Kinases. *Microbiol. Mol. Biol. Rev.* **2011**, *75*, 50–83. [CrossRef] [PubMed]
168. Owuor, E.D.; Kong, A.N.T. Antioxidants and Oxidants Regulated Signal Transduction Pathways. *Biochem. Pharmacol.* **2002**, *64*, 765–770. [CrossRef] [PubMed]
169. Liu, T.; Lv, Y.F.; Zhao, J.L.; You, Q.D.; Jiang, Z.Y. Regulation of Nrf2 by Phosphorylation: Consequences for Biological Function and Therapeutic Implications. *Free Radic. Biol. Med.* **2021**, *168*, 129–141. [CrossRef] [PubMed]
170. Wang, Y.; Cui, R.; Zhang, X.; Qiao, Y.; Liu, X.; Chang, Y.; Yu, Y.; Sun, F.; Wang, J. SIRT1 Increases YAP- and MKK3-Dependent P38 Phosphorylation in Mouse Liver and Human Hepatocellular Carcinoma. *Oncotarget* **2016**, *7*, 11284–11298. [CrossRef] [PubMed]
171. Song, D.; Lian, Y.; Zhang, L. The Potential of Activator Protein 1 (AP-1) in Cancer Targeted Therapy. *Front. Immunol.* **2023**, *14*, 1224892. [CrossRef]
172. Zipper, L.M.; Mulcahy, R.T. Inhibition of ERK and P38 MAP Kinases Inhibits Binding of Nrf2 and Induction of GCS Genes. *Biochem. Biophys. Res. Commun.* **2000**, *278*, 484–492. [CrossRef] [PubMed]
173. Gao, Z.; Ye, J. Inhibition of Transcriptional Activity of C-JUN by SIRT1. *Biochem. Biophys. Res. Commun.* **2008**, *376*, 793–796. [CrossRef] [PubMed]

174. Sies, H.; Belousov, V.V.; Chandel, N.S.; Davies, M.J.; Jones, D.P.; Mann, G.E.; Murphy, M.P.; Yamamoto, M.; Winterbourn, C. Defining Roles of Specific Reactive Oxygen Species (ROS) in Cell Biology and Physiology. *Nat. Rev. Mol. Cell Biol.* **2022**, *23*, 499–515. [CrossRef] [PubMed]
175. Weiss-Sadan, T.; Ge, M.; Hayashi, M.; Gohar, M.; Yao, C.H.; de Groot, A.; Harry, S.; Carlin, A.; Fischer, H.; Shi, L.; et al. NRF2 Activation Induces NADH-Reductive Stress, Providing a Metabolic Vulnerability in Lung Cancer. *Cell Metab.* **2023**, *35*, 487–503. [CrossRef] [PubMed]
176. Wang, Y.Y.; Chen, J.; Liu, X.M.; Zhao, R.; Zhe, H. Nrf2-Mediated Metabolic Reprogramming in Cancer. *Oxid. Med. Cell Longev.* **2018**, *2018*, 9304091. [CrossRef]
177. Hybertson, B.M.; Gao, B. Role of the Nrf2 Signaling System in Health and Disease. *Clin. Genet.* **2014**, *86*, 447–452. [CrossRef]
178. Zhang, X.; Guo, J.; Wei, X.; Niu, C.; Jia, M.; Li, Q.; Meng, D. Bach1: Function, Regulation, and Involvement in Disease. *Oxid. Med. Cell Longev.* **2018**, *2018*, 1347969. [CrossRef] [PubMed]
179. Ju, D.T.; Huang, R.F.S.; Tsai, B.C.K.; Su, Y.C.; Chiu, P.L.; Chang, Y.M.; Padma, V.V.; Ho, T.J.; Yao, C.H.; Kuo, W.W.; et al. Folic Acid and Folinic Acid Protect Hearts of Aging Triple-Transgenic Alzheimer's Disease Mice via IGF1R/PI3K/AKT and SIRT1/AMPK Pathways. *Neurotox. Res.* **2023**, *41*, 648–659. [CrossRef]
180. Benarroch, E.E. Nrf2, Cellular Redox Regulation, and Neurologic Implications. *Neurology* **2017**, *88*, 1942–1950. [CrossRef] [PubMed]
181. López-Otín, C.; Blasco, M.A.; Partridge, L.; Serrano, M.; Kroemer, G. The Hallmarks of Aging. *Cell* **2013**, *153*, 1194–1217. [CrossRef] [PubMed]
182. López-Otín, C.; Blasco, M.A.; Partridge, L.; Serrano, M.; Kroemer, G. Hallmarks of Aging: An Expanding Universe. *Cell* **2023**, *186*, 243–278. [CrossRef] [PubMed]
183. Suh, J.H.; Shenvi, S.V.; Dixon, B.M.; Liu, H.; Jaiswal, A.K.; Liu, R.M.; Hagen, T.M. Decline in Transcriptional Activity of Nrf2 Causes Age-Related Loss of Glutathione Synthesis, Which Is Reversible with Lipoic Acid. *Proc. Natl. Acad. Sci. USA* **2004**, *101*, 3381–3386. [CrossRef] [PubMed]
184. Schmidlin, C.J.; Dodson, M.B.; Madhavan, L.; Zhang, D.D. Redox Regulation by NRF2 in Aging and Disease. *Free Radic. Biol. Med.* **2019**, *134*, 702–707. [CrossRef] [PubMed]
185. Batliwala, S.; Xavier, C.; Liu, Y.; Wu, H.; Pang, I.H. Involvement of Nrf2 in Ocular Diseases. *Oxid. Med. Cell Longev.* **2017**, *2017*, 1703810. [CrossRef] [PubMed]
186. Taylor, E.L.; Collins, J.A.; Gopalakrishnan, P.; Chubinskaya, S.; Loeser, R.F. Age and Oxidative Stress Regulate Nrf2 Homeostasis in Human Articular Chondrocytes. *Osteoarthr. Cartil.* **2023**, *31*, 1214–1223. [CrossRef] [PubMed]
187. Frantz, M.C.; Rozot, R.; Marrot, L. NRF2 in Dermo-Cosmetic: From Scientific Knowledge to Skin Care Products. *BioFactors* **2023**, *49*, 32–61. [CrossRef] [PubMed]
188. Aranda-Rivera, A.K.; Cruz-Gregorio, A.; Pedraza-Chaverri, J.; Scholze, A. Nrf2 Activation in Chronic Kidney Disease: Promises and Pitfalls. *Antioxidants* **2022**, *11*, 1112. [CrossRef] [PubMed]
189. Wang, Y.; Xiong, Y.; Zhang, A.; Zhao, N.; Zhang, J.; Zhao, D.; Yu, Z.; Xu, N.; Yin, Y.; Luan, X.; et al. Oligosaccharide Attenuates Aging-Related Liver Dysfunction by Activating Nrf2 Antioxidant Signaling. *Food Sci. Nutr.* **2020**, *8*, 3872–3881. [CrossRef] [PubMed]
190. Wu, X.; Wei, J.; Yi, Y.; Gong, Q.; Gao, J. Activation of Nrf2 Signaling: A Key Molecular Mechanism of Protection against Cardiovascular Diseases by Natural Products. *Front. Pharmacol.* **2022**, *13*, 1057918. [CrossRef] [PubMed]
191. Dinkova-Kostova, A.T.; Kostov, R.V.; Kazantsev, A.G. The Role of Nrf2 Signaling in Counteracting Neurodegenerative Diseases. *FEBS J.* **2018**, *285*, 3576–3590. [CrossRef] [PubMed]
192. Chen, C.; Zhou, M.; Ge, Y.; Wang, X. SIRT1 and Aging Related Signaling Pathways. *Mech. Ageing Dev.* **2020**, *187*, 111215. [CrossRef]
193. Liu, T.; Zhu, S.; Yang, Y.; Qin, W.; Wang, Z.; Zhao, Z.; Liu, T.; Wang, X.; Duan, T.; Zhang, H.; et al. Oroxylin A Ameliorates Ultraviolet Radiation-Induced Premature Skin Aging by Regulating Oxidative Stress via the Sirt1 Pathway. *Biomed. Pharmacother.* **2024**, *171*, 116110. [CrossRef]
194. Bielach-Bazyluk, A.; Zbroch, E.; Mysliwiec, H.; Rydzewska-Rosolowska, A.; Kakareko, K.; Flisiak, I.; Hryszko, T. Sirtuin 1 and Skin: Implications in Intrinsic and Extrinsic Aging—A Systematic Review. *Cells* **2021**, *10*, 813. [CrossRef]
195. Sun, K.; Wu, Y.; Zeng, Y.; Xu, J.; Wu, L.; Li, M.; Shen, B. The Role of the Sirtuin Family in Cartilage and Osteoarthritis: Molecular Mechanisms and Therapeutic Targets. *Arthritis Res. Ther.* **2022**, *24*, 286. [CrossRef] [PubMed]
196. Kong, S.; Yeung, P.; Fang, D. The Class III Histone Deacetylase Sirtuin 1 in Immune Suppression and Its Therapeutic Potential in Rheumatoid Arthritis. *J. Genet. Genom.* **2013**, *40*, 347–354. [CrossRef]
197. Lu, J.; Wang, Z.; Cao, J.; Chen, Y.; Dong, Y. A Novel and Compact Review on the Role of Oxidative Stress in Female Reproduction. *Reprod. Biol. Endocrinol.* **2018**, *16*, 80. [CrossRef] [PubMed]
198. Valipour, J.; Taghizadeh, F.; Esfahani, R.; Ramesh, M.; Rastegar, T. Role of Nuclear Factor Erythroid 2-Related Factor 2 (Nrf2) in Female and Male Fertility. *Heliyon* **2024**, *10*, e29752. [CrossRef] [PubMed]

199. Alam, F.; Syed, H.; Amjad, S.; Baig, M.; Khan, T.A.; Rehman, R. Interplay between Oxidative Stress, SIRT1, Reproductive and Metabolic Functions. *Curr. Res. Physiol.* **2021**, *4*, 119–124. [CrossRef]
200. Khan, M.Z.; Khan, A.; Huang, B.; Wei, R.; Kou, X.; Wang, X.; Chen, W.; Li, L.; Zahoor, M.; Wang, C. Bioactive Compounds Protect Mammalian Reproductive Cells from Xenobiotics and Heat Stress-Induced Oxidative Distress via Nrf2 Signaling Activation: A Narrative Review. *Antioxidants* **2024**, *13*, 597. [CrossRef] [PubMed]
201. Li, X.; He, Y.; Wu, S.; Zhang, P.; Gan, M.; Chen, L.; Zhao, Y.; Niu, L.; Zhang, S.; Jiang, Y.; et al. Regulation of SIRT1 in Ovarian Function: PCOS Treatment. *Curr. Issues Mol. Biol.* **2023**, *45*, 2073–2089. [CrossRef] [PubMed]
202. Sotoudeh, G.; Abshirini, M. Antioxidant Capacity and Menopausal Symptoms. In *Aging: Oxidative Stress and Dietary Antioxidants*; Academic Press: Cambridge, MA, USA, 2020; pp. 125–133.
203. Ishikawa, A.; Matsushita, H.; Shimizu, S.; Morita, N.; Hanai, R.; Sugiyama, S.; Watanabe, K.; Wakatsuki, A. Impact of Menopause and the Menstrual Cycle on Oxidative Stress in Japanese Women. *J. Clin. Med.* **2023**, *12*, 829. [CrossRef] [PubMed]
204. Mendoza, C.C.C.; Zamarripa, C.A.J. Menopause Induces Oxidative Stress. In *Oxidative Stress and Chronic Degenerative Diseases—A Role for Antioxidants*; IntechOpen Limited: London, UK, 2013; pp. 289–316.
205. Bonaccorsi, G.; Piva, I.; Greco, P.; Cervellati, C. Oxidative Stress as a Possible Pathogenic Cofactor of Post-Menopausal Osteoporosis: Existing Evidence in Support of the Axis Oestrogen Deficiency-Redox Imbalance-Bone Loss. *Indian J. Med. Res.* **2018**, *147*, 341–351. [CrossRef]
206. Chandankhede, M.; Gupta, M.; Pakhmode, S. Assessment of Psychological Status and Oxidative Stress in Postmenopausal Women: A Cross-Sectional Study. *J. Menopausal Med.* **2021**, *27*, 155–161. [CrossRef] [PubMed]
207. Lv, Y.; Zhang, S.; Weng, X.; Huang, J.; Zhao, H.; Dai, X.; Bai, X.; Bao, X.; Zhao, C.; Zeng, M.; et al. Estrogen Deficiency Accelerates Postmenopausal Atherosclerosis by Inducing Endothelial Cell Ferroptosis through Inhibiting NRF2/GPX4 Pathway. *FASEB J.* **2023**, *37*, e22992. [CrossRef]
208. Zhao, F.; Guo, L.; Wang, X.; Zhang, Y. Correlation of Oxidative Stress-Related Biomarkers with Postmenopausal Osteoporosis: A Systematic Review and Meta-Analysis. *Arch. Osteoporos.* **2021**, *16*, 4. [CrossRef] [PubMed]
209. Hajializadeh, Z.; Khaksari, M. The Protective Effects of 17-β Estradiol and SIRT1 against Cardiac Hypertrophy: A Review. *Heart Fail. Rev.* **2022**, *27*, 725–738. [CrossRef] [PubMed]
210. Mendoza, C.C.C.; Zamarripa, C.A.J.; Hernández, M.E.O.; Iglesias, G.G. Possible Role of Nrf2 in Oxidative and Inflammatory Processes During Menopause. In *A Master Regulator of Oxidative Stress—The Transcription Factor Nrf2*; IntechOpen Limited: London, UK, 2016; pp. 113–129.
211. Sánchez-Rodríguez, M.A.; Zacarías-Flores, M.; Correa-Muñoz, E.; Arronte-Rosales, A.; Mendoza-Núñez, V.M. Oxidative Stress Risk Is Increased with a Sedentary Lifestyle during Aging in Mexican Women. *Oxid. Med. Cell Longev.* **2021**, *2021*, 9971756. [CrossRef] [PubMed]
212. Sharifi-Rad, M.; Anil Kumar, N.V.; Zucca, P.; Varoni, E.M.; Dini, L.; Panzarini, E.; Rajkovic, J.; Tsouh Fokou, P.V.; Azzini, E.; Peluso, I.; et al. Lifestyle, Oxidative Stress, and Antioxidants: Back and Forth in the Pathophysiology of Chronic Diseases. *Front. Physiol.* **2020**, *11*, 694. [CrossRef] [PubMed]
213. Tan, B.L.; Norhaizan, M.E.; Liew, W.P.P. Nutrients and Oxidative Stress: Friend or Foe? *Oxid. Med. Cell Longev.* **2018**, *2018*, 9719584. [CrossRef] [PubMed]

International Journal of
Molecular Sciences

MDPI

Review

Medicinal Mushrooms in Colon Cancer Therapy: Mechanisms of Action of Bioactive Compounds and Therapeutic Potential

Jinangi Bentharavithana [1,2], Tahidul Islam [1,3] and Baojun Xu [1,*]

[1] Food Science and Technology Program, Department of Life Sciences, Beijing Normal-Hong Kong Baptist University, Zhuhai 519085, China; jinangibv97@gmail.com (J.B.); nobab01@hkbu.edu.hk (T.I.)
[2] Department of Food Science and Technology, Faculty of Agriculture, University of Peradeniya, Peradeniya 20400, Sri Lanka
[3] School of Chinese Medicine, Hong Kong Baptist University, Hong Kong SAR, China
[*] Correspondence: baojunxu@uic.edu.cn; Tel.: +86-756-3620636

Abstract: Colon cancer is the second leading cause of cancer-related deaths in the world. This is commonly observed among older adults, and the occurrence of colon cancer is mainly influenced by unhealthy lifestyle factors. Edible medicinal mushrooms have been demonstrated to have anti-colon cancer effects both individually and in combination with conventional therapies, including synergistically enhancing the efficacy of chemotherapy medications such as 5-fluorouracil in preclinical models. Medicinal mushrooms such as *Lentinus edodes*, *Phellinus linteus*, *Ganoderma lucidum*, *Inonotus obliquus*, *Pleurotus ostreatus*, *Hericium erinaceus*, *Pleurotus eryngii*, *Gloeostereum incarnatum*, and *Termitomyces heimii* are emerging as promising candidates, not only because conventional treatments for colon cancer face significant limitations, including side effects, psychological impacts on patients, high cost, limited specificity toward cancer and healthy cells, and the development of drug resistance, but also due to the diverse array of bioactive compounds present within them. Therefore, there is a strong demand for innovative, affordable, and minimally invasive treatments such as medicinal mushrooms. Their bioactive compounds, including terpenoids, sterols, phenols, polysaccharides, acids, sesquiterpenes, alkaloids, lactones, metal-chelating agents, nucleotide analogs, glycoproteins, β-glucan, cerebrosides, steroids, terpenes, quinolones, anthraquinones, benzoic acid derivatives, linoleic acid, ascorbic acid, glycosides, organic acids, flavonoids, grifolin, tocopherols, proteins, indoles, lectin, and laccases, exert anti-colon cancer activities through various mechanisms, including anti-proliferative effects, cell cycle arrest, anti-inflammatory effects, antioxidant effects, induction of apoptosis, cytotoxic effects, and antimigratory effects. Further research is needed to elucidate the molecular mechanisms and confirm the safety and efficacy of medicinal mushrooms as a holistic anti-colon cancer treatment.

Keywords: apoptosis; bioactive compounds; clinical study; edible mushrooms; medicinal mushrooms

1. Introduction

Mushrooms are filamentous fungi with fruiting bodies. For a long time, they have been an essential component in the human diet as well as in medicine and pharmacology [1]. The Chinese and Japanese have utilized mushrooms for medicinal purposes for thousands of years. Nowadays, mushrooms are popular, valuable foods because they are low in calories, carbohydrates, fat, and sodium and are cholesterol-free. Besides, mushrooms provide important nutrients, including selenium, potassium, riboflavin, niacin, vitamin

D, proteins, and fiber [2]. There are many health-promoting functionalities found in mushrooms and it has been reported that more than 100 medicinal functions are produced by mushrooms and fungi, which have antioxidant, anticancer, antidiabetic, antiallergic, immunomodulating, cardiovascular protector, anticholesterolemic, antiviral, antibacterial, antiparasitic, antifungal, detoxification, and hepatoprotective effects, and also protect against tumor development and inflammatory processes [3–6]. Colorectal cancer is the third most commonly diagnosed cancer and the fourth most common cause of cancer-related mortality globally. The highest incidence and mortality rates are seen in high-income countries due to unhealthy diets [7]. However, other factors such as a history of polyps in the colon, age over 50 years, a personal history of inflammatory bowel disease, family colon cancer history, obesity, low exercise, red meat consumption, smoking, and alcohol consumption are driving factors for the growth of colon cancer. The incidence of colorectal cancer is especially increased in developing countries. It has been reported that nearly 2 million new cases and about 1 million deaths are expected in 2018 [8], which demonstrates that only 50% of colon cancer patients could survive. Treatments such as chemotherapy, surgery, radiation therapy, targeted therapy, and immunotherapy may damage healthy cells and tissues and cause side effects in the body. Therefore, researchers are exploring natural alternative methods with minimum side effects. This review focuses on medicinal mushrooms as an alternative and their different molecular mechanisms in lab, animal, and human studies against colon cancer.

2. Edible and Medicinal Mushrooms with Anti-Colon Cancer Effects

2.1. Definition of Medicinal and Edible Mushrooms

In many in vivo, in vitro, and clinical studies, medicinal mushroom consumption has shown an inverse correlation with gastrointestinal cancer occurrence [9]. According to the Oxford Dictionary, the term 'medicinal' means having healing, curative, or therapeutic attributes or used as a medicine or related to medicine. The term 'edible' means fit to be eaten. Medicinal mushrooms are macroscopic fungi belonging to the phylum Basidiomycota. These can be consumed as food items and used as extracts or powders to treat diseases. Bioactive compounds present in medicinal mushrooms enhance their medicinal and nutraceutical properties. Medicinal mushrooms are rich in a plethora of biologically active secondary metabolites that possess beneficial health effects such as anticancer, antimicrobial, antiviral, anti-inflammatory, antidiabetic, and antioxidant effects. Even though there are 140,000 mushroom species available in nature, only 2000 species are in edible form. Among the edible forms, only 25 species are commercially available [10,11]. There are around 2000 mushrooms that show medicinal properties, and over 600 have been confirmed with official data [12]. Edible mushrooms show medicinal properties through anticancer, antiviral, hepatoprotective, anti-cardiovascular disease, immunopotentive, antioxidative, and hypocholesterolemic effects. Therefore, many research techniques are involved in extracting the bioactive compounds present in edible mushrooms that show medicinal properties [13]. Edible mushrooms are used to prepare drugs and nutraceuticals that have antitumor, antioxidant, and antimicrobial effects. The low starch, low fat, and high fiber contents of edible mushrooms make them ideal food items for both obese and diabetic patients. Due to this, edible mushrooms are emerging as potent sources of pharmaceuticals [14]. *Letinula edodes*, *Ganoderma Lucidum*, *Grifola frondosa*, *Trametes versicolor*, and *Inonotus obliquus* are popular edible medicinal fungi that have been used in both traditional and modern medicine and research [12]. There are both edible medicinal mushroom species and inedible medicinal mushroom species. *Heterobasidion annosum* and *Trametes versicolor* are medicinal mushrooms that can be used as potential medicines for colorectal cancer treatment, but they are not edible [15,16].

2.2. Types of Edible and Medicinal Mushrooms with Anti-Colon Cancer Properties

Mushrooms are becoming more interesting due to their dual edible and medicinal nature [17]. The bioactive compounds present in medicinal mushrooms demonstrate anticancer activity through various mechanisms. This section classifies key edible and medicinal mushrooms (MMs) according to their demonstrated anti-colon cancer effects and usage. Figure 1 presents edible and medicinal mushrooms with anti-colon cancer effects.

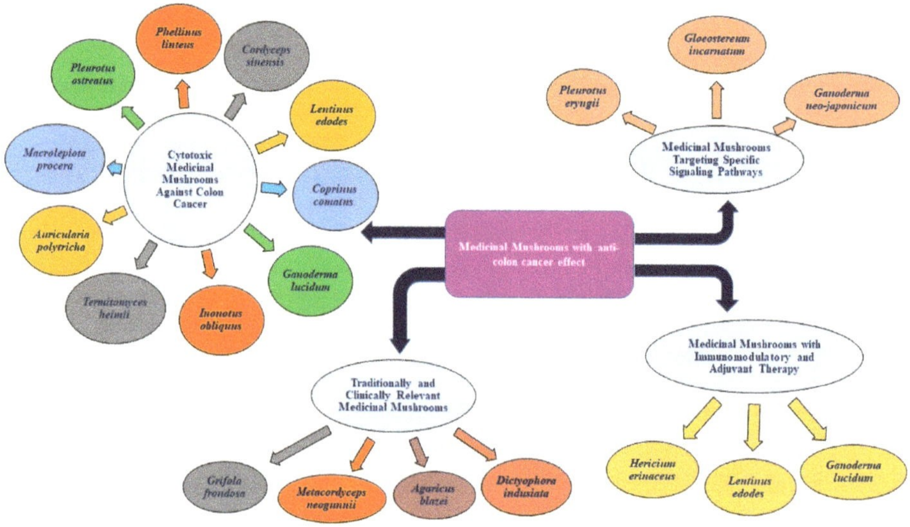

Figure 1. Edible and medicinal mushrooms with anti-colon cancer effects.

2.2.1. Potent Cytotoxic Medicinal Mushrooms Against Colon Cancer

Meshima (*Phellinus linteus* (Berk. et Curt.) Teng), caterpillar fungus (*Cordyceps sinensis* (Berk.) Sacc.), shiitake mushrooms (*Lentinus edodes* (Berk.) Pegler), shaggy ink cap (*Coprinus comatus* (O.F. Müll.) Pers.), and lingzhi/reishi (*Ganoderma lucidum* (Curtis) P. Karst) are some of the traditional and commercially available edible MMs with anti-colon cancer properties. Methanolic extracts of *Phellinus linteus* dramatically decreased the cell viability of human colorectal carcinoma cell line 116 (HCT-116) and surg pathol-derived colorectal cancer cell line 480 (SW-480). Other mushroom varieties showed selective inhibition of SW-480 cancer cell line viability. The MTT assay showed a broad spectrum of cytotoxic effects induced by *Phellinus linteus* when compared to the other four varieties [18]. Research conducted using a hot water extract of *Inonotus obliquus* (IOWE) showed that it can inhibit the growth of human colorectal adenocarcinoma cell line 29 (HT-29) in a dose-dependent manner. The most effective IOWE concentration was 1.0 mg/mL for 48 h, with a maximum inhibitory activity of 56% [6]. *Termitomyces heimii* showed an in vitro cytotoxic effect against the HCT-116 colon cancer cell line [9]. *Auricularia polytricha*, *Macrolepiota procera*, and *Pleurotus ostreatus* are edible and medicinal mushrooms with anti-colon cancer effects [19]. A study conducted on the anticancer properties of *Pleurotus ostreatus* grown in selenium (Se) showed anti-colon cancer effects against a human colorectal adenocarcinoma cell line (Caco-2) and normal human colon mucosal epithelial cancer cell line (NCM-460). These findings provide insight into the anti-colon cancer effects of selenium-rich fruiting bodies of edible medicinal mushrooms [20].

2.2.2. Medicinal Mushrooms with Immunomodulatory Effects and Adjuvant Therapy

Lentinus edodes is one of the most popular edible mushrooms used as a food in Eastern countries. The *Lentinus edodes* extract has shown its potential as a novel chemotherapeutic adjuvant agent when combined with 5-fluorouracil [21]. β-Glucan from *Lentinus edodes* showed promising results for colitis-associated colorectal cancer and the gut microbiota [22]. *Ganoderma lucidum* is another edible MM that has an inhibitory effect against colorectal cancer. In vitro co-administration of a non-toxic concentration (0.3 mg/mL) of an extract of a commercial product that contained spores and fruiting bodies (30:8 ratio) of *Ganoderma lucidum* (GLSF) and paclitaxel, a chemotherapy medication (0.125 μM), showed significant cancer cell growth inhibition and apoptosis in the CT26 murine colon carcinoma cell line and HCT-15 human colon cancer cell line. In vivo studies showed that CT26 tumor cell growth was suppressed when mice were orally administered a modified diet that contained 1.25% GLSF powder [23]. *G. lucidum* supplements are used by cancer patients to boost their immune system. This mushroom's fruiting bodies, mycelia, or spores contain anticancer chemical compounds, particularly polysaccharides and triterpenes [24]. Since ancient times in China, *Hericium erinaceus*, also known as lion's mane, has been used to treat digestive system disorder-related diseases. This medicinal mushroom studied worldwide has anticancer effects against colon cancer cells [25].

2.2.3. Medicinal Mushrooms Targeting Specific Signaling Pathways

Gloeostereum incarnatum (GI) is an edible mushroom with anti-colon cancer activity. Its water-soluble polysaccharide extract (GIPS) shows anti-colon cancer activity through the Wnt signaling pathway [26]. *Ganoderma neo-japonicum* Imazeki was identified as a potential natural anticancer agent against human colon cancer through cellular and computational models [27]. *Pleurotus eryngii*, also known as king trumpet mushroom or French horn mushroom, is an edible mushroom that showed inhibitory activity against the growth of HCT-116 colon cancer cells and an anti-inflammatory function in the RAW264.7 macrophage cell line [28].

2.2.4. Traditionally and Clinically Relevant Medicinal Mushrooms

Inonotus obliquus (Chaga) is an edible medicinal mushroom with antioxidant, anti-inflammatory, antitumor, immunomodulatory, and antimutagenic activities. The toxicity associated with improper usage of Chaga is now addressed by toxicity assessments, risk analyses, and guidelines for proper usage [29]. In vitro research conducted on n-hexane extracts of *Hericium erinaceus*, *Metacordyceps neogunnii*, and *Dictyophora indusiata* showed that these edible and medicinal mushrooms exerted anti-colon cancer effects against the HCT-116 human colon cancer cell line [30]. The raw form of *Ganoderma lucidum* medicinal mushroom is not edible, but powders, dietary supplements, and tea are commercially available products of this mushroom. This species is renowned for its pharmaceutical activities rather than its nutritional value. A water extract of sporoderm-broken spores of *Ganoderma lucidum* showed an anticarcinogenic effect against colorectal cancer [31]. *Lentinus edodes* shows its anticancer activity by activating immune responses, prohibiting colon cancer cell proliferation, suppressing tumor growth, and inducing cell apoptosis [32]. It also demonstrates antioxidant, antidiabetic, antihepatitic, hyperlipidemic, antifungal, immunomodulatory, antibacterial, antiviral, and antitumor activities [33]. The water extracts of the mycelium of *Agaricus blazei* (82.4%), *Hericeum erinaceus* (14.7%) and *Grifola frondosa* (2.9%), collectively called Andosan™, a medicinal mushroom extract, showed their potential as a natural preventive and therapeutic agent for colorectal cancer in an A/J Min/+ mice model [34,35].

2.3. Chemical Compositions of Mushrooms with Anti-Colon Cancer Properties

Edible and medicinal mushrooms are generally low in calories, lipids, and cholesterol and have high contents of protein and essential vitamins. The fruiting bodies of edible and medicinal mushrooms contain a variety of secondary [11,14]. Edible mushrooms are usually low in fat and energy while high in protein. Generally, many edible mushrooms that show medicinal properties are rich in minerals (iron, phosphorus) and vitamins (riboflavin, thiamine, ergosterol, niacin, ascorbic acid). β-Glucans and glycoproteins are the main polysaccharides present in edible and medicinal mushrooms. Terpenoids, acids, alkaloids, sesquiterpenes, lactones, sterols, metal-chelating agents, vitamins, nucleotide analogs, and polyphenolic compounds are the main secondary metabolites present in these mushrooms [13]. Chitin and starch are the main carbohydrates present in edible and medicinal mushrooms. Many medicinal mushrooms are good sources of crude fiber such as chitin, mannan, and hemicellulose. The nutrient, enzyme, bioactive compound, vitamin, and mineral contents vary with the type of edible medicinal mushroom [12]. Edible and medicinal mushrooms are rich sources of selenium, potassium, riboflavin, niacin, vitamin D, proteins, and fiber. Also, they are free from cholesterol and contain low calories and fat, sodium, and carbohydrate contents. Therefore, these edible and medicinal mushrooms are used to treat different diseases, including cancer, and to improve human health [2].

2.4. Bioactive Compounds in Edible and Medicinal Mushrooms with Anti-Colon Cancer Effects

Edible and medicinal mushrooms are good sources of biologically active compounds such as terpenoids, sterols, phenols, and polysaccharides [2,12,28]. The bioactive molecules can be categorized as high-molecular-weight compounds (polysaccharides and proteins) and low-molecular-weight compounds (indoles, terpenoids, and phenols) [9]. There are secondary metabolites, glycoproteins, and polysaccharides. The secondary metabolites include acids, terpenoids, polyphenols, sesquiterpenes, alkaloids, lactones, sterols, metal-chelating agents, nucleotide analogs, and vitamins. β-Glucan is the main polysaccharide present in edible MMs [2]. Several bioactive polysaccharides and polysaccharide–protein complexes also play significant roles in cancer treatment [36]. The bioactive compounds in *Hericium erinaceus*, such as terpenoids, polyphenols, cerebrosides, and polysaccharides, have shown a variety of potential therapeutic applications such as antioxidant, immune-modulatory, and anticancer properties [25]. The secondary metabolites in medicinal mushrooms, such as steroids, terpenes, quinolones, anthraquinones, and benzoic acid derivatives, have shown anti-carcinogenesis effects. In addition, linoleic acid, ascorbic acid, polysaccharides, glycosides, organic acids, flavonoids, phenols, grifolin, and tocopherols exert their metabolic activities by actively scavenging nitrite radicals, hydroxyl radicals, and oxygen free radicals, which aids in suppressing cancer cell proliferation. Bioactive compounds in medicinal mushrooms demonstrate their anticancer activities through multiple mechanisms, including inhibition of nuclear factor kappa-light-chain-enhancer of activated B cells (NF-κB) activation, inhibition of protein kinase B (Akt) processes, and inhibition of cyclooxygenase-2 (COX-2) expression [37].

2.4.1. Polysaccharides as Bioactive Compounds in Edible and Medicinal Mushrooms

Anti-Colon Cancer Mechanisms of Medicinal Mushroom Polysaccharides

Polysaccharides are increasingly being recognized as the main active and multifunctional bioactive compounds in medicinal mushrooms. Medicinal mushroom-derived polysaccharides are being recognized as potential anticancer agents mainly due to their role in modulating the immune system [9,17,25]. In Chinese traditional medicine, *Ganoderma lucidum* polysaccharides (GLPs) are used for cancer prevention. Sporoderm-broken spores of *G. lucidum* water extract (BSGLWE) mainly contained GLPs [31]. A study conducted on *H. erinaceus* fruiting bodies extracted and purified a novel polysaccharide called

HEFP-2b. HEFP-2b showed anti-colon cancer properties in HCT-116 and DLD1 colorectal adenocarcinoma cell line [25]. Lentinan extracted from *Lentinus edodes* is a clinically used β-1, 3-glucan polysaccharide with immunomodulatory and antitumor effects. A water-extracted polysaccharide from *Lentinus edodes* (SLNT) showed an anti-colon cancer effect in athymic nude mice by inhibiting human colon cancer cell proliferation and suppressing tumor growth [33]. Polysaccharides in *Agaricus bisporus, Antrodia camphorate, Boletus edulis, Cordyceps Sinensis, Flammulina velutipes, Lentinus edodes, Lignosus rhinoceros, Morchella esculenta,* and *Pleurotus eryngii* are active against colorectal carcinogenesis [37–39]. The polysaccharides extracted from *Agaricus bisporus* were encapsulated within alginate-kappa carrageenan microcapsules, and these microcapsules activated natural killer cells (NK cells), acting as a potential treatment for colon cancer [40]. *Agaricus blazei* Murill polysaccharide significantly enhanced the activity of CD8+ T cells, which aid in the killing of colorectal cancer cells, and effectively inhibited intraperitoneal tumors in mice. Therefore, *Agaricus blazei* Murill polysaccharide can be considered a promising therapeutic agent to treat colon cancer, particularly involving intraperitoneal dissemination [41].

Structural Diversity and Bioactivity of Polysaccharides in Medicinal Mushrooms

A homogeneous neutral polysaccharide, WAAP-1 (molecular weight of 10.1 kDa), extracted from *Agaricus bisporus* significantly inhibited HT-29 colon cancer cell proliferation. The monosaccharides present in WAAP-1 are glucose, mannose, and galactose in a molar ratio of 84.95:8.97:4.50, while the main chain is mainly composed of (1,4)-α-D-Glcp and (1,6)-β-D-Manp [17]. Polysaccharides are emerging as potential components to prevent and treat colorectal cancers. The very first research that found the *Agaricus bisporus* polysaccharide WAAP-2, with a triple-helical structure, tangled chain conformation, and a molecular weight of 121 kDa, demonstrated an anticancer colon effect against HT-29 cells. The primary structure of WAAP-2 contains mannose, glucose, and galactose at a molar ratio of 8.81:81.80:6.89 [42]. *Gloeostereum incarnatum* polysaccharide extract has shown anti-colon cancer effects in mice [43]. Anticancer research conducted using *Lentinus edodes* showed that its β-glucan fraction (lentinan) reduced inflammation, inhibited colon cancer cell proliferation, and induced colon cancer cell apoptosis. Among the three β-glucan fractions, the fraction with the lowest molecular weight showed the most promising anti-colon cancer activity by rebuilding the intestinal mucosal barrier [22].

α-Glucan, a low-molecular-weight carbohydrate, extracted from the hot water-soluble fraction of *Pleurotus ostreatus* mushroom showed anti-proliferative and pro-apoptotic effects against HT-29 colon cancer cells [44]. *Boletus edulis*, a medicinal mushroom having anti-colon cancer properties, contains carbohydrates as its major component, constituting 58.1% of the mushroom's dry weight. The extract obtained using ohmic heating demonstrated a total sugar concentration of 437.5 mg/g extract, trehalose 341 mg/g, and rhamnose, mannitol, glucose, and fructose [45,46]. Lentinan, a polysaccharide with β-glucans, demonstrated immunomodulating properties and antitumor activity against human colorectal cancer [37,47,48].

Polysaccharide-K, which is also known as Krestin or PSK, is usually found in *Trametes versicolor* and shows anti-colon cancer properties. PSK is a complex molecule containing polysaccharides and proteins. The polysaccharides primarily contain β-glucans. The amino acid profile of PSK is rich in acidic amino acids, mainly glutamic and aspartic acids [37,49–51]. Maitake D-fraction is a complex compound derived from the Maitake mushroom (*Grifola frondosa*) that shows anti-colon cancer properties. This compound also contains β-glucans as the primary polysaccharide and forms a proteoglycan complex by binding proteins into the polysaccharide [37,52–55]. Schizophyllan, derived from *Schizophyllum commune,* is another promising polysaccharide with a β-glucan structure that shows anti-colon cancer properties [37,56,57].

Future Directions, Limitations, and Suggestions Concerning the Use of Polysaccharides in Medicinal Mushrooms

The anti-colon cancer effects of polysaccharides are exerted through diverse mechanisms, demonstrating the wide potential of mushroom polysaccharides. The structural diversity of polysaccharides allows them to act through multiple mechanisms rather than relying on a single pathway. This aligns well with modern colon cancer treatment strategies, as combination approaches improve therapeutic efficacy. Monosaccharide composition, linkage type, and molecular weight influence the structure–function relationship in anti-colon cancer activity. As a future direction, modifying polysaccharide branching patterns, molecular weight, and glycosidic linkages could enhance bioavailability. Developing nanoformulations may also improve oral bioavailability and tumor targeting. Despite being multifunctional bioactive agents, polysaccharides face challenges such as structural heterogeneity, poor bioavailability, and limited clinical research in humans. Additionally, variations in extraction methods can lead to inconsistent polysaccharide compositions.

2.4.2. Proteins as Bioactive Compounds in Edible and Medicinal Mushrooms

A novel bioactive protein in *Pleurotus eryngii*, called 'PEP', exhibited anti-inflammatory and anti-colon cancer effects [58,59]. Protein extracts from *Calvatia lilacina*, *Pleurotus ostreatus*, and *Volvariella volvacea* showed anticancer effects against the SW480 and Tohoku Hospital Pediatrics-1 human monocytic leukemia (THP-1) cell lines. Each protein extract had a unique protein profile with a distinct molecular weight [60]. Low-carbohydrate proteins and peptides derived from mushrooms exhibit anticancer properties through unique mechanisms [61]. A protein fraction known as Cibacron blue affinity eluted protein (CBAEP) was isolated from *Termitomyces clypeatus*, *Pleurotus florida*, *Calocybe indica*, *Astraeus hygrometricus*, and *Volvariella volvacea*. CBAEP showed anti-proliferative activity in tumor cells, enhanced natural killer cell activity, and stimulated the proliferation of immune cells, particularly splenocytes, thymocytes, and bone marrow cells [62].

Lectin, a bioactive protein in *Agaricus bisporus* demonstrates anti-colon cancer activity by inhibiting cancer cell growth and stimulating the immune system [63–66]. Lectins in *Agrocybe aegerita* and *Grifola frondose* show anticancer activity by inducing apoptosis in cancer cells and stimulating the immune system, respectively [67]. Musarin, a polysaccharide peptide isolated from *Trametes versicolor* powder, downregulated the epidermal growth factor receptor-ras signaling pathway (EGFR-Ras signaling pathway) in colorectal cancer stem-like cells and inhibited the proliferation of CSC-like CD24+CD44+ HT29 cells [15,43]. Laccases are bioactive proteins that have shown promising anti-colon cancer activity through anti-proliferative and cytotoxic effects [5,66,68–70].

2.4.3. Phenolic Compounds as Bioactive Compounds in Edible and Medicinal Mushrooms

Unique Attributes of Phenolic Compounds in Medicinal Mushrooms

Phenolic compounds present in certain mushroom species exhibit a wide range of biological activities. These findings have significant implications for the development of novel and effective mushroom-based anticancer therapies. Phenolic compounds are abundant in some mushrooms, which help exert diverse biological activities, including anti-colon cancer effects, and have antimicrobial, antioxidant, and anti-inflammatory properties [28,71,72]. In point of fact, phenolic compounds are present in many plants as well as mushrooms. Compared to plants, fungal phenolics are unique fungal-specific metabolites. Even though fungi are underutilized compared to plants, they have superior antioxidant activity. Fungi contain a diverse range of phenolic compounds that contribute to their antioxidant properties. Among species, the phenolic profile significantly varies, indicating species-specific uniqueness in biochemical composition. In addition, many fungi exhibit comparative or may even exhibit superior antioxidant activity in free radical scavenging. The extraction methods used to extract fungi influence the antioxidant activity of fungi. Therefore, fungi could offer unique phenolics and potentially higher bioactivity [73].

Fungi do not rely specifically on shikimate and phenylpropanoid pathways for the biosynthesis of phenolic compounds and they may have modified or alternative pathways. Most importantly, fungi modify their phenolic compounds using extracellular enzymes, which allow them to synthesize new hybrid molecules. Fungi molecules often demonstrate higher bioavailability in human cells [74]. A combination of different mushroom extracts exhibited superior cytotoxic effects against human colon cancer cell lines compared to single-species extracts due to the synergistic effects of both polyphenols and polysaccharides [75]. The study demonstrated the effectiveness of the combination of mushroom extracts against colon cancer cells compared to single-species extracts. As discussed above, fungi show species-specific uniqueness in biochemical composition, and interactions among these compounds may enhance the anti-colon cancer activity.

Anti-Colon Cancer Potential of Medicinal Mushroom-Derived Phenolics

A study was conducted on the phenolic compound concentrations and antioxidant activities of five edible and five medicinal mushrooms that are commonly cultivated in Korea to determine the DPPH radical scavenging activities. *Pleurotus eryngii* showed 15% activity when the reaction time was one minute and *Agaricus bisporus* showed 78% activity when the reaction time was 30 min. There were 28 phenolic compounds present in these 10 mushroom species. The average total concentrations of phenolic compounds in edible and medicinal mushrooms were 174 μg/g and 477 μg/g, respectively. The average total concentrations of flavonoids in edible and medicinal mushrooms were 22 μg/g and 76 μg/g, respectively [71]. Another study conducted on determining the total phenolic and flavonoid contents in *Agaricus bisporus, Boletus edulis, Calocybe gambosa, Cantharellus cibarius, Craterellus cornucopioides, Hygrophorus marzuolus, Lactarius deliciosus*, and *Pleurotus ostreatus* demonstrated that the total flavonoid concentrations and total phenolic concentrations ranged between 0.9 and 3 mg per gram dry matter (DM) and 1 and 6 mg per gram (DM), respectively. All 8 varieties of mushrooms contained homogentisic acid, free phenolic acid, and flavonoids such as myricetin and catechin [76].

Termitomyces mushroom species contain gallic acid, chlorogenic acid, caffeic acid, ellagic acid, catechins, epicatechins, rutin, isoquercitrin, quercitrin, quercetin, and kaempferol as phenolic compounds [72]. A polyphenol-rich extract was isolated from *Pleurotus eryngii* to study the anti-colon cancer activity and anti-inflammatory activity. The polyphenol-rich extract contained gallic acid monohydrate, 3-(3,4-dihydroxyphenyl)-propionic acid, methyl gallate, syringic acid, ellagic acid, and catechin. The polyphenol-rich extract exhibited both anti-inflammatory effects and inhibitory effects against human colon cancer cells [28]. The butanol fraction of *Ganoderma neo-japonicum* Imazeki showed the highest antioxidant activity against colonic carcinoma cells and the highest concentration of phenolic compounds [27]. A study demonstrated that the ethyl acetate fraction of *Pleurotus tuber-regium* mushroom sclerotium was rich in phenolic compounds, particularly chlorogenic acid and syringic acid. The ethyl acetate fraction demonstrated strong antioxidant activity by effectively scavenging free radicals in DPPH, ABTS, and hydrogen peroxide assays. In addition, it showed both in vivo and in vitro anti-angiogenic effects, suggesting a connection between the antioxidant properties and anti-angiogenic effects [77]. *Boletus edulis* is a medicinal mushroom with anti-colon cancer properties and its extract contains ellagic acid, rutin, and taxifolin at 532 μg/g, 465 μg/g, and 259 μg/g, respectively [46].

In general, medicinal mushrooms are rich in species-specific phenolic compounds and flavonoids that contribute to their anticancer activities. Medicinal mushrooms contain higher concentrations of these compounds than other non-medicinal mushrooms.

2.4.4. Other Bioactive Compounds in Edible and Medicinal Mushrooms

Apart from major bioactive compounds, *Termitomyces* species have cerebrosides, ergostanes, fatty acid amides, serine, saponins, and proteases [72]. Both the hexane and chloroform fractions of *Ganoderma neo-japonicum* showed anti-colon cancer effects. These fractions were enriched with sterols and terpenoids, respectively. 1,25-Dihydroxyvitamin D3 3-glycoside, an active metabolite of vitamin D3, and stellasterol demonstrated strong binding affinities to B-cell lymphoma 2 (Bcl-2) anti-apoptotic protein and increased cell death in cancer cells [27]. The main active compound of the hexane extract of *Pleurotus sajor-caju* was the sterol ergosta-5,7,22-trien-3β-ol. In silico docking confirmed the anti-colon cancer effect by demonstrating the credible interaction between the active compound and Bcl-2 protein, a key protein that prevents apoptosis [4]. 4-Acetyl-antroquinonol B extracted from *Antrodia camphorata* inhibits colorectal cancer by suppressing tumorigenesis and cancer stem-like phenotype [78]. 4-Acetyl-antroquinonol B in mushrooms enhances the response of colorectal cancer cells to the drug cetuximab by inhibiting the KRAS signaling pathway [79,80]. Alkaloids, folate, enzymes, ergosterol, organic acids, and selenium are also bioactive compounds in mushrooms. Antroquinonol, polysaccharide, cordycepin, hispolon, lectins, krestin, sulfated polysaccharides, lentinan, and Maitake D Fraction are some of the anticancer components present in medicinal mushrooms [78,81].

3. Mechanisms of Action Against Colon Cancer

Edible and medicinal mushrooms may exert their anticancer effects through immunomodulation, antioxidation, and direct cytotoxicity in cancer cells [82]. Medicinal mushrooms have been recognized as potential agents to fight against cancer. Bioactive compounds in medicinal mushrooms interfere with the cellular pathways of cancer cells and stimulate the immune system to fight against cancer cells [83,84]. Figure 2 illustrates the mechanisms of action of edible mushrooms in colon cancer management.

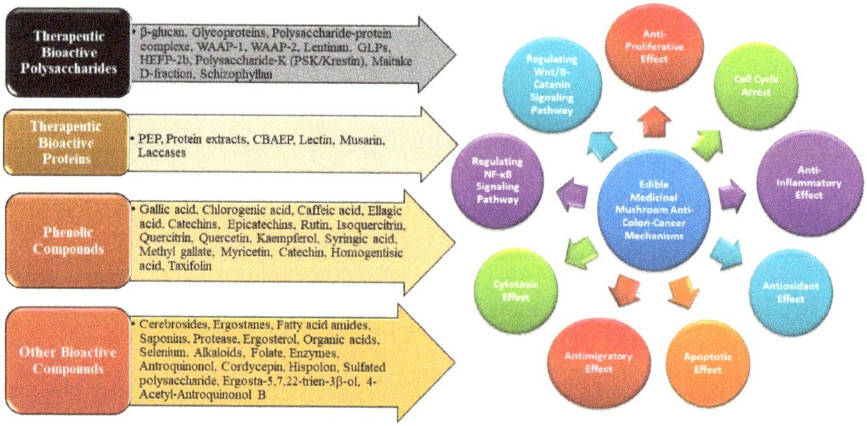

Figure 2. Anti-colon cancer mechanisms of edible and medicinal mushrooms. CBAEP—Cibacron blue affinity eluted protein; PEP—A novel bioactive protein in *Pleurotus eryngii*.

3.1. Anti-Colon Cancer Mechanism Through Anti-Proliferative Effects

3.1.1. Introduction to Anti-Proliferative Activity of Medicinal Mushrooms

Medicinal mushroom extracts and their bioactive compounds that exhibit anti-proliferative effects are emerging as alternative and complementary therapeutic strategies for treating colon cancer. Further research is required to optimize their bioactive component bioavailability and explain their molecular mechanisms.

Figure 3 illustrates the anti-proliferative effects of edible medicinal mushrooms against colon cancer. Several in vivo as well as in vitro studies further confirmed the anti-proliferative effects of edible medicinal mushrooms.

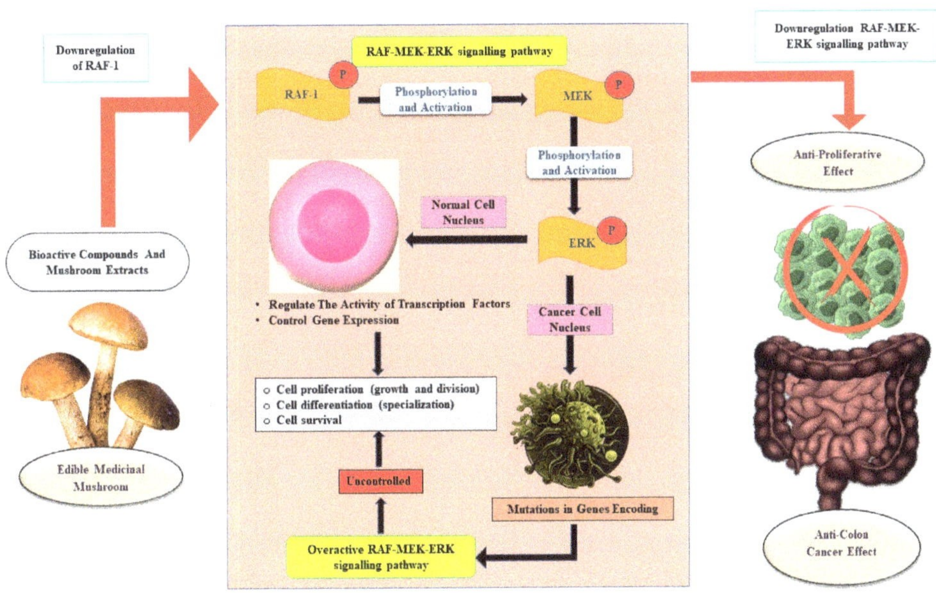

Figure 3. Anti-proliferative effects of edible and medicinal mushrooms against colon cancer. ERK—Extracellular signal-regulated kinase; MEK—Mitogen-activated protein kinase; RAF-1—RAF proto-oncogene serine/threonine-protein kinase.

3.1.2. In Vitro and In Vivo Studies of the Anti-Proliferative Activity of Medicinal Mushrooms

Aqueous extracts of *Auricularia polytricha*, *Macrolepiota procera*, and *Pleurotus ostreatus* showed an irreversible anti-proliferative effect against human colorectal adenocarcinoma cell line 205 (COLO-205) [19]. A study conducted on selenium-enriched *Pleurotus ostreatus* edible MM extract (SME) downregulated the expression of Raf-1 [20]. Raf-1 is a crucial protein kinase in the mitogen-activated protein kinase (MAPK) pathway that promotes growth, differentiation, and survival of cells. Due to gene mutation or under a constitutively active state of Raf-1, oncogenesis occurs, transforming normal cells into cancerous cells and increasing cancer cell proliferation in the colon [85]. The inhibition of colon cancer cell proliferation by Se-enriched mushrooms showed a dose-dependent nature. Purified selenomethionine mainly targeted the blocking of the RAF-MEK-ERK signaling pathway, which is involved in cell survival and proliferation [20]. Caproic acid, a medium-chain fatty acid found in *D. indusiata* edible MM, shows anti-proliferative effects against human colorectal [30]. Research conducted on the structural characterization and anti-colon cancer activity of *Agaricus bisporus* used a sequential extraction method (room temperature water, hot water, high-pressure hot water, dilute alkaline solution, and concentrated alkaline solution) to obtain five crude polysaccharides, and a homogeneous polysaccharide called 'WAAP-1' was purified from them using a DEAE Cellulose-52 column. WAAP-1 showed promising anticancer properties by significantly inhibiting the proliferation of HT-29 colon cancer cells [17].

A polyphenol-rich extract isolated from *Pleurotus eryngii* showed dose and time-dependent suppression of HCT-116 cell proliferation. The extract showed no inhibitory effect against normal human colonic myofibroblasts (CCD-18Co cells) [28]. PEP in *Pleurotus eryngii* showed dose and time-dependent suppression of HCT-116 and MC38 murine colon adenocarcinoma cell line [58]. The α-glucan fraction of *Pleurotus ostreatus* significantly decreased the growth of HT-29 cells in a dose-dependent manner [44]. A significant decrease in the number of precancerous lesions (aberrant crypt foci and microadenomas) was observed in mice fed the polysaccharide extract of *Pleurotus pulmonarius* fruiting bodies or mycelia. The reduction in precancerous lesions was due to decreased cell proliferation, increased cell death, and reduced levels of the inflammatory cytokine, tumor necrosis factor alpha (TNF-α) [86]. Aqueous extracts of *Auricularia polytricha, Macrolepiota procera,* and *Pleurotus ostreatus* demonstrated significant inhibition of the growth and proliferation of human colorectal adenocarcinoma cell line COLO-205 [19]. The methanolic extract of *Lignosus rhinocerotis* (Tiger milk mushroom) demonstrated strong anti-proliferative activity against HCT-116 cells, while the aqueous extract demonstrated weaker anti-proliferative activity. The IC_{50} values of the methanolic and aqueous extracts were 600 μg/mL and 1200 μg/mL, respectively [87]. *Boletus edulis* extracts showed dose-dependent and time-dependent reduction of Caco-2 cell viability. The IC_{50} values of Caco-2 cells at 24 h, 48 h, and 72 h incubation periods were >2000 μg/mL, 1880 $\pm$ 18 μg/mL, and 1509 $\pm$ 43 μg/mL, respectively, which demonstrated a significant anti-proliferative effect against human colon cancer cells [46].

3.1.3. Limitations and Future Directions of In Vitro and In Vivo Studies on the Anti-Proliferative Activity of Medicinal Mushrooms

Many studies are limited to in vitro and animal studies and suggest that bioactive compounds inhibit proliferation by targeting oncogenic pathways. They are not clinically validated, which raises a critical issue about translatability to humans. Furthermore, the bioavailability of these bioactive compounds remains a major hurdle. Many mushroom-derived polysaccharides and proteins demonstrate poor absorption and raise concerns related to high doses that may risk toxicity. The standardization of extraction methods is another critical issue. Using different extracts raises concerns about varied concentrations and the types of bioactive compounds extracted. Future research must prioritize human trials to determine the anti-colon cancer mechanism through anti-proliferative effects. In addition, optimizing delivery systems through nanoencapsulation and exploring synergistic combinations with chemotherapy could improve the anti-proliferative effects against colon cancer cells.

3.2. Anti-Colon Cancer Mechanism Through Cell Cycle Arrest

3.2.1. Introduction to Anti-Colon Cancer Mechanism Through Cell Cycle Arrest by Medicinal Mushrooms

Medicinal mushrooms demonstrate an anti-colon cancer mechanism through cell cycle arrest. The induction of cell cycle arrest in colon cancer cells is emerging as a promising approach due to the inability to maintain cellular homeostasis in cancer cells. Through various in vivo and in vitro studies, polysaccharides, polyphenols, proteins, and other bioactive compounds have demonstrated selective targeting of cancer cells by arresting their cell cycle progression.

3.2.2. In Vitro and In Vivo Studies on Cell Cycle Arrest by Medicinal Mushrooms

Figure 4 illustrates the cell cycle arrest mechanism of edible medicinal mushrooms in colon cancer. A polyphenol-rich extract isolated from edible *Pleurotus eryngii* induced cell cycle arrest in HCT-116 cells. The expression of cyclin B and cyclin E, which are essential proteins for cell cycle progression, was downregulated in the mitotic phase of the cancer cell cycle [28]. PEP in *Pleurotus eryngii* downregulates cyclin B, cyclin E, and Cyclin-dependent kinase (CDC-2) proteins and arrests the cell cycle, leading to apoptosis induction [58]. The *Lentinus edodes* alcohol precipitate (LAP) obtained from *L. edodes* caused cell cycle arrest at the G0/G1 phase and prevented COLO 205 xenografted cancer cell proliferation in nude mice. Only 1 mg/mL of LAP was sufficient to arrest 61.3% of COLO 205 cancer cells in the G0/G1 phase of the cell cycle [21]. BSGLWE significantly induced cell cycle arrest at the G2/M phase in HCT-116 cells. The arrested cell percentages at 5 mg/mL (23.7%) and 7.5 mg/mL (32.2%) were higher when compared with the control (12.0%). After BSGLWE treatment, the proportion of HCT-116 cells in the G0/G1 phase slightly decreased, and the proportion of cancer cells in the S phase was significantly decreased. The G2/M checkpoint was confirmed by measuring cyclin B1 and cyclin A2. The mRNA levels of cyclin B1 and cyclin A were significantly reduced while the mRNA level of cell cycle arrest protein cyclin-dependent kinase inhibitor 1 (p21) was increased [31]. HEFP-2b polysaccharide arrested the colon cancer cell cycle at its S phase and inhibited HCT-116 and DLD1 cell growth [25]. The anti-proliferative activity of SLNT against HT-29 cells was concentration-dependent. SLNT at a 1600 µg/mL concentration demonstrated the strongest activity against HT-29 cells and the cell viability was 35.5% [32]. The extract of *Pleurotus sajor-caju* prevented further dividing and multiplying of cancer cells by inducing cell cycle arrest in HCT-116 cells at the G2/M phase. In addition, the extract interfered with the p21/ tumor protein p53 (p53) cell cycle regulation pathway [4]. The extract of *Boletus edulis* interfered with the G0/G1 phase and reduced the proportion of cancer cells in the S phase [46].

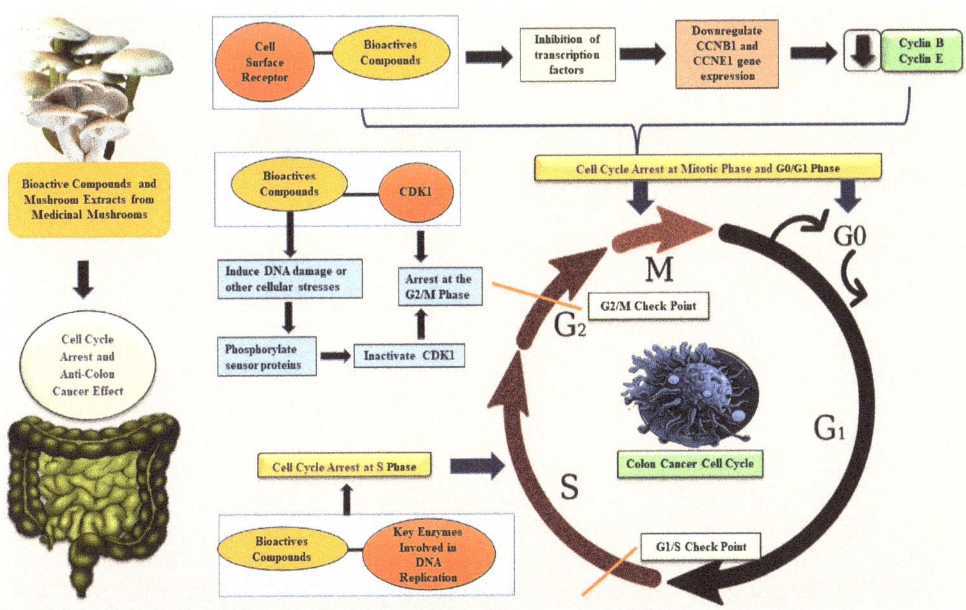

Figure 4. Cell cycle arrest induced by edible and medicinal mushrooms in colon cancer. CCNB1—Cyclin B1; CCNE1—Cyclin E1; CDK1—Cyclin-dependent kinase 1; DNA—Deoxyribonucleic acid.

3.3. Anti-Colon Cancer Mechanism Through Anti-Inflammatory Effects

Chronic inflammation, including prolonged exposure to pro-inflammatory cytokines, is a risk factor for colorectal cancer (CRC). Bioactive compounds in medicinal mushrooms suppress inflammation-related signaling pathways such as NF-κB, Wnt/β-catenin, prostaglandin E2 (PGE2), and COX-2. Further clinical research is required to capture their full potential in human oncology.

Figure 5 illustrates the anti-inflammatory effects of edible medicinal mushrooms against colon cancer. Polysaccharides purified from *Gloeostereum incarnatum* (GIPS) (30 mg/kg) significantly changed 89 of these cytokines (over 1.5-fold change). These altered cytokines were linked with inflammatory pathways, including the Wnt signaling pathway. Pro-inflammatory cytokines IL-1β, IL-6, and TNF-α were downregulated while IL-15 and IL-18 were promoted. GIPS suppressed matrix metallopeptidase-2 (MMP-2) enzyme [43,88]. The six families of interleukins (IL) (IL-1, -2, -6, -8, -10, -17) have their own ways of being involved in oncogenic or antitumor functions in colorectal cancer [89]. For instance, IL-1β promotes colon cancer cell proliferation and tumorigenesis and alters the tumor microenvironment [89–91].

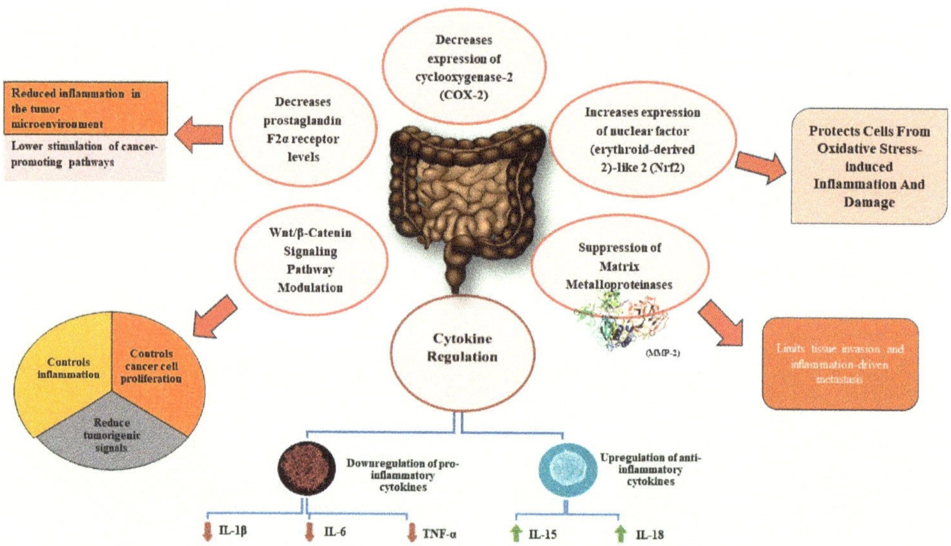

Figure 5. Anti-inflammatory effects of edible and medicinal mushrooms against colon cancer. COX-2—Cyclooxygenase-2; IL—Interleukin; MMP-2—Matrix metallopeptidase-2; Nrf2—Nuclear factor erythroid 2; TNF-α—Tumor necrosis factor alpha.

IL-6 promotes mitosis, proliferation, metastasis, migration, and angiogenesis and provides a microenvironment suitable for the metastasis of cancer cells [89]. TNF-α is associated with CRC as it promotes tumor growth, invasion, and metastasis [92,93]. IL-15 inhibits cancer cell proliferation and angiogenesis and promotes apoptosis [89,94,95]. IL-18 improves intestinal barrier integrity and improves the immune system by acting on NK cells [89,91,96]. A study investigated the anti-inflammatory effects of the extract of *Agaricus bisporus* against Caco-2 cells. The Caco-2 cells were stimulated with LPS and TNF-α. The extract decreased the cyclooxygenase-2 and prostaglandin F2α receptor levels in cells associated with inflammation and increased the expression of nuclear factor (erythroid-derived 2)-like 2 associated with cellular protection. The anti-inflammatory activity was further supported by reducing the production of IL-6 in Caco-2 cells [97].

3.4. Anti-Colon Cancer Mechanism Through Antioxidant Effects

During energy production in cells, free radicals (FR) and oxidants are formed, which are either reactive oxygen species (ROS) or reactive nitrogen species (RNS). The excess production of these FRs and oxidants could cause oxidative stress and an imbalance in the antioxidant system. This causes detrimental effects, including colon cancer [30,98–100]. Some edible MM species are rich sources of antioxidants [19]. The methanolic extracts of five different medicinal mushrooms (*Phellinus linteus, Cordyceps sinensis, L. edodes, Coprinus comatus,* and *G. lucidum)* caused a prooxidant/antioxidant imbalance in colon cancer cell lines, leading to oxidative stress. *Phellinus linteus* showed the highest phenolic content, flavonoid content, and DPPH scavenging capacity. It was determined that anti-migratory effects had a positive correlation with increased superoxide anion radical production ($O_2^{\bullet-}$) [18].

3.5. Anti-Colon Cancer Mechanism Through the NF-κB Signaling Pathway

NF-κB is a protein complex signaling pathway that may control the transcription of DNA, cytokine production, cell survival, and, in particular, inflammatory responses, cellular growth, and apoptosis, and it may be related to diseases such as cancer [101–103]. The activation of NF-κB in inflammatory bowel disease might induce cellular transformation, mediate cellular proliferation, and prevent the elimination of pre-neoplastic and fully malignant cells by upregulating anti-apoptosis proteins, increasing the risk of colorectal cancer [104]. However, NF-κB also may make a remarkable contribution to colon cancer progression through the regulation of the target genes that help in cell proliferation (Cyclin D1), angiogenesis (vascular endothelial growth factor (VEGF), IL-8, COX2), and metastasis (MMP9). It has been found that various edible and medicinal mushroom compounds have anti-colon cancer activity, which may help in cell proliferation, angiogenesis, and metastasis.

3.6. Anti-Colon Cancer Mechanism Through the Wnt/β-Catenin Signaling Pathway

The anti-colon cancer effects of GIPS were demonstrated in a mouse model (*Apc*^{MinC}/*Gpt* mice), where the growth, size, and number of tumors were reduced after an 8-week administration but body weight was not affected. GIPS (90 mg/kg) showed better results than GIPS (30 mg/kg). GIPS regulated the Wnt/β-catenin signaling pathway, which promoted cancer cell proliferation, and showed resistance to chemotherapy. GIPS downregulated the expression of β-catenin, Frizzled-7, WNT1, LRP5/6, MMP-2, and MMP-9 and upregulated the expression of DKK1 and Kremen-2. In addition, GIPS suppressed GSK-3β phosphorylation in colon cancer tissues [26]. Abnormal activation of the Wnt/β-catenin signaling pathway is one of the main features in CRC. This occurs due to the mutation of either the APC gene (Adenomatous Polyposis Coli) or the β-catenin gene. The Wnt/β-catenin signaling pathway has significant roles in maintaining stem cell homeostasis and intestinal function in the human body. Mutations in APC or the β-catenin gene promote the accumulation of β-catenin in the cell nucleus, which eventually promotes colon cancer cell proliferation and tumorigenesis. Therefore, targeting this pathway to treat colon cancer would be an effective strategy [105,106]. MMP-1 gene expression is also associated with colon cancer cell invasion and migration [107,108].

3.7. Anti-Colon Cancer Mechanism Through Apoptotic Effects

3.7.1. Initiation of Apoptotic Effects Through Intrinsic and Extrinsic Pathways

Table 1 illustrates the anti-colon cancer mechanism of edible medicinal mushrooms through apoptotic effects. Apoptosis is a well-regulated process that programs the death of cells. During colon cancer, the apoptotic pathway is dysregulated or inhibited through various mechanisms by the overexpression of antiapoptotic proteins or underexpression of proapoptotic proteins [109].The extrinsic pathway (death receptor-mediated) and intrinsic pathway (mitochondria-dependent) are the two primary pathways involved in initiating apoptosis [110].

Table 1. Anti-colon cancer mechanism of edible and medicinal mushroom through apoptotic effects.

Scientific Name	Common Name	Active Component/Fraction	Extraction Method	Experimental Model	Mechanism	References
Agaricus bisporus	White button mushroom	WAAP-1, WAAP-2	Polysaccharide extraction	HT-29 human colon cancer cells	Upregulation of Caspase-3, BAX, E-cadherin; downregulation of Bcl-2, Vimentin	[17,42]
		Microencapsulated polysaccharide extract	Polysaccharide extraction	Caco-2 human colon cancer cells	Downregulation of Bcl-2, TGF-β; upregulation of IkappaB-α; inhibits NF-κB signaling pathway	[40]
Lentinus edodes	Shiitake mushroom	LAP	Mycelium extract	COLO 205 human colon cancer cells	Upregulation of p53, p21/Cip1, p27/Kip1; downregulation of cyclins B and D3; induces apoptosis	[21]
		SLNT	Water extraction	HT-29 human colon cancer cells	Activation of mitochondrial apoptotic pathways, caspase-3 and caspase-8 activation, cytochrome c release, MMP loss	[32]
Pleurotus eryngii	King oyster mushroom	Polyphenol-rich extract, PEP (Protein)	Polyphenol extraction	HCT-116 human colon cancer cells	Upregulation of caspase-3, *caspase-8* p53, c-PARP; cleaves caspase-3; induces cell death	[28,58,59]
Pleurotus ostreatus	Oyster mushroom	α-Glucan	Polysaccharide extraction	HT-29 human colon cancer cells	Upregulation of BAX, cytochrome c; induces apoptosis	[44]
		Protein extract	Protein extraction	SW480 human colon cancer cells	Increased ROS production, glutathione depletion, loss of mitochondrial membrane potential	[60]
Pleurotus pulmonarius	Indian oyster mushroom	Polysaccharides	Fruiting body/mycelium extract	HT-29 human colon cancer cells	Regulation of Bcl-2, BAX, cytochrome C; Induces apoptosis	[86]
Volvariella volvacea	Straw mushroom	Protein extract	Protein extraction	SW480 human colon cancer cells	Increased SubG1 phase cells, loss of mitochondrial membrane potential	[60]
Agaricus blazei, Hericeum erinaceus, Grifola frondosa	-	Andosan™	Water extraction	Caco-2 human colon cancer cells	Induces early and late apoptosis	[35]
Ganoderma neo-japonicum	Reishi mushroom	Hexane and chloroform fractions	Fraction extraction	Colonic carcinoma cells	Induction of apoptosis	[27]

Table 1. *Cont.*

Scientific Name	Common Name	Active Component/Fraction	Extraction Method	Experimental Model	Mechanism	References
Ganoderma lucidum		BSGLWE	Water extraction	HCT-116 human colon cancer cells	Upregulation of NAG-1; downregulation of Bcl-2, pro-caspase-3, pro-caspase-9; cleaves PARP; caspase-8 activation	[31]
Pleurotus sajor-caju	-	PSC-hex	n-hexane extraction	-	ROS generation; p53, BAX, caspase-3 upregulation; Bcl-2 downregulation	[4]

Abbreviations: Andosan™—Water extracts of the mycelia of *Agaricus blazei* (82.4%), *Hericeum erinaceus* (14.7%), and *Grifola frondosa* (2.9%); BAX—Bcl-2-associated x, apoptosis regulator; Bcl-2—B-cell lymphoma 2; BSGLWE—Sporoderm-broken spores of *G. lucidum* water extract; Caco-2—Human colorectal adenocarcinoma cell line; COLO-205—Human colorectal adenocarcinoma cell line 205; HCT-116—Human colorectal carcinoma cell line 116; HT-29—Human colorectal adenocarcinoma cell line 29; IkappaB-α—NF-kappa-B inhibitor alpha; LAP—*Lentinus edodes* alcohol precipitate; NAG-1—Nonsteroidal anti-inflammatory drug-activated gene-1; NF-κB—Nuclear factor kappa-light-chain-enhancer of activated B cells; p21/Cip1 or CDKN1A—Cyclin-dependent kinase inhibitor 1; p27/Kip1 or CDKN1B—Cyclin-dependent kinase inhibitor 1B; p53—Tumor protein p53; ROS—Reactive oxygen species; PARP—Poly (ADP-ribose) polymerase; PSC-hex—n-hexane fraction of *Pleurotus sajor-caju*; SLNT—Water-extracted polysaccharide from *Lentinus edodes*; SW-480—Surg pathol-derived colorectal cancer cell line 480; TGF-β—Transforming growth factor beta; WAAP—A homogeneous neutral polysaccharide.

3.7.2. Apoptotic Effects of *Inonotus obliquus*, *Pleurotus ostreatus*, and *Agaricus bisporus*

Edible medicinal mushrooms have shown promising results against CRC by activating apoptotic pathways. IOWE inhibited the proliferation of HT-29 cells through apoptotic effects. IOWE decreased the level of Bcl-2, an anti-apoptotic protein, and increased the levels of Bcl-2-associated x (BAX), an apoptosis regulator, and caspase-3, triggering apoptosis [6]. Members of the Bcl-2 family of proteins are major contributors in CRC through colon cancer cell initiation, progression, and resistance to cancer treatments [111].

SME enhances the regulation of p53 and caspase-3 [20]. p53 protein regulates apoptosis. Notably, 60% of CRC patients have a mutation in this gene and show resistance to cancer treatments due to dysregulation of the apoptotic pathway [112]. Caspase-3 is a key enzyme involved in apoptosis that causes the death of cancer cells and acts as a tumor suppressor. However, recent studies have focused on the non-apoptotic function of caspase-3, which could contribute to tumor progression in the colon [113]. The α-glucan fraction of *Pleurotus ostreatus* induced apoptosis in HT-29 cells by increasing levels of BAX and cytosolic cytochrome-c proteins [44].

WAAP-1 promoted apoptosis of colon cancer cell line HT-29 and inhibited the epithelial-mesenchymal transition in HT-29 cells. The expression of caspase-3, BAX, and E-cadherin proteins was upregulated while Bcl-2 and Vimentin protein expression was downregulated [17]. WAAP-2 showed an apoptosis-inducing effect by arresting the HT-29 cell cycle at the S phase. Also, by upregulating caspase-3 and BAX protein expression and downregulating Bcl-2 protein expression, WAAP-2 promoted apoptosis. WAAP-2 induced the expression of E-cadherin and inhibited Vimentin expression. By inducing the expression of E-cadherin, the migration and invasion of colorectal cancer cells can be inhibited. By inhibiting Vimentin expression, the epithelial-mesenchymal transition can be affected, decreasing the invasive potential of CRC cells [42]. Natural killer cells treated with microencapsulated polysaccharide extract from *Agaricus bisporus* induced apoptosis in Caco-2 cells by downregulating the expression of genes that promote cell survival, such as Bcl-2 and transforming growth factor beta (TGF-*β*), and upregulating NF-kappa-B inhibitor alpha (IkappaB-α) gene expression, which inhibits the NF-κB signaling pathway [40].

3.7.3. Apoptotic Effects of *Pleurotus eryngii* and *Ganoderma lucidum*

A polyphenol-rich extract of *Pleurotus eryngii* caused apoptosis induction in HCT-116 cells. The molecular mechanisms involved downregulating cell cycle proteins and upregulating apoptosis proteins. The polyphenol extract upregulated caspase-3 and cleaved caspase-3 proteins, leading to cell death [28]. PEP upregulated apoptosis proteins such as p53 and c-poly (adp-ribose) polymerase (PARP), initiating cell death. The in vivo study conducted on PEP treatment of mice with transplanted colon cancer cells showed increasing levels of p21, p53, caspase-3, and cleaved caspase-3 in tumor tissues and suppressed tumor growth in mice. Increased cleavage of caspase-8 was observed, confirming the activation of the extrinsic apoptotic pathway. The activated caspase-8 then cleaved caspase-3 and PARP, ultimately leading to programmed cell death [58].

Apoptosis increased in colon cancer cells in a dose-dependent and time-dependent manner after BSGLWE treatment. BSGLWE treatment induced nuclear fragmentation, condensation, coagulation, and formation of apoptotic bodies in HCT-116 cells. BSGLWE upregulated survivin expression while downregulating Bcl-2 expression. In addition, Bcl-2, pro-caspase-3, pro-caspase-9, and cleaved PARP protein levels were altered. BSGLWE also induced death receptor-mediated apoptosis in HT-29 and HCT-116 cells through the activation and upregulation of caspase-8. Caspase-8 is a key initiator of the extrinsic apoptosis pathway. Subsequently, the upregulation of caspase-8 resulted in the cleavage and activation of caspase-3. This triggered programmed cell death in colon cancer cells. The expression of nonsteroidal anti-inflammatory drug-activated gene-1 (NAG-1) at both mRNA and protein levels was induced. Higher concentrations of BSGLWE (both 5 and 7.5 mg/mL) increased the secretion of NAG-1 into the cell culture medium [31]. NAG-1 induces various anticancer effects and its role in apoptosis is complex. NAG-1 acts as a tumor suppressor that inhibits cancer cell growth and promotes apoptosis. NAG-1 combined with the downregulation of Bcl-2 or related anti-apoptotic proteins enhances the apoptosis mechanism [114,115].

3.7.4. Apoptotic Effects of *Lentinus edodes* as a Promising Medicinal Mushroom

The combined therapeutic effect of *Lentinus edodes* mycelium extract (LEM) and 5-fluorouracil against human colon cancer cells in xenografted nude mice showed promising antitumor activity by inducing apoptosis. Many COLO 205 cancer cells underwent apoptosis when treated with concentrations greater than 15 mg/mL LAP for 15 h. At the same time and concentration, COLO 205 cells showed morphological changes and internucleosomal DNA degradation. Also, LAP enhanced the effectiveness of 5-fluorouracil. The synergistic effect of LEM and 5-fluorouracil increased the expression of cell cycle inhibitor proteins p53, p21/Cip1, and cyclin-dependent kinase inhibitor 1B (p27/Kip1) and decreased the expression of cyclins B and D3 in tumor cells [21].

SLNT affects tumor growth and induces apoptosis in a dose-dependent manner. SLNT at 1600 μg/mL resulted in 48.50% apoptotic cells, while the control group's value was 7.31%. AO/EB staining showed more apoptotic cells than normal cells. In DAPI staining, many SLNT-treated cells showed condensed chromatin, confirming the presence of apoptotic cells. SLNT treatment significantly reduced tumor size and weight. This mechanism was dose-dependent (0.2, 1.0, and 5.0 mg/kg inhibition percentages were 17.88%, 48.87%, and 57.90%, respectively), but the rates were lower than the positive control drug 5-FU (67.23% inhibition). SLNT activates mitochondrial apoptotic pathways, activates caspase-3, releases cytochrome c, and causes loss of mitochondrial membrane potential.

SLNT significantly increased the levels of TNF-α, a pro-inflammatory cytokine, and caspase-8 in HT-29 cells. TNF-α triggers apoptosis through an extrinsic pathway. Caspase-8 is one of the key components of the extrinsic pathway and acts as an initiator caspase.

The SLNT-induced apoptosis was dependent on capase-8 activation to some extent. This was further confirmed through using a caspase-8 inhibitor, which reduced the extent of apoptosis induced by SLNT. This suggested that caspase-8 played a crucial role in mediating the anti-colon cancer effects of the *L. edodes* polysaccharide [32].

3.7.5. Apoptotic Effects of Other Promising Medicinal Mushrooms in Anti-Colon Cancer Therapy

The hexane and chloroform fractions of *G. neo-japonicum* exert their anti-colon cancer activity against colonic carcinoma cells through the induction of apoptosis [27]. Polysaccharides extracted from *Pleurotus pulmonarius* induced dose-dependent apoptosis in human colon cancer cells. The extract regulated Bcl-2, BAX, and cytochrome C protein expression [86]. Andosan™ induced apoptosis in Caco-2 cells. Andosan™-treated groups showed significant increases in the abundance of both early and late apoptotic cells compared to untreated controls. The control group, 1.0% Andosan™, and 5.0% Andosan™ treated groups, respectively, showed 5.7% $\pm$ 1.5%, 15.3% $\pm$ 2.1%, and 35.6% $\pm$ 4.5% in early apoptosis, while late apoptosis showed 7.3% $\pm$ 2.1%, 35.6% $\pm$ 4.5%, and 39.7% $\pm$ 7.6% [35]. Protein extracts of *Calvatia lilacina, Pleurotus ostreatus*, and *Volvariella volvacea* demonstrated apoptosis induction in SW480 cells. *Pleurotus ostreatus* increased ROS production, depletion of glutathione, and loss of mitochondrial membrane potential, while other extracts dramatically increased the proportion of SW480 cells in the sub-G1 phase. *Volvariella volvacea* demonstrated a stronger apoptotic effect, increasing the proportion of SubG1 cells from 1.9% to 97.8% [60]. The n-hexane fraction of Pleurotus sajor-caju (PSC-hex) induced cancer cell apoptosis by breaking down the mitochondrial membrane potential in cancer cells, ROS generation, increasing the expression of p53, BAX, and caspase-3 proteins, and decreasing the expression of Bcl-2 [4].

3.8. Anti-Colon Cancer Mechanism Through Antimigratory Effects

Table 2 summarizes the anti-colon cancer mechanism through antimigratory effects. The antimigratory effects of methanolic extracts of *Phellinus linteus, Cordyceps sinensis, Lentinus edodes, Coprinus comatus*, and *Ganoderma lucidum* mushroom varieties were determined using the Transwell assay and immunofluorescence staining of β-catenin. *Phellinus linteus* showed significant antimigratory effects against both HCT-116 and SW-480 colon cancer cell lines. At the same time, all of the other mushroom extracts had significant anti-migratory effects against only the HCT-116 colon cancer cell line. The antimigratory effects had a positive correlation with both increased superoxide anion radical production ($O_2^{\bullet -}$) and reduced expression of β-catenin protein [18].

Table 2. Table 2. Anti-colon cancer mechanism through antimigratory effects and cytotoxic effects.

Scientific Name	Common Name	Active Compound/ Fraction	Experimental Model	Mechanism of Action	References
Phellinus linteus	N/A	Methanolic extract	HCT-116 and SW-480 colon cancer cell lines	Significant antimigratory effects against both cell lines; correlated with increased superoxide anion radical production and reduced β-catenin expression	[18]
Cordyceps sinensis	N/A	Methanolic extract	HCT-116 colon cancer cell line	Significant antimigratory effect; correlated with increased superoxide anion radical production and reduced β-catenin expression	[18]
Lentinus edodes	Shiitake mushroom	Methanolic extract	HCT-116 colon cancer cell line	Significant antimigratory effect; correlated with increased superoxide anion radical production and reduced β-catenin expression	[18]

Table 2. *Cont.*

Scientific Name	Common Name	Active Compound/ Fraction	Experimental Model	Mechanism of Action	References
Coprinus comatus	Shaggymane	Methanolic extract	HCT-116 colon cancer cell line	Significant antimigratory effect; correlated with increased superoxide anion radical production and reduced β-catenin expression	[18]
Ganoderma lucidum	Reishi mushroom	Methanolic extract	HCT-116 colon cancer cell line	Significant antimigratory effect; correlated with increased superoxide anion radical production and reduced β-catenin expression	[18]
Macrolepiota procera	Parasol mushroom	Ethanolic extract	COLO-205 human colon cancer cell line	Significant cytotoxic effect	[19]
Pleurotus ostreatus	Oyster mushroom	Aqueous extract	COLO-205 human colon cancer cell line	Significant cytotoxic effect	[19]
Hericium erinaceus	Lion's mane mushroom	n-hexane extract	HCT-116 human colon carcinoma cell line	Lower cytotoxicity level	[30]
Metacordyceps neogunnii	N/A	n-hexane extract	HCT-116 human colon carcinoma cell line	Highest anti-colon cancer effect (68.6 ± 3.6% cytotoxicity)	[30]
Dictyophora indusiata	Bamboo mushroom	n-hexane extract	HCT-116 human colon carcinoma cell line	Lower cytotoxicity level	[30]
Agaricus bisporus	Button mushroom	Microcapsulated polysaccharide extract	Caco-2 human colon cancer cell line	Increased CD16+CD56+ NK cell population; 74.09% cytotoxic activity; G0/G1 phase cell cycle arrest	[40]
Calvatia lilacina	N/A	Protein extract	SW480 and THP-1 cells	Concentration-dependent cytotoxicity; significant decrease in cell viability at high concentrations	[60]
Pleurotus ostreatus	Oyster mushroom	Protein extract	SW480 cell line	Concentration-dependent cytotoxicity; significant decrease in cell viability at high concentrations	[60]
Volvariella volvacea	Straw mushroom	Protein extract	SW480 cell line	Concentration-dependent cytotoxicity; significant decrease in cell viability at high concentrations	[60]
Ganoderma neo-japonicum	N/A	Hexane and chloroform fractions	Colonic carcinoma cells	Cytotoxic effect	[27]
Lentinus edodes	Shiitake mushroom	Alcohol precipitate	Colon epithelial cells	Selective cytotoxicity in cancerous cells	[21]
PSC-hex	N/A	Hexane extract	Colorectal cancer cells	Strong cytotoxicity (IC50 = 0.05 mg/mL)	[4]
Agaricus blazei, Hericeum erinaceus, Grifola frondosa	-	Andosan™	Caco-2 human colon cancer cell line	Concentration-dependent cytotoxic effects; cell viability reduction correlated with concentration	[35]

Abbreviations: Andosan™—Water extracts of the mycelia of Agaricus blazei (82.4%), Hericeum erinaceus (14.7%) and Grifola frondosa (2.9%); Caco-2—Human colorectal adenocarcinoma cell line; COLO-205—Human colorectal adenocarcinoma cell line 205; HCT-116—Human colorectal carcinoma cell line 116; HT-29—human colorectal adenocarcinoma cell line 29; NK cells—Natural killer cells; SW-480—Surg pathol-derived colorectal cancer cell line 480; THP-1—Tohoku Hospital Pediatrics-1 human monocytic leukemia cell line.

3.9. Anti-Colon Cancer Mechanism Through Cytotoxic Effecst

Table 2 summarizes the anti-colon cancer mechanism through cytotoxic effects. The ethanolic extract of *Macrolepiota procera* and aqueous extract of *Pleurotus ostreatus* showed significant cytotoxic effects against the COLO-205 cell line [19]. The n-hexane extracts of *Hericium erinaceus*, *Metacordyceps neogunnii*, and *Dictyophora indusiata* MMs showed dose-dependent anticancer activity. *M. neogunnii* showed the highest cytotoxicity level (68.6 ± 3.6%) while *H. erinaceus* (18.3 ± 1.7%) and *D. indusiata* (19.3 ± 3.2%) demonstrated lower cytotoxicity levels against HCT-116 cells at 100 μg/mL [30]. The microcapsulated polysaccharide extract from *A. bisporus* significantly increased the CD16+CD56+ NK cell population, showing 74.09% cytotoxic activity against the Caco-2 cell line. Many treated cancer cells were arrested at the G0/G1 phase [40].

The Andosan™ concentration showed a strong inverse correlation with Caco-2 cell viability. Even the lowest concentration, 0.5%, showed a 14% reduction while the highest concentration, 5%, showed 90% cell viability reduction [35]. Protein extracts of *Calvatia lilacina*, *Pleurotus ostreatus*, and *Volvariella volvacea* demonstrated concentration-dependent cytotoxicity against SW480 and THP-1 cells. After 24 h of treatment with 500 μg/mL, all three protein extracts significantly decreased the cell viability of SW480 cells (8%, 2%, and 7%, respectively). *Pleurotus ostreatus* showed concentration-dependent cell viability reduction. The viability decreased to 39%, 10%, and 7% at 10 and 25 μg/mL, 50 μg/mL, and 100 μg/mL, respectively. *Volvariella volvacea* decreased cell viability by 20% at 10 μg/mL. The results suggested that the type of extract and the concentration used influence the degree of cytotoxicity [60]. PSC-hex demonstrated a strong cytotoxicity effect against colorectal cancer cells, with an IC_{50} value of 0.05 mg/mL [4]. Regardless of the tumor protein p53 status, the combined effect of LEM and 5-fluorouracil enhanced the cytotoxic effect. LAP demonstrated selective cytotoxicity against colon epithelial cells and less cytotoxicity against non-cancerous colon epithelial cells [21]. The hexane and chloroform fractions of *Ganoderma neo-japonicum* exerted their anti-colon cancer activity against colonic carcinoma cells through cytotoxic effects [27].

3.10. Anti-Colon Cancer Mechanism Through Gene Modulation

The bioactive compounds in medicinal mushrooms also exert their anti-colon cancer effects through gene modulation. This mainly involves critical genes such as Baculoviral IAP Repeat-Containing 5 (BIRC5), human telomerase reverse transcriptase (hTERT), hypoxia-inducible factor-1 alpha (HIF-1α or HIF1A), and Multidrug Resistance Protein 1 (MDR1). The following sections delve into the anti-colon cancer mechanism through gene modulation.

3.10.1. Anti-Colon Cancer Mechanism Through BIRC5 (Survivin) Modulation

BIRC5, which is also known as survivin, plays a significant role as an inhibitor of the apoptosis protein (IAP) family and as an immune-related gene. It functions as an inhibitor of apoptosis (IAP) and regulates cell division by functioning as a chromosomal passenger protein. Overexpression of BIRC5 may show relationships with various types of cancer, particularly human colon cancer. BIRC5 interacts with other molecular pathways, such as Wnt/β-catenin and p53, and supports tumor growth and evasion of cell death [116,117].

Polysaccharides found in medicinal mushrooms such as *Ganoderma lucidum*, *Phellinus linteus*, and *Lentinus edodes* indirectly downregulate BIRC5 by inhibiting the NF-κB and PI3K/AKT/mTOR pathways and direct suppression at transcriptional and translational levels. Proteomic analyses conducted during research studies confirmed that mushroom extracts reduced BIRC5 levels in colon cancer cell lines. This was correlated with decreased cell proliferation and increased apoptosis. Triterpenoids present in *Ganoderma lucidum* suppressed BIRC5 expression through blocking STAT3 signaling, a protein-mediated pathway of gene expression that upregulates survivin expression. Due to this, the caspases were activated, leading to apoptosis in colon cancer cells. Lectins and proteins present in *Ganoderma lucidum* induced the p53-mediated suppression of BIRC5, promoting cancer cell cycle arrest and programmed cell death. Overall, downregulation of BIRC5 inhibits the colon cancer survival rate through inducing programmed cell death, restricting cancer cell proliferation, and enhancing chemo sensitivity [118–120]. A study conducted on the bioactive compounds present in a mixture of *Trametes versicolor*, *Ganoderma lucidum*, and *Dioscorea opposite* extract interfered with BIRC5 through proteomic and metabolic reprogramming.

The mechanism of interference with BIRC5 mainly included the downregulation of BIRC5 and indirect suppression of BIRC5 through metabolic and translational reprogramming. Direct downregulation of BIRC5 was clearly demonstrated in the tandem mass tag (TMT) proteomics analysis. BIRC5 inhibition promotes apoptosis in cancer cells. BIRC5 production was indirectly reduced through disrupting key translation-related protein synthesis such as EIF4 family proteins. The mushroom extract also created an unfavorable environment for cancer cells and suppressed the anti-apoptotic mechanism. These studies suggested that the bioactive compounds in the mushroom extract downregulated BIRC5 and promoted the apoptosis of cancer cells [121].

3.10.2. Anti-Colon Cancer Mechanism Through Telomerase Reverse Transcriptase Modulation

hTERT is the catalytic subunit of telomerase, which is responsible for maintaining the length of telomeres. This leads to shortening of telomeres during each cell division and eventually leads to apoptosis. However, in colon cancer cells, hTERT reactivates and allows proliferation. A study conducted on hTERT proteins in colorectal cancer investigated the expression and tumorigenesis activity of hTERT, which is more frequent in colorectal tumors. Certain clinicopathological features were also associated with hTERT expression. Immunohistochemistry showed that increased levels of hTERT were associated with tumor behavior in colon cancer [122,123]. A study conducted using the combined extract of both *G. lucidum* spores and *Sanghuangporus vaninii* demonstrated that TERT mediated the Wnt signaling pathway by downregulating hTERT [124]. A study conducted using *Ganoderma tsugae* and its fungal immunomodulatory protein (FIP) inhibited the activity of hTERT. Many studies have been conducted to investigate the effect of hTERT in human lung carcinoma cells rather than in human colon cancer cells. Therefore, more studies on colorectal cancer and hTERT should be conducted to identify and understand the mechanisms and specific bioactive compounds present in medicinal mushrooms. Fungal immunomodulatory proteins transcriptionally downregulated hTERT, leading to telomerase inhibition in the A549 human lung adenocarcinoma cell line [125].

3.10.3. Anti-Colon Cancer Mechanism Through Hypoxia-Inducible Factor 1 Alpha Modulation

HIF-1α plays a significant role in the cellular response to low oxygen levels. It functions as a key transcriptional regulator. HIF-1α promotes angiogenesis, metabolic adaptation, and cell survival, leading to colon cancer aggressiveness and tumor growth in the human body through hypoxia-mediated pathways and under normoxic conditions. A study was conducted to investigate HIF-1α-driven colon cancer proliferation, demonstrating that overexpression of HIF-1α enhances tumor growth in colon cancer cells [126,127].

G. lucidum is a medicinal mushroom that has triterpenes and polysaccharides as bioactive compounds. These compounds exhibit antitumor activity under hypoxic conditions by inhibiting HIF-1α. *Antrodia camphorata* also suppresses HIF-1α expression and helps to reduce tumor hypoxia adaptation. The bioactive compounds in these mushrooms had the ability to interfere with tumor hypoxia signaling, which is a critical pathway in colon cancer progression [128]. *Ganoderma lucidum* (Lingzhi or Reishi) aqueous mushroom extract (GLE) plays a role on hypoxia-induced responses that include HIF-1α modulation. The results demonstrated that the extract dramatically reduced hypoxia-induced HIF-1α stabilization in rat thymocytes. Generally, HIF-1α levels increase under hypoxic conditions. However, GLE treatment suppressed this upregulation of HIF-1α and contributed to the restoration of redox balance and reduction of inflammatory responses [129].

3.10.4. Anti-Colon Cancer Mechanism Through Multidrug Resistance Protein 1 Modulation

MDR1, also known as ABCB1 or P-glycoprotein, is an ATP-binding cassette (ABC) transporter. MDR1 functions as an efflux pump that expels chemotherapeutic agents out of cancer treated cells. MDR1 decreases intracellular drug accumulation and reduces the effectiveness of cancer treatments. Increased expression of MDR1 correlates with poor prognosis, and reduced response to therapy has become a major issue in treatments [130].

Mushroom-derived bioactive compounds downregulate MDR1 expression by inhibiting the NF-κB, AP-1, and PI3K/Akt pathways. These compounds can block MDR1 efflux activity and enhance intracellular drug accumulation. *G. lucidum* has triterpenoids such as ganoderic acids, which inhibit NF-κB and MAPK signaling and suppress MDR1. *T. versicolor* contains PSK and polysaccharide-peptide (PSP), which decrease MDR1-mediated resistance by inhibiting PI3K/Akt and immunomodulation. *Phellinus linteus* contains hispidin and polysaccharides that downregulate MDR1 via ROS-mediated pathways [78]. A study conducted on *G. lucidum* identified polysaccharides, triterpenes (ganoderic acids), and other metabolites as major bioactive components in *G. lucidum*. The bioactive compounds modulate the NF-κB and STAT3 signaling pathways. NF-κB is identified as a transcriptional regulator of MDR1 and, by suppressing the activation of NF-κB, indirectly downregulates the expression of MDR1, which may be very valuable for chemo treatments. MDR1 downregulation decreases the expression of pro-inflammatory cytokines because chronic inflammation promotes MDR1 upregulation [23]. Ganoderic acid, a bioactive triterpenoid derived from *G. lucidum*, regulated MDR1 expression in the surg pathol-derived colorectal cancer cell line 620 (SW620/Ad300) cells. These compounds inhibit NF-κB activation, which indirectly downregulates MDR1 expression [131].

4. Challenges and Limitations

Edible medicinal mushrooms in colon cancer therapy face several challenges and limitations, particularly, limited clinical evidence, variability in the composition of mushroom species, mushroom cultivation and preparation techniques, potential for drug interactions, quality control and standardization, limited understanding of mechanisms, potential side effects, and ethical considerations. While many edible medicinal mushrooms demonstrate anti-colon cancer properties, the specific mechanisms are yet to be discovered. Establishing consistent efficacy, quality, and safety is critical in mushroom-derived products. Well-designed clinical trials are crucial before using medicinal mushrooms in clinical practice. Mushroom-derived drug development and delivery systems require new modifications to improve their pharmacokinetics and therapeutic efficacy. It is necessary to develop a clear regulatory framework for both patients and healthcare providers and to address ethical concerns [36,132–134].

5. Future Directions and Research Opportunities

Globally, colon cancer has become one of the most significant causes of mortality. Conventional therapies for cancers have significant adverse impacts on patients and their limitations encourage the exploration of alternative therapies. Edible MMs have gained significant attention in recent years for their potent anticancer properties. Further research should be focused on mechanistic, translational, and clinical studies to capture their full potential to treat colon cancer. Research in this field should prioritize bioactive compound isolation, identifying their mechanisms of action, developing synergistic therapeutic effects, and standardizing delivery systems. MMs improve the quality of life of cancer patients by preventing lymph node metastasis and decreasing the side effects of chemotherapy [82]. It is important to continue the identification and characterization of different anti-colon cancer compounds in mushroom species and understand their mechanisms of action. A deeper understanding of how these bioactive compounds exert their effects is important to developing effective and safer anti-colon cancer therapeutics. Novel drugs can be formulated using these active compounds but they need to be confirmed through rigorous human clinical trials. Combining mushroom-derived compounds with conventional cancer treatments may offer a synergistic approach to fight against colon cancers [82,135,136].

As strong suggestions for future improvements in this field, AI-driven bioactive compound discovery and analysis could be used to predict synergistic mushroom compound combinations to treat colorectal cancer-specific pathways [137]. Engineered mushrooms through clustered regularly interspaced short palindromic repeats (CRISPR) gene editing technology can be used to develop mushroom species that can overexpress anti-CRC compounds [138]. Oral mushroom spore-based vaccines could be introduced as a new concept in vaccine development, where people with colorectal cancers could use spores to deliver CRC-specific antigens to the gut. Eventually this will stimulate an immune response in the gut to fight against colorectal cancer [139–141].

6. Conclusions

Colon cancer has become a major global health issue. The search for novel, affordable, and effective treatments with minimum side effects is demanding. This review comprehensively explores the anti-colon cancer properties of edible medicinal fungi. It is evident that bioactive compounds derived from medicinal mushroom species show significant effects against colon cancer. These natural compounds exert their actions through various mechanisms, demonstrating anti-proliferative, pro-apoptotic, and anti-metastatic effects in preclinical models. Also, a synergistic effect can be gained by combining medicinal mushrooms with standard chemotherapy or combining multiple mushroom species. While current research is promising, well-designed human clinical trials are crucial to confirm their effectiveness and safety. Improving extraction methods and formulations could further enhance their therapeutic benefits.

Author Contributions: J.B. and T.I.—Investigation and writing of the original draft; B.X.: Conceptualization, Funding acquisition, Supervision, writing—review and editing. All authors have read and agreed to the published version of the manuscript.

Funding: This work was jointly supported by the Guangdong Higher Education Upgrading Plan (2021–2025) under grant nos. UICR0400015-24B and UICR0400016-24B at Beijing Normal-Hong Kong Baptist University, Zhuhai, China.

Conflicts of Interest: The authors declare no conflict of interest.

Abbreviations

The following abbreviations are used in this manuscript:

ABC	ATP-binding cassette
Andosan™	Water extracts of the mycelia of *Agaricus blazei* (82.4%), *Hericeum erinaceus* (14.7%), and *Grifola frondosa* (2.9%)
BAX	Bcl-2-associated x, apoptosis regulator
Bcl-2	B-cell lymphoma 2
BIRC5	Baculoviral IAP repeat-containing 5
BSGLWE	Sporoderm-broken spores of *G. lucidum* water extract
Caco-2	Human colorectal adenocarcinoma cell line
CBAEP	Cibacron blue affinity eluted protein
CCNB1	Cyclin B1
CCNE1	Cyclin E1

CDK1	Cyclin-dependent kinase 1
COLO-205	Human colorectal adenocarcinoma cell line 205
COX-2	Cyclooxygenase-2
CRC	Colorectal cancer
CRISPR	Clustered regularly interspaced short palindromic repeats
DNA	Deoxyribonucleic acid
ERK	Extracellular signal-regulated kinase
FIP	Fungal immunomodulatory protein
GIPS	Polysaccharides purified from *Gloeostereum incarnatum*
GLE	*Ganoderma Lucidum* extract
GLP	*Ganoderma lucidum* polysaccharides
GLSF	*Ganoderma lucidum* extract containg spores and fruiting bodies (30:8 ratio)
HCT-116	Human colorectal carcinoma cell line 116
HIF-1α or HIF1A	Hypoxia-inducible factor-1 alpha
HT-29	Human colorectal adenocarcinoma cell line 29
hTERT	Human telomerase reverse transcriptase
IkappaB-α	NF-kappa-B inhibitor alpha
IL	Interleukin
LAP	*Lentinus edodes* alcohol precipitate
LEM	*Lentinus edodes* mycelium extracts
MDR1	Multidrug resistance protein 1
MEK	Mitogen-activated protein kinase
MM	Medicinal mushrooms
MMP-2	Matrix metallopeptidase-2
MRP1	Multidrug resistance-associated protein 1
NAG-1	Nonsteroidal anti-inflammatory drug-activated gene-1
NF-κB	Nuclear factor kappa-light-chain-enhancer of activated B cells
NK cells	Natural killer cells
Nrf2	Nuclear factor erythroid 2
p21/Cip1 or CDKN1A	Cyclin-dependent kinase inhibitor 1
p27/Kip1 or CDKN1B	Cyclin-dependent kinase inhibitor 1B
p53	Tumor protein p53
PARP	Poly (ADP-ribose) polymerase
PSC-hex	n-hexane fraction of *Pleurotus sajor-caju*
RAF-1	RAF proto-oncogene serine/threonine-protein kinase
ROS	Reactive oxygen species
SLNT	Water-extracted polysaccharide from *Lentinus edodes*
SW-480	Surg pathol-derived colorectal cancer cell line 480
SW620/Ad300	Surg pathol-derived colorectal cancer cell line 620
TGF-β	Transforming growth factor beta
THP-1	Tohoku Hospital Pediatrics-1 human monocytic leukemia cell line
TNF-α	Tumor necrosis factor alpha
WAAP	A homogeneous neutral polysaccharide

References

1. Chaturvedi, V.K.; Agarwal, S.; Gupta, K.K.; Ramteke, P.W.; Singh, M.P. Medicinal mushroom: Boon for therapeutic applications. *3 Biotech.* **2018**, *8*, 334. [CrossRef] [PubMed]
2. Valverde, M.E.; Hernández-Pérez, T.; Paredes-López, O. Edible mushrooms: Improving human health and promoting quality life. *Int. J. Microbiol.* **2015**, *2015*, 376387. [CrossRef] [PubMed]
3. Chang, S.T.; Wasser, S.P. The role of culinary-medicinal mushrooms on human welfare with a pyramid model for human health. *Int. J. Med. Mushrooms.* **2012**, *14*, 95–134. [CrossRef] [PubMed]

4. Finimundy, T.C.; Abreu, R.M.V.; Bonetto, N.; Scariot, F.J.; Dillon, A.J.P.; Echeverrigaray, S.; Barros, L.; Ferreira, I.C.F.R.; Henriques, J.A.P.; Roesch-Ely, M. Apoptosis induction by *Pleurotus sajor-caju* (Fr.) Singer extracts on colorectal cancer cell lines. *Food Chem. Toxicol.* **2018**, *112*, 383–392. [CrossRef]
5. Hu, D.D.; Zhang, R.Y.; Zhang, G.Q.; Wang, H.X.; Ng, T.B. A laccase with antiproliferative activity against tumor cells from an edible mushroom, white common Agrocybe cylindracea. *Phytomedicine* **2011**, *18*, 374–379. [CrossRef]
6. Lee, S.H.; Hwang, H.S.; Yun, J.W. Antitumor activity of water extract of a mushroom, *Inonotus obliquus*, against HT-29 human colon cancer cells. *Phytother. Res.* **2009**, *23*, 1784–1789. [CrossRef]
7. Torre, L.A.; Siegel, R.L.; Ward, E.M.; Jemal, A. Global cancer incidence and mortality rates and trends--an update. *Cancer Epidemiol. Biomark. Prev.* **2016**, *25*, 16–27. [CrossRef]
8. Rawla, P.; Sunkara, T.; Barsouk, A. Epidemiology of colorectal cancer: Incidence, mortality, survival, and risk factors. *Przeglad Gastroenterologiczny.* **2019**, *14*, 89–103. [CrossRef]
9. Singha, K.; Hor, P.K.; Soren, J.P.; Mondal, J.; Mondal, K.C.; Pati, B.R.; Mohapatra, P.K.D. Exploration of bioactive prospects of a polysaccharide fraction from *Termitomyces heimii* against colorectal cancer and broad-spectrum bacteria. *Bioact. Carbohydr. Diet. Fibre* **2021**, *25*, 100255. [CrossRef]
10. Abdel-Azeem, A.M.; Abdel-Azeem, M.A.; Khalil, W.F. Endophytic fungi as a new source of antirheumatoid metabolites. In *Bioactive Food as Dietary Interventions for Arthritis and Related Inflammatory Diseases*, 2nd ed.; Academic Press: Cambridge, MA, USA, 2019; pp. 355–384. [CrossRef]
11. Hamza, A.; Mylarapu, A.; Krishna, K.V.; Kumar, D.S. An insight into the nutritional and medicinal value of edible mushrooms: A natural treasury for human health. *J. Biotechnol.* **2024**, *381*, 86–99. [CrossRef]
12. Frljak, J.; Mulabećirović, A.H.; Isaković, S.; Karahmet, E.; Toroman, A.; Frljak, J.; Mulabećirović, A.H.; Isaković, S.; Karahmet, E.; Toroman, A. Biological active components of selected medical fungi. *Open J. Prev. Med.* **2021**, *11*, 9–22. [CrossRef]
13. Kumar, K.; Mehra, R.; Guiné, R.P.F.; Lima, M.J.; Kumar, N.; Kaushik, R.; Ahmed, N.; Yadav, A.N.; Kumar, H. Edible mushrooms: A comprehensive review on bioactive compounds with health benefits and processing aspects. *Foods* **2021**, *10*, 2996. [CrossRef] [PubMed]
14. Khatun, S.; Islam, A.; Cakilcioglu, U.; Chatterjee, N.C. Research on mushroom as a potential source of nutraceuticals: A review on Indian perspective. *J. Exp. Agric Int.* **2012**, *2*, 47–73. [CrossRef]
15. He, Z.; Lin, J.; He, Y.; Liu, S. Polysaccharide-peptide from *Trametes versicolor*: The potential medicine for colorectal cancer treatment. *Biomedicines* **2022**, *10*, 2841. [CrossRef]
16. Sadowska, A.; Zapora, E.; Sawicka, D.; Niemirowicz-Laskowska, K.; Surażyński, A.; Sułkowska-Ziaja, K.; Kała, K.; Stocki, M.; Wołkowycki, M.; Bakier, S.; et al. *Heterobasidion annosum* induces apoptosis in DLD-1 cells and decreases colon cancer growth in in vivo model. *Int. J. Mol. Sci.* **2020**, *21*, 3447. [CrossRef]
17. Zhang, N.; Liu, Y.; Tang, F.Y.; Yang, L.Y.; Wang, J.H. Structural characterization and in vitro anti-colon cancer activity of a homogeneous polysaccharide from *Agaricus bisporus*. *Int. J. Biol. Macromol.* **2023**, *251*, 126410. [CrossRef]
18. Šeklić, D.S.; Stanković, M.S.; Milutinović, M.G.; Topuzović, M.D.; Štajn, A.Š.; Marković, S.D. Cytotoxic, antimigratory, pro-and antioxidative activities of extracts from medicinal mushrooms on colon cancer cell lines. *Arch. Biol. Sci.* **2016**, *68*, 93–105. [CrossRef]
19. Arora, S.; Goyal, S.; Balani, J.; Tandon, S. Enhanced antiproliferative effects of aqueous extracts of some medicinal mushrooms on colon cancer cells. *Int. J. Med. Mushrooms* **2013**, *15*, 301–314. [CrossRef]
20. Fekry, T.; Salem, M.F.; Abd-Elaziz, A.A.; Muawia, S.; Naguib, Y.M.; Khalil, H. Anticancer properties of selenium-enriched oyster culinary-medicinal mushroom, *Pleurotus ostreatus* (Agaricomycetes), in colon cancer in vitro. *Int. J. Med. Mushrooms* **2022**, *24*, 1–20. [CrossRef]
21. Wu, C.H.; Wu, C.C.; Ho, Y.S. Antitumor activity of combination treatment of *Lentinus edodes* mycelium extracts with 5-Fluorouracil against human colon cancer cells xenografted in nude mice. *J. Cancer Mol.* **2007**, *3*, 15–22. [CrossRef]
22. Liu, N.; Zou, S.; Xie, C.; Meng, Y.; Xu, X. Effect of the β-glucan from *Lentinus edodes* on colitis-associated colorectal cancer and gut microbiota. *Carbohydr. Polym.* **2023**, *316*, 121069. [CrossRef] [PubMed]
23. Liu, M.M.; Liu, T.; Yeung, S.; Wang, Z.; Andresen, B.; Parsa, C.; Orlando, R.; Zhou, B.; Wu, W.; Li, X.; et al. Inhibitory activity of medicinal mushroom *Ganoderma lucidum* on colorectal cancer by attenuating inflammation. *Precis. Clin. Med.* **2021**, *4*, 231–245. [CrossRef] [PubMed]
24. Wachtel-Galor, S.; Yuen, J.; Buswell, J.A.; Benzie, I.F.F. Ganoderma lucidum (Lingzhi or Reishi). In *Herbal Medicine: Biomolecular and Clinical Aspects*, 2nd ed.; CRC Press: Boca Raton, FL, USA, 2011; pp. 175–199. Available online: https://www.ncbi.nlm.nih.gov/books/NBK92757/ (accessed on 15th March 2025).
25. Liu, J.Y.; Hou, X.X.; Li, Z.Y.; Shan, S.H.; Chang, M.C.; Feng, C.P.; Wei, Y. Isolation and structural characterization of a novel polysaccharide from *Hericium erinaceus* fruiting bodies and its arrest of the cell cycle at S-phage in colon cancer cells. *Int. J. Biol. Macromol.* **2020**, *157*, 288–295. [CrossRef]

26. He, J.; Yang, A.; Zhao, X.; Liu, Y.; Liu, S.; Wang, D. Anti-colon cancer activity of water-soluble polysaccharides extracted from *Gloeostereum incarnatum* via Wnt/β-catenin signaling pathway. *Food Sci. Hum. Well.* **2021**, *10*, 460–470. [CrossRef]
27. Lau, M.F.; Chua, K.H.; Sabaratnam, V.; Kuppusamy, U.R. In vitro and in silico anticancer evaluation of a medicinal mushroom, *Ganoderma neo-japonicum* Imazeki, against human colonic carcinoma cells. *Biotech. Appl. Biochem.* **2021**, *68*, 902–917. [CrossRef]
28. Hu, Q.; Yuan, B.; Xiao, H.; Zhao, L.; Wu, X.; Rakariyatham, K.; Zhong, L.; Han, Y.; Muinde Kimatu, B.; Yang, W. Polyphenols-rich extract from *Pleurotus eryngii* with growth inhibitory of HCT116 colon cancer cells and anti-inflammatory function in RAW264.7 cells. *Food Funct.* **2018**, *9*, 1601–1611. [CrossRef]
29. Peng, H.; Shahidi, F. Bioactive compounds and bioactive properties of Chaga (*Inonotus obliquus*) mushroom: A review. *J. Food Bioact.* **2020**, *12*, 9–75. [CrossRef]
30. Daba, G.M.; Elkhateeb, W.A.; El-Dein, A.N.; Ahmed, E.F.; El Hagrassi, A.M.; Fayad, W.; Wen, T.C. Therapeutic potentials of n-hexane extracts of the three medicinal mushrooms regarding their anti-colon cancer, antioxidant, and hypocholesterolemic capabilities. *Biodiversitas* **2020**, *21*, 2437–2445. [CrossRef]
31. Na, K.; Li, K.; Sang, T.; Wu, K.; Wang, Y.; Wang, X. Anticarcinogenic effects of water extract of sporoderm-broken spores of *Ganoderma lucidum* on colorectal cancer in vitro and in vivo. *Int. J. Oncol.* **2017**, *50*, 1541–1554. [CrossRef]
32. Wang, J.; Li, W.; Huang, X.; Liu, Y.; Li, Q.; Zheng, Z.; Wang, K. A polysaccharide from *Lentinus edodes* inhibits human colon cancer cell proliferation and suppresses tumor growth in athymic nude mice. *Oncotarget* **2017**, *8*, 610–623. [CrossRef]
33. Bisen, P.S.; Baghel, R.K.; Sanodiya, B.S.; Thakur, G.S.; Prasad, G.B.K.S. *Lentinus edodes*: A macrofungus with pharmacological activities. *Curr. Med. Chem.* **2010**, *17*, 2419–2430. [CrossRef] [PubMed]
34. Firenzuoli, F.; Gori, L.; Lombardo, G. The medicinal mushroom *Agaricus blazei* Murrill: Review of literature and pharmaco-toxicological problems. *Evid.-Based Complement. Altern. Med.* **2007**, *5*, 3–15. [CrossRef] [PubMed]
35. Hetland, G.; Eide, D.M.; Tangen, J.M.; Haugen, M.H.; Mirlashari, M.R.; Paulsen, J.E. The *Agaricus blaze*-based mushroom extract, Andosan^{TM}, protects against intestinal tumorigenesis in the A/J Min/+ mouse. *PLoS ONE* **2016**, *11*, e0167754. [CrossRef] [PubMed]
36. Wasser, S.P. Current findings, future trends, and unsolved problems in studies of medicinal mushrooms. *Appl. Microbiol. Biotechnol.* **2011**, *89*, 1323–1332. [CrossRef]
37. Macharia, J.M.; Zhang, L.; Mwangi, R.W.; Rozmann, N.; Kaposztas, Z.; Varjas, T.; Sugár, M.; Alfatafta, H.; Pintér, M.; Bence, R.L. Are chemical compounds in medical mushrooms potent against colorectal cancer carcinogenesis and antimicrobial growth? *Cancer Cell Int.* **2022**, *22*, 379. [CrossRef]
38. Liu, J.; Jia, L.; Kan, J.; Jin, C.H. In vitro and in vivo antioxidant activity of ethanolic extract of white button mushroom (*Agaricus bisporus*). *Food Chem. Toxicol.* **2013**, *51*, 310–316. [CrossRef]
39. Mwangi, R.W.; Macharia, J.M.; Wagara, I.N.; Bence, R.L. The antioxidant potential of different edible and medicinal mushrooms. *Biomed. Pharmacother.* **2022**, *147*, 112621. [CrossRef]
40. El-Deeb, N.M.; Ibrahim, O.M.; Mohamed, M.A.; Farag, M.M.S.; Farrag, A.A.; El-Aassar, M.R. Alginate/κ-carrageenan oral microcapsules loaded with *Agaricus bisporus* polysaccharides MH751906 for natural killer cells mediated colon cancer immunotherapy. *Int. J. Biol. Macromol.* **2022**, *205*, 385–395. [CrossRef]
41. Xu, Y.; Xu, T.; Huang, C.; Liu, L.; Kwame, A.W.; Zhu, Y.; Ren, J. Preventive intervention with *Agaricus blazei* murill polysaccharide exerts anti-tumor immune effect on intraperitoneal metastasis colorectal cancer. *Int. J. Biol. Macromol.* **2024**, *282 Pt 3*, 136810. [CrossRef]
42. Dong, K.; Wang, J.; Tang, F.; Liu, Y.; Gao, L. A polysaccharide with a triple helix structure from *Agaricus bisporus*: Characterization and anti-colon cancer activity. *Int. J. Biol. Macromol.* **2024**, *281 Pt 4*, 136521. [CrossRef]
43. He, Y.Y.; Liu, S.; Newburg, D.S. Musarin, a novel protein with tyrosine kinase inhibitory activity from *Trametes versicolor*, inhibits colorectal cancer stem cell growth. *Biomed. Pharmacother.* **2021**, *144*, 112339. [CrossRef] [PubMed]
44. Lavi, I.; Friesem, D.; Geresh, S.; Hadar, Y.; Schwartz, B. An aqueous polysaccharide extract from the edible mushroom *Pleurotus ostreatus* induces anti-proliferative and pro-apoptotic effects on HT-29 colon cancer cells. *Cancer Lett.* **2006**, *244*, 61–70. [CrossRef] [PubMed]
45. Chaitanya, N.S.; Devi, A.; Sahu, S.; Alugoju, P. Molecular mechanisms of action of Trehalose in cancer: A comprehensive review. *Life Sci.* **2021**, *269*, 118968. [CrossRef]
46. Quero, J.; Paesa, M.; Morales, C.; Mendoza, G.; Osada, J.; Teixeira, J.A.; Ferreira-Santos, P.; Rodríguez-Yoldi, M.J. Biological properties of *Boletus edulis* extract on Caco-2 cells: Antioxidant, anticancer, and anti-inflammatory effects. *Antioxidants* **2024**, *13*, 908. [CrossRef]
47. Park, G.S.; Shin, J.; Hong, S.; Gopal, J.; Oh, J.W. Anticarcinogenic potential of the mushroom polysaccharide lentinan on gastric and colon cancer cells: Antiproliferative, antitumorigenic, Mu-2-related death-inducing gene, MUDENG ramifications. *J. Ind. Eng. Chem.* **2024**, *135*, 122–130. [CrossRef]
48. Zhang, M.; Zhang, Y.; Zhang, L.; Tian, Q. Mushroom polysaccharide lentinan for treating different types of cancers: A review of 12 years clinical studies in China. *Prog. Mol. Biol. Translat. Sci.* **2019**, *163*, 297–328. [CrossRef]

49. Fritz, H.; Kennedy, D.A.; Ishii, M.; Fergusson, D.; Fernandes, R.; Cooley, K.; Seely, D. Polysaccharide K and *Coriolus versicolor* extracts for lung cancer: A systematic review. *Integr. Cancer Ther.* **2015**, *14*, 201–211. [CrossRef]
50. Lu, H.; Yang, Y.; Gad, E.; Wenner, C.A.; Chang, A.; Larson, E.R.; Dang, Y.; Martzen, M.; Standish, L.J.; Disis, M.L. Polysaccharide Krestin is a novel TLR2 agonist that mediates inhibition of tumor growth via stimulation of CD8 T cells and NK cells. *Clin. Cancer Res.* **2010**, *17*, 67. [CrossRef]
51. Rosendahl, A.H.; Sun, C.; Wu, D.Q.; Andersson, R. Polysaccharide-K (PSK) increases p21(WAF/Cip1) and promotes apoptosis in pancreatic cancer cells. *Pancreatology* **2012**, *12*, 467–474. [CrossRef]
52. Camilleri, E.; Blundell, R.; Baral, B.; Karpiński, T.M.; Aruci, E.; Atrooz, O.M. Unveiling the full spectrum of maitake mushrooms: A comprehensive review of their medicinal, therapeutic, nutraceutical, and cosmetic potential. *Heliyon* **2024**, *10*, e30254. [CrossRef]
53. Fontes Vieira, P.A.; Gontijo, D.C.; Vieira, B.C.; Fontes, E.A.F.; de Assunção, L.S.; Leite, J.P.V.; Oliveira, M.G.D.A.; Kasuya, M.C.M. Antioxidant activities, total phenolics and metal contents in *Pleurotus ostreatus* mushrooms enriched with iron, zinc or lithium. *LWT* **2013**, *54*, 421–425. [CrossRef]
54. Kodama, N.; Komuta, K.; Sakai, N.; Nanba, H. Effects of D-Fraction, a polysaccharide from *Grifola frondosa* on tumor growth involve activation of NK cells. *Biol. Pharm. Bull.* **2002**, *25*, 1647–1650. [CrossRef]
55. Zhao, F.; Guo, Z.; Ma, Z.R.; Ma, L.L.; Zhao, J. Antitumor activities of *Grifola frondosa* (Maitake) polysaccharide: A meta-analysis based on preclinical evidence and quality assessment. *J. Ethnopharmacol.* **2021**, *280*, 114395. [CrossRef] [PubMed]
56. Brown, G.D.; Gordon, S. Immune recognition. A new receptor for beta-glucans. *Nature* **2001**, *413*, 36–37. [CrossRef] [PubMed]
57. Lemieszek, M.; Rzeski, W. Anticancer properties of polysaccharides isolated from fungi of the Basidiomycetes class. *Contemp. Oncol.* **2012**, *16*, 285. [CrossRef]
58. Yuan, B.; Ma, N.; Zhao, L.; Zhao, E.; Gao, Z.; Wang, W.; Song, M.; Zhang, G.; Hu, Q.; Xiao, H. In vitro and in vivo inhibitory effects of a *Pleurotus eryngii* protein on colon cancer cells. *Food Funct.* **2017**, *8*, 3553–3562. [CrossRef]
59. Yuan, B.; Zhao, L.; Rakariyatham, K.; Han, Y.; Gao, Z.; Muinde Kimatu, B.; Hu, Q.; Xiao, H. Isolation of a novel bioactive protein from an edible mushroom *Pleurotus eryngii* and its anti-inflammatory potential. *Food Funct.* **2017**, *8*, 2175–2183. [CrossRef]
60. Chen, C.H.; Wu, J.Y.; Chen, C.H.; Chang, W.H.; Chung, K.T.; Liu, Y.W.; Lu, F.J. Anti-cancer effects of protein extracts from *Calvatia lilacina, Pleurotus ostreatus* and *Volvariella volvacea. Evid. Based Complement. Altern. Med.* **2011**, *2011*, 982368. [CrossRef]
61. Rezvani, V.; Pourianfar, H.R.; Mohammadnejad, S.; Madjid Ansari, A.; Farahmand, L. Anticancer potentiality and mode of action of low-carbohydrate proteins and peptides from mushrooms. *Appl. Microbiol. Biotechnol.* **2020**, *104*, 6855–6871. [CrossRef]
62. Maiti, S.; Bhutia, S.K.; Mallick, S.K.; Kumar, A.; Khadgi, N.; Maiti, T.K. Antiproliferative and immunostimulatory protein fraction from edible mushrooms. *Environ. Toxicol. Pharm.* **2008**, *26*, 187–191. [CrossRef]
63. Evans, R.; Rhodes, J.; Kinsella, A. Mushroom lectin inhibits invasion of a human colon cancer cell line into collagen gels. *Clin. Sci.* **1995**, *88*, 12P. [CrossRef]
64. Evans, R.C.; Fear, S.; Ashby, D.; Williams, E.; Van der Vliet, M.; Rhodes, J.M.; Dunstan, F.D.J. Diet and colorectal cancer: An investigation of the lectin/galactose hypothesis. *Gastroenterology* **2002**, *122*, 1784–1792. [CrossRef] [PubMed]
65. Yu, L.G.; Fernig, D.G.; White, M.R.H.; Spiller, D.G.; Appleton, P.; Evans, R.C.; Grierson, I.; Smith, J.A.; Davies, H.; Gerasimenko, O.V.; et al. Edible mushroom (*Agaricus bisporus*) lectin, which reversibly inhibits epithelial cell proliferation, blocks nuclear localization sequence-dependent nuclear protein import. *J. Biol. Chem.* **1999**, *274*, 4890–4899. [CrossRef]
66. Zhou, R.; Liu, Z.K.; Zhang, Y.N.; Wong, J.H.; Ng, T.B.; Liu, F. Research progress of bioactive proteins from the edible and medicinal mushrooms. *Curr. Prot. Peptide Sci.* **2018**, *20*, 196–219. [CrossRef]
67. Singh, S.S.; Wang, H.; Chan, Y.S.; Pan, W.; Dan, X.; Yin, C.M.; Akkouh, O.; Ng, T.B. Lectins from edible mushrooms. *Molecules* **2014**, *20*, 446. [CrossRef]
68. Camilleri, E.; Blundell, R.; Baral, B.; Karpiński, T.M.; Aruci, E.; Atrooz, O.M. A comprehensive review of the health benefits, phytochemicals, and enzymatic constituents for potential therapeutic and industrial applications of turkey tail mushrooms. *Discov. Appl. Sci.* **2024**, *6*, 257. [CrossRef]
69. Sun, J.; Chen, Q.J.; Zhu, M.J.; Wang, H.X.; Zhang, G.Q. An extracellular laccase with antiproliferative activity from the Sanghuang mushroom *Inonotus baumii. J. Mol. Catal. B Enzym.* **2014**, *99*, 20–25. [CrossRef]
70. Yuzugullu Karakus, Y.; Isik, S.; Kale Bakir, E.; Turkmenoglu, A.; Deveci Ozkan, A. Characterization of the three-phase partitioned laccase from *Trametes versicolor* strain with antiproliferative activity against breast cancer cells. *Int. J. Biol. Macromol.* **2025**, *286*, 138504. [CrossRef]
71. Kim, M.Y.; Seguin, P.; Ahn, J.K.; Kim, J.J.; Chun, S.C.; Kim, E.H.; Seo, S.H.; Kang, E.Y.; Kim, S.L.; Park, Y.J.; et al. Phenolic compound concentration and antioxidant activities of edible and medicinal mushrooms from Korea. *J. Agric. Food Chem.* **2008**, *56*, 7265–7270. [CrossRef]
72. Paloi, S.; Kumla, J.; Paloi, B.P.; Srinuanpan, S.; Hoijang, S.; Karunarathna, S.C.; Acharya, K.; Suwannarach, N.; Lumyong, S. Termite mushrooms (Termitomyces), a potential source of nutrients and bioactive compounds exhibiting human health benefits: A review. *J. Fungi* **2023**, *9*, 112. [CrossRef]

73. Silva, M.; Lageiro, M.; Ramos, A.C.; Reboredo, F.H.; Gonçalves, E.M. Cultivated mushrooms: A comparative study of antioxidant activity and phenolic content. *Biol. Life Sci. Forum* **2024**, *40*, 13. [CrossRef]
74. Chu, M.; Khan, R.D.; Zhou, Y.; Agar, O.T.; Barrow, C.J.; Dunshea, F.R.; Suleria, H.A.R. LC-ESI-QTOF-MS/MS characterization of phenolic compounds in common commercial mushrooms and their potential antioxidant activities. *Processes* **2023**, *11*, 1711. [CrossRef]
75. Durgo, K.; Koncar, M.; Komes, D.; Belscak-Cvitanovic, A.; Franekic, J.; Jakopovich, I.; Jakopovich, N.; Jakopovich, B. Cytotoxicity of blended versus single medicinal mushroom extracts on human cancer cell lines: Contribution of polyphenol and polysaccharide content. *Int. J. Med. Mushrooms* **2013**, *15*, 435–448. [CrossRef]
76. Palacios, I.; Lozano, M.; Moro, C.; D'Arrigo, M.; Rostagno, M.A.; Martínez, J.A.; García-Lafuente, A.; Guillamón, E.; Villares, A. Antioxidant properties of phenolic compounds occurring in edible mushrooms. *Food Chem.* **2011**, *128*, 674–678. [CrossRef]
77. Lin, S.; Ching, L.T.; Chen, J.; Cheung, P.C.K. Antioxidant and anti-angiogenic effects of mushroom phenolics-rich fractions. *J. Funct. Foods* **2015**, *17*, 802–815. [CrossRef]
78. Park, H.J. Current uses of mushrooms in cancer treatment and their anticancer mechanisms. *Int. J. Mol. Sci.* **2022**, *23*, 10502. [CrossRef]
79. Chang, T.C.; Yeh, C.T.; Adebayo, B.O.; Lin, Y.C.; Deng, L.; Rao, Y.K.; Huang, C.C.; Lee, W.H.; Wu, A.T.H.; Hsiao, M.; et al. 4-Acetylantroquinonol B inhibits colorectal cancer tumorigenesis and suppresses cancer stem-like phenotype. *Toxicol. Appl. Pharmacol.* **2015**, *288*, 258–268. [CrossRef]
80. Chu, Y.C.; Tsai, T.-Y.; Yadav, V.K.; Deng, L.; Huang, C.-C.; Tzeng, Y.-M.; Yeh, C.-T.; Chen, M.-Y. 4-Acetyl-Antroquinonol B improves the sensitization of cetuximab on both Kras mutant and wild type colorectal cancer by modulating the expression of Ras/Raf/miR-193a-3p signaling axis. *Int. J. Mol. Sci.* **2021**, *22*, 7508. [CrossRef]
81. Ayeka, P.A. Potential of mushroom compounds as immunomodulators in cancer immunotherapy: A review. *Evid. Based Complement. Altern. Med.* **2018**, *2018*, 7271509. [CrossRef]
82. Dan, A.; Swain, R.; Belonce, S.; Jacobs, R.J. Therapeutic effects of medicinal mushrooms on gastric, breast, and colorectal cancer: A scoping review. *Cureus* **2023**, *15*, e37574. [CrossRef]
83. De Silva, D.D.; Rapior, S.; Fons, F.; Bahkali, A.H.; Hyde, K.D. Medicinal mushrooms in supportive cancer therapies: An approach to anti-cancer effects and putative mechanisms of action. *Fungal Divers.* **2012**, *55*, 1–35. [CrossRef]
84. Figueiredo, L.; Régis, W.C.B. Medicinal mushrooms in adjuvant cancer therapies: An approach to anticancer effects and presumed mechanisms of action. *Nutrire* **2017**, *42*, 28. [CrossRef]
85. Subramanian, R.R.; Yamakawa, A. Combination therapy targeting Raf-1 and MEK causes apoptosis of HCT116 colon cancer cells. *Int. J. Oncol.* **2012**, *41*, 1855–1862. [CrossRef]
86. Lavi, I.; Nimri, L.; Levinson, D.; Peri, I.; Hadar, Y.; Schwartz, B. Glucans from the edible mushroom *Pleurotus pulmonarius* inhibit colitis-associated colon carcinogenesis in mice. *J. Gastroenterol.* **2012**, *47*, 504–518. [CrossRef]
87. Zaila, C.S.; Zuraina, M.F.; Norfazlina, M.N.; Mun, L.L.; Nurshahirah, N.; Florinsiah, L.; Rajab, N.F. Antiproliferative effect of *Lignosus rhinocerotis*, the Tiger Milk mushroom on HCT 116 human colorectal cancer cells. *Open Conf. Proc. J.* **2013**, *4*, 65–70. [CrossRef]
88. Aakif, M.; Balfe, P.; Elfaedy, O.; Awan, F.N.; Pretorius, F.; Silvio, L.; Castinera, C.; Mustafa, H. Study on colorectal cancer presentation, treatment and follow-up. *Int. J. Color. Dis.* **2016**, *31*, 1361–1363. [CrossRef]
89. Li, J.; Huang, L.; Zhao, H.; Yan, Y.; Lu, J. The role of interleukins in colorectal cancer. *Int. J. Biol. Sci.* **2020**, *16*, 2323. [CrossRef]
90. Bergman, M.; Levin, G.S.; Bessler, H.; Djaldetti, M.; Salman, H. Resveratrol affects the cross talk between immune and colon cancer cells. *Biomed. Pharmacother.* **2013**, *67*, 43–47. [CrossRef]
91. Pastille, E.; Wasmer, M.H.; Adamczyk, A.; Vu, V.P.; Mager, L.F.; Phuong, N.N.T.; Palmieri, V.; Simillion, C.; Hansen, W.; Kasper, S.; et al. The IL-33/ST2 pathway shapes the regulatory T cell phenotype to promote intestinal cancer. *Mucosal Immunol.* **2019**, *12*, 990. [CrossRef]
92. Al Obeed, O.A.; Alkhayal, K.A.; Al Sheikh, A.; Zubaidi, A.M.; Vaali-Mohammed, M.A.; Boushey, R.; Mckerrow, J.H.; Abdulla, M.H. Increased expression of tumor necrosis factor-α is associated with advanced colorectal cancer stages. *World J. Gastroenterol. WJG* **2014**, *20*, 18390. [CrossRef]
93. Zhao, P.; Zhang, Z. TNF-α promotes colon cancer cell migration and invasion by upregulating TROP-2. *Oncol. Lett.* **2018**, *15*, 3820–3827. [CrossRef] [PubMed]
94. Liu, X.; Li, Y.; Sun, X.; Muftuoglu, Y.; Wang, B.; Yu, T.; Hu, Y.; Ma, L.; Xiang, M.; Guo, G.; et al. Powerful anti-colon cancer effect of modified nanoparticle-mediated IL-15 immunogene therapy through activation of the host immune system. *Theranostics* **2018**, *8*, 3490–3503. [CrossRef] [PubMed]
95. Waldmann, T.A. The biology of interleukin-2 and interleukin-15: Implications for cancer therapy and vaccine design. *Nat. Rev. Immunol.* **2006**, *6*, 595–601. [CrossRef] [PubMed]
96. Sharma, D.; Malik, A.; Guy, C.S.; Karki, R.; Vogel, P.; Kanneganti, T.D. Pyrin inflammasome regulates tight junction integrity to restrict colitis and tumorigenesis. *Gastroenterology* **2017**, *154*, 948. [CrossRef]

97. Muszyńska, B.; Grzywacz, A.; Kała, K.; Gdula-Argasińska, J. Anti-inflammatory potential of in vitro cultures of the white button mushroom, *Agaricus bisporus* (Agaricomycetes), in Caco-2 Cells. *Int. J. Med. Mushrooms* **2018**, *20*, 129–139. [CrossRef]
98. Chaudhary, P.; Janmeda, P.; Docea, A.O.; Yeskaliyeva, B.; Abdull Razis, A.F.; Modu, B.; Calina, D.; Sharifi-Rad, J. Oxidative stress, free radicals and antioxidants: Potential crosstalk in the pathophysiology of human diseases. *Front. Chem.* **2023**, *11*, 1158198. [CrossRef]
99. Muchtaridi, M.; Az-Zahra, F.; Wongso, H.; Setyawati, L.U.; Novitasari, D.; Ikram, E.H.K. Molecular mechanism of natural food antioxidants to regulate ROS in treating cancer: A review. *Antioxidants* **2024**, *13*, 207. [CrossRef]
100. Pham-Huy, L.A.; He, H.; Pham-Huy, C. Free radicals, antioxidants in disease and health. *Int. J. Biomed. Sci.* **2008**, *4*, 89. [CrossRef]
101. Baeuerle, P.A.; Henkel, T. Function and activation of NF-KappaB in the immune system. *Annu. Rev. Immunol.* **1994**, *12*, 141–179. [CrossRef]
102. Barkett, M.; Gilmore, T.D. Control of apoptosis by Rel/NF-κB transcription factors. *Oncogene* **1999**, *18*, 6910–6924. [CrossRef]
103. Dolcet, X.; Llobet, D.; Pallares, J.; Matias-Guiu, X. NF-kB in development and progression of human cancer. *Virchows Arch.* **2005**, *446*, 475–482. [CrossRef] [PubMed]
104. Wan, F.; Lenardo, M.J. Specification of DNA binding activity of NF- B proteins. *Cold Spring Harb. Perspect. Biol.* **2009**, *1*, a000067. [CrossRef] [PubMed]
105. Chen, Y.; Chen, M.; Deng, K. Blocking the Wnt/β-catenin signaling pathway to treat colorectal cancer: Strategies to improve current therapies (Review). *Int. J. Oncol.* **2022**, *62*, 24. [CrossRef]
106. He, K.; Gan, W.J. Wnt/β-Catenin signaling pathway in the development and progression of colorectal cancer. *Cancer Manag. Res.* **2023**, *15*, 435. [CrossRef]
107. Heslin, M.J.; Yan, J.; Johnson, M.R.; Weiss, H.; Diasio, R.B.; Urist, M.M. Role of matrix metalloproteinases in colorectal carcinogenesis. *Ann. Surg.* **2001**, *233*, 786. [CrossRef]
108. Said, A.H.; Raufman, J.P.; Xie, G. The role of matrix metalloproteinases in colorectal cancer. *Cancers* **2014**, *6*, 366. [CrossRef]
109. Pfeffer, C.M.; Singh, A.T.K. Apoptosis: A target for anticancer therapy. *Int. J. Mol. Sci.* **2018**, *19*, 448. [CrossRef]
110. Xu, J.; Shen, R.; Jiao, Z.; Chen, W.; Peng, D.; Wang, L.; Yu, N.; Peng, C.; Cai, B.; Song, H.; et al. Current advancements in antitumor properties and mechanisms of medicinal components in edible mushrooms. *Nutrients* **2022**, *14*, 2622. [CrossRef]
111. Ramesh, P.; Medema, J.P. BCL-2 family deregulation in colorectal cancer: Potential for BH3 mimetics in therapy. *Apoptosis* **2020**, *25*, 305. [CrossRef]
112. Güllülü, Ö.; Hehlgans, S.; Rödel, C.; Fokas, E.; Rödel, F. Tumor suppressor protein p53 and inhibitor of apoptosis proteins in colorectal cancer—A promising signaling network for therapeutic interventions. *Cancers* **2021**, *13*, 624. [CrossRef]
113. Zhou, M.; Liu, X.; Li, Z.; Huang, Q.; Li, F.; Li, C.Y. Caspase-3 regulates the migration, invasion, and metastasis of colon cancer cells. *Int. J. Cancer* **2018**, *143*, 921–930. [CrossRef] [PubMed]
114. Baek, S.J.; Kim, J.S.; Jackson, F.R.; Eling, T.E.; McEntee, M.F.; Lee, S.H. Epicatechin gallate-induced expression of NAG-1 is associated with growth inhibition and apoptosis in colon cancer cells. *Carcinogenesis* **2004**, *25*, 2425–2432. [CrossRef] [PubMed]
115. Kim, I.Y.; Park, S.Y.; Kang, Y.; Thapa, D.; Choi, H.G.; Kim, J.A. Role of nonsteroidal anti-inflammatory drug-activated gene-1 in docetaxel-induced cell death of human colorectal cancer cells with different p53 status. *Arch. Pharm. Res.* **2011**, *34*, 323–330. [CrossRef]
116. Krieg, A.; Werner, T.A.; Verde, P.E.; Stoecklein, N.H.; Knoefel, W.T. Prognostic and clinicopathological significance of survivin in colorectal cancer: A meta-analysis. *PLoS ONE* **2013**, *8*, e65338. [CrossRef]
117. Li, G.; Wang, Y.; Wang, W.; Lv, G.; Li, X.; Wang, J.; Liu, X.; Yuan, D.; Deng, S.; You, D. BIRC5 as a prognostic and diagnostic biomarker in pan-cancer: An integrated analysis of expression, immune subtypes, and functional networks. *Front. Genet.* **2024**, *15*, 1509342. [CrossRef]
118. Gariboldi, M.B.; Marras, E.; Ferrario, N.; Vivona, V.; Prini, P.; Vignati, F.; Perletti, G. Anti-cancer potential of edible/medicinal mushrooms in breast cancer. *Int. J. Mol. Sci.* **2023**, *24*, 10120. [CrossRef]
119. Jakopovic, B.; Oršolić, N.; Jakopovich, I. Proteomic research on the antitumor properties of medicinal mushrooms. *Molecules* **2021**, *26*, 6708. [CrossRef]
120. Randeni, N.; Xu, B. New insights into signaling pathways of cancer prevention effects of polysaccharides from edible and medicinal mushrooms. *Phytomedicine* **2024**, *132*, 155875. [CrossRef]
121. Jakopovic, B.; Horvatić, A.; Klobučar, M.; Gelemanović, A.; Grbčić, P.; Oršolić, N.; Jakopovich, I.; Pavelić, S.K. Treatment with medicinal mushroom extract mixture inhibits translation and reprograms metabolism in advanced colorectal cancer animal model as evidenced by tandem mass Tags proteomics analysis. *Front. Pharmacol.* **2020**, *11*, 1202. [CrossRef]
122. Jiang, Y.A.; Luo, H.S.; Zhang, Y.Y.; Fan, L.F.; Jiang, C.Q.; Chen, W.J. Telomerase activity and cell apoptosis in colon cancer cell by human telomerase reverse transcriptase gene antisense oligodeoxynucleotide. *World J. Gastroenterol.* **2003**, *9*, 1981–1984. [CrossRef]
123. Simsek, B.C.; Pehlivan, S.; Karaoglu, A. Human telomerase reverse transcriptase expression in colorectal tumors: Correlations with immunohistochemical expression and clinicopathologic features. *Ann. Diagn. Pathol.* **2010**, *14*, 413–417. [CrossRef] [PubMed]

124. Huo, J.; Nie, K.; Yang, T.; Zhang, S.; Zhu, Z.; Peng, X.; Zhang, Y. Network pharmacology combined with transcriptomics reveals that *Ganoderma lucidum* spore and *Sanghuangporus vaninii* compound extract exerts anti-colorectal cancer effects via CYP24A1-mediated VDR pathway and TERT-mediated Wnt signaling pathway. *J. Ethnopharmacol.* **2025**, *348*, 119820. [CrossRef] [PubMed]
125. Liao, C.; Hsiao, Y.; Hsu, C.; Lin, M.; Wang, J.C.; Huang, Y.; Ko, J. Transcriptionally mediated inhibition of telomerase of fungal immunomodulatory protein from *Ganoderma tsugae* in A549 human lung adenocarcinoma cell line. *Mol. Carcinog.* **2006**, *45*, 220–229. [CrossRef]
126. Dang, D.T.; Chen, F.; Gardner, L.B.; Cummins, J.M.; Rago, C.; Bunz, F.; Kantsevoy, S.V.; Dang, L.H. Hypoxia-inducible factor-1alpha promotes nonhypoxia-mediated proliferation in colon cancer cells and xenografts. *Cancer Res.* **2006**, *66*, 1684–1936. [CrossRef]
127. Ioannou, M.; Paraskeva, E.; Baxevanidou, K.; Simos, G.; Papamichali, R.; Papacharalambous, C.; Samara, M.; Koukoulis, G. HIF-1α in colorectal carcinoma: Review of the literature. *J. Buon.* **2015**, *20*, 680–689. Available online: https://pubmed.ncbi.nlm.nih.gov/26214618 (accessed on 17th March 2025).
128. Kirdeeva, Y.; Fedorova, O.; Daks, A.; Barlev, N.; Shuvalov, O. How should the worldwide knowledge of traditional cancer healing be integrated with herbs and mushrooms into modern molecular pharmacology? *Pharmaceuticals* **2022**, *15*, 868. [CrossRef]
129. Maithani, R.; Tulsawani, R. Lingzhi or Reishi medicinal mushroom (agaricomycetes) *Ganoderma lucidum* aqueous extract reverses hypoxia induced redox imbalance and inflammatory response in rat thymocytes. *Int. J. Med. Mushrooms* **2025**, *27*, 51–63. [CrossRef]
130. Cao, D.; Qin, S.; Mu, Y.; Zhong, M. The role of MRP1 in the multidrug resistance of colorectal cancer. *Oncol. Lett.* **2017**, *13*, 2471–2476. [CrossRef]
131. Jiang, Z.; Jin, T.; Gao, F.; Liu, J.; Zhong, J.; Zhao, H. Effects of ganoderic acid Me on inhibiting multidrug resistance and inducing apoptosis in multidrug resistant colon cancer cells. *Process Biochem.* **2011**, *46*, 1307–1314. [CrossRef]
132. Chang, S.T.; Wasser, S.P. Current and future research trends in agricultural and biomedical applications of medicinal mushrooms and mushroom products (Review). *Int. J. Med. Mushrooms* **2018**, *20*, 1121–1133. [CrossRef]
133. Wasser, S.P. Medicinal mushroom science: History, current status, future trends, and unsolved problems. *Int. J. Med. Mushrooms* **2010**, *12*, 1–16. [CrossRef]
134. Jeitler, M.; Michalsen, A.; Frings, D.; Hübner, M.; Fischer, M.; Koppold-Liebscher, D.A.; Murthy, V.; Kessler, C.S. Significance of medicinal mushrooms in integrative oncology: A narrative review. *Fron. Pharmacol.* **2020**, *11*, 580656. [CrossRef] [PubMed]
135. Islam, M.R.; Akash, S.; Rahman, M.M.; Nowrin, F.T.; Akter, T.; Shohag, S.; Rauf, A.; Aljohani, A.S.M.; Simal-Gandara, J. Colon cancer and colorectal cancer: Prevention and treatment by potential natural products. *Chem. Biol. Inter.* **2022**, *368*, 110170. [CrossRef] [PubMed]
136. Patel, S.; Goyal, A. Recent developments in mushrooms as anti-cancer therapeutics: A review. *3 Biotech* **2012**, *2*, 1–15. [CrossRef]
137. Paul, D.; Sanap, G.; Shenoy, S.; Kalyane, D.; Kalia, K.; Tekade, R.K. Artificial intelligence in drug discovery and development. *Drug Discov. Today* **2021**, *26*, 80–93. [CrossRef]
138. Leal, K.; Rojas, E.; Madariaga, D.; Contreras, M.J.; Nuñez-Montero, K.; Barrientos, L.; Goméz-Espinoza, O.; Iturrieta-González, I. Unlocking fungal potential: The CRISPR-Cas system as a strategy for secondary metabolite discovery. *J. Fungi* **2024**, *10*, 748. [CrossRef]
139. Kwong, K.W.; Xin, Y.; Lai, N.C.; Sung, J.C.; Wu, K.C.; Hamied, Y.K.; Sze, E.T.; Lam, D.M. Oral Vaccines: A Better Future of Immunization. *Vaccines* **2023**, *11*, 1232. [CrossRef]
140. Pérez-Martínez, A.S.; Acevedo-Padilla, S.A.; Bibbins-Martínez, M.; Galván-Alonso, J.; Rosales-Mendoza, S. A perspective on the use of *Pleurotus* for the development of convenient fungi-made oral subunit vaccines. *Vaccine* **2015**, *33*, 25–33. [CrossRef]
141. Uddin, M.S.; Kaldis, A.; Menassa, R.; Ortiz Guluarte, J.; Barreda, D.R.; Guan, L.L.; Alexander, T.W. Mucosal immunization with spore-based vaccines against *Mannheimia haemolytica* enhances antigen-specific immunity. *Vaccines* **2024**, *12*, 375. [CrossRef]

International Journal of
Molecular Sciences

MDPI

Review

Double-Edged Sword Effect of Diet and Nutrition on Carcinogenic Molecular Pathways in Breast Cancer

Anca-Narcisa Neagu [1,*], Claudiu-Laurentiu Josan [1], Taniya M. Jayaweera [2], Krishan Weraduwage [2], Niyogushima Nuru [2] and Costel C. Darie [2,*]

1 Laboratory of Animal Histology, Faculty of Biology, "Alexandru Ioan Cuza" University of Iași, Carol I bvd. 20A, 700505 Iasi, Romania; claudiujosan22@gmail.com

2 Biochemistry & Proteomics Laboratories, Department of Chemistry and Biochemistry, Clarkson University, Potsdam, NY 13699-5810, USA; jayawetm@clarkson.edu (T.M.J.); weraduk@clarkson.edu (K.W.); nurun@clarkson.edu (N.N.)

* Correspondence: aneagu@uaic.ro (A.-N.N.); cdarie@clarkson.edu (C.C.D.); Tel.: +1-(315)-268-7763 (C.C.D.); Fax: +1-(315)-268-6610 (C.C.D.)

Abstract: Environmental exposure to a mixture of chemical xenobiotics acts as a double-edged sword, promoting or suppressing tumorigenesis and the development of breast cancer (BC). Before anything else, we are what we eat. In this review, we highlight both "the good" and "the bad" sides of the daily human diet and dietary patterns that could influence BC risk (BCR) and incidence. Thus, regularly eating new, diversified, colorful, clean, nutrient-rich, energy-boosting, and raw food, increases apoptosis and autophagy, antioxidation, cell cycle arrest, anti-inflammation, and the immune response against BC cells. Moreover, a healthy diet could lead to a reduction in or the inhibition of genomic instability, BC cell stemness, growth, proliferation, invasion, migration, and distant metastasis. We also emphasize that, in addition to beneficial compounds, our food is more and more contaminated by chemicals with harmful effects, which interact with each other and with endogenous proteins and lipids, resulting in synergistic or antagonistic effects. Thus, a healthy and diverse diet, combined with appropriate nutritional behaviors, can exert anti-carcinogenic effects and improve treatment efficacy, BC patient outcomes, and the overall quality of life of BC patients.

Keywords: breast cancer (BC); breast cancer risk (BCR); diet; nutrition; carcinogenic molecular pathways; bioactive compounds; food contaminants

1. Introduction

BC is the most diagnosed cancer worldwide, with over 2.3 million new cases and 685,000 deaths in 2020; BC diagnosis is predicted to increase to over 3 million new cases and one million deaths by 2040 [1]. Evidence suggests that environmental exposure to certain chemicals and lifestyle account for 70% to 90% of the risk factors for chronic diseases, whereas only 10% to 30% can be explained by a specific genomic landscape [2]. Both dietary compounds and daily nutrition-related habits may play key roles in preventing diseases [3] or, on the contrary, can exert carcinogenic effects, even at low levels of exposure [4]. The expression a "double-edged sword" refers to something that has both good and bad consequences. For example, food contaminants are usually chemical substances, as well as microbial or physical compounds present in edibles, which may be harmful to human health, with different levels of severity [5]. Moreover, food contaminants are generally environmental pollutants, which are not present naturally in raw food, with contamination occurring during the production, processing, distribution, storage, packaging, transportation, or preparation of food. On the contrary, a plethora of bioactive compounds are present in "functional food", dietary supplements, and plant or animal nutraceuticals, which are known to exert beneficial effects on health [6].

Phytochemistry, as well as various omics fields, such as nutrigenomics, nutriproteomics, metabolomics, interactomics, and exposomics [7], offer a wide range of opportunities for the development of personalized diets for women at risk of developing BC [3]. It was shown that certain dietary compounds, mainly phytonutrients, can modulate gene expression, acting on the main cancer hallmarks: cell growth and proliferation, genome instability and mutagenesis, angiogenesis, metabolism reprogramming, anti-apoptosis, tumor-promoting inflammation, therapy resistance, invasion, and metastasis [8]. Plant-based diets may contribute to the inhibition of BC cell proliferation, differentiation, invasion, metastasis, angiogenesis, and anti-apoptotic mechanisms, by targeting numerous molecular pathways and promoting beneficial effects in terms of the prevention and treatment of BC [9]. Moreover, plant-based dietary bioactive compounds can modulate hormone-induced reactions, genetic and epigenetic regulation, and promote the activation of oxidative stress (OS) and endoplasmic reticulum stress, triggering redox reactions, as well as the aggressive behavior of tumor cells caused by the production of reactive oxygen species (ROS) [10]. Many types of food rich in bioactive compounds, such as polyphenolic compounds, carotenoids, terpenoids, and sulfur-containing compounds, have been recommended as chemopreventive and chemotherapeutic agents, including asparagus, broccoli, Brussels sprouts, kale, and other cruciferous vegetables, carrots, grapefruit, soy, spinach, and tomatoes, mainly for ER-negative BC [11]. On the other hand, a plethora of risk factors, such as alcohol consumption [12], added sugar in foods [13], and LDL cholesterol [14], stimulate inflammation, BC cell proliferation, invasion, migration, and metastasis. Moreover, nowadays, more than ever, more food is processed and contaminated with man-made chemicals that act as endocrine disruptors, accumulate in high-fat tissues, and stimulate BC initiation and progression.

There are dietary patterns, such as the typical Western diet (WD), characterized by a high intake of pre-packaged foods, refined grains, red and processed/ultra-processed meat, candy, cookies, high-sugar drinks, fried foods, high-saturated fat and high-fructose products, dairy products, alcohol, salt, sugar, sweeteners, and other additives [15]. Thus, the WD has been linked to the development of excess weight/obesity [16], the alteration of human gastrointestinal microbiota [17,18], and carcinogenesis [19]. In postmenopausal obese women, the risk of developing BC is increased [20], while gut microbiota dysbiosis plays a role in the development of BC through estrogen-dependent pathways and microbial-derived metabolites [21]. Conversely, the Mediterranean diet (MD) involves a plant-based dietary pattern, based on a high intake of olive oil and plant-based foods, such as vegetables, non-refined cereals, legumes, and nuts, and a moderate or low level of consumption of dairy and meat-based products, alcohol, and sweets [22]. Consequently, the MD could reduce BCR and enhance survival through its anti-inflammatory effects, antioxidant characteristics, and hormone–receptor interactions [22]. Kalam et al. (2023) showed that diet and diet-related behaviors can modulate carcinogenesis, cancer progression, treatment efficacy, and recurrence [23]. Post-diagnosis plant-based diets can also benefit prognosis in cancer patients [24], due to the bioactive compounds that occur in small quantities in foods and that may have beneficial effects on health [25]. In this review, we highlight "the good" and "the bad" sides of the daily human diet and dietary patterns that could influence BC risk (BCR) and incidence. We conclude that a balanced and personalized dietary structure, combined with appropriate nutritional behaviors, can improve treatment efficacy and BC patient outcomes.

2. The Good

Circadian nutritional behaviors have been associated with BC [26] and, when disrupted, become factors for BCR [27]. Moreover, eating more frequently, reducing evening energy intake, and fasting for longer intervals, can downregulate the biomarkers of inflammation, as well as BCR [28]. Breakfast is the first meal of the day, often considered to be "the most important meal of the day" [29]; evidence shows that a good quality breakfast and breakfast-based nutrients are beneficial for human health, in association with a lower body mass index (BMI), a higher level of satisfaction in life, and a higher intellectual performance [30–37]. However, Elahy et al. (2023) concluded that there are no associations between the frequency of breakfast meals or after-dinner snack habits and BCR in postmenopausal women [38]. It is known that cyclical fasting or fasting-mimicking diets enhance antitumor chemotherapy effects in TNBC models [39], as well as the activity of endocrine therapeutics in mouse models of hormone receptor-positive BC [40]. A systematic review conducted by Anemoulis et al. (2023) concluded that intermittent fasting reduces chemotherapy-induced DNA damage in BC patients [41]. These nutritional patterns combined with specific therapeutic drugs reduce circulating insulin growth factor 1 (IGF1), insulin and leptin, and inhibit AKT/mTOR signaling through the upregulation of early growth response protein 1 (EGR1) and the phosphatase and tensin homolog deleted on chromosome 10 (PTEN), thus promoting long-lasting tumor regression and reverting the acquired resistance to chemotherapy [40]. Thus, intermittent fasting, based on different patterns of time-restricted feeding behaviors, attenuates obesity-induced TNBC progression in cell and animal models, involving multiple pathways, such as those involved in EMT reduction, cell cycle disruption, reductions in systemic glucose and cholesterol levels, and the downregulation of inflammatory factors in the TME [42]. A multicase control study in Spain showed that having breakfast at a later time of day was associated with a non-significant increase in BCR [26].

Among the diverse dietary patterns, the Mediterranean diet, rich in antioxidants and anti-inflammatory compounds; the plant-based dietary pattern, rich in fibers, phytochemicals, lignans, carotenoids, vitamins C and E, folate, and phenolic acid; the prudent dietary pattern, rich in spices, plant-based oils, low-fat dairy, and seafood; the healthy dietary pattern, rich in fruits, vegetables, legumes, seeds, and nuts; the ketogenic dietary pattern, known as a high-fat and low-carb diet; the paleolithic dietary pattern, rich in vegetables, fruits, lean meats, fish, nuts, and seeds; and the dietary approaches used to stop hypertension, were all correlated with a lower risk of developing BC [43]. Diverse breakfast patterns have been geographically identified and are based on cereal or sweetened breads and milk, eggs, sweetened beverages, sandwiches, and fruit consumption [30]. In Romania, traditional food products are much more popular than foods purchased from chain stores [44], because local consumers are attracted by traditional and authentic gastronomy [45]. Moreover, the main basic food products consumed are meat and meat products traditionally based on pork/chicken, such as pork crackling, pork fat, grilled pork neck, pork loin, smoked pork knuckle, grilled meat rolls, cabbage rolls with ground pork and rice, chicken liver, deep-fried chicken, pastrami, and sausages, milk and/or milk products (sour cream, cheese), vegetables (fries, beans, mushrooms, smashed potatoes, cabbage, pickles, garlic), and fruits, bread/bakery and pastry products (mamaliga/polenta, noodles, ice cream, fried cheese doughnuts), and fish (mainly fried crap) or fish products [44]. Fast food consumption is more popular among children and adolescents [44]. There are studies that have suggested that certain foods are consumed in excess (products high saturated fat, cholesterol, salt, sugar, refined grains, and alcohol), while there is a deficiency in regard to the intake of nutritional factors (essential amino acids, polyunsaturated fatty acids, vitamins C, A, B, D, folic acid, calcium, and iron) [46].

These meals contain complex mixtures of natural and added chemical compounds that can exert double-edged sword effects on carcinogenic molecular pathways in breast tissue. For example, milk is a complex biofluid, rich in primary nutrients and is considered a complete and basic food that is consumed by billions of people worldwide [47]. Thus, pure cow's milk and its derivative dairy products, such as cream, butter, cheese, and yogurt are a major sources of nutrition in the human diet [48]. Wajszczyk et al. (2021) showed that individual dairy products have a statistically significant, but bi-directional relationship with BCR, which was different for premenopausal and postmenopausal Polish women [49]. A cohort study conducted by these authors concluded that an increase in consumption of one serving of dairy products/week may significantly decrease the BCR, by 2%, for premenopausal women only, while cottage cheese consumption significantly reduced the BCR by 20%, following an increase of one serving/week, for postmenopausal women only [49]. A 5% decrease in BCR has been observed as a result of an increase in dairy consumption of one glass of milk/week in both strata of women during the menopause [49]. Arafat et al. (2023), after performing a systematic literature review, also concluded that dairy consumption was inversely associated with BCR, even when the effects of different types of dairy products and the dose–response relationship on BCR remain unknown [50]. More recently, Riseberg et al. (2024) showed that the overall amount of dairy consumption was not associated with BCR, but the results can vary according to the tumor subtype, and heterogeneity was observed in terms of the type of dairy food and the patient's period of life [51]. Deschasaux-Tanguy et al. (2022) found no association between the consumption of dairy products or dairy desserts high in sugar and BCR [52].

Almost 70% of the fat in milk is saturated, 25% is monounsaturated, and 2.3% is polyunsaturated, with a variable omega-6/omega-3 ratio [53]. Moreover, α-linolenic acid (ALA), an omega-3 fatty acid, induces apoptosis and inhibits invasion, metastasis and angiogenesis, and arrests the cell cycle in human BC cells by inhibiting fatty acid synthase (FASN), which is usually overexpressed in various cancers [54,55]. Conjugated linoleic acid (CLA) is known as a group of isomers of linoleic acid (LA), of which cis-9 trans-11 (c9,t11-CLA) is the one that has the highest percentage, being produced through biohydrogenation by lactic acid bacteria in the rumen or by endogenous synthesis in the mammary gland [56]. Consequently, CLA is present in the milk, dairy, and meat products of ruminants, with dairy products being the principal source of CLA in the human diet [57]. Moreover, 3 g/day of CLA is the recommended intake, with beneficial effects on human health [58], in principal due to the antitumor effects of CLA [56]. Zeng et al. (2019) demonstrated that dietary intake of c9,t11-CLA enriched from butter reduces BC progression in vivo via the downregulation of the progesterone receptor (PR) and Ki-67 expression [57]. Milk proteins, which include 17–20% whey proteins (α-lactalbumin, β-lactoglobulin, glycoproteins, lactoferrin, immunoglobulins, peptide hormones, enzymes like lactoperoxidase, and serum albumin) and 78–80% casein (αs-1, αs-2, β-, and k-casein), as well as milk peptides, exert many biological proprieties, such as antibacterial, antiviral, antifungal, and antioxidant activities [59]. Using cow's milk bottom-up proteomics, based on combinatorial SDS-PAGE profiling and trypsin digestion, followed by nanoHPLC-electrospray ionization tandem mass spectrometry (nLC-ESI-MS/MS), Vincent et al. (2016) identified 186 different major and minor proteins, including β2-microglobulin (β2-M), osteopontin (OPN), lipoprotein lipase (LPL), sulfhydryl oxidase (SOX), xanthine dehydrogenase/oxidase (XOR), β-1,4-galactosyltransferase 1 (Gal-T1), lactadherin/milk fat globule-EGF factor 8 (MFG-E8), lactotransferrin, mucins 1 and 15, α-1-acid glycoprotein, α-1B-glycoprotein, α-2-HS-glycoprotein, pancreatic secretory granule membrane major glycoprotein GP2, platelet glycoprotein 4, Zn-α-2-glycoprotein, milk glycosylation-dependent cell adhesion molecule 1/lactophorin (GlyCAM1), dystroglycan (DG), and peptidoglycan recognition protein 1 [48]. Some reports suggest that OPN may exhibit antitumorigenic characteristics in certain circumstances [60]. Orally administrated lactoferrin (LF), a natural proapoptotic iron-binding multifunctional glycoprotein from bovine milk, belonging to the transferrin family, can exert strong anticancer activities [61]. Pereira et al. (2016) showed that bovine LF preferentially induces apoptosis in the highly metastatic BC cell lines, Hs 578T and MDA-MB-231, but not in the less metastatic T47D or in the non-tumorigenic MCF10A cell lines [62]. The authors demonstrated that LF decreases the extracellular acidification rate and causes intracellular acidification in metastatic BC cells through the inhibition of plasmalemmal V-H^+-ATPase, which transports protons across cellular membranes [62].

Milk is also an abundant source of extracellular vesicles (EVs); bovine milk-derived extracellular vesicles (MEVs) are able to sensitize TNBC cells into doxorubicin by targeting metabolism and STAT signaling pathways, thus reducing the abundance of many tumorigenic proteins associated with a worse prognosis and low overall survival in TNBC [63]. Samuel et al. (2021) showed that orally administrated MEVs survive the degrading conditions in the mouse gut and can be detected in various organs, while MEVs implanted in BC cells reduced the primary tumor burden, but accelerated metastasis in BC mouse models through the induction of EMT, providing context-based and opposing roles of MEVs as metastasis promoters and suppressors [64]. Ramezani et al. (2023) showed that bovine milk lactoferrin-loaded exosomes induce selective toxicity against BC cells compared to normal cells and that incorporating lactoferrin into exosomes could have an antitumorigenic role through inducing the overexpression of the proapoptotic BH3 interacting domain death agonist (BID) protein and diminishing the expression of the anti-apoptotic protein Bcl2 in the human MDA-MB-231 BC cell line, following exoLF treatment [65]. Shariatikia et al. (2017) showed that mare, donkey, cow, and camel milk, and their casein and whey proteins, have potent cytotoxic effects against the MCF7 BC cell line in a dose-dependent manner, while sheep and goat milk and their proteins did not exert any cytotoxic activity [59].

A comprehensive meta-analysis of prospective cohort studies concluded that fermented milk-derived products, including yogurt and sour milk, were associated with lower cancer mortalities [66]. However, another meta-analysis conducted by Chen et al. (2019) concluded that low-fat/skimmed milk, whole milk, and yogurt intake had no effect on BCR [67]. To sustain this idea, a pooled analysis of 21 cohort studies conducted by Wu et al. (2021) showed that dietary calcium consumption was not clearly associated with BCR, but higher yogurt and cottage/ricotta cheese intake were found to be inversely correlated with the risk of ER-negative BC [68]. Cheese is one of the fermented dairy products rich in proteins, minerals, organic acids, bioactive peptides, oligosaccharides, and vitamins, and also contains diverse non-pathogenic microorganisms that act as probiotics [69]. Kamal and Talib (2023) showed that a combination of the ketogenic diet, which is high in fat, low in carbohydrates, and sufficient in terms of proteins, and probiotics inhibits BC in mice through the downregulation of IGF-1 and immune system modulation [70]. Ryser et al. (2022) showed that a gram-negative bacterium, *Morganella morganii*, isolated from the outer layers of raclette-type cheese, influences the formation of biogenic amines in cheese, such as cadaverine and putrescine, which is undesirable, since their consumption can cause intoxication [71]. Interestingly, treatment of BC cell lines with cadaverine according to its serum reference range reverted ETM, decreased cell motility and invasion, and inhibited cell stemness by reducing mitochondrial oxidation [72]. Moreover, Ritota et al. (2022) emphasized that bioactive compounds, such as picrocrocin/safranal and crocin, from cow and ewe cheeses made with saffron, usually used to add a natural yellow to orange color to cheeses, exert antiproliferative effects in regard to MDA-MB-231 BC cells [73].

Nondairy/plant-based substitutes for cow's milk and its derivatives, manufactured from soy, rice, coconut, and almonds, have gained in popularity in recent years, with adoption of vegetarian and vegan diets [74]. In the USA, the proportion of individuals reporting that they consumed soy milk was 1.54% in 2017–2020 [75]. Moreover, non-Hispanic Asian and Black ethnicities have significantly increased the consumption of soy milk [75]. Soy milk and its derivatives contain many natural isoflavones that are similar to estrogen hormones, known as phytoestrogens [76], so the beneficial effects of soy food intake remain controversial [77]. There is a hypothesis that suggests that soy isoflavones, genistein and daidzein, may stimulate the proliferation of ER+ BC cells, even at low concentrations [77]. However, genistein, a natural isoflavonoid, is also considered to be a potent anti-BC agent, usually present in high quantities in soybeans, inducing antiproliferative effects/arrest of the cell cycle and apoptosis, and preventing tumor angiogenesis [78]. In different brands of commercially available soy milk, the mean genistein content is 17.58 ± 8.38 μg/mL [76]. At high concentrations, genistein kills MCF7 BC cells [79] or delays TNBC tumor growth [80], inhibiting proliferation/differentiation and inducing apoptosis [81,82]. Genistein also inhibits angiogenesis [83], induces tamoxifen resistance and growth in ER+/HER2+ BC cells, and inhibits the growth of ER-/HER2+ BC cells [84]. At the molecular level, genistein suppresses the IGF-1R/p-AKT signalling pathway; decreases the Bcl-2/Bax ratio [81]; downregulates the NF-κB/Bcl-xL/TAp63 signaling pathway; induces the modification of key epigenetic cancer-associated genes, their enzymatic activities, genomic DNA, and histone methylation [80]; upregulates PI3K and MAPK signalling; and downregulates

$p27^{Kip1}$ levels in ER+/HER2+ BC cells [84]. In addition, daidzein, an isoflavone from fruits, nuts, soy beans, and soy-based products [85], inhibits BC cell proliferation, induces apoptosis [85,86], and inhibits TNF-α-induced migration and invasion [87]. The antineoplastic effects of daidzein are mediated by cell cycle arrest, the inhibition of cyclin D and CDK2/4, and increases in $p21^{Cip1}$, $p57^{Kip2}$ expression, and caspase-9 activity [85]. Moreover, daidzein generates ROS, disrupts mitochondrial function, decreases cyclin-D expression [86], inhibits TNF-α, suppresses Hedgehog/GLI1 signalling [87], upregulates Bax, downregulates Bcl-2, induces apoptosis, lowers the ERα/β ratio, and increases ROS production [88].

Meat has been a part of the human diet since the early stages of our existence, with archeological findings suggesting that food was cooked for the first time around 300,000–400,000 years ago (some 7000–14,000 generations ago) [89]. Even if meat consumption is correlated with an increased BCR, Lo et al. (2019) showed that poultry consumption may be associated with reduced risk, so substituting poultry for red meat could decrease BCR [90].

Grains are one of the most important foods consumed worldwide that contain bioactive compounds, mainly found in whole grain cereals, such as wheat, rye, oats, and barley [91,92]. A meta-analysis conducted by Xiao et al. (2018) concluded that high intake of whole grains might be inversely associated with BCR in case-control studies, but not in cohort studies [91]. However, carbohydrate-based foods with high glycemic index may influence BCR via the insulin growth factor axis [93]. The contamination of cereals and cereals-based products with mycotoxins, such as aflatoxins, has been associated with mutagenesis and carcinogenesis [94,95]. Aljazzar et al. (2022) showed that aflatoxin B1 (AFB1) causes significant toxicity in regard to MCF7 BC cells through oxidative stress (OS), as well as transcriptomic alterations in regard to drug-metabolizing enzymes, transporters, and antioxidant enzymes as a result of in xenobiotics [95]. Moreover, AFB1 upregulates pro-inflammatory markers, such as tumor necrosis factor alpha (TNFα), cytokine, and cyclooxygenase-2, correlated with the reduction in the mRNA expression of immunity-related genes, including interleukins 8 and 10 [95]. In BC, TNFα has a pro-tumorigenic role and correlates with increased proliferation, a higher malignancy grade, metastasis, and poor prognosis [96,97], due to the ability of TNFα to upregulate TAZ, a transcriptional co-activator that promotes BCSC self-renewal in human BC cell lines [98]. Dietary baker's yeast, commonly used in baking bread and other bakery products, can enhance the apoptotic ability of paclitaxel against the human MCF7 BC cell line and the metastatic murine 4T1 cell line [99]. A systematic review conducted by Grudzinska et al. (2023) emphasized that dietary sprouts may play a role in the chemoprevention of BC [100]. Thus, germinated wheat flour reduces the growth of MCF7 and MDA-MB-231 human BC cell lines, upregulating apoptosis [101].

Eggs are also frequently consumed worldwide as a nutrient-rich food, due to their protein and peptide content [102]. Due to advancements in MS-based proteomics, 167 proteins were identified in the egg white proteome, which varies based upon the storage of the eggs at different temperatures over different time spans, early embryonic development, egg varieties, and stress conditions [103]. Ovalbumin and ovotransferrin have been detected as major egg proteins responsible for multiple bioactivities [103,104]. Thus, the native ovotransferrin (OTRF/OTF)/conalbumin, an iron-binding glycoprotein from the transferrin protein family, present both in avian plasma and egg white [105], is known to have anticancer activities, with negative effects on the proliferation of MCF7 BC cells by inducing apoptosis [104,106].

Nutritional supplements are also consumed at breakfast time [30]. For example, bee pollen is an excellent dietary supplement in regard to human nutrition [107]. Pollen, also known as bee bread (BB), contains various bioactive polyphenolic compounds, such as isoflavonoids (genistein, daidzein, glycitein, biochanin A, formononetin, puerarin, coumestrol, and equol) and flavonoids, like quercetin, kaempferol, apigenin, luteolin, myricetin, hesperetin, rutin, naringenin, catechin, epicatechin, epigallocatechin, and proanthocyanidins [108]. Genistein is the major isoflavone identified in bee pollen [109], interfering with several biological processes, pathways, and genes/proteins, including PTEN, PI3K, PIP3, AKT, mTOR, Bcl-2 Bax, caspase-3, cyclin B, VEGF, HIF1α, p21, and p16, EGFR/AKT/NF-κB, DNA methylation, the ER pathway, and MMP genes [78]. At a concentration of 370 μM, genistein revealed a cytotoxic effect on MCF7, T47D, and MDA-MB-231 BC cell lines, while at a concentration of 0.37 μM, no significant effect on BC cell viability was observed [110]. Synergically, doxorubicin (DOX), cisplatin, and BB, blocked the migration of MDA-MB-231 cells and suppressed the proapoptotic BID gene, overexpressing the anti-apoptotic Bcl-2 gene, reducing the toxicity of chemotherapeutic drugs on this TNBC cell line [111]. Another study found that the synergistic effect of BB with DOX led to the suppression of the proliferation of breast tumors in 4T1 tumor-bearing BALB/C mice and inhibited the oxidative damage of DOX, increasing the expression of apoptotic genes and proteins, the p53 level, as well as serum interferon-γ (IFN-γ), and reducing the estrogen level, Ki-67 and Bcl-2 proliferation biomarkers, nitric oxide, and pro-inflammatory cytokines [108].

Leafy vegetables can be effective in terms of "green chemoprevention" and treating BC [112]. Moreover, broccoli and broccoli sprouts contain many active biochemicals, such as sulforaphane (SFN), a natural organosulfur compound known to counteract the tumorigenic effects of chemical xenobiotics in food and the environment [112]. Kaboli et al. (2020) concluded that SFN reduces NF-κB activity, downregulates apoptosis inhibitors, decreases the activity of histone deacetylases leading to cell cycle arrest, as well as increases the sensitivity of BC cells to chemotherapy [113]. Moreover, SFN is involved in the modulation of gene expression and nuclear factor-erythroid factor 2-related factor 2 (NRF2) antioxidant signaling [113]. Wu et al. (2019) indicated that NRF2 signaling has a double-edged sword effect in regard to cell survival, because the NRF2/KEAP1 pathway exerts anticancer activities, but also activates pro-survival genes and promotes cancer cell proliferation [114]. Moreover, Surh (2021) emphasized that SFN blocks the T cell-mediated immune response necessary for tumor immune surveillance [115]. Recently, Zhang et al. (2022) showed that SFN derived from broccoli, kale, cabbage, cauliflower, garden cress, and mustard [112,113], suppresses the metastasis of TNBC cells by targeting the RAF/MEK/ERK pathway to inhibit the formation of actin stress fibers and TGF-β1-induced BC cell migration, invasion, and metastasis [116].

Evidence suggests that consuming a high amount of onions (*Allium cepa*) and garlic (*Allium sativum*) is protective against BC [117]. Allicin, the major active biocompound present in freshly crushed garlic [118], is also a bioactive organo-sulfur compound [119], able to induce cell cycle arrest and apoptosis in MCF7 and MDA-MB-231 BC cell lines through tumor-suppressor p53 signaling pathway activation [120]. Moreover, Shi et al. (2024) showed that a combinatorial treatment of allicin with doxorubicin (DOX) resulted in better effects in regard to inhibiting proliferation and increasing the apoptosis of MCF7 and MDA-MB-231 DOX-resistant BC cells, than the treatment with DOX or allicin alone [118]. Thus, allicin inactivates the NRF2/HO-1 signaling pathway and improves the DOX sensitivity of BC cells [118]. Onion contains allicin, quercetin, fisetin, and other organo-sulfur compounds, such as diallyl disulfide and diallyl trisulfide [121]. Onion, one of the most popular vegetables in the world, is a major source of quercetin [122,123], a flavonoid abundantly found in plants, vegetables, and fruits, mainly in cruciferous vegetables, grapes, apples, tomatoes, and blueberries [124]. Quercetin inhibits the proliferation, migration, and invasion of 4T1 BC cells, suppressing the IL-6/JAK2/STAT3 signaling pathway and promoting the cytotoxicity of tumor immune cells in the TME [125]. Moreover, quercetin induces apoptosis and suppresses cell proliferation in MCF7 and MDA-MB-231 BC cells, changing endonuclease-G (Endo-G) and the expression of caspases 3/8/9 [124].

It was also shown that lycopene has anticancer activities and that it could be considered as a potentially effective compound in BC prevention and treatment [126,127]. The regulation of oxidative and inflammatory processes, angiogenesis, the induction of apoptosis, the inhibition of cell proliferation and metastasis formation, as well as the modulation of gap junctional intercellular communication, growth factors, and signal transduction pathways, have been the most cited mechanisms in terms of lycopene action [127,128]. Lycopene is a red-colored carotenoid pigment found in tomatoes and tomatoes-based products, red fruits, red carrots, watermelons, red grapefruits, papayas, and apricots, which is known to enhance protection against cancer [127]. Takeshima et al. (2014) showed that lycopene induces ERK1/2 activation, cyclin D1 suppression, and p21 upregulation in MCF7 (ER/PR positive), SK-BR-3 (HER2+), and MDA-MB-468 (TNBC) cell lines [129]. Lycopene inhibits AKT phosphorylation and mTOR in TNBC cells, leading to the upregulation of proapoptotic Bax [129]. Unfortunately, chlorpyrifos (CPF) is an organophosphate insecticide extensively used in the production of tomatoes [130]. Ventura et al. (2019) demonstrated that the concentration of this xenobiotic in the environment alters mammary histology and the hormonal balance in chronically exposed rats, acting as a BCR factor [131]. These authors emphasized that CPF alters HDAC1 mRNA expression, which promotes mammary tumor development [131]. Moreover, CPF acts as an endocrine disruptor that promotes migration, invasion, and the stemness phenotype in 3D cultures of MCF7 and MDA-MB-231 BC cell lines and induces the activation of many BC-related pathways, such as EMT [132].

Resveratrol is a natural dietary polyphenol that affects the expression of several cytokines, caspases, MMPs, adhesion molecules, and growth factors. It modulates the activity of several signalling pathways, such as PI3K/AKT, NF-κB, and Notch, which play crucial roles in carcinogenesis [133,134]. Moreover, resveratrol induces Bax-dependent, but p53-independent, apoptosis in MDA-MB-231 BC cells [135]. Resveratrol exists in 70 types of plants and can inhibit the migration and metastasis of MDA-MB-231 human BC cells by reversing transforming growth factor (TGF)-β1-induced EMT [136]. In vitro, resveratrol can decrease the expression levels of MMP2 and MMP9, fibronectin, α-SMA, P-PI3K, SMAD2, SMAD3, P-SMAD2, P-SMAD3, vimentin, Snail1, and Slug, increasing the expression levels of E-cadherin [136]. In vivo, resveratrol inhibits lung metastasis in mice bearing MDA-MB-231 human BC xenografts [136].

Curcumin is a polyphenol derived from turmeric (*Curcuma longa*) that inhibits breast cancer stem cell (BCSC) properties and cell proliferation and promotes apoptosis in the MCF7 BC cell line [137,138]. This polyphenol inhibits the proliferation of BCSCs through the modulation of several signaling pathways, such as NF-kB signaling, which is known as an important curcumin-regulated pathway [139]. NF-kB signaling is involved in the maintenance of a variety of stem cells [140], including BCSCs, which overexpress components of the NF-kB signaling pathway and have high NF-kB activity levels [141]. Moreover, sonic hedgehog (Shh) and WNT/β-catenin signalling pathways are also crucial in maintaining the stemness of BCSCs, with curcumin decreasing the activity of BCSCs by inhibiting tumor sphere formation and decreasing BCSC biomarkers, such as CD44, ALDHA1, Nanog, OCT4, and SOX2, thus downregulating both the Shh and WNT/β-catenin signalling pathway activities, which results in BCSC inhibition [142]. In addition, nuclear factor erythroid 2-related factor 2 (NRF2) is known to regulate oxidative stress, being involved in the development of cancer stem cells and metastasis [143]. Many authors have demonstrated or reviewed how curcumin activates the NRF2 signaling pathway, inducing cellular protection against oxidative injury [144,145]. Moreover, curcumin has very little toxicity in terms of normal stem cells, but has numerous cytotoxic effects on CSCs, due to the suppression of IL-6, IL-8, and IL-1, which stimulate CSCs [146]. Overall, curcumin could function as a cytotoxic and anti-metastasis agent for BC [142]. These chemopreventive, anticancer, and cytotoxic proprieties are also modulated through the downregulation of oncogenic RAF-1, the suppression of telomerase, and the upregulation of TNF-α and IL-8 genes [138]. In MDA-MB-231 and Hs578T TNBC cells, curcumin inhibits motility and migration, downregulating the expression of the proteins involved in EMT, such as the mTOR and PI3K/AKT signalling pathways [147]. Curcumin also downregulates the mRNA expression of vimentin, fibronectin, and β-catenin, and upregulates E-cadherin mRNA expression levels [148].

Black and green tea can have chemopreventive effects on BC development, with some authors suggesting that women with a family history of BC should drink about five cups of tea per day in order to decrease the BCR [149]. Flavan-3-ols are a subclass of flavonoids, consisting of polyphenolic phytochemicals found in a wide variety of food sources, especially in fruits and teas [150–153]. Catechins and theaflavins are both flavan-3-ols with significant bioactive properties, including anticancer, anti-mutagenic, antioxidative, and anti-inflammatory effects [154]. The beneficial effects, at the molecular level, of bioactive compounds that exert anti-BC potential are listed in Figure 1 and Table 1.

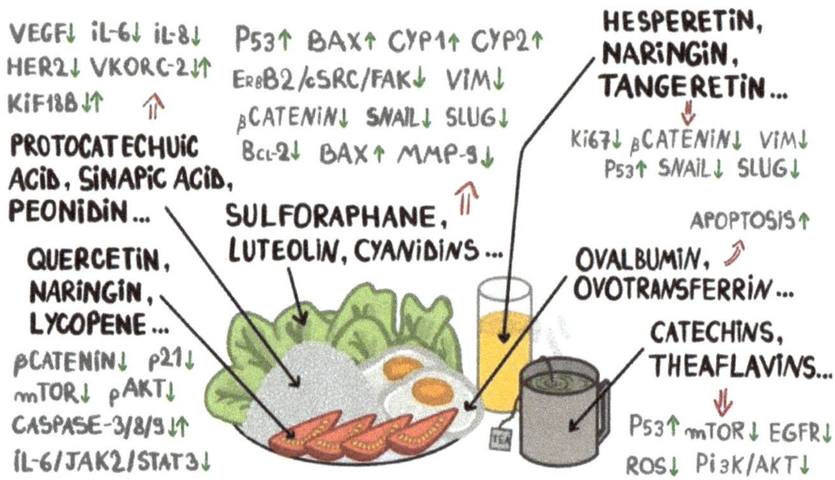

Figure 1. Biomarkers and biological pathways involved in breast cancer initiation and progression, targeted by bioactive compounds present in a daily, diverse diet.

Table 1. Potential anticarcinogenic roles of phytochemicals in regard to breast cancer.

Bioactive Dietary Compounds	Food Sources	In Vitro and In Vivo Models	Effects on Molecular Biomarkers/ Signaling Pathways	Effects on Biopathological Processes	Role in BC
Sulforaphane	Organo-sulfur compound obtained from broccoli/broccoli sprouts, kale, cabbage, cauliflower, garden cress, mustard [112,113]	MDA-MB-231 and MDA-MB-157 [116]	Targets MAPK/ERK [116]; downregulates NF-κB, AKT, and KEAP1; affects histone deacetylases involved in chromatin remodeling, and NRF2 antioxidant signaling [113]	Inhibits cell proliferation; causes apoptosis and cell cycle arrest; has antioxidant activities [113]; suppresses TGF-β1-induced migration, invasion, and metastasis of TNBC cells [116]	Chemoprotective [116], putative potential for BC treatment [113]
Allicin	Organo-sulfur compound from garlic (*Allium sativum*) [119]	MCF7 and HCC-70 [155], MCF7 and MDA-MB-231 [120]	Downregulates caspase 3/8/9 and Bcl-XL; upregulates NOXA, p21, and BAK expression [155]; induces p53 activation [120]	Decreases BC cell proliferation and viability and increases apoptosis, induces cell cycle arrest [120,155], improves DOX sensitivity [118]	Antitumor [155]

Table 1. *Cont.*

Bioactive Dietary Compounds	Food Sources	In Vitro and In Vivo Models	Effects on Molecular Biomarkers/ Signaling Pathways	Effects on Biopathological Processes	Role in BC
Quercetin	Flavonoid from fruits, vegetables (cruciferous vegetables, grapes, apples, tomatoes, blueberries), and herbal products (*Hypericum perforatum*, *Sambucus nigra*) [156]	4T1 and xenograft mouse model [125]; MCF-7 and MDA-MB-231 [124]	Suppresses IL-6/JAK2/STAT3 pathway [125], modulates the expression of caspase-3/8/9 [124]	Suppresses TNBC progression (proliferation, migration, and invasion) [125], induces apoptosis [124]	Potential anti-BC agent [124]; potential adjuvant for immune therapy in TNBC [125]
Luteolin/luteolol/ digitoflavone	Flavonoid from carrots, broccoli, celery, perilla mint leaves and seeds, apple skin, cabbages, parsley, onion leaves, thyme [157–159]	MDA-MB-231, BT549, and mouse model [157]; MDA-MB-231, MDA-MB-486, 4T1, BT549 [158]; MDA-MB-453 and MCF7 [160]	Reverses EMT through the suppression of β-catenin and VIM; stimulates E-cadherin and claudin; downregulates N-cadherin, Snail, and Slug; reorganizes F-actin [157]; inactivates AKT/mTOR and downregulates MMP9 through H3K27Ac and H3K56Ac [158]; upregulates miR-203, inhibits Ras/Raf/MEK/ ERK, downregulates Bcl-2, upregulates Bax, impedes TGFβ1-induced EMT, decreases VIM, ZEB1, and N-cadherin, and increases E-cadherin [160]	Inhibits migration and invasion of TNBC cells [157], inhibits proliferation and metastasis, and promotes apoptosis of AR+ TNBC cells [158]	Chemopreventive and potential therapeutic agent for TNBC [157], including AR+ TNBC [158]
Hesperetin	Flavanone glycoside from *Citrus* fruits (oranges and lemons) [161]	MDA-MB-231 [161]; MCF7, including mammospheres [162–164]	Inhibits the Fyn/paxillin/RhoA signalling pathway [161]; activates the ASK1/JNK pathway, initiates the accumulation of ROS [162]; modulates the expression of p53, PPARG, and Notch1 [163]; downregulates CDK2/4 and cyclins, and upregulates $p21^{Cip1}$ and $p27^{Kip1}$, stimulates the binding of CDK4 to $p21^{Cip1}$ [164]	Inhibits the migration and invasion induced by TGF-β1 [161]; exerts cytotoxic and proapoptotic effects [162]; inhibits BCSCs, exerts cytotoxicity on mammospheres, inhibits mammospheres, colony formation and migration, modulates cell cycle and induces apoptosis [163]; suppresses proliferation and stops the cell cycle in G1 [164]	Potential anti-BC agent, especially for TNBC [161]

Table 1. *Cont.*

Bioactive Dietary Compounds	Food Sources	In Vitro and In Vivo Models	Effects on Molecular Biomarkers/ Signaling Pathways	Effects on Biopathological Processes	Role in BC
Hesperidin	Flavanone from citrus fruits [165]	MCF7 cells [165], MDA-MB-231 [166], mammospheres [167], MDA-MB-231 [168], Wistar rats [169]	Suppresses AKT and NF-kB signalling, inhibits PD-L1 [166], increases p53 [167], binds to MCL-1 receptor [168], attenuates Ki67 [169]	Suppresses cell proliferation [165], inhibits cell migration and growth [166], suppresses mammospheres and colony formation, induces apoptosis [167], exerts cytotoxic effects [168]	Anti-BC activity [165,167], protective against DMBA-induced BC [169]
Naringenin	Flavanone glycoside from grapefruits, apples, onions, tea [170,171]	MDA-MB-231, Wistar rats induced with BC by DMBA [170], C57BL/6J mice induced with BC through a transplant of E0771 cells [172], Balb/c mice induced with BC through a transplant of transduced 4T1-Luc2 cells [173], MCF-7 [174]	Modulates mitochondrial-mediated pathway, upregulates caspase-3/7 [170], increases AMPK, decreases cyclin D1 [172], inhibits PKC, inhibits secretion of TGF-β1, causing its intracellular accumulation [173], inhibits PI3K and MAPK [174]	Inhibits cell proliferation and cell cycle, induces apoptosis, reduces the incidence of BC tumors [170], decreases cell viability in vivo, suppresses cell cycle progression [172], inhibits lung metastasis, increasing the survival rates of the mice [172], suppresses proliferation, impairs glucose uptake [174]	Antineoplastic agent [170], putative therapeutic option for TGF-β1 modulation [173], antiproliferative agent [174]
Naringin	Flavanone glycoside from tomatoes, grapefruits, and other *Citrus* fruits [175]	MDA-MB-231, MDA-MB-468, BT-549 [175]	Increases p21, decreases survivin/BIRC5, suppresses β-catenin pathway [176]	Inhibits cell proliferation, stimulates apoptosis [176]	Potential treatment agent for BC [176]
Apigenin	Flavone from parsley, onions, chamomile, oranges, wheat sprouts [177], celery, green peppers [178], thyme [179]	MDA-MB-453 and BT-474 [177], SK-BR-3 [177,178], MCF-7 [177,180], HBL-100 [177], MCF7-T, MCF7-F [179], MDA-MB-231, A549, SK-Hep1, nude mice [181]	Depletes HER2/neu and disrupts HER2/neu-GRP94 complex [177], modulates CDK1, $p21^{Cip1}$, and p53 [178], induces degradation of ERα and AIB1 [179], blocks PI3K/AKT pathway [177,181] and β4 integrin function, inhibits pAKT, inhibits cell motility, migration, and invasion [181], inhibits AKT/FOXM pathway, suppresses FOXM1, and modulates ER signalling [180]	Suppresses BC cell growth, induces apoptosis [177], inhibits proliferation [180], ref. [177], activates p53-induced apoptosis [177], inhibits growth of ERα+ BC cells [179], inhibits metastasis [181]	Potential anticancer treatment [177]

Table 1. *Cont.*

Bioactive Dietary Compounds	Food Sources	In Vitro and In Vivo Models	Effects on Molecular Biomarkers/ Signaling Pathways	Effects on Biopathological Processes	Role in BC
Tangeretin	Flavone from lemons, oranges [182], other *Citrus* fruits [183]	MCF7, MDA-MB-468, MDA-MB-231, nude mice injected with MDA-MB-231 cells [182,184], Sprague-Dawley rats induced with DMBA [185], Wistar rats induced with DMBA [183]	Inhibits STAT3 and SOX2 pathways, decreases STAT3-DNA binding, reduces STAT3 in BCSCs [182], induces CYP1A1/CYP1B1 activity [184], decreases ROS and pro-inflammatory factors, protects against LPO [185], upregulates p53/p21, suppresses MMP2/9 and VEGF, reduces PCNA and COX2 [183]	Inhibits proliferation [182–184] and metastasis [183], inhibits BCSC formation, induces apoptosis, inhibits mammospheres and colony formation [182], decreases tumorigenicity and OS levels, boosts antioxidant levels [185]	Anti-BC effects
Daidzein	Isoflavone from fruits, nuts, soy beans and soy-based products [85]	MCF-7 [85,86,186], MDA-MB-453 [85], T47D [186], MCF-10DCIS [87]	Induces cell cycle arrest, inhibits cyclin D, CDK2/4, increases $p21^{Cip1}$ and $p57^{Kip2}$ expression, increases caspase-9 activity [85]; generates ROS, disrupts mitochondrial function [86]; inhibits TNF-α and suppresses hedgehog/Gli1 signalling [87]; upregulates Bax and downregulates Bcl-2, induces apoptosis and lowers ERα/β ratio and ROS outbursts [88]	Inhibits cell proliferation, induces apoptosis [85,86]; inhibits migration and invasion [87]	Anti-BC potential [88]
Genistein	Phytoestrogenic soy (*Glycine max*)-derived compound [81] from soy nuts, soy powder, soy milk, tofu, miso, natto [187], lupin, fava beans, kudzu, and psoralea [80]; exerts tyrosine kinase-modulating activities [84]	MCF-7 [79,81,82]; MCF7 and MDA-MB-435 transfected with human HER2 [84]; PDX mouse models for TNBC [80]	Suppresses IGF-1R/p-AKT and decreases Bcl-2/Bax [81]; downregulates NF-κB/Bcl-xL/TAp63, influences key epigenetic associated genes, genomic DNA, and histone methylation [80]; upregulates PI3K and MAPK signalling, downregulates $p27^{Kip1}$ levels in ER+/HER2+ BC cells [84]	High concentrations kill MCF7 BC cells [79] or delay TNBC tumor growth [80]; inhibits proliferation/differentiation, induces apoptosis [81,82]; inhibits angiogenesis [83]; induces tamoxifen resistance and growth in ER+/HER2+ BC cells and inhibits growth of ER-/HER2+ BC cells [84]	Exhibits anticancer effects on various cancers [188]; chemoprevention in terms of BC carcinogenesis is concentration-, exposure time-, and BC subtype-dependent

Table 1. *Cont.*

Bioactive Dietary Compounds	Food Sources	In Vitro and In Vivo Models	Effects on Molecular Biomarkers/ Signaling Pathways	Effects on Biopathological Processes	Role in BC
Genistin	A glucoside form of genistein, readily absorbed in the intestine, found in soy beans and soy-derived foods, some legumes, and vegetables [187,189]	MCF-7, MDA-MB-231 [189]	Docks to ERα, ERβ, lowers CA 15-3 levels [190]; induces negative regulation of ERα signalling pathway, suppresses expression of oncogenic biomarkers [189]	Stimulates cell cycle arrest and apoptosis, reduces BC cell growth, proliferation, and angiogenesis [189]	Chemoprevention and therapy in terms of ER+ BCs [189]; useful for potential new drug discovery for BC management and treatment [190]
Lycopene	Major carotenoid found in tomatoes, red fruits, red carrots, watermelons, grapefruits, papayas [127]	MCF7, SK-BR-3, MDA-MB-468 [129]	Inhibits pAKT and mTOR signalling pathways, upregulates Bax [129]	Inhibits cell proliferation and cell cycle progression, initiates apoptosis [128]	Chemopreventive for TNBC [129]
Gallic acid	Hydroxybenzoic acid in fruits, vegetables, medicinal plants, such as grapes, gallnuts, pomegranates, hawthorn, tea leaves, capers [191–193], honey [194]	MCF7 [191], HCC1806 [195], MDA-MB-231 [196]	Suppresses PI3K/AKT/EGFR, nuclear accumulation of β-catenin [191,195], activates mitochondrial apoptosis pathways [195]	Inhibits survival of acidity-adapted BC cells and reduces metastatic characteristics induced by acidity [191], suppresses proliferation, promotes apoptosis [195] and ferroptosis [196]	Promising therapeutic agent for metastatic BC [191], antioxidant [193], suppresses TNBC progression [195]
Vanillic acid	Hydroxybenzoic acid in medicinal plants (e.g., *Angelica sinensis*), olives, cereals, whole grains, fruits, green tea, juices, berries, wines [197]	MCF7 [198]	Affects ROS pathway [198]	Generates ROS, promotes apoptosis [198]	Antiproliferative effects [198]
Protocatechuic acid	Hydroxybenzoic acid in olives (*Olea europaea*)/olive oil, hibiscus, white grape (*Vitis vinifera*) wine [199], purple rice bran extract [200], edible mushrooms (*Hydnum repandum*) [201], potatoes, onions, wheat [202]	MCF7 [199]	Reduces IL-6, IL-8, and suppresses VEGF [199]	Induces apoptosis and limits invasion and metastasis [199]	Potent anticancer agent [199], antioxidant [202]

Table 1. *Cont.*

Bioactive Dietary Compounds	Food Sources	In Vitro and In Vivo Models	Effects on Molecular Biomarkers/ Signaling Pathways	Effects on Biopathological Processes	Role in BC
Syringic acid	Hydroxybenzoic acid from olive oil, dates, grapes [203], foxtail millet bran (*Setaria italica*) [204]	MCF7, MDA-MB-231 [204]	Downregulates GRP78/SERBP-1/SCD1 signalling axis [204]	Antiproliferative activities [203,204]	Anti-BC agent [204], antioxidant [203]
Ellagic acid	Hydroxybenzoic acid from fruits, seeds, nuts, pomegranates, raspberries, strawberries, black raspberries, almonds, and walnuts [205]	MCF7 [205]	Regulates TGF-β/SMAD3 signalling axis, inhibits CDK6, binds to ACTN4 and induces its degradation via ubiquitin–proteasome pathway, reduces VEGFR-2 [205]	Suppresses BC cell growth, migration, invasion, metastasis, stimulates apoptosis, inhibits angiogenesis [205]	Anti-BC activities [205]
Caffeic acid	Hydroxycinnamic acid from fruits, green and roasted coffee, vegetables, tea, oils, spices [206,207], honey, and propolis extracts [194,208]	MCF7 [206]	Stimulates p53 and p21 genes, inhibits CDK2 [206], inhibits DNA methylation [208]	Induces apoptosis, cytotoxic effects, morphological changes in BC cells [206]	Putative antitumor agent [206]
Cinnamic acid	Hydroxycinnamic acid from cinnamon, grapes, tea, cocoa, spinach, celery [209]	MDA-MB-231, HEK293 [209]	Increases TNF-α-TNFR1 apoptotic pathway and caspases 8/3 [209]	Increases apoptosis and DNA damage [209]	Anti-BC agent [209]
p-Coumaric acid	Hydroxycinnamic acid from whole cereal grains, fruit, vegetables, Brazilian green propolis extracts [210]	MCF7 [210], BT20, BT549, MDA-MB-231, MDA-MB-436 TNBC [208]	Inhibits iNOS, COX-2, IL-1β, TNF-α, suppresses p-IκB, ERK1/2, blocks NF-κB and MAPKs pathways [211]	Reduces cell viability/cytotoxic effects, reverts the epigenetic silencing of the tumor suppressor *RASSF1A* [208], supports anti-inflammatory and immunomodulatory mechanisms [211]	Putative antiproliferative/anticancer agent [210,212]
Ferulic acid	Hydroxycinnamic acid from plants: ferulic (*Ferula foetida*), angelica, jujube kernel, rice bran, wheat bran [213], nuts, seeds [214]	MDA-MB-231 [215], MCF7 [216]	Regulates EMT [215]	Decreases viability and proliferation, increases apoptosis via activation of caspase-8 and -9, suppresses migration and metastasis [215,216]	Antitumor agent [215], antioxidant agent that protects DNA from OS [214,217]

Table 1. *Cont.*

Bioactive Dietary Compounds	Food Sources	In Vitro and In Vivo Models	Effects on Molecular Biomarkers/ Signaling Pathways	Effects on Biopathological Processes	Role in BC
Sinapic acid	Hydroxycinnamic acid from citrus fruits (oranges, grapefruits, lemons), berries; herbs (canola, mustard seed, rapeseed); cereals, wheat, rice, spices, oil seeds, vegetables, vinegar, *Salvia officinalis*, *Myristica fragrans* [218]	MCF7, T47D, MDA-MB-468, SK-BR-3 [219]	Downregulation of *VKORC1* and *KIF18B* [219]	Induces apoptosis [219]	Cytotoxic agent in regard to luminal A BC cell lines [219]
Rosmarinic acid	Hydroxycinnamic acid from medicinal plants, herbs, spices (*Boraginaceae*, *Lamiaceae*, *Labiatae*) [220]	MDA-MB-231, MDA-MB-468 TNBC [220]	Upregulates TNF, GADD45A, BNIP3, HRK, TNFRSF25, inhibits BIRC5/survivin, MARK4, hedgehog pathway and hippo signalling, decreases proliferation and migration via Bcl-2/BAX signalling pathway, inhibits NF-κB signalling [220,221]	Antiproliferation and migration/cell cycle arrest, apoptosis [220]	Anti-BC agent, antioxidant [220]
Chlorogenic acid	Hydroxycinnamic acid from fruits (apples, plums), vegetables (potatoes, eggplants), olive oil, spices, wine, coffee beans [222–224], honey [194]	Subcutaneous tumor mouse model of 4T1 cells [222]	Inhibits NF-κB/EMT signalling pathways [222]	Induces apoptosis, inhibits pulmonary metastasis, and improves anti-BC immunity [222]	Potential candidate for therapy of BC [222]
Avenanthramides (AVN-A, B, C)	Phenolic alkaloids found in oats (*Avena sativa*, *Poaceae*) [225]	MDA-MB-231 [226]	Activates caspase 3/7 [226]	Activates apoptosis and senescence, blocks cell proliferation, inhibits EMT and metastasis [225]	Anticancer effects [225]

Table 1. *Cont.*

Bioactive Dietary Compounds	Food Sources	In Vitro and In Vivo Models	Effects on Molecular Biomarkers/ Signaling Pathways	Effects on Biopathological Processes	Role in BC
Cyanidins/ cyanidin 3-O-glucoside	Water-soluble anthocyanins found in leaves, petals, flowers, red fruits, blackberries, cranberries, grapes, cherries, apples, raspberries, peaches, plums, beans, red cabbage, red onions, purple sweet potatoes, carrots, avocadoes, olives [227,228]	BT474, MDA-MB-231, MCF7 [228,229]	Increases the expression of p53, Bax, caspase 3, *CYP1*, *CYP2*, and decreases *Bcl2* [228], blocks ERBB2/cSRC/FAK pathway [229]	Proapoptotic and cytotoxic effects [228], inhibits invasion and metastasis [230], anti-mutagenic and anticarcinogenic effects [231]	Anticancer agent [228]
Delphinidin	Polyphenolic natural pigment occurring in berries, eggplant, wine [232]	MDA-MB-231, BT474 [233]	Induces protective autophagy via suppression of mTOR and activation of AMPK pathway in HER2+ BC cells [233]	Inhibits proliferation [108], promotes apoptosis and autophagy [233], exerts anti-mutagenic and anticarcinogenic effects [231]	Anticancer effects [233], antioxidant [232]
Malvidin/ malvidin-3-O-glucoside	Abundant anthocyanin in red wine, red grapes (*Vitis vinifera*), the skin of colored fruits, blueberries (*Vaccinium corymbosum*), blackberries (*Rubus* sp.) bilberries (*Vaccinium myrtillus*), red raspberries (*Rubus idaeus*), black raspberries (*Rubus occidentalis*), cranberries (*Vaccinium macrocarpon*), strawberries (*Fragaria ananassa*) [234]	MCF7	Increases p21, caspases 3/8/9, Bax/Bcl-2, inhibits NF-κB, PI3K, TNF-α, STAT3, MMP2/9, IL-6, WNT, Notch1, and cyclin D1 [234]	Induces cell cycle arrest, antioxidation, anti-inflammation, autophagy, and apoptosis; inhibits proliferation, metastasis/cell invasion [234]	Anticarcinogenic potential [234]
Peonidin	Anthocyanidin found in purple sweet potatoes (*Ipomoea batatas*) [235], pigmented rice (red, black, dark purple) [230]	In silico [235]	Inhibits the overexpression of HER2 protein [235]	Proapoptotic, antiproliferative, anti-metastasis role [230]	Anti-BC activity [235]

Table 1. *Cont.*

Bioactive Dietary Compounds	Food Sources	In Vitro and In Vivo Models	Effects on Molecular Biomarkers/ Signaling Pathways	Effects on Biopathological Processes	Role in BC
Resveratrol	Non-flavonoid polyphenol from blueberries, grapes, red wine, raspberries, mulberries, apples, pomegranates, soy beans, peanuts	MDA-MB-231 [135]	Modulates PI3K/AKT, NF-κB, and Notch signalling pathways [133,134]	Induces Bax-dependent, but p53-independent, apoptosis [135]	Chemopreventive and putative therapeutic agent [236]
Curcumin	Polyphenol derived from turmeric (*Curcuma longa*)	MCF7 [137,138], MDA-MB-231, and Hs 578T [147]	Modulates NF-κB [137]; downregulates oncogenic RAF-1, suppresses telomerase, upregulates TNF-α and IL-8 genes [138]; inhibits EMT through downregulation of mTOR and PI3K/AKT signaling [147]	Inhibits cell stemness [148], proliferation, and promotes apoptosis [137]; suppresses motility and metastasis in TNBC [147]	Chemopreventive agent [137], anticancer and cytotoxic properties [138]; potential therapeutic agent [147]
Epicatechin	Flavan-3-ol from green tea, cocoa, grapes, apricots, green algae [237]	4T1 [237], [238], TNBC mice model [150], MCF-7 [239] [240], MDA-MB-468 [153], MDA-MB-231 [240]	Increases Bax/Bcl-2 ratio, increases the expression of CDH1, MTSS1, PTEN, BMRS, FAT1, and SMAD4 [237], modulates the AMPK and Akt/mTOR pathways [238]	Decreases cell growth [150,238], inhibits metastasis-associated proliferation, reduces migration [150], cytostatic effects at lower concentrations [239], inhibits proliferation [238], proapoptotic [153,240]	Antiproliferative agent, similar effects to doxorubicin in terms of tumor growth inhibition and survival rates [238], could be used as an inhibitor for BC progression (anti-metastatic, anti-migratory, anti-invasion) [150]
Catechins	Flavan-3-ols found in black grapes, strawberries, cider, red algae, green algae, red wines, kiwis, green tea, gooseberries [241,242]	4T1 [241]	Downregulates EGFR, APP, Bcl-2, DNMT, HIF1a, and PSMB5; upregulates caspase 3 and GADD45b [241]	Suppresses proliferation, stimulates apoptosis [241]	Antiproliferative agent [241,242]
Epigallocatechin gallate	Flavan-3-ol from green and black tea, apples, cherries, red algae, other fruits and vegetables [151,208,242]	BT20, BT549, MDA-MB-436, MDA-MB-231, MCF7, T47D, Hs 578T, allograft Balb/c model [151,208]	Downregulates mTOR, PI3K/AKT, p53/Bcl-2, EGFR, VEGF, STAT3, NF-κB, SCUBE2, TIMP3, DNMT, ERα; activates JNK, caspases 9/3 [151,208]	Decreases cell growth, increases apoptosis, prevents DNA damage and proliferation, inhibits invasion, reduces cell viability, has cytotoxic effects [208]	Antiproliferative and anti-invasion agent, hypomethylating agent [208]

Table 1. *Cont.*

Bioactive Dietary Compounds	Food Sources	In Vitro and In Vivo Models	Effects on Molecular Biomarkers/ Signaling Pathways	Effects on Biopathological Processes	Role in BC
Theaflavin	Antioxidant polyphenol found in black tea [152,243]	T47D, MDA-MB-231 [244], MCF-7, ZR-75-1 [245]	Upregulates Fas/caspase 8, downregulates pAKT/pBAD pathway [244]; increases p53, Bax, activates caspase 6/7/9, increases ROS, stimulates p53, downregulates MMP2 and MMP9, inhibits the translocation of NF-kB/p65 to the nucleus [245]	Induces apoptosis [244], probably in a p-53-dependent manner [152], reduces cell viability, inhibits cell migration, could induce p53 phosphorylation of the Ser15 residue [245]	Proapoptotic agent [244]
Theaflavin-3-gallate	Polyphenol from fruits and veggies	MCF-7, MCF-10A [246]	Downregulates HSP90, MMP9, VEGFA, and SPP1 genes [246]	Inhibits cell proliferation, no cytotoxic effects on non-malignant breast cells (MCF-10A), induces apoptosis by stopping the cell cycle in the G2/M phase, decreases migration and colony formation [246]	Potentiate other BC therapies [246]
Phlorizin	Bioactive chalcone found in *Asteraceae*, *Ericaceae*, *Saxifragaceae*, *Proteaceae*, *Rosaceae*, *Rutaceae*, *Fabaceae*, *Lamiaceae*, *Plantaginaceae*, *Pyrus communis* [247]	MDA-MB-231, T47D [248], MCF7 [249]	Inhibits ERα signalling pathway, increases apoptotic caspase 3 via p53 [249]	Stimulates apoptosis, induces cytotoxicity/genotoxicity [248,249], antioxidant, anti-inflammatory, affects the composition of gut microbiota and development [250]	Anti-BC potential [247]
Kaempferol	Flavonoid from plants, fruits, vegetables, onions, apples, berries, tea [124,251]	MCF7 [252–254], MDA-MB-231, xenograft models [124,255]	Suppresses cyclin D1, p21, TWIST, and p38 MAPK [252], downregulates SNAI2, PLAU, CSF1, inhibits IGF1/IGF1R-mediated EMT [255]	Induces apoptosis, inhibits growth, migration, and proliferation of BC cells [124,252,256]	Anticancer effects [124,252], potentiates sensitivity to chemotherapy drugs [253,254]

3. The Bad

The pro-inflammatory diet, based on a high intake of red, processed meat and alcohol [257], was associated with an increased BCR [258], while a long-term anti-inflammatory diet can improve the survival of BC patients [259]. The Western dietary pattern, rich in hydrogenated fat, soft drinks, animal fat, fast food, refined cereals, sweets, and processed meat, and the unhealthy dietary pattern, rich in sugars, processed juices, soft drinks, potato chips and mayonnaise, desserts, solid oils, red and processed meat, and high salt intake, are also associated with increased BCR [43]. Eating breakfast at a later time was associated with increased BCR among premenopausal women [26] and skipping breakfast was associated with an elevated risk in terms of all-cause and cancer-related mortality [260] and seems to be a bad idea for patients with cancer [261]. Moreover, poor diets that include refined sugar, saturated and trans fats, as well as a low level of natural antioxidants and fiber intake, were linked to increased BCR through the modulation of inflammation-related pathways and biomarkers [262] that play a key role in BC initiation and progression [263]. To exemplify this, nuclear factor-kappa B (NF-kB) is a pro-inflammatory nuclear transcription factor and the activation of the NF-kB signaling pathway is common in BC [264]. Alcohol exposure activates the NF-kB pathway and enhances the transcription of NF-kB-targeted genes [12]. Moreover, Starek-Swiechowich et al. (2023) showed that alcohol, even at low concentrations, as well as its major metabolite, acetaldehyde, causes TNBC cell proliferation, migration, and invasion via the activation of p38 MAPK and JNK phosphorylation [12]. Kansestani et al. (2019) showed that high glucose intake increased MCF7 BC cell proliferation, viability, VEGF secretion, and Bcl-2 expression, decreasing apoptosis, and stimulating angiogenesis, due to the activation of the NF-kB pathway by increasing reactive oxygen species [265].

BC progression can be also affected by systemic nutrients [14]. For example, meat consumption is annually rising at a global level, because meat is an important source of animal-based proteins, with superior anabolic potential when compared with plant-derived proteins [266], but it also enhances the risk of several types of cancer and other chronic diseases [267]. Lo et al. (2019) emphasized that the increased consumption of red meat was associated with invasive BC risk, which was also correlated with certain meat-cooking practices [90]. Inoue-Choi et al. (2015) showed that a higher intake of processed meat was associated with a 27% higher risk of localized postmenopausal BC and a 19% higher risk of distant BC [268]. Moreover, higher nitrite intake from processed meat was positively associated with localized cancer, whereas heme iron intake from red meat was positively correlated with BCR overall and all cancer stages [268]. Kim and Shin (2021) showed that processed meat consumption increases the risk of hypercholesterolemia, hypertriglyceridemia, and dyslipidemia, whereas red meat consumption also increases the risk of hypercholesterolemia, hyper-LDL cholesterolemia, and dyslipidemia [269], which have been linked to BC incidence [270]. Brindisi et al. (2022) demonstrated that cholesterol activates the estrogen-related receptor alpha (ERRα) pathway, promoting EMT in MCF7 and MDA-MB-231 BC cells [271]. Moreover, these authors concluded that BC cells exposed to high cholesterol levels promoted an increase in macrophage infiltration with the induction of the M2 phenotype, known as the tumor growth promoter [272], angiogenesis, and the induction of the cancer-associated fibroblast phenotype [271]. Recently, Magalhães et al. (2024) demonstrated that a high-cholesterol diet promotes phenotypic changes in BC cells and their intravasation through the LDL–LDLR axis, contributing to BC progression and metastasis in vitro and in various animal models [14]. Moreover, LDL also increases the serine proteinase inhibitor, clade E member 2 (SERPINE2) expression, which is known to be overexpressed in invasive ductal carcinoma of the breast [273]. Krawczynska et al. (2024) showed that neutrophils exposed to a cholesterol metabolite become able to secrete extracellular vesicles that promote EMT and stem cell characteristics in BC cells through the activation of the WNT/β-catenin signaling pathway, influencing BC progression [274]. Additionally, the low HDL cholesterol level was related to an increased BCR [270]. However, a cohort study in Japanese women, conducted by Narii et al. (2023), showed that triglycerides were not associated with BCR, while HDL cholesterol was inversely associated with BCR only in women over 50 years old [275].

Several pieces of analyses suggest that a higher intake of fish can be associated with higher incidence rates of ER+ BC [276], while oily fish intake has been found to be negatively correlated with the incidence of total BC, mainly in the cases of ER- BC [277]. Other authors have confirmed the protective effect of omega-3 fatty acids as a result of fish consumption against BC in Asian patients [278]. However, there are serious risks involving the consumption of fish contaminated with toxins, such as methylmercury, polychlorinated biphenyls (PCBs), dioxins, pesticides, and plastic waste [279]. Thus, food of aquatic origin is an important source of human exposure to methylmercury [280] and a low level of exposure to mercury can induce cancer cell proliferation by the estrogen receptor (ER), extracellular signal-regulated kinases 1/2 (ERK1/2), c-JUN NH(2)-terminal kinase (JNK), NADPH-oxidase, and nuclear factor erythroid 2-related factor 2 (NRF2) signaling, combined with anti-apoptotic and pro-survival signaling, the accumulation of DNA modifications, and the inhibition of DNA repair machinery [281]. Microplastics also accumulate in aquatic organisms and trigger various endocrine and metabolic pathways [282]. A study performed by Park et al. (2023) showed that polypropylene microplastics enhance metastasis-related gene expression and cytokines in BC cells [283]. Fish absorb dioxins and PCBs from their environment and these toxins enter the human body through food, including the consumption of fatty fish [284]. Invasive BCR was positively associated with dioxin exposure [285].

The various transformations involving processed meat can lead to the formation of harmful and potentially carcinogenic compounds [286]. Thus, processed meat contains polycyclic aromatic hydrocarbons (PAHs), heterocyclic aromatic amines (HAAs), residues, such as nitrosamines, biogenic amines, and a wide variety of other contaminants, such as antibiotics and other veterinary medicines and growth promoters [287–290]. Low-dose PAH-enriched mixtures have been shown to upregulate aryl hydrocarbon receptor (AhR) expression and cytochrome P450 (CYP) activity in ER+ BC cells, thus increasing cell proliferation and stimulating the expression of anti-apoptotic proteins [291]. Furthermore, a cohort study concluded that PAH exposure during pregnancy could interact with tobacco smoke, thus impacting the breast tissue in mothers and daughters, potentially influencing BCR across many generations [292]. HAAs are formed during high-temperature meat cooking and are linked to BC initiation [293–295]. Moreover, aromatic amines are generally highly lipophilic and can accumulate in the fatty tissue of breasts [296]. Nitrosamines are potent carcinogens formed in processed meats, when nitrites and nitrates interact with amines during food processing [297]. Studies have shown a positive association between BC and the consumption of processed meats containing these carcinogens [298]. Research indicates that consuming processed meat more than once a week is linked with a higher risk of BC, especially in women who also consume alcohol [299–301]. Moreover, a meta-analysis of prospective studies found that processed meat intake is associated with a 6% increase in BC development [302], while another study revealed that the said consumption is linked to a 9% higher BCR [303].

Nevertheless, milk consumption is more often associated with benefits than harm, but milk intake can be associated with a higher risk of hormone-related cancers [304]. A comprehensive meta-analysis of prospective cohort studies concluded that a high amount of whole/high-fat milk consumption can be associated with higher cancer mortality [66]. The intake of milk and dairy products has been related to a higher risk of breast and prostate cancers, due to the positive association with systemic levels of insulin-like growth factor 1 (IGF-1), insulin and estrogen signaling [305,306], and the IGF signaling pathway, implicated in the regulation of breast cancer stem cells (BCSCs), EMT, local migration and invasion, angiogenesis, and chemotherapy resistance [307]. More than 400 fatty acids (FAs) have been detected in milk; however, saturated fatty acids (SFAs) are present in a greater concentration [308]. Recently, based on a systematic review and meta-analysis, Mei et al. (2024) emphasized that high total SFA levels have been correlated with increased BCR [309]. Additionally, Jiang et al. (2024) also studied the association between dietary intake of SFAs and BCR, emphasizing that this impact depends on the carbon chain lengths of SFAs, attributable to the dietary sources and biological activities of such compounds [310].

An increased level of intake of omega-3 fatty acids associated with a decrease in omega-6, resulting in a higher omega-3/omega-6 ratio, is inversely associated with BCR [311]. Linoleic acid (LA), an omega-6 acid, predominantly present in oil seeds (soy bean, sunflower, rapeseed, and cotton), as well as α-linolenic acid (ALA), an omega-3 fatty acid found in flaxseed and fresh forage, are the FA precursors of conjugated linoleic acid (CLA) synthesis [56]. Espinosa-Neira et al. (2011) previously demonstrated that LA, an essential and the major polyunsaturated fatty acid (PUFA) in most diets, induces an EMT-like process in the mammary epithelial cells, MCF10A [312]. The authors showed that LA promotes a decrease in E-cadherin expression, as well as in increase in Snail1, Snail2, Twist1, Twist2, Sip1, vimentin, and N-cadherin expression [312]. Moreover, LA induces focal adhesion kinase (FAK) and NF-κB activation, and promotes an increase in MMP2 and MMP9 secretion, cell migration, and invasion in MCF10A cells [312]. Serna-Marquez et al. (2017) demonstrated that LA induces AKT2 activation, invasion, increases NF-κB-DNA binding activity, miR34a upregulation, and miR9 downregulation in the MDA-MB-231 BC cell line [313].

Milk also contains a plethora of proteins that may develop malignant activities. Thus, lactadherin/milk fat globule-EGR factor 8 (MFG-E8) is a glycoprotein associated with the milk fat globule membrane that plays crucial roles in cell adhesion and the promotion of angiogenesis, thus overexpressed lactadherin has been associated with poor prognosis and low survival in BC and other types of malignancies [314]. MFG-E8 can be considered as a potential biomarker and therapeutic target for breast carcinoma, emphasizing the decreased expression in ER+ and HER2+ BCs, while highly expressed in TNBC cell lines and patient sera [315]. Consequently, MFG-E8 downregulation was associated with cell cycle arrest and cell apoptosis, also inhibiting the expression of MMPs and EMT-associated proteins [315]. Osteopontin (OPN), a secreted multifunctional phosphorylated protein [60] present in milk from cows, buffalos, sheep, goats, and yaks [316], has also been identified to be highly expressed in the tumor tissue and plasma of BC patients, including TNBC tissues and cells, in association with a poor clinical prognosis and reduced survival, while OPN downregulation inhibits BC skeletal metastasis in vitro [317,318]. Recently, Guo et al. (2024) showed that OPN promotes tumor growth and metastasis and GPX4-mediated anti-lipid peroxidation in TNBC by activating the PI3K/Akt/mTOR pathway, elevating cell proliferation, invasive and migratory abilities, tumor sphere formation and angiogenesis [318]. Jia et al. (2024) showed that cottage/ricotta cheese intake was causally associated with luminal A-like BC, while the risk of ER-negative BC decreased [69]. Moreover, cow's milk is a rich source of growth factors, including transforming growth factor (TGF)-β [319], a pluripotent cytokine with a key role in EMT, invasion, migration, and apoptosis, which acts as a growth inhibitor in early BC and a growth promoter in advanced stages of the disease [320].

Beyond all these negative effects, milk and milk-based products may contain pesticides that act as endocrine disruptors, with carcinogenic potential [67]. For example, atrazine, a herbicide that can bioaccumulate over time, was detected over the permissible limit for human consumption in bovine milk samples obtained from dairy farms [321]. Atrazine promotes 4T1 TNBC cell proliferation and migration, suppresses local and systemic immune function, and upregulates the expression of matrix metalloproteinase MMP2, MMP7, and MMP9 [322]. Moreover, atrazine induced IL-4 overexpression, while IFN-γ and TNF-α were found to be decreased in TME and serum [322]. Dairy products, including raw milk, ultra-high temperature milk, pasteurized milk, pasteurized and traditional butter, and pasteurized and traditional cheese, may also contain cypermethrin, deltamethrin, and hexachlorobenzene, over the maximum limit set by the EU [323]. Cypermethrin is a synthetic pyrethroid frequently used in agriculture and households for insect control [324]. Flumethrin, another pyrethroid pesticide, induces genotoxicity in MCF7 BC cells, even at low concentrations [325]. Hexachlorobenzene (HCB) is an organochlorine compound, which is able to bioaccumulate in high-fat tissues, that binds to the aryl hydrocarbon receptor, activating the membrane and nuclear pathways involved in BC development, such as ERα signaling, and insulin-like growth factor-1, epidermal growth factor, and transforming growth factor beta 1 receptors [326]. Thus, HCB stimulates epithelial cell proliferation, migration, invasion, and angiogenesis [326]. Unfortunately, even in the case of vegan milk substitutes, such as soy milk and its derivatives, genetically modified (GM) glyphosate-tolerant soy beans (GT), which dominate the soy bean market throughout the world, introduce thousands of tons of herbicides into the food chain [327] and bioaccumulate glyphosate (GLY), as well as its major degradation product aminomethylphosphonic acid (AMPA), which may itself act as an endocrine disruptor, mimicking 17β-estradiol that promotes ERα activity in BC cells [328]. The most common crops associated with the use of GLY include soy bean (*Glycine max*), corn (*Zea mays*), canola (*Brassica napus*), sugar beet (*Beta vulgaris*), and wheat (*Triticum aestivum*) [329]. Both GLY and AMPA have been shown to act as potential endocrine disruptors, while also exhibiting cytotoxic effects on BC cell lines. A study performed on MCF-7 and MDA-MB-468 cells highlighted the effects of these compounds on well-known signaling pathways, concluding that GLY and AMPA can dysregulate hedgehog, TGF-β, NOTCH, JAK-STAT, WNT, RAS, MAPK, and PI3K-AKT pathways, also affecting DNA repair processes, the cell cycle, and apoptosis [330], or increasing BC cell proliferation rates [328].

A literature search conducted by Keum et al. (2015) concluded that consuming more than five eggs/week was significantly associated with an increased BCR compared with no egg consumption [102]. As well as dairy products, home-grown eggs could be exposed to pesticides, such as hexachlorocyclohexane, aldrin, and malathion, more than commercial eggs, due to the direct interactions between eggs and the polluted environment [331]. Hexachlorocyclohexane has been associated with an increased BCR [332].

Augustin et al. (2013) emphasized that BCR revealed strong positive associations with bread and pasta consumption in women [93]. Pogurschi et al. (2021) analyzed the presence of acrylamide, which forms in several heated products consumed almost daily, such as coffee beverages, potato chips, pasta, pizza bases, cereal flakes and breakfast cereals, pancakes, pretzels, bread and pastries [333,334]. A large cohort study suggested a positive association between dietary acrylamide and BCR, mainly in premenopausal women, while other authors showed that acrylamide consumption did not increase the BCR [335,336]. A hormonal mode of action for acrylamide has been hypothesized to decipher the tumorigenesis in mammary glands based on acrylamide-induced DNA adduction and its mutagenesis-related potential [337].

Recently, a study performed by Lara-Castor et al. (2024) found that the intake of sugar-sweetened beverages among children and adolescents in 185 countries increased by 23% from 1990 to 2018 [338]. The consumption of sugar-sweetened beverages was associated with a slightly higher BCR among lean women, but no significant increases in BCR were reported overall, as well as for the consumption of artificially sweetened beverages [339]. However, there is evidence based on large-scale population-based studies that suggests that there is a positive association between a higher intake of artificial sweeteners, such as aspartame and acesulfame-K, and increased BCR [340]. Other findings, based on a systematic review and meta-analysis, confirmed that there is no association between the exposure to artificial sweeteners and the incidence of BC [341]. However, many studies sustain a direct association between sugar-sweetened beverages and weight gain, being overweight, and obesity in children and adolescents [342], whereas obesity can increase the amount of circulating proinflammatory cytokines, promote tumor angiogenesis, and stimulate cancer cell stemness, increasing BC growth, invasion, and metastasis [343]. The harmful effects, at the molecular level, of dietary compounds that exert pro-tumorigenic effects in BC are illustrated in Figure 2.

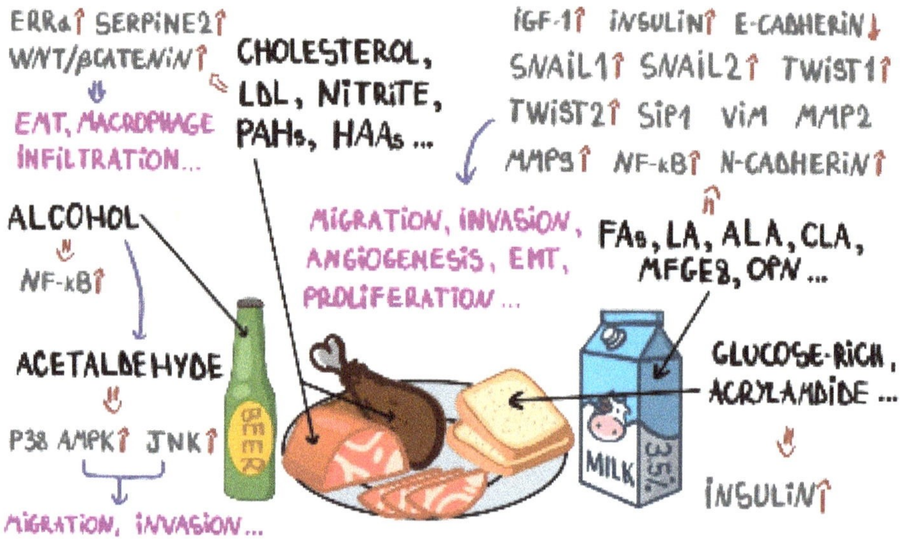

Figure 2. Biomarkers and molecular pathways involved in BC development, targeted by harmful compounds present in a daily diet.

4. Outlook

Evidence suggests that environmental exposure to chemicals and lifestyle account for 70% to 90% of the risk factors for chronic diseases, whereas only 10% to 30% can be explained by the specific genomic landscape [2]. In this review, we highlighted both "the good" and "the bad" sides of the daily human diet or dietary patterns and behaviors that influence BC risk (BCR) and incidence. Thus, pro-inflammatory diets, based on a high intake of red, processed meat, and alcohol, have been associated with increased BCR, while diets rich in antioxidants and anti-inflammatory compounds were all correlated with a lower risk of developing BC.

Milk, meat, eggs, and bread, including their derivatives, are complex foods, rich in nutrients, considered complete and basic ingredients in almost every meal and specific dietary pattern worldwide. However, milk, bread, eggs, and meat bi-directionally impact BCR. Moreover, milk contains proteins and lipids that can induce apoptosis and BC cell cycle arrest, inhibiting invasion, metastasis, and tumor angiogenesis. However, several milk proteins, such as osteopontin, may exhibit antitumorigenic characteristics, but only in certain circumstances. In addition, osteopontin can also promote tumor growth and metastasis in TNBC, by activating the PI3K/AKT/mTOR pathway, elevating BC cell proliferation, invasion, and migratory abilities, tumor sphere formation, and angiogenesis. Lactoferrin, a natural proapoptotic iron-binding multifunctional glycoprotein from bovine milk, can exert strong anticancer activities, inducing apoptosis in highly metastatic BC cell lines. Bovine milk-derived extracellular vesicles (EVs) are able to sensitize TNBC cells to doxorubicin, but many proteins from EVs provide context-based and opposing roles, acting both as BC metastasis promoters and suppressors. The intake of milk and dairy products has been related to a higher risk of breast and prostate cancers, due to their positive association with systemic levels of insulin-like growth factor 1 (IGF-1), insulin and estrogen signaling, which are implicated in the regulation of breast cancer stem cells (BCSCs), EMT, local migration and invasion, angiogenesis, and chemotherapy resistance. Linoleic acid (LA), an essential and the major polyunsaturated fatty acid (PUFA) in most diets, induces an EMT-like process in mammary epithelial cells. Lactadherin, a glycoprotein associated with the milk fat globule membrane, plays crucial roles in cell adhesion and the promotion of angiogenesis, so overexpressed lactadherin has been associated with a poor prognosis and a low survival rate in BC. Cottage/ricotta cheese intake is causally associated with luminal A-like BC and its consumption decreases the risk of ER-negative BC, while several additives derived for saffron and used to color cheese may exert antiproliferative effects on TNBC cells. Atrazine, a herbicide that can bioaccumulate over time, detected over the permissible limit for human consumption in bovine milk, promotes TNBC cells proliferation and migration, suppresses local and systemic immune function, and upregulates the expression of matrix metalloproteinases MMP2/7/9. Hexachlorobenzene (HCB), an organochlorine compound, is able to bioaccumulate in high-fat tissues, impacting ERα signaling, insulin-like growth factor-1, the epidermal growth factor, and transforming growth factor beta 1 receptors.

Meat consumption is annually rising at the global level, but red and processed meat are associated with invasive BC risk, which is also correlated with certain meat-cooking practices. High-cholesterol-based dietary patterns promote the intravasation of BC cells, as well as the progression of BC, due to the activation of ERα signaling pathway and the development an inflammatory TME. Food of aquatic origin is an important source of human exposure to methylmercury and other endocrine disruptors and even a low level of exposure can induce BC cell proliferation through the activation of ER, ERK1/2, JNK, NADPH-oxidase, and NRF2 signaling, combined with anti-apoptotic and pro-survival signaling, the accumulation of DNA modifications, and the inhibition of DNA repair machinery. Microplastics accumulated in aquatic organisms also trigger various endocrine and metabolic pathways. Low-dose polycyclic aromatic hydrocarbon (PAH)-enriched mixtures have been shown to increase BC cell proliferation and stimulate the expression of anti-apoptotic proteins. Moreover, PAH exposure during pregnancy impacts the breast tissue in mothers and daughters, potentially influencing BCR across many generations. Heterocyclic aromatic amines (HAAs) are formed during high-temperature meat cooking, are generally highly lipophilic, accumulate in the fatty tissue of the breasts, and were linked to BC initiation. Ethanol intake, alone or combined with red and processed meat, even at low concentrations, as well as its major metabolite, acetaldehyde, causes TNBC cell proliferation, migration and invasion, as well as high levels of glucose, which increases BC cell proliferation and viability, while decreasing apoptosis and stimulating tumor angiogenesis and ROS production.

Grains are one of the most important foods consumed worldwide, which contain bioactive compounds, but the contamination of cereals and cereals-based products with mycotoxins, such as aflatoxins, has been associated with mutagenesis and carcinogenesis. Common crops are associated with the use of glyphosate (GLY) and its major degradation product, aminomethylphosphonic acid (AMPA). Even in case of vegan milk or meat substitutes, such as soy-based derivatives, genetically modified glyphosate-tolerant soy beans (GT), which dominate the soy bean market throughout the world, introduce thousands of tons of herbicides into the food chain. GLY and AMPA can dysregulate hedgehog, TGF-β, NOTCH, JAK-STAT, WNT, RAS, MAPK, and PI3K-AKT pathways, also affecting DNA repair processes, the cell cycle, and apoptosis of BC cells. Acrylamide, which forms in several heated products consumed almost daily, such as coffee beverages, potato chips, pasta, pizza bases, cereal flakes and breakfast cereals, pancakes, pretzels, bread and pastries, has a hormonal mode of action, enhancing tumorigenesis in mammary glands based on acrylamide-induced DNA adduction and its mutagenesis-related potential. Ovotransferrin, an iron-binding glycoprotein present in both avian plasma and egg white, has anticancer activities, with negative effects on the proliferation of BC cells by inducing apoptosis. However, home-grown eggs could be exposed to pesticides, such as hexachlorocyclohexane, aldrin, and malathion, more than commercial eggs, due to the direct interactions between hens or eggs and the polluted environment. Many studies claim that there is a direct association between sugar-sweetened beverages and weight gain, being overweight, and obesity in children and adolescents, whereas obesity can increase the amount of circulating proinflammatory cytokines, promote tumor angiogenesis, and stimulate cancer cell stemness, increasing BC growth, invasion, and metastasis.

However, regularly eating new, diversified, colorful, clean, nutrient-rich, energy-boosting, and raw food, enables increased apoptosis and autophagy, antioxidation, cell cycle arrest, anti-inflammation, and immune response against BC cells, the reduction or inhibition of genomic instability, BC cell stemness, growth, proliferation, invasion, migration, and metastasis. Uncontrolled growth is a hallmark of cancer and is an important event during BC development and progression. Dietary compounds can induce cell cycle arrest through the upregulation of p21, and the downregulation of CDK2/4/6, survivin/BIRC5, PCNA, cyclin D1, MARK4, and RASSF1A. For example, survivin, also known as the BIRC5 protein, is overexpressed in TNBC [344], associated with a high proliferative capacity of tumor cells, so targeting survivin/BIRC5 may be an option for patients with TNBC. A diet containing tomatoes, *Citrus* fruits, medicinal plants, and spices, rich in naringin and Rosmarinic acid, could help the induction of BC cell cycle arrest. The development of cyclin-dependent kinase (CDK1/2/4/6) inhibitors with reduced toxicity is important for improving BC patient survival outcomes. Thus, hesperetin, apigenin, daidzein, ellagic, and caffeic acids, all present in herbal teas, *Citrus* fruits, nuts, berries, spices, vegetables, green and roasted coffee, oils, honey, and propolis extracts, downregulate CDK 1/2/4/6 that, as well as cyclins, play an important role in BC cell cycle progression [345]. The cyclin-dependent kinase inhibitor 1A (p21/CIP1) also has an important role in the cell cycle by inhibiting the activity of CDKs [346]. Allicin found in garlic, hesperidin from Citrus fruits, naringin, apigenin, tangerine, daidzein, caffeic acid, malvidin, an abundant anthocyanin found in red wine and colored fruits, such as all types of berries, and kaempferol, have the ability to induce cell cycle arrest. Microtubule-affinity regulating kinase 4 (MARK4) controls the early step in cell division and migration through the Hippo signaling pathway and is considered to be a potential drug target. Rosmarinic acid shows a significant binding affinity to MARK4, inhibiting its activity [347].

Inducing apoptosis is a key strategy to control excessive BC cell proliferation and natural products possess this property, stimulating proapoptotic mechanisms, including mitochondrial functions, PI3K/AKT, ROS, and MAPK-mediated pathways [348]. Thus, bioactive compounds increase ROS, caspases 3/7/8/9, the Bax/Bcl-2 ratio, BAK, NOXA, p53, and downregulate STAT3, pAKT, mTOR, JAK2, and survivin/BIRC5. Thus, all the bioactive compounds listed in Table 1 are able to induce apoptosis. Consequently, a plant-based diet and "eating the rainbow" is important for the induction of cellular death in BC. As the hallmark of BC neoplastic behavior, metastasis is a multistep process that includes cell migration and the colonization of distant niches into target organs. The epithelial–mesenchymal transition (EMT) confers BC cells with enhanced stem cells, and invasive and metastatic proprieties, so EMT blocking is crucial to avoid the spread of BC cells [349]. Thus, everyday plant-based meals contain the phytonutrients able to downregulate mesenchymal markers and EMT-associated pathways β-catenin, vimentin, N-cadherin, MMP 2/9, PI3K/AKT, mTOR, ERK1/2, Snail1, Slug, ZEB1, overexpressing epithelial biomarkers, such as claudin, E-cadherin, and PTEN. Dietary compounds also target antioxidation through activation of the NRF/KEAP1 signaling pathway with anti-BC effects, and anti-inflammation, through the downregulation of pro-inflammatory interleukins 1/6/8, NF-kB, TNF-α, COX2, and iNOS. Autophagy can be stimulated by the upregulation of p-AMPK and the downregulation of mTOR.

Eating more frequently, reducing evening energy intake, consuming early breakfast and dinner also reduce systemic inflammation and, consequently, BCR. It is important to carry out complex studies to understand, mainly at the molecular level, the bioavailability, absorption in the small intestine, distribution, metabolism, bioaccumulation, degradation, interaction with colonic microflora, elimination, and toxicity of harmful, as well as beneficial dietary compounds, on their own or in chemical mixtures, in order to translate the results from experiments involving BC cell lines, systemic reviews, meta-analysis, and cohort studies, into the most appropriate and personalized dietary patterns for each BC patient, approaching the disease in a holistic manner. Thus, we suggest that a balanced dietary structure combined with personalized nutritional behaviors can really improve treatment efficacy and BC patient outcomes.

5. Conclusions

Here, we highlighted both "the good" and "the bad" sides of daily human diet or dietary patterns and behaviors that influence BC risk and incidence, and provided an outlook on the major factors that can influence the onset and/or progression of BC. Environmental exposure to chemicals and lifestyle account for 70% to 90% of the risk factors for chronic diseases, whereas only 10% to 30% can be explained by the specific genomic landscape. So, regularly eating new food allows for increased apoptosis and autophagy, antioxidation, cell cycle arrest, anti-inflammation, and immune response against BC cells, the reduction or inhibition of genomic instability, BC cell stemness, growth, proliferation, invasion, migration, and metastasis. In addition, eating more frequently, reducing evening energy intake, consuming early breakfast and dinner also reduce systemic inflammation and, consequently, BC risk. Thus, a balanced and personalized dietary structure combined with appropriate nutritional behaviors can improve treatment efficacy and BC patient outcomes.

Author Contributions: Conceptualization, A.-N.N. and C.C.D.; literature search, A.-N.N., C.-L.J., T.M.J., K.W., N.N. and C.C.D.; writing—original draft preparation, A.-N.N., C.-L.J., T.M.J., K.W., N.N. and C.C.D.; writing—review and editing, A.-N.N., C.-L.J., T.M.J., K.W., N.N. and C.C.D.; project administration, C.C.D.; funding acquisition, C.C.D. All authors have read and agreed to the published version of the manuscript.

Funding: This publication was supported in part by the National Cancer Institute of the National Institutes of Health under Award Number R15CA260126. The content is solely the responsibility of the authors and does not necessarily represent the official views of the National Institutes of Health.

Acknowledgments: The authors thank the members of the Biochemistry and Proteomics Laboratories for the pleasant working environment. This project was supported in part by the National Cancer Institute of the National Institutes of Health under Award Number R15CA260126. The content is solely the responsibility of the authors and does not necessarily represent the official views of the National Institutes of Health.

Conflicts of Interest: The authors declare that there are no conflicts of interest.

Abbreviations

AMPK—monophosphate-activated protein kinase; APP—amyloid-beta precursor; ASK1—apoptosis signal-regulating kinase 1; BAD—Bcl-2-associated death promoter; Bax—Bcl-2-associated X protein, apoptosis regulator; Bcl-2—B-cell lymphoma 2 protein; BCSCs—breast cancer stem cells; BNIP3—Bcl-2 interacting protein 3; CDKs—cyclin-dependent kinases; COX2—cyclooxygenase 2; CSF1—colony-stimulating factor 1; DNMT1—DNA (cytosine-5)-methyltransferase 1; EGF—epidermal growth factor; EGFR—epidermal growth factor receptor; EMT—epithelial-mesenchymal transition; ERα—estrogen receptor alpha; ERK1/2—extracellular signal-regulated protein kinase 1/2; GADD45b—growth arrest and DNA damage-inducible beta; GRP78—glucose regulated protein 78; HER2—epidermal growth factor receptor 2; HIF1—hypoxia-inducible factor 1; HRK—harakiri; IGF1—insulin growth factor 1; IGF1R—insulin-like growth factor-1 receptor; IκB—IkappaB kinase; IL-6—interleukin 6; iNOS—inducible nitric oxide synthase; JAK2—Janus kinase 2; JNK—c-Jun N-terminal kinase ; KIF18B—kinesin family member 18B; KEAP1—Kelch-like ECH-associated protein 1; MAPK—mitogen-activated protein kinase; MARK4—microtubule affinity-regulating kinase 4; MCL1—myeloid cell leukemia 1; MMP-2—matrix metalloproteinase 2; MMP-9—matrix metalloproteinase 9; mTOR—mammalian target of rapamycin; MTSS1—metastasis suppressor protein 1; NF-κB—nuclear factor kappa light chain enhancer of activated B cells; NOTCH1—neutrogenic locus notch homolog protein 1; NRF2—nuclear factor erythroid 2-related factor 2; PCNA—proliferating-cell nuclear antigen; PD-L1—programmed death-ligand 1; pAKT—phosphorylated serine/threonine-specific protein kinase; pERK-phosphorylated-extracellular signal-regulated kinase; PI3K –phosphatidylinositol-4,5-bisphosphate 3-kinase; PPARG—peroxisome proliferator-activated receptor gamma; PSMB5—

proteasome subunit beta type-5; p27/Kip1—cyclin-dependent kinase inhibitor 1B; PDX—patient-derived xenograft; PLAU—plasminogen activator, urokinase; PTEN—phosphatase and TENsin homolog deleted on chromosome 10; RASSF1A—Ras association domain family protein 1 isoform A; ROS—reactive oxygen species; SCD—stearoyl-CoA 9-desaturase; SCUBE2—signal peptide, CUB domain and EGF-like domain containing 2; SERBP1—SERPINE1 mRNA-binding protein 1; SLUG—zinc finger protein SNAI2; STAT3—signal transducer and activator of transcription; TGF-β1—transforming growth factor beta 1; TIMP3—TIMP metallopeptidase inhibitor 3; TME—tumor microenvironment; TNBC—triple negative breast cancer; TNF—tumor necrosis factor; TNFRSF25—tumor necrosis factor receptor superfamily 25; VEGF—vascular endothelial growth factor; VIM—vimentin; VKORC1—vitamin K epoxide reductase complex subunit 1, WNT—wingless-related interaction site.

References

1. Arnold, M.; Morgan, E.; Rumgay, H.; Mafra, A.; Singh, D.; Laversanne, M.; Vignat, J.; Gralow, J.R.; Cardoso, F.; Siesling, S.; et al. Current and future burden of breast cancer: Global statistics for 2020 and 2040. *Breast* **2022**, *66*, 15–23. [CrossRef] [PubMed]
2. Bucher, M.L.; Anderson, F.L.; Lai, Y.; Dicent, J.; Miller, G.W.; Zota, A.R. Exposomics as a tool to investigate differences in health and disease by sex and gender. *Exposome* **2023**, *3*, osad003. [CrossRef] [PubMed]
3. Sellami, M.; Bragazzi, N.L. Nutrigenomics and Breast Cancer: State-of-Art, Future Perspectives and Insights for Prevention. *Nutrients* **2020**, *12*, 512. [CrossRef] [PubMed]
4. Kobets, T.; Smith, B.P.C.; Williams, G.M. Food-Borne Chemical Carcinogens and the Evidence for Human Cancer Risk. *Foods* **2022**, *11*, 2828. [CrossRef]
5. Rather, I.A.; Koh, W.Y.; Paek, W.K.; Lim, J. The Sources of Chemical Contaminants in Food and Their Health Implications. *Front. Pharmacol.* **2017**, *8*, 830. [CrossRef]
6. Sorrenti, V.; Burò, I.; Consoli, V.; Vanella, L. Recent Advances in Health Benefits of Bioactive Compounds from Food Wastes and By-Products: Biochemical Aspects. *Int. J. Mol. Sci.* **2023**, *24*, 2019. [CrossRef]
7. Neagu, A.-N.; Jayaweera, T.; Corrice, L.; Johnson, K.; Darie, C.C. Breast Cancer Exposomics. *Life* **2024**, *14*, 402. [CrossRef]
8. Hanahan, D. Hallmarks of Cancer: New Dimensions. *Cancer Discov.* **2022**, *12*, 31–46. [CrossRef]
9. Bhattacharya, T.; Dutta, S.; Akter, R.; Rahman, H.; Karthika, C.; Nagaswarupa, H.P.; Murthy, H.C.A.; Fratila, O.; Brata, R.; Bungau, S. Role of Phytonutrients in Nutrigenetics and Nutrigenomics Perspective in Curing Breast Cancer. *Biomolecules* **2021**, *11*, 1176. [CrossRef]
10. Mecca, M.; Sichetti, M.; Giuseffi, M.; Giglio, E.; Sabato, C.; Sanseverino, F.; Marino, G. Synergic Role of Dietary Bioactive Compounds in Breast Cancer Chemoprevention and Combination Therapies. *Nutrients* **2024**, *16*, 1883. [CrossRef]
11. Yuan, M.; Zhang, G.; Bai, W.; Han, X.; Li, C.; Bian, S. The Role of Bioactive Compounds in Natural Products Extracted from Plants in Cancer Treatment and Their Mechanisms Related to Anticancer Effects. *Oxidative Med. Cell. Longev.* **2022**, *2022*, 1429869. [CrossRef] [PubMed]
12. Starek-Świechowicz, B.; Budziszewska, B.; Starek, A. Alcohol and breast cancer. *Pharmacol. Rep.* **2023**, *75*, 69–84. [CrossRef] [PubMed]
13. Epner, M.; Yang, P.; Wagner, R.W.; Cohen, L. Understanding the Link between Sugar and Cancer: An Examination of the Preclinical and Clinical Evidence. *Cancers* **2022**, *14*, 6042. [CrossRef] [PubMed]
14. Magalhães, A.; Cesário, V.; Coutinho, D.; Matias, I.; Domingues, G.; Pinheiro, C.; Serafim, T.; Dias, S. A high-cholesterol diet promotes the intravasation of breast tumor cells through an LDL–LDLR axis. *Sci. Rep.* **2024**, *14*, 9471. [CrossRef]
15. Clemente-Suárez, V.J.; Beltrán-Velasco, A.I.; Redondo-Flórez, L.; Martín-Rodríguez, A.; Tornero-Aguilera, J.F. Global Impacts of Western Diet and Its Effects on Metabolism and Health: A Narrative Review. *Nutrients* **2023**, *15*, 2749. [CrossRef]
16. Malinowska, A.M.; Mlodzik-Czyzewska, M.A.; Chmurzynska, A. Dietary patterns associated with obesity and overweight: When should misreporters be included in analysis? *Nutrition* **2020**, *70*, 110605. [CrossRef]
17. López-Taboada, I.; González-Pardo, H.; Conejo, N.M. Western Diet: Implications for Brain Function and Behavior. *Front. Psychol.* **2020**, *11*, 564413. [CrossRef]
18. Heras, V.L.; Melgar, S.; MacSharry, J.; Gahan, C.G. The Influence of the Western Diet on Microbiota and Gastrointestinal Immunity. *Annu. Rev. Food Sci. Technol.* **2022**, *13*, 489–512. [CrossRef]
19. Newsome, R.; Yang, Y.; Jobin, C. Western diet influences on microbiome and carcinogenesis. *Semin. Immunol.* **2023**, *67*, 101756. [CrossRef]
20. Dehesh, T.; Fadaghi, S.; Seyedi, M.; Abolhadi, E.; Ilaghi, M.; Shams, P.; Ajam, F.; Mosleh-Shirazi, M.A.; Dehesh, P. The relation between obesity and breast cancer risk in women by considering menstruation status and geographical variations: A systematic review and meta-analysis. *BMC Women's Health* **2023**, *23*, 392. [CrossRef]
21. Ruo, S.W.; Alkayyali, T.; Win, M.; Tara, A.; Joseph, C.; Kannan, A.; Srivastava, K.; Ochuba, O.; Sandhu, J.K.; Went, T.R.; et al. Role of Gut Microbiota Dysbiosis in Breast Cancer and Novel Approaches in Prevention, Diagnosis, and Treatment. *Cureus* **2021**, *13*, e17472. [CrossRef] [PubMed]

22. Chen, G.; Leary, S.; Niu, J.; Perry, R.; Papadaki, A. The Role of the Mediterranean Diet in Breast Cancer Survivorship: A Systematic Review and Meta-Analysis of Observational Studies and Randomised Controlled Trials. *Nutrients* **2023**, *15*, 2099. [CrossRef] [PubMed]
23. Kalam, F.; James, D.L.; Li, Y.R.; Coleman, M.F.; A Kiesel, V.; Feliciano, E.M.C.; Hursting, S.D.; Sears, D.D.; Kleckner, A.S. Intermittent fasting interventions to leverage metabolic and circadian mechanisms for cancer treatment and supportive care outcomes. *JNCI Monogr.* **2023**, *2023*, 84–103. [CrossRef] [PubMed]
24. Hardt, L.; Mahamat-Saleh, Y.; Aune, D.; Schlesinger, S. Plant-Based Diets and Cancer Prognosis: A Review of Recent Research. *Curr. Nutr. Rep.* **2022**, *11*, 695–716. [CrossRef] [PubMed]
25. Teodoro, A.J. Bioactive Compounds of Food: Their Role in the Prevention and Treatment of Diseases. *Oxid. Med. Cell Longev.* **2019**, *2019*, 3765986. [CrossRef]
26. Palomar-Cros, A.; Harding, B.N.; Espinosa, A.; Papantoniou, K.; Pérez-Gómez, B.; Straif, K.; Ardanaz, E.; Villa, T.F.; Amiano, P.; Gómez-Acebo, I.; et al. Association of time of breakfast and nighttime fasting duration with breast cancer risk in the multicase-control study in Spain. *Front. Nutr.* **2022**, *9*, 941477. [CrossRef]
27. D'cunha, K.; Park, Y.; Protani, M.M.; Reeves, M.M. Circadian rhythm disrupting behaviours and cancer outcomes in breast cancer survivors: A systematic review. *Breast Cancer Res. Treat.* **2022**, *198*, 413–421. [CrossRef]
28. Marinac, C.R.; Sears, D.D.; Natarajan, L.; Gallo, L.C.; Breen, C.I.; Patterson, R.E. Frequency and circadian timing of eating may influence biomarkers of inflammation and insulin resistance associated with breast cancer risk. *PLoS ONE* **2015**, *10*, e0136240. [CrossRef]
29. Spence, C. Breakfast: The most important meal of the day? *Int. J. Gastron. Food Sci.* **2017**, *8*, 1–6. [CrossRef]
30. Gibney, M.J.; Barr, S.I.; Bellisle, F.; Drewnowski, A.; Fagt, S.; Livingstone, B.; Masset, G.; Moreiras, G.V.; Moreno, L.A.; Smith, J.; et al. Breakfast in Human Nutrition: The International Breakfast Research Initiative. *Nutrients* **2018**, *10*, 559. [CrossRef]
31. Ferrer-Cascales, R.; Sánchez-SanSegundo, M.; Ruiz-Robledillo, N.; Albaladejo-Blázquez, N.; Laguna-Pérez, A.; Zaragoza-Martí, A. Eat or Skip Breakfast? The Important Role of Breakfast Quality for Health-Related Quality of Life, Stress and Depression in Spanish Adolescents. *Int. J. Environ. Res. Public Health* **2018**, *15*, 1781. [CrossRef] [PubMed]
32. López-Gil, J.F.; Tully, M.A.; Cristi-Montero, C.; Brazo-Sayavera, J.; Gaya, A.R.; Calatayud, J.; López-Bueno, R.; Smith, L. Is the frequency of breakfast consumption associated with life satisfaction in children and adolescents? A cross-sectional study with 154,151 participants from 42 countries. *Nutr. J.* **2024**, *23*, 78. [CrossRef] [PubMed]
33. A Betts, J.; Richardson, J.D.; A Chowdhury, E.; Holman, G.D.; Tsintzas, K.; Thompson, D. The causal role of breakfast in energy balance and health: A randomized controlled trial in lean adults. *Am. J. Clin. Nutr.* **2014**, *100*, 539–547. [CrossRef] [PubMed]
34. Adolphus, K.; Lawton, C.L.; Dye, L. The effects of breakfast on behavior and academic performance in children and adolescents. *Front. Hum. Neurosci.* **2013**, *7*, 425. [CrossRef] [PubMed]
35. Tolfrey, K.; Zakrzewski, J.K. Breakfast, glycaemic index and health in young people. *J. Sport Health Sci.* **2012**, *1*, 149–159. [CrossRef]
36. Tang, Z.; Zhang, N.; Liu, A.; Luan, D.; Zhao, Y.; Song, C.; Ma, G. The effects of breakfast on short-term cognitive function among Chinese white-collar workers: Protocol for a three-phase crossover study. *BMC Public Health* **2017**, *17*, 92. [CrossRef] [PubMed]
37. Bellisle, F.; Hébel, P.; Salmon-Legagneur, A.; Vieux, F. Breakfast Consumption in French Children, Adolescents, and Adults: A Nationally Representative Cross-Sectional Survey Examined in the Context of the International Breakfast Research Initiative. *Nutrients* **2018**, *10*, 1056. [CrossRef]
38. Elahy, V.; Thomson, C.; Neuhouser, M.L.; Jiang, L.; Lee, S.; Pan, K.; Vitolins, M.; Chlebowski, R.; Lane, D.; Odegaard, A.O. Frequency of Consuming Breakfast Meals and After-Dinner Snacks Is not Associated with Postmenopausal Breast Cancer Risk: Women's Health Initiative Observational Study. *J. Nutr.* **2023**, *153*, 1089–1100. [CrossRef]
39. Ligorio, F.; Fucà, G.; Vingiani, A.; Iannelli, F.; Lobefaro, R.; Ferraris, C.; Belfiore, A.; Scaperrotta, G.; Depretto, C.; Martinetti, A.; et al. 248P Targeting triple-negative breast cancer metabolism with neoadjuvant chemotherapy plus fasting-mimicking diet plus/minus metformin: The BREAKFAST trial. *Ann. Oncol.* **2023**, *34*, S282–S283. [CrossRef]
40. Caffa, I.; Spagnolo, V.; Vernieri, C.; Valdemarin, F.; Becherini, P.; Wei, M.; Brandhorst, S.; Zucal, C.; Driehuis, E.; Ferrando, L.; et al. Fasting-mimicking diet and hormone therapy induce breast cancer regression. *Nature* **2020**, *583*, 620–624. [CrossRef]
41. Anemoulis, M.; Vlastos, A.; Kachtsidis, V.; Karras, S.N. Intermittent Fasting in Breast Cancer: A Systematic Review and Critical Update of Available Studies. *Nutrients* **2023**, *15*, 532. [CrossRef] [PubMed]
42. Son, D.-S.; Done, K.A.; Son, J.; Izban, M.G.; Virgous, C.; Lee, E.-S.; Adunyah, S.E. Intermittent Fasting Attenuates Obesity-Induced Triple-Negative Breast Cancer Progression by Disrupting Cell Cycle, Epithelial–Mesenchymal Transition, Immune Contexture, and Proinflammatory Signature. *Nutrients* **2024**, *16*, 2101. [CrossRef] [PubMed]
43. Zamzam, S.; Said, S.; Yaghi, J.; Faisal, F.S.; Hassan, D.; Majeed, S.A.; Al Rajabi, A.; Tayyem, R. Dietary Patterns Associated with Breast Cancer in the Middle East: A Scoping Review. *Nutrients* **2024**, *16*, 579. [CrossRef] [PubMed]
44. Soare, I.; Zugravu, C.L.; Zugravu, G.A. Research on Consumer Perception Regarding Traditional Food Products of Romania. *Foods* **2023**, *12*, 2723. [CrossRef]
45. Lequeux-Dincă, A.-I.; Preda, M.; Vijulie, I. Authentic Romanian Gastronomy—A Landmark of Bucharest's City Center. *Tour. Hosp.* **2024**, *5*, 251–275. [CrossRef]

46. Costandache, M.; Nenciu, D.-S. The structure of Romanian's food consumption and its implications on health condition and quality of life. *Rev. Romana Stat.* **2013**, *12*, 41–51.
47. Antunes, I.C.; Bexiga, R.; Pinto, C.; Roseiro, L.C.; Quaresma, M.A.G. Cow's Milk in Human Nutrition and the Emergence of Plant-Based Milk Alternatives. *Foods* **2022**, *12*, 99. [CrossRef]
48. Vincent, D.; Ezernieks, V.; Elkins, A.; Nguyen, N.; Moate, P.J.; Cocks, B.G.; Rochfort, S. Milk Bottom-Up Proteomics: Method Optimization. *Front. Genet.* **2016**, *6*, 360. [CrossRef]
49. Wajszczyk, B.; Charzewska, J.; Godlewski, D.; Zemła, B.; Nowakowska, E.; Kozaczka, M.; Chilimoniuk, M.; Pathak, D.R. Consumption of Dairy Products and the Risk of Developing Breast Cancer in Polish Women. *Nutrients* **2021**, *13*, 4420. [CrossRef]
50. Arafat, H.M.; Omar, J.; Shafii, N.; Naser, I.A.; Al Laham, N.A.; Muhamad, R.; Al-Astani, T.A.D.; Shaqaliah, A.J.; Shamallakh, O.M.; Shamallakh, K.M.; et al. The association between breast cancer and consumption of dairy products: A systematic review. *Ann. Med.* **2023**, *55*, 2198256. [CrossRef]
51. Riseberg, E.; Wu, Y.; Lam, W.C.; Eliassen, A.H.; Wang, M.; Zhang, X.; Willett, W.C.; A Smith-Warner, S. Lifetime dairy product consumption and breast cancer risk: A prospective cohort study by tumor subtypes. *Am. J. Clin. Nutr.* **2024**, *119*, 302–313. [CrossRef] [PubMed]
52. Deschasaux-Tanguy, M.; Barrubés Piñol, L.; Sellem, L.; Debras, C.; Srour, B.; Chazelas, E.; Wendeu-Foyet, G.; Hercberg, S.; Galan, P.; Kesse-Guyot, E.; et al. Dairy product consumption and risk of cancer: A short report from the NutriNet-Santé prospective cohort study. *Int. J. Cancer* **2022**, *150*, 1978–1986. [CrossRef] [PubMed]
53. Månsson, H.L. Fatty acids in bovine milk fat. *Food Nutr. Res.* **2008**, *52*, 1821. [CrossRef]
54. Huang, W.; Guo, X.; Wang, C.; Alzhan, A.; Liu, Z.; Ma, X.; Shu, Q. α-Linolenic acid induces apoptosis, inhibits the invasion and metastasis, and arrests cell cycle in human breast cancer cells by inhibiting fatty acid synthase. *J. Funct. Foods* **2022**, *92*, 105041. [CrossRef]
55. Yan, H.; Zhang, S.; Yang, L.; Jiang, M.; Xin, Y.; Liao, X.; Li, Y.; Lu, J. The Antitumor Effects of α-Linolenic Acid. *J. Pers. Med.* **2024**, *14*, 260. [CrossRef]
56. Balcazar, I.C.A.; Rivera, L.D.G.; Chavira, J.S.; Drouaillet, B.E.; Albarrán, M.R.; Martínez, Y.B. Relationship between the Composition of Lipids in Forages and the Concentration of Conjugated Linoleic Acid in Cow's Milk: A Review. *Animals* **2022**, *12*, 1621. [CrossRef] [PubMed]
57. Zeng, Y.; Liu, P.; Yang, X.; Li, H.; Li, H.; Guo, Y.; Meng, X.; Liu, X. The dietary c9,t11-conjugated linoleic acid enriched from butter reduces breast cancer progression in vivo. *J. Food Biochem.* **2020**, *44*, e13163. [CrossRef]
58. Nasrollahzadeh, A.; Tavani, S.M.; Arjeh, E.; Jafari, S.M. Production of conjugated linoleic acid by lactic acid bacteria; important factors and optimum conditions. *Food Chem. X* **2023**, *20*, 100942. [CrossRef]
59. Shariatikia, M.; Behbahani, M.; Mohabatkar, H. Anticancer activity of cow, sheep, goat, mare, donkey and camel milks and their caseins and whey proteins and in silico comparison of the caseins. *Mol. Biol. Res. Commun.* **2017**, *6*, 57–64.
60. Tan, Y.; Zhao, L.; Yang, Y.-G.; Liu, W. The Role of Osteopontin in Tumor Progression Through Tumor-Associated Macrophages. *Front. Oncol.* **2022**, *12*, 953283. [CrossRef]
61. Superti, F. Lactoferrin from Bovine Milk: A Protective Companion for Life. *Nutrients* **2020**, *12*, 2562. [CrossRef] [PubMed]
62. Pereira, C.S.; Guedes, J.P.; Gonçalves, M.; Loureiro, L.; Castro, L.; Gerós, H.; Rodrigues, L.R.; Côrte-Real, M. Lactoferrin selectively triggers apoptosis in highly metastatic breast cancer cells through inhibition of plasmalemmal V-H+-ATPase. *Oncotarget* **2016**, *7*, 62144–62158. [CrossRef]
63. Sanwlani, R.; Kang, T.; Gummadi, S.; Nedeva, C.; Ang, C.; Mathivanan, S. Bovine milk-derived extracellular vesicles enhance doxorubicin sensitivity in triple negative breast cancer cells by targeting metabolism and STAT signalling. *Proteomics* **2023**, *23*, e2200482. [CrossRef]
64. Samuel, M.; Fonseka, P.; Sanwlani, R.; Gangoda, L.; Chee, S.H.; Keerthikumar, S.; Spurling, A.; Chitti, S.V.; Zanker, D.; Ang, C.-S.; et al. Oral administration of bovine milk-derived extracellular vesicles induces senescence in the primary tumor but accelerates cancer metastasis. *Nat. Commun.* **2021**, *12*, 3950. [CrossRef] [PubMed]
65. Ramezani, R.; Mohammadian, M.; Hosseini, E.S.; Zare, M. The effect of bovine milk lactoferrin-loaded exosomes (exoLF) on human MDA-MB-231 breast cancer cell line. *BMC Complement. Med. Ther.* **2023**, *23*, 228. [CrossRef] [PubMed]
66. Jin, S.; Je, Y. Dairy Consumption and Total Cancer and Cancer-Specific Mortality: A Meta-Analysis of Prospective Cohort Studies. *Adv. Nutr. Int. Rev. J.* **2021**, *13*, 1063–1082. [CrossRef] [PubMed]
67. Chen, L.; Li, M.; Li, H. Milk and yogurt intake and breast cancer risk: A meta-analysis. *Medicine* **2019**, *98*, e14900. [CrossRef] [PubMed]
68. Wu, Y.; Huang, R.; Wang, M.; Bernstein, L.; Bethea, T.N.; Chen, C.; Chen, Y.; Eliassen, A.H.; Freedman, N.D.; Gaudet, M.M.; et al. Dairy foods, calcium, and risk of breast cancer overall and for subtypes defined by estrogen receptor status: A pooled analysis of 21 cohort studies. *Am. J. Clin. Nutr.* **2021**, *114*, 450–461. [CrossRef]
69. Jia, Y.; Feng, C.; Sun, S.; Li, C.; Li, J.; Yao, P.; Wei, X.; Lv, W.; Wang, W.; Zhang, Y.; et al. Cheese intake, probiotics and breast cancer: A Mendelian randomization analysis. *J. Funct. Foods* **2024**, *119*, 106352. [CrossRef]
70. Kamal, A.K.S.; Talib, W.H. Combination of ketogenic diet and probiotics inhibits breast cancer in mice by immune system modulation and reduction of Insulin growth factor-1. *Pharmacia* **2023**, *70*, 1411–1422. [CrossRef]

71. Ryser, L.T.; Arias-Roth, E.; Berthoud, H.; Delbès-Paus, C.; Chassard, C.; Bruggmann, R.; Irmler, S. Cadaverine, putrescine, and histamine formation of Morganella morganii in raclette-type cheese. *Int. Dairy J.* **2022**, *129*, 105362. [CrossRef]
72. Kovács, T.; Mikó, E.; Vida, A.; Sebő, É.; Toth, J.; Csonka, T.; Boratkó, A.; Ujlaki, G.; Lente, G.; Kovács, P.; et al. Cadaverine, a metabolite of the microbiome, reduces breast cancer aggressiveness through trace amino acid receptors. *Sci. Rep.* **2019**, *9*, 1300. [CrossRef] [PubMed]
73. Ritota, M.; Comitato, R.; Manzi, P. Cow and Ewe Cheeses Made with Saffron: Characterization of Bioactive Compounds and Their Antiproliferative Effect in Cervical Adenocarcinoma (HeLa) and Breast Cancer (MDA-MB-231) Cells. *Molecules* **2022**, *27*, 1995. [CrossRef] [PubMed]
74. Woodside, J.V.; Brennan, S.; Cantwell, M. Are Soy-Milk Products Viable Alternatives to Cow's Milk. In *Beverage Impacts on Health and Nutrition*, 2nd ed.; Wilson, T., Temple, N.J., Eds.; Springer International Publishing: Cham, Switzerland, 2016; pp. 151–162.
75. Storz, M.A.; Brommer, M.; Lombardo, M.; Rizzo, G. Soy Milk Consumption in the United States of America: An NHANES Data Report. *Nutrients* **2023**, *15*, 2532. [CrossRef] [PubMed]
76. Freddo, N.; Nardi, J.; Bertol, C.D.; Dallegrave, E.; Leal, M.B.; Barreto, F.; Frizzo, I.B.; Rossato-Grando, L.G. Isoflavone Quantitation in Soymilk: Genistein Content and Its Biological Effect. *CyTA-J. Food* **2019**, *17*, 20–24. [CrossRef]
77. Poschner, S.; Maier-Salamon, A.; Zehl, M.; Wackerlig, J.; Dobusch, D.; Pachmann, B.; Sterlini, K.L.; Jäger, W. The Impacts of Genistein and Daidzein on Estrogen Conjugations in Human Breast Cancer Cells: A Targeted Metabolomics Approach. *Front. Pharmacol.* **2017**, *8*, 699. [CrossRef]
78. Bhat, S.S.; Prasad, S.K.; Shivamallu, C.; Prasad, K.S.; Syed, A.; Reddy, P.; Cull, C.A.; Amachawadi, R.G. Genistein: A Potent Anti-Breast Cancer Agent. *Curr. Issues Mol. Biol.* **2021**, *43*, 1502–1517. [CrossRef]
79. Pawlicka, M.A.; Zmorzyński, S.; Popek-Marciniec, S.; Filip, A.A. The Effects of Genistein at Different Concentrations on *MCF-7* Breast Cancer Cells and *BJ* Dermal Fibroblasts. *Int. J. Mol. Sci.* **2022**, *23*, 12360. [CrossRef]
80. Sharma, M.; Arora, I.; Chen, M.; Wu, H.; Crowley, M.R.; Tollefsbol, T.O.; Li, Y. Therapeutic Effects of Dietary Soybean Genistein on Triple-Negative Breast Cancer via Regulation of Epigenetic Mechanisms. *Nutrients* **2021**, *13*, 3944. [CrossRef]
81. Chen, J.; Duan, Y.; Zhang, X.; Ye, Y.; Ge, B.; Chen, J. Genistein induces apoptosis by the inactivation of the IGF-1R/p-Akt signaling pathway in MCF-7 human breast cancer cells. *Food Funct.* **2015**, *6*, 995–1000. [CrossRef]
82. Choi, E.J.; Jung, J.Y.; Kim, G.-H. Genistein inhibits the proliferation and differentiation of MCF-7 and 3T3-L1 cells via the regulation of ERα expression and induction of apoptosis. *Exp. Ther. Med.* **2014**, *8*, 454–458. [CrossRef] [PubMed]
83. Varinska, L.; Gal, P.; Mojzisova, G.; Mirossay, L.; Mojzis, J. Soy and Breast Cancer: Focus on Angiogenesis. *Int. J. Mol. Sci.* **2015**, *16*, 11728–11749. [CrossRef] [PubMed]
84. Yang, X.; Yang, S.; McKimmey, C.; Liu, B.; Edgerton, S.M.; Bales, W.; Archer, L.T.; Thor, A.D. Genistein induces enhanced growth promotion in ER-positive/erbB-2-overexpressing breast cancers by ER-erbB-2 cross talk and p27/kip1 downregulation. *Carcinogenesis* **2010**, *31*, 695–702. [CrossRef] [PubMed]
85. Choi, E.J.; Kim, G.-H. Daidzein causes cell cycle arrest at the G1 and G2/M phases in human breast cancer MCF-7 and MDA-MB-453 cells. *Phytomedicine* **2008**, *15*, 683–690. [CrossRef] [PubMed]
86. Jin, S.; Zhang, Q.Y.; Kang, X.M.; Wang, J.X.; Zhao, W.H. Daidzein induces MCF-7 breast cancer cell apoptosis via the mitochondrial pathway. *Ann. Oncol.* **2009**, *21*, 263–268. [CrossRef]
87. Bao, C.; Namgung, H.; Lee, J.; Park, H.-C.; Ko, J.; Moon, H.; Ko, H.W.; Lee, H.J. Daidzein Suppresses Tumor Necrosis Factor-α Induced Migration and Invasion by Inhibiting Hedgehog/Gli1 Signaling in HumanBreast Cancer Cells. *J. Agric. Food Chem.* **2014**, *62*, 3759–3767. [CrossRef]
88. Kumar, V.; Chauhan, S.S. Daidzein Induces Intrinsic Pathway of Apoptosis along with ER α/β Ratio Alteration and ROS Production. *Asian Pac. J. Cancer Prev.* **2021**, *22*, 603–610. [CrossRef]
89. Furness, J.B.; Bravo, D.M. Humans as cucinivores: Comparisons with other species. *J. Comp. Physiol. B* **2015**, *185*, 825–834. [CrossRef]
90. Lo, J.J.; Park, Y.-M.; Sinha, R.; Sandler, D.P. Association between meat consumption and risk of breast cancer: Findings from the Sister Study. *Int. J. Cancer* **2019**, *146*, 2156–2165. [CrossRef]
91. Xiao, Y.; Ke, Y.; Wu, S.; Huang, S.; Li, S.; Lv, Z.; Yeoh, E.-K.; Lao, X.; Wong, S.; Kim, J.H.; et al. Association between whole grain intake and breast cancer risk: A systematic review and meta-analysis of observational studies. *Nutr. J.* **2018**, *17*, 87. [CrossRef]
92. Pajari, A.M.; Freese, R.; Kariluoto, S.; Lampi, A.M.; Piironen, V. Bioactive compounds in whole grains and their implications for health. In *Whole Grains and Health*, 2nd ed.; Landberg, R., Scheers, N., Eds.; John Wiley & Sons Ltd.: Chichester, UK, 2021; pp. 301–336. [CrossRef]
93. Augustin, L.S.A.; Malerba, S.; Lugo, A.; Franceschi, S.; Talamini, R.; Serraino, D.; Jenkins, D.J.A.; La Vecchia, C. Associations of bread and pasta with the risk of cancer of the breast and colorectum. *Ann. Oncol.* **2013**, *24*, 3094–3099. [CrossRef] [PubMed]
94. Kumar, P.; Gupta, A.; Mahato, D.K.; Pandhi, S.; Pandey, A.K.; Kargwal, R.; Mishra, S.; Suhag, R.; Sharma, N.; Saurabh, V.; et al. Aflatoxins in Cereals and Cereal-Based Products: Occurrence, Toxicity, Impact on Human Health, and Their Detoxification and Management Strategies. *Toxins* **2022**, *14*, 687. [CrossRef] [PubMed]
95. Aljazzar, A.; El-Ghareeb, W.R.; Darwish, W.S.; Abdel-Raheem, S.M.; Ibrahim, A.M.; Hegazy, E.E.; Mohamed, E.A. Effects of aflatoxin B1 on human breast cancer (MCF-7) cells: Cytotoxicity, oxidative damage, metabolic, and immune-modulatory transcriptomic changes. *Environ. Sci. Pollut. Res.* **2022**, *30*, 13132–13140. [CrossRef] [PubMed]

96. Mercogliano, M.F.; Bruni, S.; Elizalde, P.V.; Schillaci, R. Tumor Necrosis Factor α Blockade: An Opportunity to Tackle Breast Cancer. *Front. Oncol.* **2020**, *10*, 584. [CrossRef] [PubMed]
97. Cruceriu, D.; Baldasici, O.; Balacescu, O.; Berindan-Neagoe, I. The dual role of tumor necrosis factor-alpha (TNF-α) in breast cancer: Molecular insights and therapeutic approaches. *Cell. Oncol.* **2020**, *43*, 1–18. [CrossRef]
98. Liu, W.; Lu, X.; Shi, P.; Yang, G.; Zhou, Z.; Li, W.; Mao, X.; Jiang, D.; Chen, C. TNF-α increases breast cancer stem-like cells through up-regulating TAZ expression via the non-canonical NF-κB pathway. *Sci. Rep.* **2020**, *10*, 1804. [CrossRef]
99. Badr El-Din, N.K.; Mahmoud, A.Z.; Hassan, T.A.; Ghoneum, M. Baker's Yeast Sensitizes Metastatic Breast Cancer Cells to Paclitaxel In Vitro. *Integr. Cancer Ther.* **2018**, *17*, 542–550. [CrossRef]
100. Grudzińska, M.; Galanty, A.; Paśko, P. Can edible sprouts be the element of effective chemopreventive strategy?—A systematic review of in vitro and in vivo study. *Trends Food Sci. Technol.* **2023**, *139*, 104130. [CrossRef]
101. Cho, K.; Lee, C.W.; Ohm, J. In Vitro Study on Effect of Germinated Wheat on Human Breast Cancer Cells. *Cereal Chem.* **2016**, *93*, 647–649. [CrossRef]
102. Keum, N.; Lee, D.H.; Marchand, N.; Oh, H.; Liu, H.; Aune, D.; Greenwood, D.C.; Giovannucci, E.L. Egg intake and cancers of the breast, ovary and prostate: A dose-response meta-analysis of prospective observational studies. *Br. J. Nutr.* **2015**, *114*, 1099–1107. [CrossRef]
103. Mann, K. Chapter 16—Proteomics of Egg White. In *Proteomics in Food Science*; Colgrave, M.L., Ed.; Academic Press: Cambridge, MA, USA, 2017; pp. 261–276.
104. Rathnapala, E.C.N.; Ahn, D.U.; Abeyrathne, S. Functional properties of ovotransferrin from chicken egg white and its derived peptides: A review. *Food Sci. Biotechnol.* **2021**, *30*, 619–630. [CrossRef]
105. Giansanti, F.; Leboffe, L.; Angelucci, F.; Antonini, G. The Nutraceutical Properties of Ovotransferrin and Its Potential Utilization as a Functional Food. *Nutrients* **2015**, *7*, 9105–9115. [CrossRef]
106. Ibrahim, H.R.; Kiyono, T. Novel Anticancer Activity of the Autocleaved Ovotransferrin against Human Colon and Breast Cancer Cells. *J. Agric. Food Chem.* **2009**, *57*, 11383–11390. [CrossRef]
107. Kostić, A.; Milinčić, D.D.; Barać, M.B.; Shariati, M.A.; Tešić, L.; Pešić, M.B. The Application of Pollen as a Functional Food and Feed Ingredient—The Present and Perspectives. *Biomolecules* **2020**, *10*, 84. [CrossRef]
108. Li, J. Bee Pollen and Doxorubicin by Synergistic Effects Inhibit the Proliferation of Breast Tumors in 4T1 Tumor-bearing BALB/c Mice: A Biochemical, Immunohistochemical, and Molecular Approach. *Pharmacogn. Mag.* **2023**, *20*, 159–178. [CrossRef]
109. El Ghouizi, A.; Bakour, M.; Laaroussi, H.; Ousaaid, D.; El Menyiy, N.; Hano, C.; Lyoussi, B. Bee Pollen as Functional Food: Insights into Its Composition and Therapeutic Properties. *Antioxidants* **2023**, *12*, 557. [CrossRef]
110. Jurzak, M.; Ramos, P.; Pilawa, B.; Bednarek, I.A. Comparative EPR Studies on the Influence of Genistein on Free Radicals in Non-Irradiated and UV-Irradiated MCF7, T47D and MDA-MB-231 Breast Cancer Cells. *Biomedicines* **2024**, *12*, 518. [CrossRef]
111. Caner, A.; Onal, M.G.; Silici, S. The effect of bee bread (Perga) with chemotherapy on MDA-MB-231 cells. *Mol. Biol. Rep.* **2021**, *48*, 2299–2306. [CrossRef]
112. Nandini, D.; Rao, R.S.; Deepak, B.; Reddy, P.B. Sulforaphane in broccoli: The green chemoprevention!! Role in cancer prevention and therapy. *J. Oral Maxillofac. Pathol.* **2020**, *24*, 405. [CrossRef]
113. Kaboli, P.J.; Khoshkbejari, M.A.; Mohammadi, M.; Abiri, A.; Mokhtarian, R.; Vazifemand, R.; Amanollahi, S.; Sani, S.Y.; Li, M.; Zhao, Y.; et al. Targets and mechanisms of sulforaphane derivatives obtained from cruciferous plants with special focus on breast cancer—Contradictory effects and future perspectives. *Biomed. Pharmacother.* **2019**, *121*, 109635. [CrossRef]
114. Wu, S.; Lu, H.; Bai, Y. Nrf2 in cancers: A double-edged sword. *Cancer Med.* **2019**, *8*, 2252–2267. [CrossRef]
115. Surh, Y.-J. Nrf2 paradox: Can cancer patients eat broccoli? *Food Front.* **2021**, *2*, 25–28. [CrossRef]
116. Zhang, Y.; Lu, Q.; Li, N.; Xu, M.; Miyamoto, T.; Liu, J. Sulforaphane suppresses metastasis of triple-negative breast cancer cells by targeting the RAF/MEK/ERK pathway. *npj Breast Cancer* **2022**, *8*, 40. [CrossRef]
117. Desai, G.; Schelske-Santos, M.; Nazario, C.M.; Rosario-Rosado, R.V.; Mansilla-Rivera, I.; Ramírez-Marrero, F.; Nie, J.; Myneni, A.A.; Zhang, Z.-F.; Freudenheim, J.L.; et al. Onion and Garlic Intake and Breast Cancer, a Case-Control Study in Puerto Rico. *Nutr. Cancer* **2019**, *72*, 791–800. [CrossRef]
118. Shi, G.; Li, X.; Wang, W.; Hou, L.; Yin, L.; Wang, L. Allicin Overcomes Doxorubicin Resistance of Breast Cancer Cells by Targeting the Nrf2 Pathway. *Cell Biochem. Biophys.* **2024**, *82*, 659–667. [CrossRef]
119. Talib, W.H.; Baban, M.M.; Azzam, A.O.; Issa, J.J.; Ali, A.Y.; AlSuwais, A.K.; Allala, S.; AL Kury, L.T. Allicin and Cancer Hallmarks. *Molecules* **2024**, *29*, 1320. [CrossRef]
120. Maitisha, G.; Aimaiti, M.; An, Z.; Li, X. Allicin induces cell cycle arrest and apoptosis of breast cancer cells in vitro via modulating the p53 pathway. *Mol. Biol. Rep.* **2021**, *48*, 7261–7272. [CrossRef]
121. Dorrigiv, M.; Zareiyan, A.; Hosseinzadeh, H. Onion (*Allium cepa*) and its Main Constituents as Antidotes or Protective Agents against Natural or Chemical Toxicities: A Comprehensive Review. *Iran. J. Pharm. Res. IJPR* **2021**, *20*, 3–26. [CrossRef]
122. Nishimura, M.; Muro, T.; Kobori, M.; Nishihira, J. Effect of Daily Ingestion of Quercetin-Rich Onion Powder for 12 Weeks on Visceral Fat: A Randomised, Double-Blind, Placebo-Controlled, Parallel-Group Study. *Nutrients* **2019**, *12*, 91. [CrossRef]
123. Savitha, S.; Bhatkar, N.; Chakraborty, S.; Thorat, B.N. Onion quercetin: As immune boosters, extraction, and effect of dehydration. *Food Biosci.* **2021**, *44*, 101457. [CrossRef]

124. Dilek, B.; Meltem, Ö. Quercetin suppresses cell proliferation using the apoptosis pathways in MCF-7 and MDA-MB-231 human breast carcinoma cells in monolayer and spheroid model cultures. *S. Afr. J. Bot.* **2023**, *162*, 259–270. [CrossRef]
125. Liao, Y.; Xie, X.; Zhang, C.; Zhong, H.; Shan, L.; Yu, P.; Xu, L. Quercetin exerts anti-tumor immune mechanism by regulating IL-6/JAK2/STAT3 signaling pathway to deplete Treg cells. *Toxicon* **2024**, *243*, 107747. [CrossRef]
126. Sesso, H.D.; Buring, J.E.; Zhang, S.M.; Norkus, E.P.; Gaziano, J.M. Dietary and Plasma Lycopene and the Risk of Breast Cancer. *Cancer Epidemiol. Biomark. Prev.* **2005**, *14*, 1074–1081. [CrossRef]
127. Kapała, A.; Szlendak, M.; Motacka, E. The Anti-Cancer Activity of Lycopene: A Systematic Review of Human and Animal Studies. *Nutrients* **2022**, *14*, 5152. [CrossRef]
128. Ozkan, G.; Günal-Köroğlu, D.; Karadag, A.; Capanoglu, E.; Cardoso, S.M.; Al-Omari, B.; Calina, D.; Sharifi-Rad, J.; Cho, W.C. A mechanistic updated overview on lycopene as potential anticancer agent. *Biomed. Pharmacother.* **2023**, *161*, 114428. [CrossRef]
129. Takeshima, M.; Ono, M.; Higuchi, T.; Chen, C.; Hara, T.; Nakano, S. Anti-proliferative and apoptosis-inducing activity of lycopene against three subtypes of human breast cancer cell lines. *Cancer Sci.* **2014**, *105*, 252–257. [CrossRef]
130. Ma, P.; Wang, L.; Xu, L.; Li, J.; Zhang, X.; Chen, H. Rapid quantitative determination of chlorpyrifos pesticide residues in tomatoes by surface-enhanced Raman spectroscopy. *Eur. Food Res. Technol.* **2019**, *246*, 239–251. [CrossRef]
131. Ventura, C.; Zappia, C.D.; Lasagna, M.; Pavicic, W.; Richard, S.; Bolzan, A.D.; Monczor, F.; Nunez, M.; Cocca, C. Effects of the pesticide chlorpyrifos on breast cancer disease. Implication of epigenetic mechanisms. *J. Steroid Biochem. Mol. Biol.* **2019**, *186*, 96–104. [CrossRef]
132. Lasagna, M.; Ventura, C.; Hielpos, M.; Mardirosian, M.; Martín, G.; Miret, N.; Randi, A.; Núñez, M.; Cocca, C. Endocrine disruptor chlorpyrifos promotes migration, invasion, and stemness phenotype in 3D cultures of breast cancer cells and induces a wide range of pathways involved in cancer progression. *Environ. Res.* **2021**, *204*, 111989. [CrossRef]
133. Ghafouri-Fard, S.; Bahroudi, Z.; Shoorei, H.; Hussen, B.M.; Talebi, S.F.; Baig, S.G.; Taheri, M.; Ayatollahi, S.A. Disease-associated regulation of gene expression by resveratrol: Special focus on the PI3K/AKT signaling pathway. *Cancer Cell Int.* **2022**, *22*, 298. [CrossRef]
134. Jang, J.Y.; Im, E.; Kim, N.D. Mechanism of Resveratrol-Induced Programmed Cell Death and New Drug Discovery against Cancer: A Review. *Int. J. Mol. Sci.* **2022**, *23*, 13689. [CrossRef]
135. Gogada, R.; Prabhu, V.; Amadori, M.; Scott, R.; Hashmi, S.; Chandra, D. Resveratrol induces p53-independent, X-linked inhibitor of apoptosis protein (XIAP)-mediated Bax protein oligomerization on mitochondria to initiate cytochrome c release and caspase activation. *J. Biol. Chem.* **2011**, *286*, 28749–28760. [CrossRef]
136. Sun, Y.; Zhou, Q.-M.; Lu, Y.-Y.; Zhang, H.; Chen, Q.-L.; Zhao, M.; Su, S.-B. Resveratrol Inhibits the Migration and Metastasis of MDA-MB-231 Human Breast Cancer by Reversing TGF-β1-Induced Epithelial-Mesenchymal Transition. *Molecules* **2019**, *24*, 1131. [CrossRef]
137. Liu, J.L.; Pan, Y.Y.; Chen, O.; Luan, Y.; Xue, X.; Zhao, J.J.; Liu, L.; Jia, H.Y. Curcumin inhibits MCF-7 cells by modulating the NF-κB signaling pathway. *Oncol. Lett.* **2017**, *14*, 5581–5584. [CrossRef]
138. Fawzy, R.M.; Abdel-Aziz, A.A.; Bassiouny, K.; Fayed, A.M. Phytocompounds-based therapeutic approach: Investigating curcumin and green tea extracts on MCF-7 breast cancer cell line. *J. Genet. Eng. Biotechnol.* **2024**, *22*, 100339. [CrossRef]
139. Liu, H.-T.; Ho, Y.-S. Anticancer effect of curcumin on breast cancer and stem cells. *Food Sci. Hum. Wellness* **2018**, *7*, 134–137. [CrossRef]
140. Rinkenbaugh, A.L.; Baldwin, A.S. The NF-κB Pathway and Cancer Stem Cells. *Cells* **2016**, *5*, 16. [CrossRef]
141. Vazquez-Santillan, K.; Melendez-Zajgla, J.; Jimenez-Hernandez, L.E.; Gaytan-Cervantes, J.; Muñoz-Galindo, L.; Piña-Sanchez, P.; Martinez-Ruiz, G.; Torres, J.; Garcia-Lopez, P.; Gonzalez-Torres, C.; et al. NF-kappaB-inducing kinase regulates stem cell phenotype in breast cancer. *Sci. Rep.* **2016**, *6*, 37340. [CrossRef]
142. Li, X.; Wang, X.; Xie, C.; Zhu, J.; Meng, Y.; Chen, Y.; Li, Y.; Jiang, Y.; Yang, X.; Wang, S.; et al. Sonic hedgehog and Wnt/β-catenin pathways mediate curcumin inhibition of breast cancer stem cells. *Anti-Cancer Drugs* **2018**, *29*, 208–215. [CrossRef] [PubMed]
143. Kumar, H.; Kumar, R.M.; Bhattacharjee, D.; Somanna, P.; Jain, V. Role of Nrf2 Signaling Cascade in Breast Cancer: Strategies and Treatment. *Front. Pharmacol.* **2022**, *13*, 720076. [CrossRef] [PubMed]
144. Ashrafizadeh, M. Curcumin Activates the Nrf2 Pathway and Induces Cellular Protection Against Oxidative Injury. *Curr. Mol. Med.* **2020**, *20*, 116–133. [PubMed]
145. Shahcheraghi, S.H.; Salemi, F.; Peirovi, N.; Ayatollahi, J.; Alam, W.; Khan, H.; Saso, L. Nrf2 Regulation by Curcumin: Molecular Aspects for Therapeutic Prospects. *Molecules* **2021**, *27*, 167. [CrossRef] [PubMed]
146. Sordillo, P.P.; Helson, L. Curcumin and cancer stem cells: Curcumin has asymmetrical effects on cancer and normal stem cells. *Anticancer. Res.* **2015**, *35*, 599–614.
147. Chen, Z.M.; Lu, P.M.; Li, M.M.; Zhang, Q.M.; He, T.; Gan, L. Curcumin suppresses metastasis of triple-negative breast cancer cells by modulating EMT signaling pathways: An integrated study of bioinformatics analysis. *Medicine* **2024**, *103*, e37264. [CrossRef]
148. Hu, C.; Li, M.; Guo, T.; Wang, S.; Huang, W.; Yang, K.; Liao, Z.; Wang, J.; Zhang, F.; Wang, H. Anti-metastasis activity of curcumin against breast cancer via the inhibition of stem cell-like properties and EMT. *Phytomedicine* **2018**, *58*, 152740. [CrossRef]
149. Zhang, D.; Nichols, H.B.; Troester, M.; Cai, J.; Bensen, J.T.; Sandler, D.P. Tea consumption and breast cancer risk in a cohort of women with family history of breast cancer. *Int. J. Cancer* **2020**, *147*, 876–886. [CrossRef]

150. Pérez-Durán, J.; Luna, A.; Portilla, A.; Martínez, P.; Ceballos, G.; Ortíz-Flores, M.Á.; Solis-Paredes, J.M.; Nájera, N. (-)-Epicatechin Inhibits Metastatic-Associated Proliferation, Migration, and Invasion of Murine Breast Cancer Cells In Vitro. *Molecules* **2023**, *28*, 298. [CrossRef]
151. Al Mamun, A.; Sharker, B.; Bhuiyan, M.R.; Sadikuj Jaman, M. Anticancer activity and therapeutic uses of catechins on breast, prostate and lung cancer: Future perspective and clinical proofs. *J. Clin. Case Rep. Med. Images Health Sci.* **2023**, *3*, 2023. [CrossRef]
152. O'neill, E.J.; Termini, D.; Albano, A.; Tsiani, E. Anti-Cancer Properties of Theaflavins. *Molecules* **2021**, *26*, 987. [CrossRef] [PubMed]
153. Thomas, P.; Dong, J. (-)-Epicatechin acts as a potent agonist of the membrane androgen receptor, ZIP9 (SLC39A9), to promote apoptosis of breast and prostate cancer cells. *J. Steroid Biochem. Mol. Biol.* **2021**, *211*, 105906. [CrossRef]
154. Sen, A.K.; Sen, D.B.; Zanwar, A.S.; Balaraman, R.; Shah, U.; Maheshwari, R.A. Catechins and Theaflavins: An Overview on Therapeutic Application. *J. Nat. Remedies* **2022**, *22*, 330–346. [CrossRef]
155. Rosas-González, V.C.; Téllez-Bañuelos, M.C.; Hernández-Flores, G.; Bravo-Cuellar, A.; Aguilar-Lemarroy, A.; Jave-Suárez, L.F.; Haramati, J.; Solorzano-Ibarra, F.; Ortiz-Lazareno, P.C. Differential effects of alliin and allicin on apoptosis and senescence in luminal A and triple-negative breast cancer: Caspase, $\Delta\psi$m, and pro-apoptotic gene involvement. *Fundam. Clin. Pharmacol.* **2020**, *34*, 671–686. [CrossRef] [PubMed]
156. Wach, A.; Pyrzyńska, K.; Biesaga, M. Quercetin content in some food and herbal samples. *Food Chem.* **2005**, *100*, 699–704. [CrossRef]
157. Lin, D.; Kuang, G.; Wan, J.; Zhang, X.; Li, H.; Gong, X.; Li, H. Luteolin suppresses the metastasis of triple-negative breast cancer by reversing epithelial-to-mesenchymal transition via downregulation of β-catenin expression. *Oncol. Rep.* **2016**, *37*, 895–902. [CrossRef] [PubMed]
158. Wu, H.-T.; Lin, J.; Liu, Y.-E.; Chen, H.-F.; Hsu, K.-W.; Lin, S.-H.; Peng, K.-Y.; Lin, K.-J.; Hsieh, C.-C.; Chen, D.-R. Luteolin suppresses androgen receptor-positive triple-negative breast cancer cell proliferation and metastasis by epigenetic regulation of MMP9 expression via the AKT/mTOR signaling pathway. *Phytomedicine* **2020**, *81*, 153437. [CrossRef]
159. Ahmed, S.; Khan, H.; Fratantonio, D.; Hasan, M.M.; Sharifi, S.; Fathi, N.; Ullah, H.; Rastrelli, L. Apoptosis induced by luteolin in breast cancer: Mechanistic and therapeutic perspectives. *Phytomedicine* **2019**, *59*, 152883. [CrossRef]
160. Gao, G.; Ge, R.; Li, Y.; Liu, S. Luteolin exhibits anti-breast cancer property through up-regulating miR-203. *Artif. Cells Nanomed. Biotechnol.* **2019**, *47*, 3265–3271. [CrossRef]
161. Lu, Q.; Lai, Y.; Zhang, H.; Ren, K.; Liu, W.; An, Y.; Yao, J.; Fan, H. Hesperetin Inhibits TGF-β1-Induced Migration and Invasion of Triple Negative Breast Cancer MDA-MB-231 Cells via Suppressing Fyn/Paxillin/RhoA Pathway. *Integr. Cancer Ther.* **2022**, *21*, 15347354221086900. [CrossRef]
162. Palit, S.; Kar, S.; Sharma, G.; Das, P.K. Hesperetin Induces Apoptosis in Breast Carcinoma by Triggering Accumulation of ROS and Activation of ASK1/JNK Pathway. *J. Cell. Physiol.* **2015**, *230*, 1729–1739. [CrossRef]
163. Hermawan, A.; Ikawati, M.; Khumaira, A.; Putri, H.; Jenie, R.I.; Angraini, S.M.; Muflikhasari, H.A. Bioinformatics and *In Vitro* Studies Reveal the Importance of p53, PPARG and Notch Signaling Pathway in Inhibition of Breast Cancer Stem Cells by Hesperetin. *Adv. Pharm. Bull.* **2021**, *11*, 351–360. [CrossRef]
164. Choi, E.J. Hesperetin Induced G1-Phase Cell Cycle Arrest in Human Breast Cancer MCF-7 Cells: Involvement of CDK4 and p21. *Nutr. Cancer* **2007**, *59*, 115–119. [CrossRef] [PubMed]
165. Lee, C.J.; Wilson, L.; Jordan, M.A.; Nguyen, V.; Tang, J.; Smiyun, G. Hesperidin suppressed proliferations of both Human breast cancer and androgen-dependent prostate cancer cells. *Phytotherapy Res.* **2010**, *24*, S15–S19. [CrossRef]
166. Kongtawelert, P.; Wudtiwai, B.; Shwe, T.H.; Pothacharoen, P.; Phitak, T. Inhibitory Effect of Hesperidin on the Expression of Programmed Death Ligand (PD-L1) in Breast Cancer. *Molecules* **2020**, *25*, 252. [CrossRef] [PubMed]
167. Hermawan, A.; Khumaira, A.; Ikawati, M.; Putri, H.; Jenie, R.I.; Angraini, S.M.; Muflikhasari, H.A. Identification of key genes of hesperidin in inhibition of breast cancer stem cells by functional network analysis. *Comput. Biol. Chem.* **2020**, *90*, 107427. [CrossRef]
168. Taghizadeh, M.S.; Niazi, A.; Moghadam, A.; Afsharifar, A. Experimental, molecular docking and molecular dynamic studies of natural products targeting overexpressed receptors in breast cancer. *PLoS ONE* **2022**, *17*, e0267961. [CrossRef]
169. Patel, P.; Shah, J. Protective effects of hesperidin through attenuation of Ki67 expression against DMBA-induced breast cancer in female rats. *Life Sci.* **2021**, *285*, 119957. [CrossRef] [PubMed]
170. Zhao, Z.; Jin, G.; Ge, Y.; Guo, Z. Naringenin inhibits migration of breast cancer cells via inflammatory and apoptosis cell signaling pathways. *Inflammopharmacology* **2019**, *27*, 1021–1036. [CrossRef]
171. Sharma, M.; Dwivedi, P.; Rawat, A.K.S.; Dwivedi, A.K. 3—Nutrition nutraceuticals: A proactive approach for healthcare. In *Nutraceuticals*; Grumezescu, A.M., Ed.; Academic Press: Cambridge, MA, USA, 2016; pp. 79–116.
172. Ke, J.-Y.; Banh, T.; Hsiao, Y.-H.; Cole, R.M.; Straka, S.R.; Yee, L.D.; Belury, M.A. Citrus flavonoid naringenin reduces mammary tumor cell viability, adipose mass, and adipose inflammation in obese ovariectomized mice. *Mol. Nutr. Food Res.* **2017**, *61*, 1600934. [CrossRef] [PubMed]
173. Zhang, F.; Dong, W.; Zeng, W.; Zhang, L.; Zhang, C.; Qiu, Y.; Wang, L.; Yin, X.; Zhang, C.; Liang, W. Naringenin prevents TGF-β1 secretion from breast cancer and suppresses pulmonary metastasis by inhibiting PKC activation. *Breast Cancer Res.* **2016**, *18*, 38. [CrossRef]

174. Harmon, A.W.; Patel, Y.M. Naringenin Inhibits Glucose Uptake in MCF-7 Breast Cancer Cells: A Mechanism for Impaired Cellular Proliferation. *Breast Cancer Res. Treat.* **2004**, *85*, 103–110. [CrossRef]
175. Bharti, S.; Rani, N.; Krishnamurthy, B.; Arya, D.S. Preclinical Evidence for the Pharmacological Actions of Naringin: A Review. *Planta Medica* **2014**, *80*, 437–451. [CrossRef] [PubMed]
176. Li, H.; Yang, B.; Huang, J.; Xiang, T.; Yin, X.; Wan, J.; Luo, F.; Zhang, L.; Li, H.; Ren, G. Naringin inhibits growth potential of human triple-negative breast cancer cells by targeting β-catenin signaling pathway. *Toxicol. Lett.* **2013**, *220*, 219–228. [CrossRef] [PubMed]
177. Way, T.D.; Kao, M.C.; Lin, J.K. Apigenin induces apoptosis through proteasomal degradation of HER2/neu in HER2/neu-overexpressing breast cancer cells via the phosphatidylinositol 3-kinase/Akt-dependent pathway. *J. Biol. Chem.* **2004**, *279*, 4479–4489. [CrossRef] [PubMed]
178. Choi, E.J.; Kim, G.H. Apigenin causes G(2)/M arrest associated with the modulation of p21(Cip1) and Cdc2 and activates p53-dependent apoptosis pathway in human breast cancer SK-BR-3 cells. *J. Nutr. Biochem.* **2009**, *20*, 285–290. [CrossRef]
179. Long, X.; Fan, M.; Bigsby, R.M.; Nephew, K.P. Apigenin inhibits antiestrogen-resistant breast cancer cell growth through estrogen receptor-α-dependent and estrogen receptor-α-independent mechanisms. *Mol. Cancer Ther.* **2008**, *7*, 2096–2108. [CrossRef]
180. Pham, T.H.; Le Page, Y.; Percevault, F.; Ferrière, F.; Flouriot, G.; Pakdel, F. Apigenin, a Partial Antagonist of the Estrogen Receptor (ER), Inhibits ER-Positive Breast Cancer Cell Proliferation through Akt/FOXM1 Signaling. *Int. J. Mol. Sci.* **2021**, *22*, 470. [CrossRef]
181. Lee, W.J.; Chen, W.K.; Wang, C.J.; Lin, W.L.; Tseng, T.H. Apigenin inhibits HGF-promoted invasive growth and metastasis involving blocking PI3K/Akt pathway and beta 4 integrin function in MDA-MB-231 breast cancer cells. *Toxicol. Appl. Pharmacol.* **2008**, *226*, 178–191. [CrossRef]
182. Ko, Y.-C.; Choi, H.S.; Liu, R.; Kim, J.-H.; Kim, S.-L.; Yun, B.-S.; Lee, D.-S. Inhibitory Effects of Tangeretin, a Citrus Peel-Derived Flavonoid, on Breast Cancer Stem Cell Formation through Suppression of Stat3 Signaling. *Molecules* **2020**, *25*, 2599. [CrossRef]
183. Arivazhagan, L.; Pillai, S.S. Tangeretin, a citrus pentamethoxyflavone, exerts cytostatic effect via p53/p21 up-regulation and suppresses metastasis in 7,12-dimethylbenz(alpha)anthracene-induced rat mammary carcinoma. *J. Nutr. Biochem.* **2014**, *25*, 1140–1153. [CrossRef]
184. Surichan, S.; Arroo, R.R.; Tsatsakis, A.M.; Androutsopoulos, V.P. Tangeretin inhibits the proliferation of human breast cancer cells via CYP1A1/CYP1B1 enzyme induction and CYP1A1/CYP1B1-mediated metabolism to the product 4′ hydroxy tangeretin. *Toxicol In Vitro* **2018**, *50*, 274–284. [CrossRef]
185. Periyasamy, K.; Baskaran, K.; Ilakkia, A.; Vanitha, K.; Selvaraj, S.; Sakthisekaran, D. Antitumor efficacy of tangeretin by targeting the oxidative stress mediated on 7,12-dimethylbenz(a) anthracene-induced proliferative breast cancer in Sprague-Dawley rats. *Cancer Chemother. Pharmacol.* **2015**, *75*, 263–272. [CrossRef] [PubMed]
186. Montalesi, E.; Cipolletti, M.; Cracco, P.; Fiocchetti, M.; Marino, M. Divergent Effects of Daidzein and Its Metabolites on Estrogen-Induced Survival of Breast Cancer Cells. *Cancers* **2020**, *12*, 167. [CrossRef] [PubMed]
187. Sharifi-Rad, J.; Quispe, C.; Imran, M.; Rauf, A.; Nadeem, M.; Gondal, T.A.; Ahmad, B.; Atif, M.; Mubarak, M.S.; Sytar, O.; et al. Genistein: An Integrative Overview of Its Mode of Action, Pharmacological Properties, and Health Benefits. *Oxidative Med. Cell. Longev.* **2021**, *2021*, 3268136. [CrossRef]
188. Hsieh, P.-L.; Liao, Y.-W.; Hsieh, C.-W.; Chen, P.-N.; Yu, C.-C. Soy Isoflavone Genistein Impedes Cancer Stemness and Mesenchymal Transition in Head and Neck Cancer through Activating miR-34a/RTCB Axis. *Nutrients* **2020**, *12*, 1924. [CrossRef]
189. Hwang, S.T.; Yang, M.H.; Baek, S.H.; Um, J.Y.; Ahn, K.S. Genistin attenuates cellular growth and promotes apoptotic cell death breast cancer cells through modulation of ERalpha signaling pathway. *Life Sci.* **2020**, *263*, 118594. [CrossRef] [PubMed]
190. Elkhalifa, A.E.O.; Al-Shammari, E.; Kuddus, M.; Adnan, M.; Sachidanandan, M.; Awadelkareem, A.M.; Qattan, M.Y.; Khan, M.I.; Abduljabbar, S.I.; Baig, M.S.; et al. Structure-Based Multi-Targeted Molecular Docking and Dynamic Simulation of Soybean-Derived Isoflavone Genistin as a Potential Breast Cancer Signaling Proteins Inhibitor. *Life* **2023**, *13*, 1739. [CrossRef] [PubMed]
191. Hong, R.; Lim, S.C.; Lee, T.B.; Han, S.I. Anticancer Effect of Gallic Acid on Acidity-Induced Invasion of MCF7 Breast Cancer Cells. *Nutrients* **2023**, *15*, 3596. [CrossRef]
192. Yang, K.; Zhang, L.; Liao, P.; Xiao, Z.; Zhang, F.; Sindaye, D.; Xin, Z.; Tan, C.; Deng, J.; Yin, Y.; et al. Impact of Gallic Acid on Gut Health: Focus on the Gut Microbiome, Immune Response, and Mechanisms of Action. *Front. Immunol.* **2020**, *11*, 580208. [CrossRef]
193. Xiang, Z.; Guan, H.; Zhao, X.; Xie, Q.; Xie, Z.; Cai, F.; Dang, R.; Li, M.; Wang, C. Dietary gallic acid as an antioxidant: A review of its food industry applications, health benefits, bioavailability, nano-delivery systems, and drug interactions. *Food Res. Int.* **2024**, *180*, 114068. [CrossRef]
194. Cheung, Y.; Meenu, M.; Yu, X.; Xu, B. Phenolic acids and flavonoids profiles of commercial honey from different floral sources and geographic sources. *Int. J. Food Prop.* **2019**, *22*, 290–308. [CrossRef]
195. Lin, S.; Qin, H.-Z.; Li, Z.-Y.; Zhu, H.; Long, L.; Xu, L.-B. Gallic acid suppresses the progression of triple-negative breast cancer HCC1806 cells via modulating PI3K/AKT/EGFR and MAPK signaling pathways. *Front. Pharmacol.* **2022**, *13*, 1049117. [CrossRef] [PubMed]

196. Khorsandi, K.; Kianmehr, Z.; Hosseinmardi, Z.; Hosseinzadeh, R. Anti-cancer effect of gallic acid in presence of low level laser irradiation: ROS production and induction of apoptosis and ferroptosis. *Cancer Cell Int.* **2020**, *20*, 18. [CrossRef] [PubMed]
197. Kaur, J.; Gulati, M.; Singh, S.K.; Kuppusamy, G.; Kapoor, B.; Mishra, V.; Gupta, S.; Arshad, M.F.; Porwal, O.; Jha, N.K.; et al. Discovering multifaceted role of vanillic acid beyond flavours: Nutraceutical and therapeutic potential. *Trends Food Sci. Technol.* **2022**, *122*, 187–200. [CrossRef]
198. Shanmugam, M. Vanillic acid exhibits potent antiproliferative and free radical scavenging effects under in vitro conditions. *Int. J. Nutr. Pharmacol. Neurol. Dis.* **2023**, *13*, 188–198.
199. Yin, M.-C.; Lin, C.-C.; Wu, H.-C.; Tsao, S.-M.; Hsu, C.-K. Apoptotic Effects of Protocatechuic Acid in Human Breast, Lung, Liver, Cervix, and Prostate Cancer Cells: Potential Mechanisms of Action. *J. Agric. Food Chem.* **2009**, *57*, 6468–6473. [CrossRef]
200. Punvittayagul, C.; Luangsuphabool, T.; Wongpoomchai, R. Protocatechuic acid as a potent anticarcinogenic compound in purple rice bran against diethylnitrosamine-initiated rat hepatocarcinogenesis. *Sci. Rep.* **2022**, *12*, 10548. [CrossRef]
201. Sułkowska-Ziaja, K.; Muszyńska, B.; Szewczyk, A. Antioxidant components of selected indigenous edible mushrooms of the obsolete order Aphyllophorales. *Rev. Iberoam. De Micol.* **2015**, *32*, 99–102. [CrossRef]
202. Cadena-Iñiguez, J.; Santiago-Osorio, E.; Sánchez-Flores, N.; Salazar-Aguilar, S.; Soto-Hernández, R.M.; Riviello-Flores, M.d.l.L.; Macías-Zaragoza, V.M.; Aguiñiga-Sánchez, I. The Cancer-Protective Potential of Protocatechuic Acid: A Narrative Review. *Molecules* **2024**, *29*, 1439. [CrossRef]
203. Jha, S.; Prabakaran, A.; Sahoo, R.K.; Batheja, S.; Gupta, U.; Alexander, A. Antiproliferative activity of syringic acid-loaded nanostructured lipid carriers against MCF-7 human breast carcinoma cells. *J. Drug Deliv. Sci. Technol.* **2024**, *98*, 105902. [CrossRef]
204. Zhang, L.; La, X.; Tian, J.; Li, H.; Li, A.; Liu, Y.; Wu, C.; Li, Z. The phytochemical vitexin and syringic acid derived from foxtail fillet bran inhibit breast cancer cells proliferation via GRP78/SREBP-1/SCD1 signaling axis. *J. Funct. Foods* **2021**, *85*, 104620. [CrossRef]
205. Golmohammadi, M.; Zamanian, M.Y.; Jalal, S.M.; Noraldeen, S.A.M.; Ramírez-Coronel, A.A.; Oudaha, K.H.; Obaid, R.F.; Almulla, A.F.; Bazmandegan, G.; Kamiab, Z. A comprehensive review on Ellagic acid in breast cancer treatment: From cellular effects to molecular mechanisms of action. *Food Sci. Nutr.* **2023**, *11*, 7458–7468. [CrossRef] [PubMed]
206. Rezaei-Seresht, H.; Cheshomi, H.; Falanji, F.; Movahedi-Motlagh, F.; Hashemian, M.; Mireskandari, E. Cytotoxic activity of caffeic acid and gallic acid against MCF-7 human breast cancer cells: An in silico and in vitro study. *Avicenna J. Phytomedicine* **2019**, *9*, 574–586.
207. Alam, M.; Ahmed, S.; Elasbali, A.M.; Adnan, M.; Alam, S.; Hassan, I.; Pasupuleti, V.R. Therapeutic Implications of Caffeic Acid in Cancer and Neurological Diseases. *Front. Oncol.* **2022**, *12*, 860508. [CrossRef] [PubMed]
208. Assumpção, J.H.M.; Takeda, A.A.S.; Sforcin, J.M.; Rainho, C.A. Effects of Propolis and Phenolic Acids on Triple-Negative Breast Cancer Cell Lines: Potential Involvement of Epigenetic Mechanisms. *Molecules* **2020**, *25*, 1289. [CrossRef]
209. Pal, A.; Tapadar, P.; Pal, R. Exploring the Molecular Mechanism of Cinnamic Acid-Mediated Cytotoxicity in Triple Negative MDA-MB-231 Breast Cancer Cells. *Anti-Cancer Agents Med. Chem.* **2021**, *21*, 1141–1150. [CrossRef]
210. Rodrigues, D.M.; Portapilla, G.B.; de Sicco, G.S.; da Silva, I.F.R.; de Albuquerque, S.; Bastos, J.K.; Campo, V.L. Novel synthetic derivatives of cinnamic and *p*-coumaric acids with antiproliferative effect on breast MCF-7 tumor cells. *Nat. Prod. Res.* **2023**, *37*, 4210–4220. [CrossRef]
211. Zhao, Y.; Liu, J. Anti-Inflammatory Effects of p-coumaric Acid in LPS-Stimulated RAW264.7 Cells: Involvement of NF-κB and MAPKs Pathways. *Med. Chem.* **2016**, *6*, 327–330. [CrossRef]
212. Kaur, J.; Kaur, R. *p*-Coumaric Acid: A Naturally Occurring Chemical with Potential Therapeutic Applications. *Curr. Org. Chem.* **2022**, *26*, 1333–1349. [CrossRef]
213. Bao, X.; Li, W.; Jia, R.; Meng, D.; Zhang, H.; Xia, L. Molecular mechanism of ferulic acid and its derivatives in tumor progression. *Pharmacol. Rep.* **2023**, *75*, 891–906. [CrossRef]
214. Gupta, A.; Singh, A.K.; Loka, M.; Pandey, A.K.; Bishayee, A. Chapter Eight—Ferulic acid-mediated modulation of apoptotic signaling pathways in cancer. In *Advances in Protein Chemistry and Structural Biology*; Donev, R., Ed.; Academic Press: Cambridge, MA, USA, 2021; pp. 215–257.
215. Zhang, X.; Lin, D.; Jiang, R.; Li, H.; Wan, J.; Li, H. Ferulic acid exerts antitumor activity and inhibits metastasis in breast cancer cells by regulating epithelial to mesenchymal transition. *Oncol. Rep.* **2016**, *36*, 271–278. [CrossRef]
216. ElKhazendar, M.; Chalak, J.; El-Huneidi, W.; Vinod, A.; Abdel-Rahman, W.M.; Abu-Gharbieh, E. Antiproliferative and proapoptotic activities of ferulic acid in breast and liver cancer cell lines. *Trop. J. Pharm. Res.* **2019**, *18*, 2571–2576.
217. Serafim, T.L.; Milhazes, N.; Borges, F.; Oliveira, P.J. Chapter 74—Caffeic and Ferulic Acid Derivatives: Use in Breast Cancer. In *Coffee in Health and Disease Prevention*; Preedy, V.R., Ed.; Academic Press: San Diego, CA, USA, 2015; pp. 663–671.
218. Pandi, A.; Kalappan, V.M. Pharmacological and therapeutic applications of Sinapic acid—An updated review. *Mol. Biol. Rep.* **2021**, *48*, 3733–3745. [CrossRef] [PubMed]
219. Dwivedi, P.S.R.; Shastry, C.S. The cytotoxic potential of sinapic acid on luminal A breast cancer; a computational and experimental pharmacology approach. *J. Biomol. Struct. Dyn.* **2023**, 1–16. [CrossRef] [PubMed]
220. Konstantinou, E.K.; Panagiotopoulos, A.A.; Argyri, K.; Panoutsopoulos, G.I.; Dimitriou, M.; Gioxari, A. Molecular Pathways of Rosmarinic Acid Anticancer Activity in Triple-Negative Breast Cancer Cells: A Literature Review. *Nutrients* **2023**, *16*, 2. [CrossRef] [PubMed]

221. Messeha, S.S.; Zarmouh, N.O.; Asiri, A.; Soliman, K.F. Rosmarinic acid-induced apoptosis and cell cycle arrest in triple-negative breast cancer cells. *Eur. J. Pharmacol.* **2020**, *885*, 173419. [CrossRef] [PubMed]
222. Zeng, A.; Liang, X.; Zhu, S.; Liu, C.; Wang, S.; Zhang, Q.; Zhao, J.; Song, L. Chlorogenic acid induces apoptosis, inhibits metastasis and improves antitumor immunity in breast cancer via the NF-κB signaling pathway. *Oncol. Rep.* **2021**, *45*, 717–727. [CrossRef]
223. Veljkovic, E.; Xia, W.; Phillips, B.; Wong, E.T.; Ho, J.; Oviedo, A. Chapter 10—Other Compounds From Tobacco With Potential Impact on Neurodegenerative Diseases. In *Nicotine and Other Tobacco Compounds in Neurodegenerative and Psychiatric Diseases*; Academic Press: Cambridge, MA, USA, 2018; pp. 83–97.
224. Sakai, E.; Farhana, F.; Yamaguchi, Y.; Tsukuba, T. Chapter 1—Potentials of natural antioxidants from plants as antiosteoporotic agents. In *Studies in Natural Products Chemistry*; Attaur, R., Ed.; Elsevier: Amsterdam, The Netherlands, 2022; pp. 1–28.
225. Turrini, E.; Maffei, F.; Milelli, A.; Calcabrini, C.; Fimognari, C. Overview of the Anticancer Profile of Avenanthramides from Oat. *Int. J. Mol. Sci.* **2019**, *20*, 4536. [CrossRef]
226. Hastings, J.; Kenealey, J. Avenanthramide-C reduces the viability of MDA-MB-231 breast cancer cells through an apoptotic mechanism. *Cancer Cell Int.* **2017**, *17*, 93. [CrossRef]
227. Posadino, A.M.; Giordo, R.; Ramli, I.; Zayed, H.; Nasrallah, G.K.; Wehbe, Z.; Eid, A.H.; Gürer, E.S.; Kennedy, J.F.; Aldahish, A.A.; et al. An updated overview of cyanidins for chemoprevention and cancer therapy. *Biomed. Pharmacother.* **2023**, *163*, 114783. [CrossRef]
228. Mirmalek, S.; Faraji, S.; Ranjbaran, S.; Aryan, H.; Arani, H.; Jangholi, E.; Marzouni, H.; Salimi-Tabatabaee, S. Cyanidin 3-glycoside induced apoptosis in MCF-7 breast cancer cell line. *Arch. Med Sci.* **2020**, *16*, 1092–1098. [CrossRef]
229. Xu, M.; Bower, K.A.; Wang, S.; Frank, J.A.; Chen, G.; Ding, M.; Wang, S.; Shi, X.; Ke, Z.; Luo, J. Cyanidin-3-Glucoside inhibits ethanol-induced invasion of breast cancer cells overexpressing ErbB2. *Mol. Cancer* **2010**, *9*, 285. [CrossRef] [PubMed]
230. Nafisah, W.; Nugraha, A.P.; Nugroho, A.; Sakinah, A.I.; Nusantara, D.S.; Philia, J.; Kurniawinata, M.I.; Aini, W.; Herlina, V.T. Benefit of Asian pigmented rice bioactive compound and its implication in breast cancer: A systematic review [version 1; peer review: 2 approved, 1 approved with reservations, 1 not approved]. *F1000Research* **2023**, *12*, 371. [CrossRef] [PubMed]
231. Omenn, G.S. What accounts for the association of vegetables and fruits with lower incidence of cancers and coronary heart disease? *Ann. Epidemiol.* **1995**, *5*, 333–335. [CrossRef] [PubMed]
232. Husain, A.; Chanana, H.; Alam Khan, S.; Dhanalekshmi, U.M.; Ali, M.; Alghamdi, A.A.; Ahmad, A. Chemistry and Pharmacological Actions of Delphinidin, a Dietary Purple Pigment in Anthocyanidin and Anthocyanin Forms. *Front. Nutr.* **2022**, *9*, 746881. [CrossRef]
233. Chen, J.; Zhu, Y.; Zhang, W.; Peng, X.; Zhou, J.; Li, F.; Han, B.; Liu, X.; Ou, Y.; Yu, X. Delphinidin induced protective autophagy via mTOR pathway suppression and AMPK pathway activation in HER-2 positive breast cancer cells. *BMC Cancer* **2018**, *18*, 342. [CrossRef]
234. Sood, R.; Sanjay; Choi, H.-K.; Lee, H.-J. Potential anti-cancer properties of malvidin and its glycosides: Evidence from in vitro and in vivo studies. *J. Funct. Foods* **2024**, *116*, 106191. [CrossRef]
235. Laksmiani, N.P.L.; I Widiastari, M.; Reynaldi, K.R. The inhibitory activity of peonidin purple sweet potato in human epidermal receptor-2 receptor (her-2) expression by in silico study. *J. Physics. Conf. Ser.* **2018**, *1040*, 012010. [CrossRef]
236. Sinha, D.; Sarkar, N.; Biswas, J.; Bishayee, A. Resveratrol for breast cancer prevention and therapy: Preclinical evidence and molecular mechanisms. *Semin. Cancer Biol.* **2016**, *40–41*, 209–232. [CrossRef] [PubMed]
237. Khezrian, A.; Shojaeian, A. Therapeutic Opportunities in Breast Cancer by Targeting Macrophage Migration Inhibitory Factor as a Pleiotropic Cytokine. *Breast Cancer Basic Clin. Res.* **2024**, *18*, 11782234241276310. [CrossRef]
238. Almaguer, G.; Ortiz-Vilchis, P.; Cordero, P.; Martinez-Vega, R.; Perez-Durán, J.; Meaney, E.; Villarreal, F.; Ceballos, G.; Nájera, N. Anticancer potential of (−)-epicatechin in a triple-negative mammary gland model. *J. Pharm. Pharmacol.* **2021**, *73*, 1675–1682. [CrossRef]
239. Kiran, S.; Patra, A.; Verma, P.; Purkait, S.; Chhabra, G.; Guttula, P.K.; Ghosh, A. Restoration of Altered Oncogenic and Tumor Suppressor microRNA Expression in Breast Cancer and Colorectal Cancer Cell using Epicatechin. *Curr. Mol. Pharmacol.* **2023**, *16*, 915–926. [CrossRef] [PubMed]
240. Pereyra-Vergara, F.; Olivares-Corichi, I.M.; Perez-Ruiz, A.G.; Luna-Arias, J.P.; García-Sánchez, J.R. Apoptosis Induced by (−)-Epicatechin in Human Breast Cancer Cells is Mediated by Reactive Oxygen Species. *Molecules* **2020**, *25*, 1020. [CrossRef]
241. Ndacyayisenga, J.; Tolo, F.M.; Wamunyokoli, F.; Maina, E.N. Effects of tea catechin extracts from BB35 and purple (TRFK 306) tea clones on the gene expression of Egfr, App, Bcl2, Dnmt, Casp3, Hif1a, Gadd45b and Psmb5 genes involved in triple negative breast cancer diseases: In silico and in vitro study. *Inform. Med. Unlocked* **2024**, *46*, 101469. [CrossRef]
242. Delgado, L.; Fernandes, I.; González-Manzano, S.; de Freitas, V.; Mateus, N.; Santos-Buelga, C. Anti-proliferative effects of quercetin and catechin metabolites. *Food Funct.* **2014**, *5*, 797–803. [CrossRef] [PubMed]
243. Zhang, L.; Meng, S.; Yan, B.; Chen, J.; Zhou, L.; Shan, L.; Wang, Y. Anti-Proliferative, Pro-Apoptotic, Anti-Migrative and Tumor-Inhibitory Effects and Pleiotropic Mechanism of Theaflavin on B16F10 Melanoma Cells. *OncoTargets Ther.* **2021**, *14*, 1291–1304. [CrossRef]
244. Lahiry, L.; Saha, B.; Chakraborty, J.; Adhikary, A.; Mohanty, S.; Hossain, D.M.S.; Banerjee, S.; Das, K.; Sa, G.; Das, T. Theaflavins target Fas/caspase-8 and Akt/pBad pathways to induce apoptosis in p53-mutated human breast cancer cells. *Carcinogenesis* **2009**, *31*, 259–268. [CrossRef] [PubMed]

245. Adhikary, A.; Mohanty, S.; Lahiry, L.; Hossain, D.M.S.; Chakraborty, S.; Das, T. Theaflavins retard human breast cancer cell migration by inhibiting NF-kappaB via p53-ROS cross-talk. *FEBS Lett.* **2010**, *584*, 7–14. [CrossRef]
246. Bhadresha, K.; Kumar, S.P.; Brahmbhatt, J.; Patel, C.; Pandya, P.; Jain, N.; Rawal, R. Theaflavin-3-gallate, a natural antagonist for Hsp90: In-silico and in-vitro approach. *Chem. Interact.* **2022**, *353*, 109774. [CrossRef]
247. Nahar, L.; Al-Groshi, A.; Kumar, A.; Sarker, S.D. Arbutin: Occurrence in Plants, and Its Potential as an Anticancer Agent. *Molecules* **2022**, *27*, 8786. [CrossRef]
248. Avelino-Flores, M.d.C.; Cruz-López, M.D.C.; Jimenez-Montejo, F.E.; Reyes-Leyva, J. Cytotoxic Activity of the Methanolic Extract of Turnera diffusa Willd on Breast Cancer Cells. *J. Med. Food* **2014**, *18*, 299–305. [CrossRef]
249. Hazman, Ö.; Sarıova, A.; Bozkurt, M.F.; Ciğerci, İ.H. The anticarcinogen activity of β-arbutin on MCF-7 cells: Stimulation of apoptosis through estrogen receptor-α signal pathway, inflammation and genotoxicity. *Mol. Cell. Biochem.* **2021**, *476*, 349–360. [CrossRef] [PubMed]
250. Ma, J.; Chen, S.; Li, Y.; Wu, X.; Song, Z. Arbutin improves gut development and serum lipids via Lactobacillus intestinalis. *Front. Nutr.* **2022**, *9*, 948573. [CrossRef] [PubMed]
251. Chinnikrishnan, P.; Ibrahim, I.A.A.; Alzahrani, A.R.; Shahzad, N.; Sivaprakasam, P.; Pandurangan, A.K. The Role of Selective Flavonoids on Triple-Negative Breast Cancer: An Update. *Separations* **2023**, *10*, 207. [CrossRef]
252. Ranganathan, S.; Halagowder, D.; Sivasithambaram, N.D. Quercetin Suppresses Twist to Induce Apoptosis in MCF-7 Breast Cancer Cells. *PLoS ONE* **2015**, *10*, e0141370. [CrossRef]
253. Mawalizadeh, F.; Mohammadzadeh, G.; Khedri, A.; Rashidi, M. Quercetin potentiates the chemosensitivity of MCF-7 breast cancer cells to 5-fluorouracil. *Mol. Biol. Rep.* **2021**, *48*, 7733–7742. [CrossRef]
254. Aljabr, B.A.; Zihlif, M.; Abu-Dahab, R.; Zalloum, H. Effect of quercetin on doxorubicin cytotoxicity in sensitive and resistant human MCF7 breast cancer cell lines. *Biomed. Rep.* **2024**, *20*, 58. [CrossRef]
255. Chen, W.J.; Tsai, J.H.; Hsu, L.S.; Lin, C.L.; Hong, H.M.; Pan, M.H. Quercetin Blocks the Aggressive Phenotype of Triple Negative Breast Cancer by Inhibiting IGF1/IGF1R-Mediated EMT Program. *J. Food Drug Anal.* **2021**, *29*, 98–112. [CrossRef]
256. Maugeri, A.; Calderaro, A.; Patanè, G.T.; Navarra, M.; Barreca, D.; Cirmi, S.; Felice, M.R. Targets Involved in the Anti-Cancer Activity of Quercetin in Breast, Colorectal and Liver Neoplasms. *Int. J. Mol. Sci.* **2023**, *24*, 2952. [CrossRef] [PubMed]
257. Adams, C.; Conigrave, J.H.; Lewohl, J.; Haber, P.; Morley, K.C. Alcohol use disorder and circulating cytokines: A systematic review and meta-analysis. *Brain, Behav. Immun.* **2020**, *89*, 501–512. [CrossRef]
258. Gholamalizadeh, M.; Afsharfar, M.; Fathi, S.; Tajadod, S.; Mohseni, G.K.; Shekari, S.; Vahid, F.; Doaei, S.; Kachaei, H.S.; Majidi, N.; et al. Relationship between breast cancer and dietary inflammatory index; a case–control study. *Clin. Nutr. ESPEN* **2022**, *51*, 353–358. [CrossRef]
259. Wang, K.; Sun, J.-Z.; Wu, Q.-X.; Li, Z.-Y.; Li, D.-X.; Xiong, Y.-F.; Zhong, G.-C.; Shi, Y.; Li, Q.; Zheng, J.; et al. Long-term anti-inflammatory diet in relation to improved breast cancer prognosis: A prospective cohort study. *npj Breast Cancer* **2020**, *6*, 36. [CrossRef] [PubMed]
260. Helo, D.; Appiah, L.; Bhende, K.M.; Byrd, T.L.; Appiah, D. The association of skipping breakfast with cancer-related and all-cause mortality in a national cohort of United States adults. *Cancer Causes Control* **2021**, *32*, 505–513. [CrossRef] [PubMed]
261. Fillon, M. Skipping breakfast is a bad idea for patients with cancer. *CA Cancer J. Clin.* **2021**, *71*, 363–365. [CrossRef]
262. Tapan, T.K.; Iyigun, Z.E.; Ilgun, S.; Ozmen, V. Evaluation of the eating habits of breast cancer patients. *Pak. J. Med. Sci.* **2020**, *36*, 1562–1566. [CrossRef]
263. Danforth, D.N. The Role of Chronic Inflammation in the Development of Breast Cancer. *Cancers* **2021**, *13*, 3918. [CrossRef]
264. Wang, W.; Nag, S.A.; Zhang, R. Targeting the NFκB Signaling Pathways for Breast Cancer Prevention and Therapy. *Curr. Med. Chem.* **2014**, *22*, 264–289. [CrossRef] [PubMed]
265. Kansestani, A.N.; Mansouri, K.; Hemmati, S.; Zare, M.E.; Moatafaei, A. High Glucose-reduced Apoptosis in Human Breast Cancer Cells Is Mediated by Activation of NF-κB. *Iran. J. Allergy Asthma Immunol.* **2019**, *18*, 153–162. [CrossRef]
266. Ford, K.L.; Arends, J.; Atherton, P.J.; Engelen, M.P.; Gonçalves, T.J.; Laviano, A.; Lobo, D.N.; Phillips, S.M.; Ravasco, P.; Deutz, N.E.; et al. The importance of protein sources to support muscle anabolism in cancer: An expert group opinion. *Clin. Nutr.* **2021**, *41*, 192–201. [CrossRef] [PubMed]
267. Godfray, H.C.J.; Aveyard, P.; Garnett, T.; Hall, J.W.; Key, T.J.; Lorimer, J.; Pierrehumbert, R.T.; Scarborough, P.; Springmann, M.; Jebb, S.A. Meat consumption, health, and the environment. *Science* **2018**, *361*, eaam5324. [CrossRef]
268. Inoue-Choi, M.; Sinha, R.; Gierach, G.L.; Ward, M.H. Red and processed meat, nitrite, and heme iron intakes and postmenopausal breast cancer risk in the NIH-AARP Diet and Health Study. *Int. J. Cancer* **2016**, *138*, 1609–1618. [CrossRef]
269. Kim, S.-A.; Shin, S. Red meat and processed meat consumption and the risk of dyslipidemia in Korean adults: A prospective cohort study based on the Health Examinees (HEXA) study. *Nutr. Metab. Cardiovasc. Dis.* **2021**, *31*, 1714–1727. [CrossRef] [PubMed]
270. Nouri, M.; Mohsenpour, M.A.; Katsiki, N.; Ghobadi, S.; Jafari, A.; Faghih, S.; Banach, M.; Mazidi, M. Effect of Serum Lipid Profile on the Risk of Breast Cancer: Systematic Review and Meta-Analysis of 1,628,871 Women. *J. Clin. Med.* **2022**, *11*, 4503. [CrossRef] [PubMed]
271. Brindisi, M.; Frattaruolo, L.; Fiorillo, M.; Dolce, V.; Sotgia, F.; Lisanti, M.P.; Cappello, A.R. New insights into cholesterol-mediated ERRα activation in breast cancer progression and pro-tumoral microenvironment orchestration. *FEBS J.* **2022**, *290*, 1481–1501. [CrossRef]

272. Strizova, Z.; Benesova, I.; Bartolini, R.; Novysedlak, R.; Cecrdlova, E.; Foley, L.K.; Striz, I. M1/M2 macrophages and their overlaps—Myth or reality? *Clin. Sci.* **2023**, *137*, 1067–1093. [CrossRef]
273. Yang, Y.; Xin, X.; Fu, X.; Xu, D. Expression pattern of human SERPINE2 in a variety of human tumors. *Oncol. Lett.* **2018**, *15*, 4523–4530. [CrossRef]
274. Krawczynska, N.; Wang, Y.; Lim, K.; Das Gupta, A.; Lenczowski, A.; Abughazaleh, M.; Bendre, S.V.; Kockaya, L.I.; Schane, C.P.; Fei, Y.; et al. Neutrophils exposed to a cholesterol metabolite secrete extracellular vesicles that promote epithelial-mesenchymal transition and stemness in breast cancer cells. *bioRxiv* **2024**, bioRxiv:2024.08.02.606061. [CrossRef]
275. Narii, N.; Zha, L.; Komatsu, M.; Kitamura, T.; Sobue, T.; Ogawa, T. Cholesterol and breast cancer risk: A cohort study using health insurance claims and health checkup databases. *Breast Cancer Res. Treat.* **2023**, *199*, 315–322. [CrossRef] [PubMed]
276. Tjønneland, A.; Stripp, C.; Christensen, J.; Thomsen, B.L.; Olsen, A.; Overvad, K.; Møller, S. Fish Intake Is Positively Associated with Breast Cancer Incidence Rate. *J. Nutr.* **2003**, *133*, 3664–3669. [CrossRef]
277. Wu, S.; Ma, X.; Zhang, X.; Shi, C.; Cao, M.; Yang, C.; Qi, Y.; Liu, Y. Relationship between oily fish intake and breast cancer based on estrogen receptor status: A Mendelian randomization study. *Breast Cancer Res. Treat.* **2023**, *203*, 145–152. [CrossRef]
278. Nindrea, R.D.; Aryandono, T.; Lazuardi, L.; Dwiprahasto, I. Protective Effect of Omega-3 Fatty Acids in Fish Consumption Against Breast Cancer in Asian Patients: A Meta-Analysis. *Asian Pac. J. Cancer Prev.* **2019**, *20*, 327–332. [CrossRef]
279. Abera, B.D.; Adimas, M.A. Health benefits and health risks of contaminated fish consumption: Current research outputs, research approaches, and perspectives. *Heliyon* **2024**, *10*, e33905. [CrossRef] [PubMed]
280. Brodziak-Dopierała, B.; Fischer, A. Analysis of the Mercury Content in Fish for Human Consumption in Poland. *Toxics* **2023**, *11*, 717. [CrossRef] [PubMed]
281. Skalny, A.V.; Aschner, M.; Sekacheva, M.I.; Santamaria, A.; Barbosa, F.; Ferrer, B.; Aaseth, J.; Paoliello, M.M.; Rocha, J.B.; Tinkov, A.A. Mercury and cancer: Where are we now after two decades of research? *Food Chem. Toxicol.* **2022**, *164*, 113001. [CrossRef] [PubMed]
282. Yedier, S.; Yalçınkaya, S.K.; Bostancı, D. Exposure to polypropylene microplastics via diet and water induces oxidative stress in Cyprinus carpio. *Aquat. Toxicol.* **2023**, *259*, 106540. [CrossRef]
283. Park, J.H.; Hong, S.; Kim, O.-H.; Kim, C.-H.; Kim, J.; Kim, J.-W.; Hong, S.; Lee, H.J. Polypropylene microplastics promote metastatic features in human breast cancer. *Sci. Rep.* **2023**, *13*, 6252. [CrossRef]
284. Polak-Juszczak, L.; Waszak, I.; Szlinder-Richert, J.; Wójcik, I. Levels, time trends, and distribution of dioxins and polychlorinated biphenyls in fishes from the Baltic Sea. *Chemosphere* **2022**, *306*, 135614. [CrossRef]
285. VoPham, T.; Bertrand, K.A.; Jones, R.R.; Deziel, N.C.; DuPré, N.C.; James, P.; Liu, Y.; Vieira, V.M.; Tamimi, R.M.; Hart, J.E.; et al. Dioxin exposure and breast cancer risk in a prospective cohort study. *Environ. Res.* **2020**, *186*, 109516. [CrossRef]
286. Meurillon, M.; Engel, E. Potential chemical hazards linked to meat processing. In *Encyclopedia of Meat Sciences*; CRC Press: Boca Raton, FL, USA, 2022.
287. Bulanda, S.; Janoszka, B. Consumption of Thermally Processed Meat Containing Carcinogenic Compounds (Polycyclic Aromatic Hydrocarbons and Heterocyclic Aromatic Amines) versus a Risk of Some Cancers in Humans and the Possibility of Reducing Their Formation by Natural Food Additives—A Literature Review. *Int. J. Environ. Res. Public Health* **2022**, *19*, 4781. [CrossRef] [PubMed]
288. Biswas, A.K.; Jairath, G.; Mendiratta, S.K.; Kumar, D.; Bauer, F. Residues associated with meat production and processing. In *Encyclopedia of Meat Sciences (Third Edition)*; Elsevier: Amsterdam, The Netherlands, 2022. [CrossRef]
289. Muthukumar, M.; Banerjee, R.; Naveena, B.M. Potential chemical hazards associated with meat. In *Encyclopedia of Meat Sciences*, 3rd ed.; Dikeman, M., Ed.; Elsevier: Oxford, UK, 2024; pp. 582–590. [CrossRef]
290. Reig, M.; Toldrá, F. Chapter 11—Residues of harmful chemicals and their detection techniques. In *Meat Quality Analysis*; Biswas, A.K., Mandal, P.K., Eds.; Academic Press: Cambridge, MA, USA, 2020; pp. 173–183.
291. Gearhart-Serna, L.M.; Davis, J.B.; Jolly, M.K.; Jayasundara, N.; Sauer, S.J.; Di Giulio, R.T.; Devi, G.R. A polycyclic aromatic hydrocarbon-enriched environmental chemical mixture enhances AhR, antiapoptotic signaling and a proliferative phenotype in breast cancer cells. *Carcinog.* **2020**, *41*, 1648–1659. [CrossRef]
292. Kehm, R.D.; Walter, E.J.; Oskar, S.; White, M.L.; Tehranifar, P.; Herbstman, J.B.; Perera, F.; Lilge, L.; Miller, R.L.; Terry, M.B. Exposure to polycyclic aromatic hydrocarbons during pregnancy and breast tissue composition in adolescent daughters and their mothers: A prospective cohort study. *Breast Cancer Res.* **2022**, *24*, 47. [CrossRef]
293. Sheikh, I.A.; Jiffri, E.H.; Kamal, M.A.; Ashraf, G.M.; Beg, M.A. Lactoperoxidase, an Antimicrobial Milk Protein, as a Potential Activator of Carcinogenic Heterocyclic Amines in Breast Cancer. *Anticancer. Res.* **2017**, *37*, 6415–6420. [CrossRef] [PubMed]
294. Rahman, U.U.; Sahar, A.; Khan, M.I.; Nadeem, M. Production of heterocyclic aromatic amines in meat: Chemistry, health risks and inhibition. A review. *LWT* **2014**, *59*, 229–233. [CrossRef]
295. Bellamri, M.; Walmsley, S.J.; Turesky, R.J. Metabolism and biomarkers of heterocyclic aromatic amines in humans. *Genes Environ.* **2021**, *43*, 29. [CrossRef] [PubMed]
296. Sheikh, I.A.; Beg, M.A.; Yasir, M. Molecular Interactions of Carcinogenic Aromatic Amines, 4-Aminobiphenyl and 4,4′-Diaminobiphenyl, with Lactoperoxidase—Insight to Breast Cancer. *Anticancer. Res.* **2017**, *37*, 6245–6249. [CrossRef] [PubMed]

297. Deveci, G.; Tek, N.A. N-Nitrosamines: A potential hazard in processed meat products. *J. Sci. Food Agric.* **2024**, *104*, 2551–2560. [CrossRef]
298. Nabizadeh, S.; Aeini, K.; Barzegar, F.; Arabameri, M.; Hosseini, H.; Kamankesh, M.; Mohammadi, A. Volatile N-nitrosamines in processed meat products: An approach for monitoring dietary exposure, assessing human risk, and evaluating variable correlations by principal component analysis and heat map. *Food Chem. Toxicol.* **2024**, *188*, 114649. [CrossRef]
299. Diallo, A.; Deschasaux, M.; Latino-Martel, P.; Hercberg, S.; Galan, P.; Fassier, P.; Allès, B.; Guéraud, F.; Pierre, F.H.; Touvier, M. Red and processed meat intake and cancer risk: Results from the prospective NutriNet-Santé cohort study. *Int. J. Cancer* **2018**, *142*, 230–237. [CrossRef]
300. Cauchi, J.P.; Camilleri, L.; Scerri, C. Environmental and lifestyle risk factors of breast cancer in Malta—A retrospective case-control study. *EPMA J.* **2016**, *7*, 20. [CrossRef]
301. Zhu, P.; Zhang, Y.; Zou, S.; Yu, X.; Song, M.; Lin, M.; Yang, H. Consumption of processed meat and its interactions with alcohol drinking and polygentic risk scores on breast cancer risk: A cohort study in the UK biobank. *medRxiv* **2022**, medRxiv:2022.08.30.22279400. [CrossRef]
302. Farvid, M.S.; Sidahmed, E.; Spence, N.D.; Angua, K.M.; Rosner, B.A.; Barnett, J.B. Consumption of red meat and processed meat and cancer incidence: A systematic review and meta-analysis of prospective studies. *Eur. J. Epidemiol.* **2021**, *36*, 937–951. [CrossRef]
303. Farvid, M.S.; Stern, M.C.; Norat, T.; Sasazuki, S.; Vineis, P.; Weijenberg, M.P.; Wolk, A.; Wu, K.; Stewart, B.W.; Cho, E. Consumption of red and processed meat and breast cancer incidence: A systematic review and meta-analysis of prospective studies. *Int. J. Cancer* **2018**, *143*, 2787–2799. [CrossRef] [PubMed]
304. Zhang, X.; Chen, X.; Xu, Y.; Yang, J.; Du, L.; Li, K.; Zhou, Y. Milk consumption and multiple health outcomes: Umbrella review of systematic reviews and meta-analyses in humans. *Nutr. Metab.* **2021**, *18*, 7. [CrossRef] [PubMed]
305. Ventura, E.R.; Konigorski, S.; Rohrmann, S.; Schneider, H.; Stalla, G.K.; Pischon, T.; Linseisen, J.; Nimptsch, K. Association of dietary intake of milk and dairy products with blood concentrations of insulin-like growth factor 1 (IGF-1) in Bavarian adults. *Eur. J. Nutr.* **2020**, *59*, 1413–1420. [CrossRef] [PubMed]
306. Melnik, B.C.; John, S.M.; Carrera-Bastos, P.; Cordain, L.; Leitzmann, C.; Weiskirchen, R.; Schmitz, G. The Role of Cow's Milk Consumption in Breast Cancer Initiation and Progression. *Curr. Nutr. Rep.* **2023**, *12*, 122–140. [CrossRef]
307. Lee, J.-S.; Tocheny, C.E.; Shaw, L.M. The Insulin-like Growth Factor Signaling Pathway in Breast Cancer: An Elusive Therapeutic Target. *Life* **2022**, *12*, 1992. [CrossRef] [PubMed]
308. Reis, L.G.; Silva, T.H.D. Effect of cow's milk with different PUFA n-6: N-3 ratios on performance, serum lipid profile, and blood parameters of grower gilts. *PLoS ONE* **2022**, *17*, e0258629. [CrossRef]
309. Mei, J.; Qian, M.; Hou, Y.; Liang, M.; Chen, Y.; Wang, C.; Zhang, J. Association of saturated fatty acids with cancer risk: A systematic review and meta-analysis. *Lipids Health Dis.* **2024**, *23*, 32. [CrossRef]
310. Jiang, Y.; Li, L.-T.; Hou, S.-H.; Chen, L.-N.; Zhang, C.-X. Association between dietary intake of saturated fatty acid subgroups and breast cancer risk. *Food Funct.* **2024**, *15*, 2282–2294. [CrossRef]
311. de Lorgeril, M.; Salen, P. Helping women to good health: Breast cancer, omega-3/omega-6 lipids, and related lifestyle factors. *BMC Med.* **2014**, *12*, 54. [CrossRef]
312. Espinosa-Neira, R.; Mejia-Rangel, J.; Cortes-Reynosa, P.; Salazar, E.P. Linoleic acid induces an EMT-like process in mammary epithelial cells MCF10A. *Int. J. Biochem. Cell Biol.* **2011**, *43*, 1782–1791. [CrossRef]
313. Serna-Marquez, N.; Diaz-Aragon, R.; Reyes-Uribe, E.; Cortes-Reynosa, P.; Salazar, E.P. Linoleic acid induces migration and invasion through FFAR4- and PI3K-/Akt-dependent pathway in MDA-MB-231 breast cancer cells. *Med. Oncol.* **2017**, *34*, 111. [CrossRef] [PubMed]
314. Durán-Jara, E.; Vera-Tobar, T.; Lobos-González, L.D.L. Lactadherin: From a Well-Known Breast Tumor Marker to a Possible Player in Extracellular Vesicle-Mediated Cancer Progression. *Int. J. Mol. Sci.* **2022**, *23*, 3855. [CrossRef] [PubMed]
315. Yang, Y.; Li, J.; Song, Q.; Zhu, K.; Yu, X.; Tian, Y.; Zhang, J. Reduction in milk fat globule-EGF factor 8 inhibits triple-negative breast cancer cell viability and migration. *Oncol. Lett.* **2019**, *17*, 3457–3465. [CrossRef]
316. Sørensen, E.S.; Christensen, B. Milk Osteopontin and Human Health. *Nutrients* **2023**, *15*, 2423. [CrossRef] [PubMed]
317. Kovacheva, M.; Zepp, M.; Schraad, M.; Berger, S.; Berger, M.R. Conditional Knockdown of Osteopontin Inhibits Breast Cancer Skeletal Metastasis. *Int. J. Mol. Sci.* **2019**, *20*, 4918. [CrossRef]
318. Guo, M.; Liu, M.; Li, W.; Wang, C.; Zhang, L.; Zhang, H. Osteopontin promotes tumor growth and metastasis and GPX4-mediated anti-lipid peroxidation in triple-negative breast cancer by activating the PI3k/Akt/mTOR pathway. *J. Cancer Res. Clin. Oncol.* **2024**, *150*, 155. [CrossRef]
319. Panahipour, L.; Stähli, A.; Haiden, N.; Gruber, R. TGF-β activity in cow milk and fermented milk products: An in vitro bioassay with oral fibroblasts. *Arch. Oral Biol.* **2018**, *95*, 15–21. [CrossRef]
320. Shukla, N.; Naik, A.; Moryani, K.; Soni, M.; Shah, J.; Dave, H. TGF-β at the crossroads of multiple prognosis in breast cancer, and beyond. *Life Sci.* **2022**, *310*, 121011. [CrossRef]
321. Urseler, N.; Bachetti, R.; Biolé, F.; Morgante, V.; Morgante, C. Atrazine pollution in groundwater and raw bovine milk: Water quality, bioaccumulation and human risk assessment. *Sci. Total. Environ.* **2022**, *852*, 158498. [CrossRef]

322. Wang, M.; Chen, J.; Zhao, S.; Zheng, J.; He, K.; Liu, W.; Zhao, W.; Li, J.; Wang, K.; Wang, Y.; et al. Atrazine promotes breast cancer development by suppressing immune function and upregulating MMP expression. *Ecotoxicol. Environ. Saf.* **2023**, *253*, 114691. [CrossRef]
323. Ranjbar, F.; Movassaghghazani, M. Cypermethrin, deltamethrin, and hexachlorobenzene contents in milk and dairy products in Tehran, Iran. *Int. J. Environ. Stud.* **2023**, *80*, 1808–1819. [CrossRef]
324. El-Danasoury, H.M.; Khozimy, A.M. Toxic Effects of Administration of Cypermethrin, Vitamin E, Zinc and Their Mixtures on the Hormonal Levels of Thyroid, Kidney Functions and some Biochemical Parameters in Male Mice. *J. Appl. Plant Prot.* **2020**, *9*, 39–49.
325. Kara-Ertekin, S.; Yazar, S.; Erkan, M. In vitro toxicological assessment of flumethrin's effects on MCF-7 breast cancer cells. *Hum. Exp. Toxicol.* **2021**, *40*, 2165–2177. [CrossRef] [PubMed]
326. Miret, N.V.; Pontillo, C.A.; Zárate, L.V.; de Pisarev, D.K.; Cocca, C.; Randi, A.S. Impact of endocrine disruptor hexachlorobenzene on the mammary gland and breast cancer: The story thus far. *Environ. Res.* **2019**, *173*, 330–341. [CrossRef] [PubMed]
327. Bøhn, T.; Millstone, E. The Introduction of Thousands of Tonnes of Glyphosate in the food Chain—An Evaluation of Glyphosate Tolerant Soybeans. *Foods* **2019**, *8*, 669. [CrossRef]
328. Muñoz, J.P.; Araya-Osorio, R.; Mera-Adasme, R.; Calaf, G.M. Glyphosate mimics 17β-estradiol effects promoting estrogen receptor alpha activity in breast cancer cells. *Chemosphere* **2023**, *313*, 137201. [CrossRef] [PubMed]
329. Ridley, W.P.; Sidhu, R.S.; Pyla, P.D.; Nemeth, M.A.; Breeze, M.L.; Astwood, J.D. Comparison of the Nutritional Profile of Glyphosate-Tolerant Corn Event NK603 with That of Conventional Corn (*Zea mays* L.). *J. Agric. Food Chem.* **2002**, *50*, 7235–7243. [CrossRef]
330. Stur, E.; Aristizabal-Pachon, A.F.; Peronni, K.C.; Agostini, L.P.; Waigel, S.; Chariker, J.; Miller, D.M.; Thomas, S.D.; Rezzoug, F.; Detogni, R.S.; et al. Glyphosate-based herbicides at low doses affect canonical pathways in estrogen positive and negative breast cancer cell lines. *PLoS ONE* **2019**, *14*, e0219610. [CrossRef]
331. Alaboudi, A.R.; Osaili, T.M.; Alrwashdeh, A. Pesticides (Hexachlorocyclohexane, Aldrin, and Malathion) Residues in Home-Grown Eggs: Prevalence, Distribution, and Effect of Storage and Heat Treatments. *J. Food Sci.* **2019**, *84*, 3383–3390. [CrossRef]
332. Liu, H.; Sun, Y.; Ran, L.; Li, J.; Shi, Y.; Mu, C.; Hao, C. Endocrine-disrupting chemicals and breast cancer: A meta-analysis. *Front. Oncol.* **2023**, *13*, 1282651. [CrossRef]
333. Pogurschi, E.N.; Zugravu, C.A.; Ranga, I.N.; Trifunschi, S.; Munteanu, M.F.; Popa, D.C.; Tudorache, M.; Custura, I. Determination of Acrylamide in Selected Foods from the Romanian Market. *Foods* **2021**, *10*, 2110. [CrossRef] [PubMed]
334. Mihalache, O.A.; Dall'asta, C. The burden of disease due to dietary exposure to acrylamide in Italy: A risk assessment-based approach. *Food Chem. Toxicol.* **2024**, *188*, 114699. [CrossRef] [PubMed]
335. Bellicha, A.; Wendeu-Foyet, G. Dietary exposure to acrylamide and breast cancer risk: Results from the NutriNet-Santé cohort. *Am. J. Clin. Nutr.* **2022**, *116*, 911–919. [CrossRef] [PubMed]
336. Atabati, H.; Abouhamzeh, B.; Abdollahifar, M.-A.; Javadinia, S.S.; Bajestani, S.G.; Atamaleki, A.; Raoofi, A.; Fakhri, Y.; Oliveira, C.A.; Khaneghah, A.M. The association between high oral intake of acrylamide and risk of breast cancer: An updated systematic review and meta-analysis. *Trends Food Sci. Technol.* **2020**, *100*, 155–163. [CrossRef]
337. Besaratinia, A.; Pfeifer, G.P. A review of mechanisms of acrylamide carcinogenicity. *Carcinogenesis* **2007**, *28*, 519–528. [CrossRef] [PubMed]
338. Lara-Castor, L.; Micha, R.; Cudhea, F.; Miller, V.; Shi, P.; Zhang, J.; Sharib, J.R.; Erndt-Marino, J.; Cash, S.B.; Barquera, S.; et al. Intake of sugar sweetened beverages among children and adolescents in 185 countries between 1990 and 2018: Population based study. *BMJ* **2024**, *386*, e079234. [CrossRef]
339. Romanos-Nanclares, A.; Collins, L.C.; Hu, F.B.; Willett, W.C.; A Rosner, B.; Toledo, E.; Eliassen, A.H. Sugar-Sweetened Beverages, Artificially Sweetened Beverages, and Breast Cancer Risk: Results From 2 Prospective US Cohorts. *J. Nutr.* **2021**, *151*, 2768–2779. [CrossRef] [PubMed]
340. Debras, C.; Chazelas, E.; Srour, B.; Druesne-Pecollo, N.; Esseddik, Y.; Szabo de Edelenyi, F.; Agaësse, C.; De Sa, A.; Lutchia, R.; Gigandet, S.; et al. Artificial sweeteners and cancer risk: Results from the NutriNet-Santé population-based cohort study. *PLoS Med.* **2022**, *19*, e1003950. [CrossRef]
341. Ye, X.; Zhang, Y.; He, Y.; Sheng, M.; Huang, J.; Lou, W. Association between Consumption of Artificial Sweeteners and Breast Cancer Risk: A Systematic Review and Meta-Analysis of Observational Studies. *Nutr. Cancer* **2023**, *75*, 795–804. [CrossRef]
342. Keller, A.; Bucher Della Torre, S. Sugar-Sweetened Beverages and Obesity among Children and Adolescents: A Review of Systematic Literature Reviews. *Child. Obes.* **2015**, *11*, 338–346. [CrossRef]
343. Picon-Ruiz, M.; Morata-Tarifa, C.; Valle-Goffin, J.J.; Friedman, E.R.; Slingerland, J.M. Obesity and adverse breast cancer risk and outcome: Mechanistic insights and strategies for intervention. *CA Cancer J. Clin.* **2017**, *67*, 378–397. [CrossRef] [PubMed]
344. Garlapati, C.; Joshi, S.; Bhattarai, S.; Krishnamurthy, J.; Turaga, R.C.; Nguyen, T.; Li, X.; Aneja, R. PLK1 and AURKB phosphorylate survivin differentially to affect proliferation in racially distinct triple-negative breast cancer. *Cell Death Dis.* **2023**, *14*, 12. [CrossRef]
345. Izadi, S.; Nikkhoo, A.; Hojjat-Farsangi, M.; Namdar, A.; Azizi, G.; Mohammadi, H.; Yousefi, M.; Jadidi-Niaragh, F. CDK1 in Breast Cancer: Implications for Theranostic Potential. *Anti-Cancer Agents Med. Chem.* **2020**, *20*, 758–767. [CrossRef] [PubMed]
346. Manousakis, E.; Miralles, C.M.; Esquerda, M.G.; Wright, R.H.G. CDKN1A/p21 in Breast Cancer: Part of the Problem, or Part of the Solution? *Int. J. Mol. Sci.* **2023**, *24*, 17488. [CrossRef]

347. Anwar, S.; Shamsi, A.; Shahbaaz, M.; Queen, A.; Khan, P.; Hasan, G.M.; Islam, A.; Alajmi, M.F.; Hussain, A.; Ahmad, F.; et al. Rosmarinic Acid Exhibits Anticancer Effects via MARK4 Inhibition. *Sci. Rep.* **2020**, *10*, 10300. [CrossRef] [PubMed]
348. Yuan, L.; Cai, Y.; Zhang, L.; Liu, S.; Li, P.; Li, X. Promoting Apoptosis, a Promising Way to Treat Breast Cancer With Natural Products: A Comprehensive Review. *Front. Pharmacol.* **2022**, *12*, 801662. [CrossRef]
349. Sun, X.; Wang, M.; Wang, M.; Yao, L.; Li, X.; Dong, H.; Li, M.; Li, X.; Liu, X.; Xu, Y. Exploring the Metabolic Vulnerabilities of Epithelial–Mesenchymal Transition in Breast Cancer. *Front. Cell Dev. Biol.* **2020**, *8*, 655. [CrossRef]

MDPI AG
Grosspeteranlage 5
4052 Basel
Switzerland
Tel.: +41 61 683 77 34

International Journal of Molecular Sciences Editorial Office
E-mail: ijms@mdpi.com
www.mdpi.com/journal/ijms

www.ingramcontent.com/pod-product-compliance
Lightning Source LLC
Chambersburg PA
CBHW040311270926
51850CB00007BC/1025
9783725854479